Theory and Analysis of Flight Structures

Theory and Analysis of Flight Structures

Robert M. Rivello
Professor of Aerospace Engineering
University of Maryland

McGraw-Hill Book Company

New York　*St. Louis*　*San Francisco*　*Toronto*　*London*　*Sydney*

Theory and Analysis of Flight Structures

Library of Congress Catalog Card Number 68-25662

07-052985-X

7 8 9 10   KPKP   7832109

to the memory of my son
Stephen Charles

Preface

The term *flight vehicle* is used to describe a broad class of crafts for transporting payloads through the atmosphere and space. It includes aircraft, missiles, and spacecraft. While the configurations of these vehicles differ greatly, their structures have notable similarities. Minimum structural weight is essential in achieving high performance in these aerospace vehicles, and, as a result, they frequently employ similar types of construction and materials. Rather than attempting to give detailed descriptions of the structural design of the various types of flight vehicles, for these descriptions would soon be obsolete, this text deals with the structural theory that is common to all of these crafts.

The book is essentially self-contained and assumes only an elementary knowledge of differential equations, matrix algebra, and mechanics of materials, subjects usually taken in the first two years of undergraduate study in engineering. It is intended for advanced undergraduate or beginning graduate students and for the practicing aerospace engineer. However, the emphasis is upon fundamentals that are applicable to a broad class of structural problems, and, as a result, the text should also be useful to those involved in the analysis of machine, civil, or naval structures.

Little of the material is original; instead, an attempt is made to present the subject matter in a unified manner which brings out the basic principles underlying the entire field of structural theory. The theoretical developments proceed from the general to the specific. While it is admittedly more difficult to understand the general theory without first treating simple special cases, the approach has the advantages of establishing a frame of reference for the special cases and of forcing an examination of the assumptions in each problem.

To limit the size of the book, relatively little is given on empirical methods. While it is realized that these methods are important and often necessary, their application usually requires little formal training, and their limitations are best understood by those with a sound theoretical background. Furthermore, no attempt has been made to provide a handbook of results. Instead, the author has tried to provide the reader with sufficient background for the independent study of more advanced texts and the technical literature.

Chapter 1 describes the load and temperature environment of the aerospace vehicle and outlines the steps that are involved in the analysis and design of the vehicle structure. Chapters 2 to 4 give an introduction to what is commonly known as the theory of elasticity. In addition to providing relationships and concepts that are used in later chapters, this material is felt to be essential in providing a better understanding of the approximate theories that follow. Exact solutions to the differential equations of the theory of elasticity are seldom possible. Finite-difference and energy methods are introduced in Chaps. 5 and 6 to provide a means for obtaining approximate solutions in these cases.

Chapters 7 to 9 deal with the theories of bending, extension, torsion, and shear of slender beams without structural discontinuities. Despite limitations, these theories are useful in preliminary design and for determining the basic and redundant-stress systems that are used in the analysis of statically indeterminate structures.

The energy theorems introduced in Chap. 6 are applied to the deflection and stress analysis of complex structures in Chaps. 10 and 11 and are used to develop the matrix methods of structural analysis in Chap. 12. The essentials of plate theory are given in Chap. 13, and the remaining chapters, 14 to 16, deal with the buckling and failure of columns, plates, and stiffened panels.

Thermal stresses and deflections are considered simultaneously with those due to loads throughout the book, rather than as an appendage to the theory. Where practical, the theory is also generalized to include the effects of nonhomogeneity, which is frequently present in the composite structures used in flight vehicles.

I am indebted to Dr. John Dugundji of the Department of Aeronautics and Astronautics of the Massachusetts Institute of Technology for his careful review of the manuscript and for his many helpful suggestions. I am grateful to Drs. Richard S. Reilly and Bruce K. Donaldson, who taught from drafts of the manuscript, for pointing out many typographical errors. I would also like to express gratitude to those who helped to prepare the manuscript: Mrs. Edna Brothers, Miss Sydney Phillips, and Miss Joan M. Duvall for typing the manuscript, and Messrs. Peter P. Ostrowski and Milton R. Grimsley, who prepared most of the illustrations.

Robert M. Rivello

Contents

**CHAPTER 16   INSTABILITY AND FAILURE OF THIN-WALLED COLUMNS
AND STIFFENED PLATES   467**

Theory and Analysis
of Flight Structures

1
Introduction

1-1 STEPS IN STRUCTURAL DESIGN

The process of design and analysis of flight structures may be divided into the following steps:

1. The determination of the critical combinations of applied loads and temperatures to which the structure is subjected.
2. The layout of the design in which the arrangement, size, and materials of the component parts of the structure are tentatively decided upon.
3. The determination of the actual stresses and deformations in the structure due to the applied loads and temperatures.
4. The determination of the allowable stresses or deformations of the structure.
5. The comparison of steps 3 and 4 to determine whether the design of step 2 is adequate and efficient. If the design is either inadequate or overdesigned (and therefore inefficient), steps 2 to 5 must be repeated until a satisfactory design is obtained.

These steps form a successive-approximation procedure, for the loads and temperatures of step 1 are functions of the details of the structural design, which in turn depend upon the loads and temperatures. In the early stages of the design process weights, loads, and temperatures are often based upon crude estimates. These are continuously refined as the design progresses and more accurate information becomes available. During the early phases, the methods of structural analysis are usually based upon simplified theories, as the expense and time necessary for more elaborate methods are not justified until the loads and temperatures are known more accurately. The structural analyst must therefore be capable of covering the range from educated guesses to sophisticated analyses. The design that finally evolves is a compromise involving structural, aerodynamic, fabrication, maintenance, and operational considerations.

1

The substantiation of the final design is usually documented by the following comprehensive reports, which are submitted to the agency which is procuring or certifying the vehicle:

1. A *weight and balance report*, which gives the weights, centers of gravity, mass moments of inertia, and weight distributions of the vehicle and each of its major components.
2. A *loads report*, which contains the aerodynamic, weight, and inertial-force distributions for each of the critical load conditions. Shear, bending-moment, torque, and axial-load curves are also given for major components.
3. A *structural-temperature report*, which gives the temperature distributions that occur simultaneously with the critical load conditions.
4. A *stress-analysis report*, which substantiates the actual and allowable stresses and deflections for each of the critical load-temperature conditions for all components of the structure.
5. An *aeroelastic report*, which gives the predicted speeds at which flutter, divergence, and control reversal will occur. The effects that structural deformations have upon air loads and control effectiveness are also contained in this report.

The structural-analysis group usually prepares the stress-analysis report and assists in the preparation of the other reports.

1-2 APPLIED LOADS AND TEMPERATURES

The loads imposed upon the structure may be divided into two classes, those encountered on the ground and those in flight. *Ground loads* are those loads imposed during fabrication, assembly, shipping, storage, and handling. In the case of missiles they include launch operations, while for aircraft they involve the loads imposed by taxiing and landing. *Flight loads* are those loads applied to the structure during its flight phase and include the loads imposed by maneuver, gusts, and wind shear. In missiles they also involve the forces encountered during boost and staging operations. Temperatures are usually not significant in the ground-operations phase, but during the flight phase they are often of equal or greater importance than the loads. This is especially true for flight in the supersonic or hypersonic regimes. In some cases the structure may have to withstand the aerodynamic loads imposed by passing through the subsonic, transonic, supersonic, and hypersonic phases of flight. At the same time it may be subjected to temperatures ranging from the extreme lows of cryogenic fuels and radiation to space, to the highs associated with aerodynamic heating, heat from the propulsion unit, and radiation from the sun.

Loads may also be categorized according to how they act upon the structure. *Surface forces* are those forces which act upon the surfaces of the structure, e.g., aerodynamic or hydrodynamic pressures, aerostatic or hydrostatic pressures, or contact pressures from other bodies. *Body forces* are those forces which act over the volume of the structure, e.g., gravitational and inertial forces.

No attempt will be made here to define the loads and temperatures for flight structures quantitatively, since several volumes would be required to cover the environmental conditions for airplanes, helicopters, missiles, spacecraft, etc. Such information can be found in Refs. 1 to 18 at the end of the chapter. In some cases, the applied loads which the structure must withstand are specified by the procuring or certifying agency, based upon statistical data obtained from operating experience with similar craft. In other cases, especially if the design and its environment are unconventional, it is part of the contractor's responsibility to obtain rational loads and thermal criteria.

A few definitions of terms relating to loads should be mentioned at this point because of their repeated use in the analysis of flight structures. *Limit loads* are the largest loads which it is anticipated that the structure will be subjected to during its lifetime. It is usually impossible to specify the largest loads that a particular vehicle will be subjected to, but it is often possible to predict statistically the number of times that an average vehicle will encounter certain load levels. In specifying the limit loads, it is usually impractical to set the loads at such a high level that none of the vehicles will ever have a structural failure. Such a design would be inefficient from a weight standpoint. It is therefore necessary to set the limit loads at a level which results in an acceptable low level of failure. The failure rate for inhabited vehicles must, of course, be much lower than that for uninhabited ones.

The limit loads are often prescribed by giving a *limit-load factor*, or the factor by which basic loads are multiplied to obtain limit loads. As an example, the loads for $1g$ level flight are often taken as a basic load condition for aircraft. In a maneuver that imposes inertial and gravitational forces upon the structure that are six times greater than those caused by the gravitational force in level unaccelerated flight, the limit-load factor n_{lim} would be 6.

In order to provide for a separation between the limit loads and the load at which the structure fails, a *factor of safety* is specified. This factor, which may vary according to the mission of the vehicle, is usually 1.5 for inhabited craft and may be as low as 1.25 for missiles. These factors are considerably lower than those used in civil or machine structures. The use of such low factors of safety requires considerable substantiation by analysis and test.

The *ultimate load* (sometimes known as the *design load*) is defined as the product of the limit load and the factor of safety. The *failing load* (ultimate strength) of the structure should be only slightly greater than the ultimate load. It should be noted that in flight structures the limit load is conventionally multiplied by the factor of safety. On the other hand, in civil and machine structures the ultimate strength is usually divided by the factor of safety to give a working strength. Both methods, of course, give the same result. The ultimate load is often specified by giving an *ultimate-load factor* n_{ult}, which is equal to the product of the limit-load factor and the factor of safety. The ultimate loads are then obtained by multiplying the basic loads by the ultimate-load factor.

1-3 ACTUAL STRESSES AND DEFLECTIONS

The major portion of this book is devoted to methods of analysis for predicting the stresses and deflections of structural components under applied loads and temperatures. In the mechanics of deformable bodies it is usually necessary to introduce simplifying assumptions to arrive at a solution to the problem. The results achieved by using these assumptions must therefore be regarded as approximate, and it is possible to assess the degree of approximation only by knowing the nature and significance of the assumptions. Considerable effort is made in this text to underscore the assumptions and limitations of the theories discussed. In practice it is seldom that all the assumptions will be fulfilled, but it is only by an intimate knowledge of the development of the theories that the equations can be intelligently applied to situations which do not precisely follow the conditions of the theory.

The approximations can be divided into *physical* and *mathematical* categories. Physical approximations are simplifying assumptions regarding the mechanical behavior of the material, the shape and proportions of the body, the manner in which it deforms (or how the stresses are distributed), and the nature of the loading. Mathematical approximations are often necessary in order to arrive at simple solutions or, in some cases, to obtain a solution at all. In many cases these mathematical approximations will also imply physical limitations. For instance, if it is assumed for mathematical convenience that the sine of an angle may be replaced by the angle, the results will be acceptable only for small angles.

In the physical approximations we usually replace the real deformable body and loads by a simple *conceptional model* which embodies the significant characteristics of behavior of the real system. For instance, we may idealize the force-displacement behavior of the material by one of the methods discussed in Chap. 3. We may also make assumptions on the mode of deformation of the body. For examples, in Chap. 7 we assume that plane cross sections of a beam remain plane and normal to the axis of a beam as it bends, and in Chap. 13 we assume that normals to the midsurface remain normal to that surface as a plate deforms.   In the study of stiffened-shell structures we shall find it convenient to replace the actual structure by an idealized one having longitudinal stiffeners which resist only axial forces and thin webs which resist only shear forces. To evaluate the accuracy of these assumptions it is necessary to compare the results with those of more accurate theories or with experiments.

In complex structures we shall subdivide the structure into simpler elements for which methods of analysis exist.   We view the composite structure as an assemblage of beams, shear webs, plates, shells, etc., and develop methods of analysis for these simpler structural shapes.

Whenever possible we shall treat the stresses and deformations associated with thermal distributions along with those resulting from applied forces.   We shall include these effects from the outset, rather than treating them as an appendage to the theory, so that the stresses and deflections resulting from loads or temperatures alone will be special cases of the more general theory, which includes both.

1-4 ALLOWABLE STRESSES OR DEFLECTIONS

The usual criteria for the allowable loads of flight structures are:

1. The load which produces a collapse of the structure
2. The load which produces a limiting *permanent* deformation in the structure after removal of the load
3. The load which produces a limiting *total* deformation of the loaded structure

For the first criterion it is required that the stresses imposed by the ultimate loads should not result in a failure of the structure.   Such a failure could be the result of rupture of the material or buckling instability of the structure.   The latter mode of failure usually establishes the design of the major portion of flight structures because of the thin-shell construction that is commonly used.   It is for this reason that a large portion of this book is devoted to the study of the buckling of structural elements.

The second criterion is usually interpreted to mean that the stresses imposed by the limit loads should not exceed the 0.2 percent offset yield stress of the material (Sec. 3-2).   Such a criterion limits the permanent strains in the structure to 0.002.   As this requirement is arbitrary, it is often waived for uninhabited craft, and the only requirement on deformations is then the total-deformation criterion.

The third criterion requires that deflections at the limit loads shall not be excessive.   Excessive deflections are those which interfere with the mission of the vehicle, e.g., those which prevent the free motion of moving parts or produce adverse dynamic or aeroelastic effects.

It is usually specified that the material properties used in determining the allowable stresses and deflections be taken from Ref. 19 or that the properties be substantiated by tests made by the materials manufacturer or the contractor. These properties should reflect the temperature of the structure and the duration of the load.

1-5 COMPARISON OF APPLIED AND ALLOWABLE STRESSES AND DEFLECTIONS

As mentioned earlier, it is necessary to compare the applied and allowable stresses and deflections to determine whether the structure is efficiently designed. This is done by computing the *margin of safety*, defined as

$$\text{MS} = \frac{\text{allowable load}}{\text{applied load}} - 1 \tag{1-1}$$

For the limit-load condition this becomes

$$\text{Limit MS} = \frac{\text{yield load}}{\text{applied limit load}} - 1 \tag{1-2}$$

and for the ultimate-load condition

$$\text{Ultimate MS} = \frac{\text{collapse load}}{\text{applied ultimate load}} - 1 \tag{1-3}$$

The smaller of these two margins of safety controls the design.

In many cases the stresses are directly proportional to the loads (or are assumed so in the linear theories), and the word "load" can therefore be replaced by "stress" in these equations. It is seen from Eq. (1-1) that an efficiently designed structure would have a very small positive margin of safety. However, in some cases it may be desirable to have relatively large positive margins of safety to provide a growth potential for the craft so that increased performance or payload could be accommodated without redesigning the structure.

It has been pointed out that the design process is one of successive approximations until a satisfactory margin of safety is reached. In most cases, and especially if the structure is unconventional, tests are performed to substantiate the analysis and prove the strength and stiffness of the structure. A reduction in the structural weight of a flight vehicle permits an increase in payload or performance. It is therefore economically feasible to use expensive materials and fabrication methods and to expend many manhours of analysis and testing if it results in a decrease in structural weight.

1-6 SUMMARY

The preceding discussion gives only a cursory introduction to the considerations that enter into the evolution of a structural design. More complete descriptions of the process may be found in Refs. 7 to 11. The remainder of this text will address itself to the third and fourth steps described in Sec. 1-1, i.e., the determination of the actual and the allowable stresses and deflections. Even with this limited scope, it is impossible to give more than an introduction to the theoretical methods that are used in structural analysis. Additional references will be given at the end of each chapter, but even here completeness is not possible. Inevitably, the structural analyst finds that he must refer to the technical journals of the professional engineering societies or to the reports of research organizations to obtain the solutions to his problems. It is hoped that this text will provide the reader with an introduction to basic theory sufficient to permit him to read and understand the more advanced theories that are found in the technical literature.

REFERENCES

1. Airworthiness Standards: Normal, Utility, and Acrobatic Category Airplanes, *Federal Aviation Agency Rept.* 23, Feb. 1, 1965.
2. Airworthiness Standards: Transport Category Airplanes, *Federal Aviation Agency Rept.* 25, Feb. 1, 1965.
3. Airworthiness Standards: Normal Category Rotorcraft, *Federal Aviation Agency Rept.* 27, Feb. 1, 1965.
4. Airworthiness Standards: Transport Category Rotorcraft, *Federal Aviation Agency Rept.*, 29, Feb. 1, 1965.
5. General Specification for Airplane Strength and Rigidity, *Military Specification* MIL-A-8860(ASG), May 18, 1960.
6. Tye, W.: Structural Airworthiness, in "Handbook of Aeronautics, no. 1, Structural Principles and Data, pt. 1," 4th ed., Pitman Publishing Corporation, New York, 1952.
7. Bruhn, E. F.: "Analysis and Design of Flight Vehicle Structures," Tri-state Offset Co., Cincinnati, Ohio, 1965.
8. Osgood, C. C.: "Spacecraft Structures," Prentice-Hall, Inc., Englewood Cliffs, N.J., 1966.
9. Bonney, E. A., C. W. Zucrow, and C. W. Besserer: "Principles of Guided Missile Design, Aerodynamics, Propulsion, Structures, and Design Practice," D. Van Nostrand Company, New York, 1956.
10. Chin, S. S.: "Guided Missile Configuration Design," McGraw-Hill Book Company, New York, 1961.
11. Abraham, L. H.: "Structural Design of Missiles and Spacecraft," McGraw-Hill Book Company, New York, 1962.
12. Hoff, N. J.: "High Temperature Effects in Aircraft Structures," Pergamon Press, New York, 1958.
13. Truitt, R. W.: "Fundamentals of Aerodynamic Heating," The Ronald Press Company, New York, 1960.

14. Glaser, P. E.: "Aerodynamically Heated Structures," Prentice-Hall, Inc., Engle-
    wood Cliffs, N.J., 1962.
15. Bisplinghoff, R. L., H. Ashley, and R. L. Halfman: "Aeroelasticity," Addison-
    Wesley Publishing Company, Inc., Reading, Mass., 1955.
16. Fung, Y. C.: "An Introduction to the Theory of Aeroelasticity," John Wiley &
    Sons, Inc., New York, 1955.
17. "Manual on Aeroelasticity," NATO Advisory Group for Aeronautical Research
    and Development, 1959.
18. Bisplinghoff, R. L., and H. A. Ashley: "Principles of Aeroelasticity," John Wiley
    & Sons, Inc., New York, 1962.
19. Metallic Materials and Elements for Flight Vehicle Structures, *Military Hand-
    book* MIL-HDBK-5A, Feb. 8, 1966.

PROBLEMS

1-1. A 600-lb satellite is mounted in the upper stage of a launch vehicle. During the
boosted vertical-flight phase, a peak *acceleration* of $9g$ is reached. The satellite is
mated to the booster by four bolts loaded in shear, each of which has an ultimate
shear strength of 2126 lb. The specified factor of safety is 1.25. Determine (a) the
limit load per bolt, (b) the ultimate load per bolt, and (c) the ultimate margin of
safety. [*Ans.* (a) 1500 lb; (b) 1875 lb; (c) 0.135.]

1-2. The fuel tank of a vertically launched rocket contains kerosene (specific gravity
0.8) and is pressurized to 100 psig at a sea-level pressure of 14.7 psia. The peak boost
acceleration of $9g$ occurs at an altitude where the ambient pressure is 5 psia and at
a time when the depth of the unexpended fuel is 100 in. Determine the ultimate
bursting pressure at the bottom of the tank at this time assuming an ultimate factor
of safety of 1.25. [*Ans.* $p_{ult} = 173$ psi.]

1-3. The nose of a cargo airplane is at a body station (BS) of 0 in. The loaded plane
weighs 150,000 lb, and its center of gravity is at BS 250. The centers of pressure of
the aerodynamic forces on the wing and tail are respectively at BS 200 and 550. The
fuselage is 600 in. long and together with its contents weighs a constant 150 lb/in.
The tail weighs 2000 lb and has a center of gravity at BS 560. Determine the ultimate shear and bending moment in the fuselage at BS 300 for a limit trimmed (no
pitching acceleration) maneuver load factor of $n = 3g$ *including gravity*. Assume
a 1.5 factor of safety. [*Ans.* $V_{ult} = 115,000$ lb, $M_{ult} = 8.7 \times 10^6$ in.-lb.]

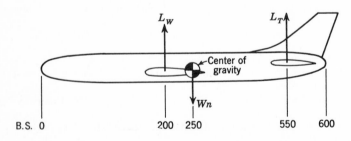

Fig. P1-3

1-4. A 96,600-lb transport airplane has a mass moment of inertia of 48,300,000 lb-in.-
$\sec^2$ about a pitch axis passing through its center of gravity. During landing, when

the aerodynamic lift is 0.9 times the weight, it is subjected to the ground loads shown. Determine (a) the limit-load factor in the vertical direction at the center of gravity and (b) the limit pitching acceleration in radians per second per second. [*Ans.* (a) $n_{lim} = 3.5g$; (b) $\ddot{\theta} = 0.889$ rad/sec².]

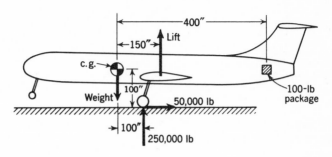

Fig. P1-4 and 1-5

1-5. An electronics package weighing 100 lb is located 400 in. aft of the center of gravity in the airplane of Prob. 1-4. Determine the ultimate vertical load that the package support brackets are subjected to during landing. [*Ans.* 664 lb.]

2
Stress and Strain

2-1 INTRODUCTION

In this chapter we consider the physical quantities of stress and strain. A thorough understanding of the definitions of these quantities and the relations which they must satisfy is fundamental to the remainder of the book. We defer until the next chapter a discussion of the relationships between stresses and strains so that the presentation in this chapter does not presuppose a material with a particular stress-strain law. It will be noted, for instance, that the equations for stresses and the equilibrium relationships which they must satisfy are the same as those for a viscous fluid, and the strain-displacement equations are applicable to any medium which deforms in a continuous manner.

2-2 STRESS: DEFINITIONS AND NOTATIONS

Consider the arbitrarily shaped body shown in Fig. 2-1, which is acted upon by an equilibrium distribution of surface forces applied to its boundary and body forces acting over its volume. If the body is deformable, the forces will be transmitted to all parts of its volume, and internal forces will be developed between the particles which make up the body. To determine the nature of these internal forces, imagine that the body is

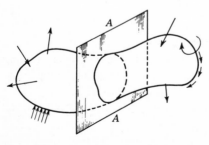

Fig. 2-1 Deformable body with applied loads.

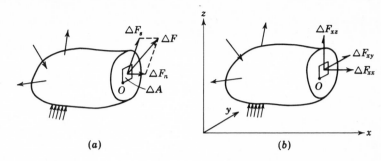

Fig. 2-2 Internal force components at O. (a) Resultant normal and shear components; (b) components parallel to x, y, z axes.

divided into two parts by the cutting plane AA. The surface and body forces on either side of the cutting plane are maintained in equilibrium by the internal forces which act across the plane. The force ΔF acting upon a small area ΔA at a point O in the cutting plane is shown in Fig. 2-2a. The *stress* at the point O is defined as the intensity of the internal force at the point and is given by the equation

$$\text{Stress} = \lim_{\Delta A \to 0} \frac{\Delta F}{\Delta A} \qquad (2\text{-}1)$$

In general, the force ΔF is not normal to the plane AA and it is customary to resolve it into a component ΔF_n normal to the plane and a component ΔF_s lying in the plane, as shown in Fig. 2-2a. The two components of the force produce different effects upon the body; the normal component causes linear deformations parallel to its direction and the in-plane force† produces shearing (angular) deformations. For this reason it is convenient to define two types of stress, a *normal stress*, given by

$$\sigma_n = \lim_{\Delta A \to 0} \frac{\Delta F_n}{\Delta A} \qquad (2\text{-}2)$$

and a *shear stress*, given by

$$\sigma_s = \lim_{\Delta A \to 0} \frac{\Delta F_s}{\Delta A} \qquad (2\text{-}3)$$

In many cases it is desirable to refer the stresses to an orthogonal set of axes x, y, and z, as shown in Fig. 2-2b. In this case we take the cutting plane perpendicular to one of the axes, e.g., the x axis, as shown in Fig. 2-2b. In such cases we resolve the in-plane force into components ΔF_{xy} and ΔF_{xz}, parallel to the y and z axes, respectively. Adopting a double-subscript

† Vectors with a single-sided arrowhead will be used to indicate shearing forces and stresses in illustrations.

notation for the stresses such that the first subscript gives the axis to which the cutting plane is perpendicular and the second subscript gives the axis to which the stress component is parallel, we have

$$\sigma_{xx} = \lim_{\Delta A \to 0} \frac{\Delta F_{xx}}{\Delta A} \qquad \sigma_{xy} = \lim_{\Delta A \to 0} \frac{\Delta F_{xy}}{\Delta A} \qquad \sigma_{xz} = \lim_{\Delta A \to 0} \frac{\Delta F_{xz}}{\Delta A} \qquad (2\text{-}4)$$

For such a notation, stresses of the form σ_{ii} are normal stresses and those of the form σ_{ij} $(i \neq j)$ are shearing stresses. The resultant shear stress in the plane is obtained from the components by the equation

$$\sigma_s = \sqrt{\sigma_{xy}{}^2 + \sigma_{xz}{}^2} \qquad (2\text{-}5)$$

Since we can pass the reference plane through point O in any direction, there are an infinite number of sets of a normal stress and two shear stresses which may represent the stresses at the point O. However, we shall see later that if we know the stresses referred to three mutually perpendicular planes passing through the point O, we can compute the stresses on any arbitrary plane through O.

2-3 EQUATIONS OF EQUILIBRIUM

The body of Fig. 2-1 and every portion of it must be in equilibrium under the action of the applied surface and body forces and the stresses. In this section we shall consider the relationships which these forces must satisfy at points lying within the body and on its surface in order for equilibrium to exist.

Consider first an infinitesimal parallelepiped of dimensions dx, dy, and dz cut from within the body at the point O (Fig. 2-3). In general, there will be a normal stress and two components of shear stress acting upon each of the faces of the element. On the faces of the parallelepiped which contain O there will be a total of nine stress components, given in the following array:

$$\sigma_{ij} = \begin{bmatrix} \sigma_{xx} & \sigma_{xy} & \sigma_{xz} \\ \sigma_{yx} & \sigma_{yy} & \sigma_{yz} \\ \sigma_{zx} & \sigma_{zy} & \sigma_{zz} \end{bmatrix} \qquad (2\text{-}6)$$

The stresses acting upon the faces which do not pass through O will usually differ slightly from those on the planes through O. Consider, for example, the shear stress in the y direction on the right-hand face. It may be expressed in terms of the stress on the left face by the first two terms of a Taylor's series expansion which gives $\sigma_{xy} + (\partial \sigma_{xy}/\partial x)\,dx$. Applying the same method to the other faces which do not contain O, we get the stresses shown in Fig. 2-3. The stresses are defined positive in the directions shown, so that positive normal stresses are tensile, and positive shear

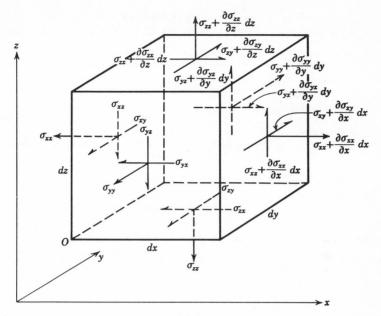

Fig. 2-3 Stresses on faces of a parallelepiped at point O.

stresses act in the direction of the positive coordinate axes on those faces which do not pass through O. We shall assume for generality that the element may also be subjected to a body force per unit volume having components X, Y, Z taken positive in the directions of the x, y, z axes, respectively. To simplify the drawing, these forces are not shown in Fig. 2-3.

Since the element is in equilibrium, the sum of the forces and moments which act upon it are zero. Taking moments about an axis passing through the center of the element and parallel to the z axis, we obtain

$$\sigma_{xy} \, dy \, dz \, \frac{dx}{2} + \left(\sigma_{xy} + \frac{\partial \sigma_{xy}}{\partial x} \, dx \right) dy \, dz \, \frac{dx}{2} - \sigma_{yx} \, dx \, dz \, \frac{dy}{2}$$
$$- \left(\sigma_{yx} + \frac{\partial \sigma_{yx}}{\partial y} \, dy \right) dx \, dz \, \frac{dy}{2} = 0$$

Dividing by $dx \, dy \, dz$ and taking the limit as dx and dy approach zero, we obtain the equation

$$\sigma_{xy} = \sigma_{yx} \tag{2-7a}$$

Similarly for moments about the other axes

$$\sigma_{yz} = \sigma_{zy} \tag{2-7b}$$

$$\sigma_{zx} = \sigma_{xz} \tag{2-7c}$$

We see that there are only six independent stress components at the point O, instead of nine as indicated in Eq. (2-6), because of the relationships (2-7).

If we consider the equilibrium of forces in the x direction in Fig. 2-3, we find

$$-\sigma_{xx}\, dy\, dz + \left(\sigma_{xx} + \frac{\partial \sigma_{xx}}{\partial x}\, dx\right) dy\, dz - \sigma_{yx}\, dx\, dz + \left(\sigma_{yx} + \frac{\partial \sigma_{yx}}{\partial y}\, dy\right) dx\, dz$$

$$- \sigma_{zx}\, dx\, dy + \left(\sigma_{zx} + \frac{\partial \sigma_{zx}}{\partial z}\, dz\right) dx\, dy + X\, dx\, dy\, dz = 0$$

Simplifying this equation and using Eqs. (2-7), we obtain

$$\frac{\partial \sigma_{xx}}{\partial x} + \frac{\partial \sigma_{xy}}{\partial y} + \frac{\partial \sigma_{xz}}{\partial z} + X = 0 \qquad (2\text{-}8a)$$

Similarly by summing forces parallel to the y and z axes we find

$$\frac{\partial \sigma_{yx}}{\partial x} + \frac{\partial \sigma_{yy}}{\partial y} + \frac{\partial \sigma_{yz}}{\partial z} + Y = 0 \qquad (2\text{-}8b)$$

$$\frac{\partial \sigma_{zx}}{\partial x} + \frac{\partial \sigma_{zy}}{\partial y} + \frac{\partial \sigma_{zz}}{\partial z} + Z = 0 \qquad (2\text{-}8c)$$

Equations (2-7) and (2-8) are the *equations of equilibrium* which must be satisfied at all interior points in a deformable body with a three-dimensional force system.

If we consider the surface S of the body, it may be divided into two parts. On part of the surface S_1, forces are prescribed; and on the remaining part S_2 displacements are given. On a free surface the forces are prescribed and are of course zero. If we consider a typical differential element on the surface S_1 (Fig. 2-4), the internal forces must be in equilibrium with the applied surface forces and the body forces. Assume that the intensity of the surface force is described by $\bar{X}$, $\bar{Y}$, and $\bar{Z}$, the forces per unit area that are parallel to the x, y, and z axes, respectively.

If we designate the area of the differential surface element as dA, the areas of the other faces of the tetrahedron dA_x, dA_y, and dA_z (perpendicular to the x, y, and z axes, respectively) are

$$dA_x = dA\, l \qquad dA_y = dA\, m \qquad dA_z = dA\, n \qquad (2\text{-}9)$$

where l, m, and n are the direction cosines of the angles that a normal to the plane dA makes with the x, y, and z axes respectively.

Summing forces in the x direction gives

$$\bar{X}\, dA - \sigma_{xx}\, dA_x - \sigma_{yx}\, dA_y - \sigma_{zx}\, dA_z + \tfrac{1}{3}X\, dA_x\, dx = 0$$

which by using Eqs. (2-7) and (2-9) and taking the limit as dx approaches

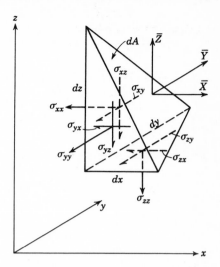

Fig. 2-4 Forces acting on an element at the surface.

zero becomes

$$\bar{X} = \sigma_{xx}l + \sigma_{xy}m + \sigma_{xz}n \qquad (2\text{-}10a)$$

Similarly, by summing forces parallel to the y and z axes we obtain

$$\bar{Y} = \sigma_{yx}l + \sigma_{yy}m + \sigma_{yz}n \qquad (2\text{-}10b)$$

$$\bar{Z} = \sigma_{zx}l + \sigma_{zy}m + \sigma_{zz}n \qquad (2\text{-}10c)$$

Equations (2-10) are the *equilibrium boundary conditions* which must be satisfied over all portions of the surface S_1. For a free surface $\bar{X}$, $\bar{Y}$, and $\bar{Z}$ are zero. Equations (2-7), (2-8), and (2-10) are the equilibrium conditions for the most general three-dimensional state of stress that can exist in a body.

In most flight vehicles the structures are thin-walled, and the stresses in the thin direction are usually negligible compared with those in the other directions. Assume, for instance, that this is the z direction; the state of stress where σ_{xz}, σ_{yz}, and σ_{zz} are negligible compared to the other stresses is defined as *plane stress*. For this case the internal equilibrium equations (2-7) and (2-8) become

$$\sigma_{xy} = \sigma_{yx} \qquad \frac{\partial \sigma_{xx}}{\partial x} + \frac{\partial \sigma_{xy}}{\partial y} + X = 0 \qquad \frac{\partial \sigma_{yx}}{\partial x} + \frac{\partial \sigma_{yy}}{\partial y} + Y = 0 \quad (2\text{-}11)$$

and the surface equilibrium conditions Eqs. (2-10) reduce to

$$\bar{X} = \sigma_{xx}l + \sigma_{xy}m \qquad \bar{Y} = \sigma_{yx}l + \sigma_{yy}m \qquad (2\text{-}12)$$

We note again that the derived relationships do not depend upon the force-displacement behavior of the material. They are therefore applicable whether the body is undergoing elastic, plastic, or creep deformation.

In fact they are applicable to a viscous fluid which can sustain shear forces. However, it should be pointed out that the changes in the lengths and directions of dx, dy, and dz that occur as the body deforms have been neglected in deriving the equations. Because of this, the resulting linear differential equations are correct only for small deformations.

2-4 STRESS TRANSFORMATIONS FOR ROTATION OF AXES

In Sec. 2-2 it was stated that the stresses acting upon any plane passing through the point O could be determined from a knowledge of the three normal- and three shear-stress components which act upon a set of mutually perpendicular planes passing through O. We shall now prove this statement, thereby showing that the state of stress at a point is completely defined by six components of stress. In the next section we shall discuss the orientation that the reference planes must have for the normal and shear stresses to have their maximum and minimum values.

Let us return to the three planes through O which are perpendicular to the x, y, z axes. We now take as our element of the body the tetrahedron formed by the three planes through O and a fourth plane in an arbitrary skew direction, as shown in Fig. 2-5. Consider a second set of orthogonal axes x', y', z' having the same origin as the x, y, z axes but with the x' axis perpendicular to the skew plane of the tetrahedron. The y' axis is in an arbitrary direction perpendicular to the x' axis. The direction of the x' axis is specified by the direction cosines l_1, m_1, n_1 of the angles that this axis respectively makes with the x, y, z axes.

Assume that the stresses acting on the planes perpendicular to the x, y, z, axes are known and the normal stress on the plane perpendicular to

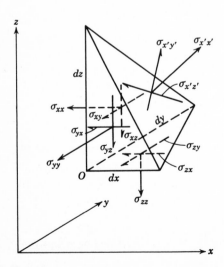

Fig. 2-5 Components of stress on an oblique plane.

the x' axis is desired.   Let the areas of the sides of the tetrahedron which are perpendicular to the x, y, z axes be dA_x, dA_y, dA_z, respectively, and the area of the side perpendicular to the x' axis be dA.   Summing forces parallel to the x' axis, we obtain

$$\begin{aligned}
\sigma_{x'x'}\, dA = {}& (\sigma_{xx}\, dA_x)l_1 + (\sigma_{xy}\, dA_x)m_1 + (\sigma_{xz}\, dA_x)n_1 \\
&+ (\sigma_{yx}\, dA_y)l_1 + (\sigma_{yy}\, dA_y)m_1 + (\sigma_{yz}\, dA_y)n_1 \\
&+ (\sigma_{zx}\, dA_z)l_1 + (\sigma_{zy}\, dA_z)m_1 + (\sigma_{zz}\, dA_z)n_1 \\
&- \tfrac{1}{3}(Xl_1 + Ym_1 + Zn_1)\, dA_x\, dx
\end{aligned}$$

By using Eqs. (2-9) (with a subscript 1 appended to the direction cosines) and taking the limit as dx approaches zero, we find

$$\begin{aligned}
\sigma_{x'x'} = {}& \sigma_{xx}l_1{}^2 + \sigma_{yy}m_1{}^2 + \sigma_{zz}n_1{}^2 + 2(\sigma_{xy}l_1m_1 \\
&+ \sigma_{yz}m_1n_1 + \sigma_{zx}n_1l_1)
\end{aligned} \qquad (2\text{-}13a)$$

so that the normal stress in the arbitrary x' direction has been found in terms of the stress components relative to the x, y, z axes.

Next consider the shear stress $\sigma_{x'y'}$ in the plane normal to the x' axis and parallel to the y' axis (which has direction cosines l_2, m_2, n_2 relative to the x, y, z axes).   Summing forces parallel to the y' axis gives

$$\begin{aligned}
\sigma_{x'y'}\, dA = {}& (\sigma_{xx}\, dA_x)l_2 + (\sigma_{xy}\, dA_x)m_2 + (\sigma_{xz}\, dA_x)n_2 \\
&+ (\sigma_{yx}\, dA_y)l_2 + (\sigma_{yy}\, dA_y)m_2 + (\sigma_{yz}\, dA_y)n_2 \\
&+ (\sigma_{zx}\, dA_z)l_2 + (\sigma_{zy}\, dA_z)m_2 + (\sigma_{zz}\, dA_z)n_2 \\
&- \tfrac{1}{3}(Xl_2 + Ym_2 + Zn_2)\, dA_x\, dx
\end{aligned}$$

which as dx approaches zero simplifies to

$$\begin{aligned}
\sigma_{x'y'} = {}& \sigma_{xx}l_1l_2 + \sigma_{yy}m_1m_2 + \sigma_{zz}n_1n_2 \\
&+ (l_1m_2 + m_1l_2)\sigma_{xy} + (m_1n_2 + n_1m_2)\sigma_{yz} \\
&+ (n_1l_2 + l_1n_2)\sigma_{zx}
\end{aligned} \qquad (2\text{-}13b)$$

Similarly for the shear stress parallel to the z' axis

$$\begin{aligned}
\sigma_{x'z'} = {}& \sigma_{xx}l_1l_3 + \sigma_{yy}m_1m_3 + \sigma_{zz}n_1n_3 \\
&+ (l_1m_3 + m_1l_3)\sigma_{xy} + (m_1n_3 + n_1m_3)\sigma_{yz} \\
&+ (n_1l_3 + l_1n_3)\sigma_{zx}
\end{aligned} \qquad (2\text{-}13c)$$

Since the orientation of the x', y', z' axes has been chosen arbitrarily, we have proved that the state of stress at the point is completely determined if the six stress components referred to the x, y, z axes are known.

For the case of plane stress Eq. (2-13a) becomes

$$\sigma_{x'x'} = \sigma_{xx}\cos^2\alpha + \sigma_{yy}\sin^2\alpha + 2\sigma_{xy}\sin\alpha\cos\alpha \qquad (2\text{-}14a)$$

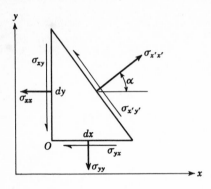

Fig. 2-6 Stress components on an oblique plane in plane stress.

where α is the angle that the x' axis makes with the x axis, as shown in Fig. 2-6. Equation (2-13b) becomes

$$\sigma_{x'y'} = -(\sigma_{xx} - \sigma_{yy}) \sin \alpha \cos \alpha + \sigma_{xy}(\cos^2 \alpha - \sin^2 \alpha) \quad (2\text{-}14b)$$

2-5 PRINCIPAL STRESSES AND MAXIMUM SHEAR STRESSES

In the last section it was shown that the normal and shear stresses at a point depend upon the orientation of the plane upon which the stresses act. It is natural to inquire what the orientations of the plane should be for stresses to have their maximum values and what the magnitudes of these stresses are. To simplify the discussion, we restrict the analysis to the case of plane stresses as shown in Fig. 2-6 (the general case for three-dimensional stresses is treated in Ref. 1). By using the trigonometric identities

$$\sin^2 \alpha = \tfrac{1}{2}(1 - \cos 2\alpha) \qquad \cos^2 \alpha = \tfrac{1}{2}(1 + \cos 2\alpha)$$

$$2 \sin \alpha \cos \alpha = \sin 2\alpha$$

we may rewrite Eqs. (2-14) as

$$\sigma_{x'x'} = \frac{\sigma_{xx} + \sigma_{yy}}{2} + \frac{\sigma_{xx} - \sigma_{yy}}{2} \cos 2\alpha + \sigma_{xy} \sin 2\alpha \quad (2\text{-}15a)$$

$$\sigma_{x'y'} = \frac{\sigma_{yy} - \sigma_{xx}}{2} \sin 2\alpha + \sigma_{xy} \cos 2\alpha \quad (2\text{-}15b)$$

If we replace α in Eq. (2-15a) by $\alpha + \pi/2$, we obtain the normal stress in the y' direction

$$\sigma_{y'y'} = \frac{\sigma_{xx} + \sigma_{yy}}{2} - \frac{\sigma_{xx} - \sigma_{yy}}{2} \cos 2\alpha - \sigma_{xy} \sin 2\alpha \quad (2\text{-}15c)$$

and by adding Eqs. (2-15a) and (2-15c) we find

$$\sigma_{x'x'} + \sigma_{y'y'} = \sigma_{xx} + \sigma_{yy} \quad (2\text{-}16)$$

Since the angle α is arbitrary, we have proved that the sum of the normal stresses in any two perpendicular directions is an invariant quantity. Placing the derivative of Eq. (2-15a) with respect to α equal to zero, we find that α for a maximum or minimum normal stress is given by

$$\tan 2\alpha = \frac{2\sigma_{xy}}{\sigma_{xx} - \sigma_{yy}} \qquad (2\text{-}17)$$

Equation (2-17) has two solutions which differ by 90°, since

$$\tan 2\alpha = \tan 2(\alpha + \pi/2)$$

Let us designate these two angles as α_1 and α_2, and the normal stresses in these directions as σ_{11} and σ_{22}. It was previously shown that the sum of the normal stresses in perpendicular directions is a constant; therefore one of these stresses is a maximum, and the other is a minimum. Substituting the trigonometric relationships from the triangle of Fig. 2-7a into Eq. (2-15a), we find the maximum and minimum normal stresses to be

$$\sigma_{11,22} = \frac{\sigma_{xx} + \sigma_{yy}}{2} \pm \sqrt{\left(\frac{\sigma_{xx} - \sigma_{yy}}{2}\right)^2 + \sigma_{xy}{}^2} \qquad (2\text{-}18)$$

These are referred to as the *principal stresses,* and the axes defined by α_1 and α_2 are called the *principal axes of stress.*

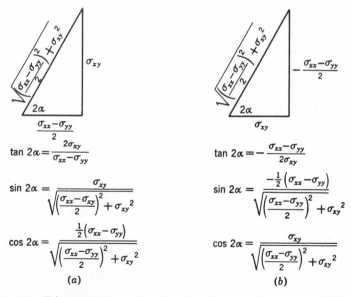

Fig. 2-7 Trigonometric relationships for (a) principal stresses and (b) maximum shear stress.

Next consider the shear-stress equation (2-15b). We see that the angle for zero shear stress is also given by Eq. (2-17). Therefore the shear stresses are zero in the directions of the principal axes. To determine the maximum and minimum shear stresses we differentiate Eq. (2-15b) with respect to α and set the result equal to zero. This gives

$$\tan 2\alpha = -\frac{\sigma_{xx} - \sigma_{yy}}{2\sigma_{xy}} \tag{2-19}$$

which is the negative reciprocal of Eq. (2-17). The angles 2α which satisfy Eqs. (2-17) and (2-19) differ by 90°, or the angles α differ by 45°. The maximum and minimum shear stresses therefore occur on planes which make angles of 45° with the principal axes.

The magnitude of the maximum and minimum shear stresses can be obtained by substituting the trigonometric relationships given for the triangle of Fig. 2-7b into Eq. (2-15b), which upon simplification gives

$$\sigma_{s,\max,\min} = \pm\sqrt{\left(\frac{\sigma_{xx} - \sigma_{yy}}{2}\right)^2 + \sigma_{xy}^2} \tag{2-20}$$

An alternate method for solving the equations in this section and Sec. 2-4 is a graphical solution given by O. Mohr and known as *Mohr's circle*.[1]† It might also be pointed out that stress is a tensor quantity, and as a result the mathematical nature of the equations is the same as those encountered in the study of moments of inertia if the normal stresses are replaced by moments of inertia and the shear stresses are replaced by products of inertia. It will also be observed in Chap. 13 that the same is true for the curvatures and twist of a surface. Any of these problems is therefore amenable to solution by Mohr's circle.

2-6 DEFLECTIONS AND STRAINS

The preceding sections of this chapter dealt with the equilibrium conditions which the stresses must satisfy at every point within the body and at all points on the surface S_1. We have seen that by the use of Eqs. (2-7) the number of unknown stress components at each point in the body is reduced to six. We found that there remain only three differential equations of equilibrium [Eqs. (2-8)], which are insufficient to solve for the six unknown stresses. However, we have not considered the deformation of the body. We shall find that these deformations lead to additional conditions which must be satisfied to obtain solutions. These conditions are based upon the physical requirement that the body deform in a continuous manner, so that no voids form and that two different points cannot

† Superior numbers key to references at the end of each chapter.

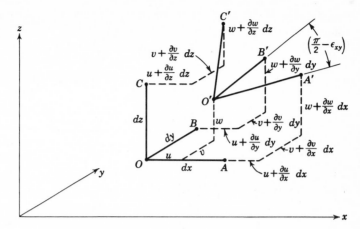

Fig. 2-8 Displacements of points O, A, B, and C parallel to the coordinate axes.

deform into the same point. In short, the deformation of each element of the body must be compatible with the deformation of its neighboring elements.

Let us consider a deformable body which is restrained against rigid-body motion and focus our attention upon four points O, A, B, and C (Fig. 2-8), which are in close proximity. It is assumed that the lines OA, OB, and OC are initially parallel to the x, y, z axes and are of lengths dx, dy, and dz. The coordinates of point O in the undeformed body are (x,y,z), so that those of A, B, and C are $(x + dx, y, z)$, $(x, y + dy, z)$, and $(x, y, z + dz)$, respectively. As a result of the application of surface forces, body forces, and temperature changes the points will deform to O', A', B', and C', as shown in Fig. 2-8. The displacements of O parallel to the x, y, z axes will be designated u, v, w. The displacements of A, B, and C can be found by Taylor's series expansions about the point O. As an example, the displacements of A are $u + (\partial u/\partial x)\, dx$, $v + (\partial v/\partial x)\, dx$, and $w + (\partial w/\partial x)\, dx$. The other displacements are shown in Fig. 2-8. We see from this figure that two types of relative motion of the points O, A, B, and C occur: (1) the distances between the points have changed, and (2) the angles between the lines OA, OB, and OC have changed, so that the lines are no longer perpendicular. When relative motions of points in a body occur, the body is said to be in a *strained* state. Strains associated with changes in length are called *longitudinal strains*, and those related to changes in angle are referred to as *shearing strains*.

The longitudinal strain in a given direction at a point is obtained by studying the length of a line segment passing through the point and aligned in the given direction.   The longitudinal strain is then defined as

$$\epsilon_{long} = \lim_{L \to 0} \frac{\Delta L}{L}$$

where L is the original length of the line and ΔL is the change in length of this line as a result of deformation.   Consider the longitudinal strain which occurs at the point O in the x direction.   From Fig. 2-8 the length of the line segment OA is dx, and

$$O'A' = L + \Delta L = \sqrt{\left(dx + \frac{\partial u}{\partial x} dx\right)^2 + \left(\frac{\partial v}{\partial x} dx\right)^2 + \left(\frac{\partial w}{\partial x} dx\right)^2} \quad (2\text{-}21)$$

The strain in the x direction is then

$$\epsilon_{xx} = \frac{dx \sqrt{(1 + \partial u/\partial x)^2 + (\partial v/\partial x)^2 + (\partial w/\partial x)^2} - dx}{dx}$$

$$= \left[1 + 2\frac{\partial u}{\partial x} + \left(\frac{\partial u}{\partial x}\right)^2 + \left(\frac{\partial v}{\partial x}\right)^2 + \left(\frac{\partial w}{\partial x}\right)^2\right]^{\frac{1}{2}} - 1$$

The radical in this equation can be rewritten by means of the binomial expansion, which is of the general form

$$(1 + x)^n = 1 + nx + \frac{n(n-1)}{2!} x^2 + \cdots$$

and is convergent for $x^2 < 1$.   By using this expansion for $n = \frac{1}{2}$ the strain may be written

$$\epsilon_{xx} = 1 + \frac{1}{2}\left[2\frac{\partial u}{\partial x} + \left(\frac{\partial u}{\partial x}\right)^2 + \left(\frac{\partial v}{\partial x}\right)^2 + \left(\frac{\partial w}{\partial x}\right)^2\right] + \cdots -1$$

The derivatives of u, v, and w with respect to x are small, and if powers of these derivatives higher than the second are neglected, the strain becomes

$$\epsilon_{xx} = \frac{\partial u}{\partial x} + \frac{1}{2}\left[\left(\frac{\partial u}{\partial x}\right)^2 + \left(\frac{\partial v}{\partial x}\right)^2 + \left(\frac{\partial w}{\partial x}\right)^2\right] \quad (2\text{-}22a)$$

In a similar fashion

$$\epsilon_{yy} = \frac{\partial v}{\partial y} + \frac{1}{2}\left[\left(\frac{\partial u}{\partial y}\right)^2 + \left(\frac{\partial v}{\partial y}\right)^2 + \left(\frac{\partial w}{\partial y}\right)^2\right] \quad (2\text{-}22b)$$

$$\epsilon_{zz} = \frac{\partial w}{\partial z} + \frac{1}{2}\left[\left(\frac{\partial u}{\partial z}\right)^2 + \left(\frac{\partial v}{\partial z}\right)^2 + \left(\frac{\partial w}{\partial z}\right)^2\right] \quad (2\text{-}22c)$$

In another form of Eqs. (2-22) the $\partial u/\partial x$, $\partial v/\partial y$, and $\partial w/\partial z$ terms in the square brackets are neglected.   The accuracy of the various approximations for the strains is discussed by Parks and Durelli.[2]

The shearing strain at a point with respect to a given pair of orthogonal axes is defined as the change (as a result of deformation) in the 90° angle between lines through the point parallel to the axes.† Referring to Fig. 2-8, the cosine of the $A'O'B'$ is

$$\cos A'O'B' = \cos\left(\frac{\pi}{2} - \epsilon_{xy}\right) = \sin \epsilon_{xy}$$

where ϵ_{xy} is the shearing strain referred to the x and y axes. For small angles the sine of the angle may be replaced by the angle; therefore

$$\cos A'O'B' = \epsilon_{xy}$$

From analytic geometry the cosine of the angle between two lines is equal to the sum of the products of their direction cosines, or

$$\epsilon_{xy} = \cos A'O'B' = l_1 l_2 + m_1 m_2 + n_1 n_2$$

where l_1, m_1, n_1 are the direction cosines of $O'A'$ and l_2, m_2, n_2 are the direction cosines of $O'B'$. From Fig. 2-8

$$l_1 = \frac{(1 + \partial u/\partial x)\, dx}{O'A'} \qquad m_1 = \frac{\partial v/\partial x\, dx}{O'A'} \qquad n_1 = \frac{\partial w/\partial x\, dx}{O'A'}$$

$$l_2 = \frac{\partial u/\partial y\, dy}{O'B'} \qquad m_2 = \frac{(1 + \partial v/\partial y)\, dy}{O'B'} \qquad n_2 = \frac{\partial w/\partial y\, dy}{O'B'}$$

so that

$$\epsilon_{xy} = \left[\left(1 + \frac{\partial u}{\partial x}\right)\frac{\partial u}{\partial y} + \frac{\partial v}{\partial x}\left(1 + \frac{\partial v}{\partial y}\right) + \frac{\partial w}{\partial x}\frac{\partial w}{\partial y}\right]\frac{dx\, dy}{(O'A')(O'B')}$$

where $O'A'$ is found from Eq. (2-21) and

$$O'B' = dy\sqrt{1 + 2\frac{\partial v}{\partial y} + \left(\frac{\partial u}{\partial y}\right)^2 + \left(\frac{\partial v}{\partial y}\right)^2 + \left(\frac{\partial w}{\partial y}\right)^2}$$

The derivatives of u, v, and w are small compared to unity, so that $O'A' \approx dx$ and $O'B' \approx dy$, in which case

$$\epsilon_{xy} = \frac{\partial v}{\partial x} + \frac{\partial u}{\partial y} + \frac{\partial u}{\partial x}\frac{\partial u}{\partial y} + \frac{\partial v}{\partial x}\frac{\partial v}{\partial y} + \frac{\partial w}{\partial x}\frac{\partial w}{\partial y} \qquad (2\text{-}22d)$$

In a similar manner

$$\epsilon_{yz} = \frac{\partial w}{\partial y} + \frac{\partial v}{\partial z} + \frac{\partial u}{\partial y}\frac{\partial u}{\partial z} + \frac{\partial v}{\partial y}\frac{\partial v}{\partial z} + \frac{\partial w}{\partial y}\frac{\partial w}{\partial z} \qquad (2\text{-}22e)$$

$$\epsilon_{zx} = \frac{\partial u}{\partial z} + \frac{\partial w}{\partial x} + \frac{\partial u}{\partial z}\frac{\partial u}{\partial x} + \frac{\partial v}{\partial z}\frac{\partial v}{\partial x} + \frac{\partial w}{\partial z}\frac{\partial w}{\partial x} \qquad (2\text{-}22f)$$

† In some texts, one-half of the change in the 90° angle is used as a definition of the shear strain.

Equations (2-22) are the so called *large-deflection* or *finite-strain-displacement equations*. It is observed that these equations are nonlinear, and considerable mathematical complexity results when they are used.

In most instances the derivatives of the displacements are small enough so that their products may be neglected, and if this is done, Eqs. (2-22) become

$$\epsilon_{xx} = \frac{\partial u}{\partial x} \qquad \epsilon_{yy} = \frac{\partial v}{\partial y} \qquad \epsilon_{zz} = \frac{\partial w}{\partial z}$$

$$\epsilon_{xy} = \frac{\partial v}{\partial x} + \frac{\partial u}{\partial y} \qquad \epsilon_{yz} = \frac{\partial w}{\partial y} + \frac{\partial v}{\partial z} \qquad \epsilon_{zx} = \frac{\partial u}{\partial z} + \frac{\partial w}{\partial x}$$

(2-23)

These linearized equations are referred to as the *small-deflection* or *infinitesimal-strain-displacement relationships;* unless otherwise stated, these relations will be used throughout the book. However, there are instances where the equations are not adequate, and the large-deflection equations must be used. Examples of this occur in the large deflection of cables, beams with longitudinally restrained ends, membranes, plates, and shells.

2-7 STRAIN-TRANSFORMATION EQUATIONS

In the last section we found the strain-displacement equations for a set of axes x, y, z. In this section we shall show that these six strains completely define the state of strain at the point, so that if we take another set of orthogonal axes x', y', z', we can determine the strain for the new axes in terms of the original strains. For simplicity we consider the case of infinitesimal strains given by Eqs. (2-23); however, the results can be shown to apply to finite strains as well. The angles between the two sets of axes are given by the direction cosines in the following table:

	x	y	z
x'	l_1	m_1	n_1
y'	l_2	m_2	n_2
z'	l_3	m_3	n_3

The following coordinate transformation equations may then be written:

$$x = x'l_1 + y'l_2 + z'l_3$$
$$y = x'm_1 + y'm_2 + z'm_3$$
$$z = x'n_1 + y'n_2 + z'n_3$$

(2-24)

The relationships between the displacements u', v', w' (referred to the x', y', z' axes) and u, v, w are

$$u' = ul_1 + vm_1 + wn_1$$
$$v' = ul_2 + vm_2 + wn_2 \tag{2-25}$$
$$w' = ul_3 + vm_3 + wn_3$$

Noting that $u' = u'(u,v,w)$, the longitudinal strain in the x' direction can be obtained by the chain rule of differentiation

$$\epsilon_{x'x'} = \frac{\partial u'}{\partial x'} = \frac{\partial u'}{\partial u}\frac{\partial u}{\partial x'} + \frac{\partial u'}{\partial v}\frac{\partial v}{\partial x'} + \frac{\partial u'}{\partial w}\frac{\partial w}{\partial x'}$$

and upon expanding $\partial u/\partial x'$, $\partial v/\partial x'$, and $\partial w/\partial x'$,

$$\epsilon_{x'x'} = \frac{\partial u'}{\partial u}\left(\frac{\partial u}{\partial x}\frac{\partial x}{\partial x'} + \frac{\partial u}{\partial y}\frac{\partial y}{\partial x'} + \frac{\partial u}{\partial z}\frac{\partial z}{\partial x'}\right)$$
$$+ \frac{\partial u'}{\partial v}\left(\frac{\partial v}{\partial x}\frac{\partial x}{\partial x'} + \frac{\partial v}{\partial y}\frac{\partial y}{\partial x'} + \frac{\partial v}{\partial z}\frac{\partial z}{\partial x'}\right)$$
$$+ \frac{\partial u'}{\partial w}\left(\frac{\partial w}{\partial x}\frac{\partial x}{\partial x'} + \frac{\partial w}{\partial y}\frac{\partial y}{\partial x'} + \frac{\partial w}{\partial z}\frac{\partial z}{\partial x'}\right)$$

Making use of Eqs. (2-24) and (2-25), we have

$$\epsilon_{x'x'} = l_1\left(l_1\frac{\partial u}{\partial x} + m_1\frac{\partial u}{\partial y} + n_1\frac{\partial u}{\partial z}\right)$$
$$+ m_1\left(l_1\frac{\partial v}{\partial x} + m_1\frac{\partial v}{\partial y} + n_1\frac{\partial v}{\partial z}\right)$$
$$+ n_1\left(l_1\frac{\partial w}{\partial x} + m_1\frac{\partial w}{\partial y} + n_1\frac{\partial w}{\partial z}\right)$$

By collecting terms and using Eqs. (2-23) we finally find

$$\epsilon_{x'x'} = \epsilon_{xx}l_1^2 + \epsilon_{yy}m_1^2 + \epsilon_{zz}n_1^2 + \epsilon_{xy}l_1m_1 + \epsilon_{yz}m_1n_1 + \epsilon_{zx}n_1l_1 \tag{2-26a}$$

In a similar manner, it can be shown that

$$\epsilon_{x'y'} = 2l_1l_2\epsilon_{xx} + 2m_1m_2\epsilon_{yy} + 2n_1n_2\epsilon_{zz}$$
$$+ (l_1m_2 + m_1l_2)\epsilon_{xy} + (m_1n_2 + n_1m_2)\epsilon_{yz}$$
$$+ (n_1l_2 + l_1n_2)\epsilon_{zz} \tag{2-26b}$$

Since the directions x' and y' were chosen arbitrarily, we see that the state of strain for any set of axes is known if the strains for one set of axes are known.

The mathematical similarity of Eqs. (2-26) to Eqs. (2-13a) and (2-13b) is immediately obvious; if we replace ϵ_{ii} by σ_{ii} and ϵ_{ij} ($i \neq j$) by

$2\sigma_{ij}$ in Eqs. (2-26), we obtain Eqs. (2-13a) and (2-13b). Thus all that has been said about stresses in Eqs. (2-13) to (2-20) applies to strain components if the above substitutions are made. The principal strains and maximum shear strains can be found from equations similar to those given for stress or by the construction of a Mohr's circle. A detailed treatment of this may be found in Ref. 3.

Corresponding to the case of plane stress, there is a state, defined as *plane strain*, in which $w = 0$ while u and v are functions of x and y only. With these deflections Eqs. (2-23) give $\epsilon_{zz} = \epsilon_{zy} = \epsilon_{zz} = 0$, which is analogous to plane stress in which $\sigma_{zz} = \sigma_{zy} = \sigma_{zz} = 0$. While the plane-stress condition applies to thin bodies, the plane-strain state occurs in elongated bodies (Sec. 4-4).

2-8 COMPATIBILITY EQUATIONS

In Sec. 2-7 we saw that if we know the six components of strain at a point referred to an orthogonal set of axes, we have completely defined the state of strain at the point. We also note that the six strains are related to the three displacements u, v, and w by Eqs. (2-23). It has been further pointed out that u, v, and w must be continuous, single-valued functions if continuity is to be preserved in the deformed body.

Since the six strains are defined in terms of three displacement functions, the strains can not be prescribed arbitrarily, and there must then be additional relationships which exist among the strains. To find these for infinitesimal strains we start by differentiating ϵ_{xy} from Eqs. (2-23) with respect to x and y, which gives

$$\frac{\partial^2 \epsilon_{xy}}{\partial x\, \partial y} = \frac{\partial^2}{\partial x\, \partial y} \frac{\partial v}{\partial x} + \frac{\partial^2}{\partial x\, \partial y} \frac{\partial u}{\partial y}$$

If the functions u and v are continuous, we may reorder the derivatives to give

$$\frac{\partial^2 \epsilon_{xy}}{\partial x\, \partial y} = \frac{\partial^2}{\partial x^2} \frac{\partial v}{\partial y} + \frac{\partial^2}{\partial y^2} \frac{\partial u}{\partial x}$$

which by using Eqs. (2-23) may be written

$$\frac{\partial^2 \epsilon_{xy}}{\partial x\, \partial y} = \frac{\partial^2 \epsilon_{yy}}{\partial x^2} + \frac{\partial^2 \epsilon_{xx}}{\partial y^2} \tag{2-27a}$$

showing that the strains ϵ_{xy}, ϵ_{xx}, and ϵ_{yy} are not independent functions.

For the two-dimensional problems of plane stress or plane strain Eq. (2-27a) is the only *compatibility equation* which is required, but in the three-dimensional case additional relationships must be found. Two of these can be derived in the same manner as Eq. (2-27a), to give

$$\frac{\partial^2 \epsilon_{yz}}{\partial y \, \partial z} = \frac{\partial^2 \epsilon_{yy}}{\partial z^2} + \frac{\partial^2 \epsilon_{zz}}{\partial y^2} \tag{2-27b}$$

$$\frac{\partial^2 \epsilon_{zx}}{\partial z \, \partial x} = \frac{\partial^2 \epsilon_{zz}}{\partial x^2} + \frac{\partial^2 \epsilon_{xx}}{\partial z^2} \tag{2-27c}$$

Three additional relationships can be found. If we differentiate ϵ_{xy} with respect to x and z and add the result to ϵ_{zx} differentiated with respect to y and x, we obtain

$$\frac{\partial^2 \epsilon_{xy}}{\partial x \, \partial z} + \frac{\partial^2 \epsilon_{zx}}{\partial y \, \partial x} = \frac{\partial^2}{\partial x \, \partial z} \left(\frac{\partial u}{\partial y} + \frac{\partial v}{\partial x} \right) + \frac{\partial^2}{\partial y \, \partial x} \left(\frac{\partial w}{\partial x} + \frac{\partial u}{\partial z} \right)$$

which by reordering the derivatives and using Eq. (2-23) can be simplified to

$$2 \frac{\partial^2 \epsilon_{xx}}{\partial y \, \partial z} = \frac{\partial}{\partial x} \left(-\frac{\partial \epsilon_{yz}}{\partial x} + \frac{\partial \epsilon_{zx}}{\partial y} + \frac{\partial \epsilon_{xy}}{\partial z} \right) \tag{2-27d}$$

In a similar manner we find that

$$2 \frac{\partial^2 \epsilon_{yy}}{\partial z \, \partial x} = \frac{\partial}{\partial y} \left(\frac{\partial \epsilon_{yz}}{\partial x} - \frac{\partial \epsilon_{zx}}{\partial y} + \frac{\partial \epsilon_{xy}}{\partial z} \right) \tag{2-27e}$$

$$2 \frac{\partial^2 \epsilon_{zz}}{\partial x \, \partial y} = \frac{\partial}{\partial z} \left(\frac{\partial \epsilon_{yz}}{\partial x} + \frac{\partial \epsilon_{zx}}{\partial y} - \frac{\partial \epsilon_{xy}}{\partial z} \right) \tag{2-27f}$$

Equations (2-27a) to (2-27f) are known as the *strain-compatibility equations*.

In addition to the above compatibility conditions, which must be satisfied at all points within the interior of the body, compatibility must also exist over the surface S_2 where displacements are specified. The requirement here is that the displacement functions u, v, and w equal prescribed values of these displacements on those parts of the surface where forces are not specified. Such conditions are referred to as *displacement boundary conditions*.

2-9 SUMMARY

It is well to review what we have learned in this chapter. It has been seen that there are six unknown stress components at each point in a deformable body but only three equations of equilibrium; therefore we cannot solve for the stresses without resorting to additional physical conditions. We turn to the three unknown deformations u, v, and w and define six components of strain in terms of these three unknowns. We then derive six auxiliary strain-compatibility equations which the strains must satisfy if the deformations are to be continuous. At this point we have six unknown stresses, six unknown strains, and three unknown displacements, or a total of 15 unknowns for the three-dimensional case. We have three equilibrium equations and six strain-displacement equations, or a total of nine equations. We thus need six additional equations if we are to obtain a solution. We have noted, however, that thus far we have made no assumptions regarding the force-temperature-displacement relationships of the material of the body. It is this material behavior which will provide us with the six additional equations we require. In the next chapter we shall discuss the strain-stress-temperature-time behavior of structural materials. We shall find that even in the case where the stress is uniaxial the general behavior is complex and not likely to be expressible by uncomplicated relationships. The situation is even more difficult in the three-dimensional case, where for mathematical simplicity we shall restrict ourselves to linearly elastic materials. The equilibrium, strain-displacement, and compatibility equations of this chapter will be combined with the linearly elastic stress-strain equations of Chap. 3 to formulate the theory for the stress analysis of elastic bodies in Chap. 4.

REFERENCES

1. Wang, C. T.: "Applied Elasticity," McGraw-Hill Book Company, New York, 1953.
2. Parks, V. J., and A. J. Durelli: Various Forms of the Strain-displacement Relations Applied to Experimental Stress Analysis, *Proc. Soc. Expl. Stress Anal.*, **21**(1): 37–47 (1964).
3. Crandall, S. H., and N. C. Dahl (eds.): "An Introduction to the Mechanics of Solids," McGraw-Hill Book Company, New York, 1959.
4. Timoshenko, S. P., and J. N. Goodier: "Theory of Elasticity," 2d ed., McGraw-Hill Book Company, New York, 1951.

PROBLEMS

2-1. The missile fuel tank shown is pressurized to 100 psig and is subjected to a torque of 5×10^6 in.-lb during a roll maneuver.

(a) Use equations from strength of materials to find the hoop and axial stresses due to the pressure and the shear stress due to torque. [*Ans.* $\sigma_{\text{axial}} = 25,000$ psi; $\sigma_{\text{hoop}} = 50,000$ psi; $\sigma_s = 25,500$ psi.]

(b) Find the principal and maximum shear stresses and the angle to each. [*Ans.* $\sigma_{11} = 65,800$ psi; $\sigma_{22} = 9100$ psi, $\sigma_{s,\text{max}} = 28,400$ psi.]

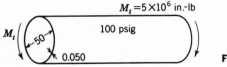

$M_t = 5 \times 10^6$ in.-lb

100 psig

M_t

50

0.050

Fig. P2-1

2-2. Show that in polar coordinates the two-dimensional equations of equilibrium in the radial and circumferential directions are

$$\frac{\partial \sigma_{rr}}{\partial r} + \frac{1}{r}\frac{\partial \sigma_{r\theta}}{\partial \theta} + \frac{\sigma_{rr} - \sigma_{\theta\theta}}{r} + R = 0$$

$$\frac{\partial \sigma_{\theta r}}{\partial r} + \frac{1}{r}\frac{\partial \sigma_{\theta\theta}}{\partial \theta} + 2\frac{\sigma_{r\theta}}{r} + \Theta = 0$$

where R and Θ are body forces in the r and θ directions.

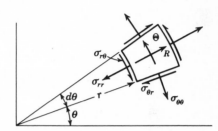

Fig. P2-2

2-3. From the results of Prob. 2-2 show that the equation of equilibrium of a circular disk rotating about its axis at a constant angular velocity Ω is

$$r\frac{d\sigma_{rr}}{dr} + \sigma_{rr} - \sigma_{\theta\theta} + \frac{r^2\omega\Omega^2}{g} = 0$$

where ω is the specific weight of the disk.

2-4. Show that the two-dimensional infinitesimal strain-displacement equations in polar coordinates are

$$\epsilon_{rr} = \frac{\partial u}{\partial r} \qquad \epsilon_{\theta\theta} = \frac{u}{r} + \frac{1}{r}\frac{\partial v}{\partial \theta} \qquad \epsilon_{r\theta} = \frac{1}{r}\frac{\partial u}{\partial \theta} + \frac{\partial v}{\partial r} - \frac{v}{r}$$

where u and v are the radial and tangential components of the displacement of a point with coordinates (r,θ).

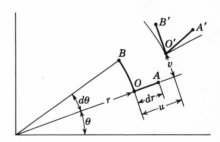

Fig. P2-4

2-5. For unrestrained thermal expansion of a body the strain components are

$$\epsilon_{xx} = \epsilon_{yy} = \epsilon_{zz} = \alpha T \qquad \epsilon_{xy} = \epsilon_{yz} = \epsilon_{zx} = 0$$

where α is the coefficient of expansion and $T = T(x,y,z)$ is the *change* in temperature at the point (x,y,z). Prove that this state of strain can occur only when T is a linear function of x, y, z.

2-6. The elementary theory of beams from strength of materials gives the stresses

$$\sigma_{xx} = -\frac{My}{I} \qquad \sigma_{yy} = 0 \qquad \sigma_{xy} = -\frac{VQ}{Ib}$$

where

$$Q = \frac{b}{2}\left[\left(\frac{h}{2}\right)^2 - y^2\right]$$

for the rectangular beam shown. Do these satisfy the equations of equilibrium and the boundary conditions for a cantilever support at $x = L$ for (a) a shear force P at $x = 0$, (b) a uniform pressure on the upper surface, (c) the beam under the action of its own weight?

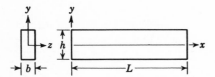

Fig. P2-6

2-7. Write the equations for (a) principal strains, (b) maximum shear strain, (c) the angles to parts (a) and (b), and (d) the strain-transformation equations for rotation of axes in plane strain.

2-8. A strain-gage rosette measures the longitudinal strains ϵ_{xx}, ϵ_{yy}, and $\epsilon_{x'x'}$ in the three directions shown. By using rotation-of-axes equations show that

$$\epsilon_{11,22} = \frac{\epsilon_{xx} + \epsilon_{yy}}{2} \pm \frac{1}{\sqrt{2}}\sqrt{(\epsilon_{xx} - \epsilon_{x'x'})^2 + (\epsilon_{yy} - \epsilon_{x'x'})^2}$$

$$\epsilon_{s,\max} = \sqrt{2}\sqrt{(\epsilon_{xx} - \epsilon_{x'x'})^2 + (\epsilon_{yy} - \epsilon_{x'x'})^2}$$

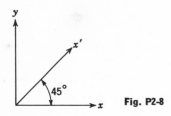

Fig. P2-8

2-9. Derive Eq. (2-26b).

3
Mechanical Behavior
of Materials

3-1 INTRODUCTION

In the preceding chapter we saw that the equilibrium conditions are generally insufficient to determine the stresses uniquely. An exception occurs in those cases where we know that some of the stress components are zero and the remaining components are equal in number to the equilibrium equations which do not identically vanish. The stresses can then be determined from equilibrium conditions alone, and the problem is referred to as being *statically determinate*. In most cases the body is *statically indeterminate*, and we must also use the compatibility conditions to obtain a solution. When this is the situation, stress-strain relationships are required for a solution. Even in determinate cases it is advisable to check the solution by substituting it into the compatibility equations to be certain that the stresses which were assumed to be zero are truly so. In either case, stress-strain relationships are required if deflections are required.

Ideally, the relationships should apply for all stress and temperature levels and include time effects; but if such relations were known (and in general they are not) they would be so complex that the mathematical formulation of the problem would be too complicated to permit a solution. For this reason we idealize the behavior of the material and in doing so restrict the applicability of the results to those cases where the assumed behavior very nearly describes the actual response of the material. Three-dimensional stress-strain-temperature laws are complicated, and we therefore begin our discussion with simple loading cases in which all stresses except one are zero. In the development of our equations we shall use the so called *phenomenological approach*, wherein empirical mathematical relationships are used to describe the experimentally observed phenomena, rather than attempting to derive relationships based upon the atomic and crystalline structure of the material.

The effects that elevated temperature and cyclic loading have upon

mechanical properties are briefly described.   Structural weight is strongly
dependent upon the mechanical properties of the materials from which
the vehicle is fabricated.   Because of this, an introduction to material
selection for minimum weight is given.

Only one type of three-dimensional stress-strain law will be considered in this chapter, namely, that for a linearly elastic isotropic
material.   Three-dimensional stress-strain relationships for plastic buckling of plates are briefly discussed in Sec. 15-8.   Further information on
three-dimensional plastic stress-strain laws can be found in Refs. 1 to 3.

3-2   THE TENSILE TEST

We begin by considering the behavior of a long straight member with a
uniform cross section in its central portion that is subjected to an axial
tensile load P, as shown in Fig. 3-1.   In the center all stresses are zero
except the normal stress in the axial direction, which is constant across
the cross section and is therefore given by $\sigma_n = P/A$, where A is the
original cross-sectional area before the application of load.   If ΔL, the
change in length between two points originally a distance L apart, is
measured, the longitudinal strain (which is constant along the reduced
section) can be computed from $\epsilon_{long} = \Delta L/L$.   The stress-strain relationship for the material can then be determined by plotting the measured
strains against the stresses for increasing loads.

For some materials, such as mild steel, the resulting curve will take
the form shown in Fig. 3-2.   We observe that initially a linear (or very
nearly linear) relationship exists between the strains and stresses.   This
proportionality eventually terminates at a stress referred to as the
proportional limit.   After this stress is reached, the slope of the curve
decreases until a stress is reached at which the strain increases with no
increase in stress (in some cases it increases with a decrease in stress, as
shown by the dotted line).   The stress at which this occurs is called the
yield point.   Eventually, it becomes necessary to raise the stress to

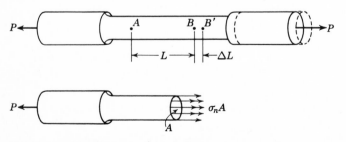

Fig. 3-1   Tensile test specimen.

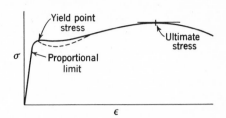

Fig. 3-2 Stress-strain curve for a material with a yield point.

increase the strain, and we refer to this effect as *strain hardening*. As the stress is further increased, a point is reached where the curve has a second horizontal tangent; thereafter strain increases while stress decreases until rupture occurs. The highest stress that is reached is referred to as the *ultimate tensile strength*. The strain at rupture multiplied by 100 is defined as the *percent elongation*.

Most materials used in flight structures do not exhibit a well-defined yield point but instead have a stress-strain curve of the shape shown in Fig. 3-3. Up to a certain stress, a linear relationship again exists between stress and strain. However, it is noted that there is no yield point, and a horizontal tangent does not exist until the ultimate stress is reached.

In Figs. 3-2 and 3-3 the strain in the linear region can be obtained from the stress by the equation

$$\epsilon_{\text{long}} = \frac{1}{E}\,\sigma_n \tag{3-1}$$

where the slope of the curve E is referred to as *Young's modulus* or the *modulus of elasticity*. Pure aluminum is more corrosion-resistant than alloyed aluminum; because of this it is often used as a cladding to protect alloyed sheet materials. In clad materials there are two linear regions in the stress-strain curve. One occurs below the proportional limit of the face layers (which have a lower strength than the core), and the other extends from this stress to the point where the core reaches its proportional limit. The first slope is known as the *primary modulus* and the other the *secondary modulus*. In each of the cases described the propor-

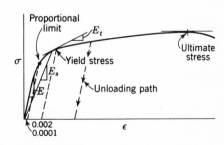

Fig. 3-3 Stress-strain curve for a material with no yield point.

tional limit is difficult to obtain accurately, and it is conventional to define this quantity as the stress at the point of intersection between the curve and a straight line drawn parallel to the linear portion of the curve but offset from the origin by a strain of 0.0001 (Fig. 3-3).

At low stress levels if the load is removed, the material will return to the zero stress condition along the same.stress-strain curve that it followed during the loading cycle, and no permanent deformations will occur. Within this range the material is said to be *elastic,* and the stress that fixes the upper limit of the range is defined as the *elastic limit.* Below the elastic limit, strain is a single-valued function of stress. For most structural materials the elastic limit very nearly coincides with the proportional limit, although the two are defined by entirely different physical considerations. It is possible, for instance, to have a material in which the strains are not directly proportional to the stresses at any stress level, but the material may still behave in an elastic manner. Such a material is referred to as being *nonlinearly elastic.* A material having both a proportional limit and an elastic limit is said to be *linearly elastic* if the stress does not exceed the lower of these two limits.

Above the elastic limit the material no longer unloads along the same stress-strain curve that it followed on the loading cycle but will usually unload along a line parallel to the linear portion of the stress-strain curve, as shown in Fig. 3-3. In this case strain is not a single-valued function of stress, for at a given stress level there is one strain for loading and another for unloading. Furthermore, the strain during unloading depends upon the highest stress reached in the loading cycle.

For materials that behave similar to Fig. 3-3, it is conventional to define an *offset yield stress* as the stress which results in a permanent strain of 0.002. This stress is at the point of intersection of the curve and a line drawn parallel to the linear portion of the curve but offset from the origin by a strain of 0.002 (Fig. 3-3). As stated in Chap. 1, one of the design criteria is often a requirement that the limit loads should not exceed this offset yield stress. In addition to the modulus of elasticity two other moduli are useful. The modulus of elasticity was defined as the slope of the stress-strain curve below the proportional limit. Above the proportional limit the slope of the curve, which is no longer a constant, is defined as the *tangent modulus* E_t, so that

$$E_t = \frac{d\sigma}{d\epsilon} \tag{3-2}$$

Below the proportional limit the modulus of elasticity may also be defined as the stress divided by the strain. Above the proportional limit this ratio, which is not constant, is defined as the *secant modulus* E_s, so that

$$E_s = \frac{\sigma}{\epsilon} \tag{3-3}$$

The tangent and secant moduli are functions of stress level, and below the proportional limit they both become equal to the modulus of elasticity.

If cross-sectional dimensions are measured during a tension test, it is found that longitudinal elongation is accompanied by a transverse contraction. The strains in the two directions are related by the equation

$$\epsilon_{trans} = -\nu\epsilon_{long} \tag{3-4}$$

where ν is referred to as *Poisson's ratio*. This ratio is constant below the proportional limit, and for most structural materials it is in the range of 0.25 to 0.33. Unless it is known more accurately, it is commonly assumed to be 0.3 for structural materials. Above the proportional limit ν gradually increases and approaches a value of 0.5 at large plastic strains.

3-3 COMPRESSION AND SHEAR TESTS

The test described in the last section may also be performed with compressive instead of tensile forces if the specimen is supported against lateral buckling. For materials having a tensile stress-strain curve similar to Fig. 3-3, the shape of the compressive curve resembles that of the tensile test. It is usually found that the compressive modulus of elasticity differs only slightly from the tensile value but that significant differences may exist in the yield stresses. The ultimate compressive stress for a ductile material is difficult to measure because the cross section increases as the length shortens, so that no maximum stress may exist. In such cases it is customary to arbitrarily take the ultimate compressive stress equal to the ultimate tensile stress.

Although standard methods have been adopted[4] for obtaining tensile and compressive properties of metals, no standard has been established for measuring shear properties. One procedure used for obtaining such data is to test a thin-walled circular tube in torsion, as shown in Fig. 3-4. In this case all stresses are zero except for a tangential shear stress σ_s, which also acts upon planes passing through the axis of the tube. We refer to this case where no other stress components are present as *pure shear*. It is known from elementary mechanics of materials, and it will be shown in Chap. 8, that this shear stress is given by $\sigma_s = M_t r/I_p$, where M_t is the applied torque, r is the radius to the point where the stress is desired, and I_p is the polar moment of inertia of the cross section of the tube. Since the wall is thin, r is essentially constant across the section, and therefore the shear stress is very nearly constant throughout the

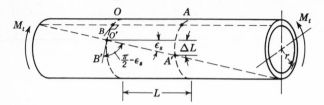

Fig. 3-4 Torsion test specimen.

tube.   Lines which were originally generators of the cylinder deflect into helices when the torque is applied, so that the angle AOB shown in Fig. 3-4 distorts to $A'O'B'$.   The change in this angle is then the shear strain ϵ_s, which for small deflections is given by $\epsilon_s = \Delta L/L$.   If this angle is determined as a function of the shear stress and the results are plotted, we obtain a shear-stress–shear-strain curve.   The shape of this curve resembles the tensile curve of the material, and so for most materials used in flight structures the appearance of the shear-stress–shear-strain curve is similar to that shown in Fig. 3-3.   As in the tensile test, a linear range exists wherein the shear strains are directly proportional to the shear stresses.   The relationship within this range may be expressed as

$$\epsilon_s = \frac{1}{G}\,\sigma_s \tag{3-5}$$

where G is known as the *shear modulus of elasticity* or the *modulus of rigidity*.

3-4 IDEALIZATIONS OF THE STRESS–STRAIN CURVE

In the two previous sections we have seen that the experimental stress-strain curve has the characteristic shape of Fig. 3-3 for uniaxial tension and compression and for pure shear.   This curve is not in a form useful for analytical work, and it is desirable to express the relationship mathematically.   As the degree of accuracy with which we mathematically approximate the stress-strain curve increases, the complexity of the mathematical equation also increases, and it is therefore desirable to use the simplest relationship which embodies the significant behavior of the material.   For this reason we idealize the stress-strain characteristics of the material.

Several idealizations have been suggested, and the choice of them depends upon the material and the stress and temperature level.   Some of the idealized models are shown in Fig. 3-5.   In Fig. 3-5a is shown the behavior of a *rigid* body in which there is no strain as load is increased or decreased. Of course no such material exists, but in many cases the deformations of a body has a negligible effect upon the analysis. This assumption forms the basis for rigid-body mechanics. In Fig. 3-5b a material is shown which behaves in a *nonlinearly elastic* manner; i.e., the material loads and unloads along the same curve. A *linearly elastic* material is shown in Fig. 3-5c; Eqs. (3-1) and (3-5) respectively describe this behavior for normal and shear stresses.

   Idealizations for materials behaving in a plastic manner are shown in Fig. 3-5d and e. Unlike the elastic models, in these cases the *plastic strains* are not recoverable. In Fig. 3-5d a *rigid, perfectly plastic* material is shown, in which no strain occurs until the yield point is reached, after which it continues to deform with no increase in stress. Upon unloading the entire strain remains as permanent set. A *rigid-strain-hardening plastic* material is illustrated in Fig. 3-5e. Again no strain occurs until the yield point is reached, but an increasing stress is required to continue the straining. The materials shown in Fig. 3-5f and g are *elastoplastic*. In Fig. 3-5f the material behaves in a linearly elastic manner until the yield point is reached, after which it behaves like a perfectly plastic body. When the stress is relieved, the elastic component of the strain is recovered, but the plastic strain remains as permanent set. A *linearly elastic-strain-*

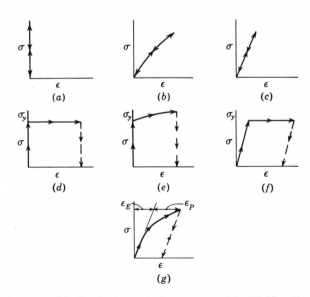

Fig. 3-5 Idealized stress-strain curves. (*a*) Rigid; (*b*) nonlinearly elastic; (*c*) linearly elastic; (*d*) rigid–perfectly plastic; (*e*) rigid–strain-hardening plastic; (*f*) linearly elastic–perfectly plastic; (*g*) linearly elastic–strain-hardening plastic.

hardening plastic material is illustrated in Fig. 3-5*g*. The behavior is linearly elastic until the proportional limit (assumed equal to the elastic limit) is reached. Above this stress there is a strain-hardening plastic component of strain ϵ_P in addition to the elastic strain ϵ_E. The elastic-strain component is recovered upon the removal of the stress. This model accurately describes the uniaxial behavior of most structural materials used in flight structures.

3-5 THREE-PARAMETER REPRESENTATIONS OF STRESS-STRAIN CURVES

We have seen that structural materials behave in a linearly elastic-strain-hardening plastic manner. Useful methods for approximating this behavior by simple three-parameter relationships have been suggested by Ramberg and Osgood[5] and by Hill.[6]

Referring to Fig. 3-5*g*, we assume that the strain can be divided into a linearly elastic component and a plastic component. Thus

$$\epsilon = \epsilon_E + \epsilon_P \tag{3-6}$$

where, as indicated by Eq. (3-1), the elastic strain is equal to σ/E. It has been found that the plastic component can be considered proportional to the stress level raised to a power which depends upon the material. Equation (3-6) may then be written as

$$\epsilon = \frac{\sigma}{E} + \beta \left(\frac{\sigma}{E}\right)^n \tag{3-7}$$

where the parameters E, β, and n are material constants which must be obtained from tests. We note that for $\sigma = 0$, Eq. (3-7) gives $d\sigma/d\epsilon = E$. Thus the slope is equal to the modulus of elasticity at the origin, and this parameter is obtained from the experimental curve in the usual manner. The remaining two parameters are determined by requiring the empirical

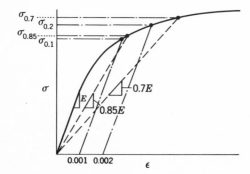

Fig. 3-6 Construction for Ramberg-Osgood and Hill parameters.

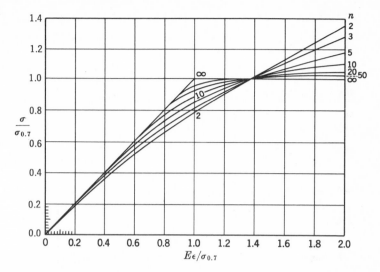

Fig. 3-7 Ramberg-Osgood nondimensionalized stress-strain curves (Ref. 5).

curve, given by Eq. (3-7), to coincide with the experimental curve at two arbitrary points.

In the Ramberg-Osgood method the points are selected so that the curves coincide at secant moduli of $0.7E$ and $0.85E$, as shown in Fig. 3-6. It was found that the point with a secant modulus of $0.7E$ is close to the 0.2 percent offset yield stress for most aircraft materials. Noting that when the stress is equal to $\sigma_{0.7}$, the strain is equal to $\sigma_{0.7}/0.7E$, and substituting the coordinates of this point into Eq. (3-7), we find that $\beta = \frac{3}{7}(E/\sigma_{0.7})^{n-1}$.

Substituting this result into Eq. (3-7) and multiplying by $E/\sigma_{0.7}$, we obtain

$$\frac{\epsilon E}{\sigma_{0.7}} = \frac{\sigma}{\sigma_{0.7}}\left[1 + \frac{3}{7}\left(\frac{\sigma}{\sigma_{0.7}}\right)^{n-1}\right] \tag{3-8}$$

Next requiring that the empirical curve pass through the point $(\sigma = \sigma_{0.85}, \epsilon = \sigma_{0.85}/0.85E)$, we find

$$n = 1 + \frac{\ln 17/7}{\ln (\sigma_{0.7}/\sigma_{0.85})} \tag{3-9}$$

Equations (3-8) and (3-9) are plotted in Figs. 3-7 and 3-8.

The three parameters in the Ramberg-Osgood method are then E, $\sigma_{0.7}$, and n. The properties E, $\sigma_{0.7}$, and $\sigma_{0.85}$ can be obtained from an experimental stress-strain curve by the graphical construction shown in Fig. 3-6. The parameter n is then found from Eq. (3-9) or Fig. 3-8.

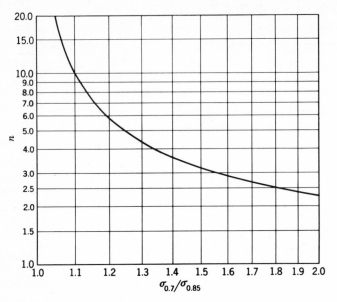

Fig. 3-8 Shape parameter n as a function of $\sigma_{0.7}/\sigma_{0.85}$ (Ref. 5).

Curves giving the dependency of these parameters upon temperature for a wide variety of materials may be found in Ref. 7. Tabulations of these parameters are also given in Refs. 8 and 9.

Equation (3-8) can be used to obtain simple analytic expressions for the tangent and secant moduli; from Eq. (3-2) we find

$$E_t = \frac{E}{1 + (3n/7)(\sigma/\sigma_{0.7})^{n-1}} \tag{3-10}$$

and from Eq. (3-3)

$$E_s = \frac{E}{1 + \tfrac{3}{7}(\sigma/\sigma_{0.7})^{n-1}} \tag{3-11}$$

These equations will be used in Chaps. 14 to 16 to predict the buckling of columns and plates in the plastic range.

In the Hill method, the two points at which the empirical equation and the actual stress-strain curve coincide are those having 0.1 and 0.2 percent offset yield stresses (Fig. 3-6). Designating these as $\sigma_{0.1}$ and $\sigma_{0.2}$, respectively, it can be shown that

$$\epsilon = \frac{\sigma}{E} + 0.002 \left(\frac{\sigma}{\sigma_{0.2}}\right)^{n'} \tag{3-12}$$

where

$$n' = \frac{0.301}{\log\,(\sigma_{0.2}/\sigma_{0.1})} \tag{3-13}$$

The parameters E, $\sigma_{0.2}$, and n' have been determined as a function of temperature for a number of aircraft structural materials and are given in Ref. 7.

Both these methods give an accurate representation of the stress-strain curve to slightly above the 0.2 percent offset yield stress. They cannot be expected to give satisfactory results at much higher stresses. It should also be pointed out that the relationships are correct only for stresses which are increasing in magnitude, since unloading occurs along a line parallel to the linear portion of the stress-strain curve.

3-6 EFFECT OF TEMPERATURE UPON SHORT–TIME STATIC PROPERTIES

The preceding sections dealt with the room-temperature behavior of materials. In addition to causing thermal expansion, temperature affects the mechanical properties of materials. Although there are exceptions, such as some forms of graphite, most materials exhibit a reduction in both strength and stiffness with an increase in temperature. This is seen in Fig. 3-9, which shows the effect of temperature upon the stress-strain

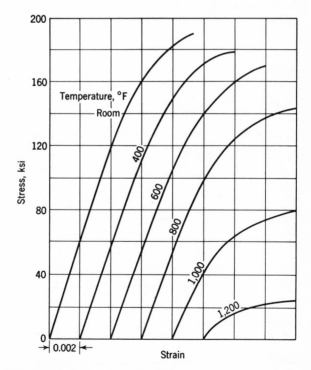

Fig. 3-9 Compressive stress-strain curves for 17-7 PH stainless-steel sheet (*NACA Tech. Note* 4074).

curve of a stainless steel.   It is observed that the modulus of elasticity, yield stress, and ultimate stress are reduced with an increase in temperature.   If the temperatures are not excessive, i.e., if they are well below the melting point, and the duration of loading is short, the given methods and relationships are applicable.   However, the material properties used in the equations must be obtained from a stress-strain curve of the material at the desired temperature.   In most cases the length of time that the material has been at a temperature prior to testing also has an effect upon the stress-strain curve and therefore upon the properties.

3-7   CREEP

The stress-strain equations discussed thus far have not included time as a parameter.   These relationships accurately describe the behavior of most structural materials at low temperatures and at moderate temperatures if the duration of loading is short.   Some materials, such as plastics, rubber, and solid propellants, continue to strain with time when subjected to a constant stress even at room temperature.   Structural materials exhibit this same behavior at high temperatures and at moderate temperatures if the duration of loading is long.   The time-dependent portion of the strain is referred to as *creep*.

Consider the case of a uniaxially loaded tensile specimen at elevated temperature, loaded by a force maintained constant with time.   The typical strain-time response observed in such a test is shown in Fig. 3-10. It is seen that there is an essentially instantaneous strain due to elastic, thermal, and (if the stresses are high enough) plastic deformation.   If the axial force is held constant, the strain will continue to increase with time but at a decreasing rate until the strain rate becomes constant.   The time interval during the constant creep rate is referred to as the *second*

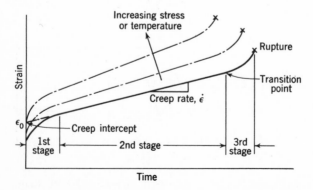

Fig. 3-10   Typical creep curves.

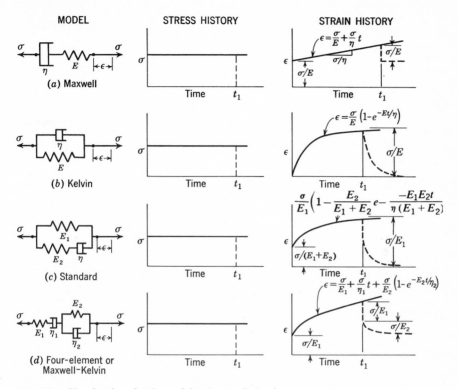

Fig. 3-11 Simple viscoelastic models of creep behavior.

stage of creep, and the interval preceding it is known as the *first stage*. The second stage is terminated at a point known as the *transition point*. The *third stage* of creep is that period of increasing creep rate which is terminated by rupture. When creep is present, failure may occur at stresses which are much below the short-time elevated-temperature ultimate tensile strength. As indicated in the figure, increasing either the stress or the temperature increases the strains and strain rates. At high temperatures and stresses no stage of constant strain rate may be discernable.

Within the second stage, if one exists, the strain may be found from the simple relationship

$$\epsilon = \epsilon_0 + \dot{\epsilon}t \tag{3-14}$$

where ϵ_0 is the *creep intercept,* obtained by extending the linear portion of the curve back to $t = 0$, and $\dot{\epsilon}$ is the second-stage *creep rate.* Additional empirical uniaxial creep relationships and three-dimensional equations may be found in Ref. 3.

The time-dependent strain suggests a viscous behavior of the material, and this has led to conceptual models containing viscous elements. Examples of these are shown in Fig. 3-11, together with the uniaxial strain equations for constant stress. Materials which can be described by models containing elastic springs and viscous dashpots are referred to as *viscoelastic*. It is seen from Fig. 3-11 that the more complicated the model, the more accurately it duplicates the strain-time curves of Fig. 3-10. In addition to increased mathematical complexity in the more complicated models, more material constants must be determined. If the material constants (represented by the elastic and viscous constants) do not depend upon stress, the material is said to be *linearly viscoelastic*. On the other hand, if the constants are functions of stress, the materials are said to be *nonlinearly viscoelastic*.

Viscoelastic models have proved to be very useful in analyzing the behavior of many plastic and rubberlike materials,[11] and an extensive literature exists on the theory of viscoelastic bodies. Since structural materials are usually nonlinearly viscoelastic, the analysis of bodies of these materials is mathematically complicated. In many cases, however, a linear solution may be useful in arriving at approximations that show underlying trends.

3-8 FATIGUE

In the preceding discussions the loads were assumed static or slowly changing with time. In many cases the loading on a structure undergoes a cyclic variation. Consider a rotating shaft which is subjected to an axial load and a bending moment applied in a plane which does not rotate with the shaft. The stress at any point on the surface of the shaft will consist of a constant stress due to the axial load plus a bending stress which varies sinusoidally as the shaft rotates. The stress history at the point will then appear as shown in Fig. 3-12. Even though the maximum

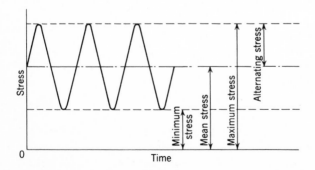

Fig. 3-12 Typical cyclic loading.

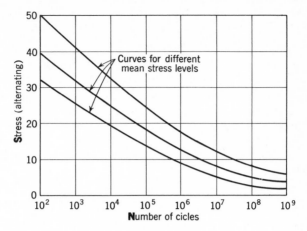

Fig. 3-13 Typical SN fatigue curves.

stresses are below the ultimate strength and may actually be less than the proportional limit, it is possible for the shaft to fail after the elapse of a sufficiently large number of load cycles. When this occurs, the part is said to have undergone a *fatigue failure*. The number of cycles to failure depends upon the material, the mean stress, the alternating stress, the type of loading which produced the stresses, the temperature, the shape of the body, the surface finish, and the corrosive effect of the environment.

Fatigue data are often presented in the form of a curve of alternating stress versus the logarithm of the number of cycles to failure for a given mean stress, as shown in Fig. 3-13. Such a curve is spoken of as an *SN curve*. For most materials the alternating stress for failure continues to decrease as the number of cycles increases. Other materials have an *SN* curve with a horizontal asymptote. This asymptotic value of stress is known as the *endurance limit*, and if the alternating stress is maintained below this limit, fatigue failure will not occur regardless of the number of cycles of load application. As indicated in Fig. 3-13, the *SN* curve is a function of the mean stress. Although the data in Fig. 3-13 are represented by lines, in reality they should be a broad bands, because of the large amounts of scatter that are present in fatigue test data.

Another method used to plot fatigue data is the *Soderberg diagram* shown in Fig. 3-14. These curves are merely a cross plot of the data in a family of *SN* curves, using the number of cycles as the parameter instead of mean stress. All points lying along a line passing through the origin have the same value of the ratio A, the alternating stress divided by the mean stress. Along the horizontal axis $A = 0$, and the loading is static.

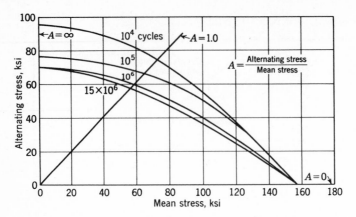

Fig. 3-14 Typical Soderberg fatigue diagram (Ref. 16).

All curves then intersect on the horizontal axis at a mean stress equal to the static strength. All points on the vertical axis have a mean stress equal to zero, so that in this case the loading is completely reversed.

In flight structures the stress history is seldom as simple as that of Fig. 3-12, usually having the random appearance shown in Fig. 3-15. It is desirable to have a method for predicting the fatigue lifetime for a random loading using experimental data from constant-amplitude tests. The simplest of several proposed methods, and one which is in reasonable agreement with experimental evidence, is that suggested by Miner.[12] In this method it is hypothesized that fatigue damage is accumulated at a constant rate, so that if at a given alternating stress σ_{a_1} and mean stress σ_{m_1} the number of cycles to failure is N_1 and the actual number of cycles run at this level is n_1, then the portion of 100 percent damage is n_1/N_1 (Fig. 3-16). If n_2 cycles are now accumulated at a second set of stresses σ_{a_2} and σ_{m_2} and the number of cycles to failure for this condition is N_2, the additional increment of damage is n_2/N_2. The total damage will then be $n_1/N_1 + n_2/N_2$. If this sum is greater than or equal to 1, failure is

Fig. 3-15 Typical random stress history.

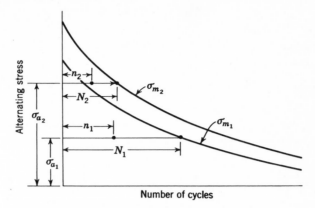

Fig. 3-16   Miner's theory of cumulative damage.

anticipated.   In general, if there are n different sets of stress amplitudes and mean stresses, the criterion for failure will be

$$\sum_{i=1}^{n} \frac{n_i}{N_i} = 1 \qquad (3\text{-}15)$$

It is possible to obtain cumulative fatigue-damage criteria which are more accurate than Eq. (3-15) for specific materials and load histories.   For instance, if the alternating stress is increased as the test proceeds, a phenomenon known as *coaxing* occurs in some materials, and it is found that the summation of n_i/N_i for failure is greater than 1.   In flight structures the loading is usually not applied in a specific pattern but occurs in a random fashion, and in this case Miner's theory has been found to be reasonably accurate.   Because of the uncertainty involved in extrapolating fatigue data from laboratory specimens to structural members with complex geometries, it is advisable to substantiate analyses on critical components by full-scale tests.

References 13 to 15 contain further information on the fatigue problems of aerospace vehicles.

3-9   ALLOWABLE MECHANICAL PROPERTIES

A large number of mechanical properties have been defined and discussed. The values of these properties allowed in the design and manufacture of aerospace vehicles are established by the procuring or certifying agency. Almost without exception, flight vehicles are built for the military services, the National Aeronautics and Space Agency, or commercial use. In the latter case their safety is certified by the Federal Aviation Agency. It is seen that in all cases the design allowables are regulated by governmental agencies, and in the interest of uniformity these organizations have adopted a set of standards which, with a few exceptions, are acceptable to all. These allowable properties are contained in Ref. 16. The design specification for an aerospace vehicle usually contains a paragraph which states that allowable properties of materials will be in accordance with this document and that specific approval must be obtained from the certifying or procuring agency for any data which are not contained in the publication. Reference 16 therefore becomes a legally binding document in the design of the vehicle.

As stated earlier, the factor of safety is applied to the loads and not to the material properties. The values used are those obtained from tests, and, except for allowances for statistical variation, they contain no factors of safety to reduce them to working stresses. Mechanical-property data in Ref. 16 are on one of the following four bases:[17]

1. The *A-basis* property value is the value above which at least 99 percent of the property values are expected to fall with a confidence of 95 percent.
2. The *B-basis* property value is the value above which at least 90 percent of the property values are expected to fall with a confidence of 95 percent.
3. The *S-basis* value is the minimum value specified by the federal, military, or SAE Aerospace Material Specification.
4. The *typical-basis* value is an average value with no statistical assurance.

Instead of presenting both A- and S-basis values, Ref. 16 gives only the smaller of these values. The use of B-basis values is permitted by the governmental agencies subject to certain limitations specified by each of the agencies. In general, these values are permitted in multiple-load-path structures, wherein the failure of one member will cause a transfer of load to the remaining members and no catastrophic failure will result. For single-load-path structures, A- or S-basis values are required. The stress-strain-tangent modulus data of Ref. 16 are typical results. They must be reduced to the A, S, or B basis before they may be used in design. Methods for accomplishing this are described in Ref. 17.

In the rapidly changing aerospace field, it is inevitable that the designer will wish to use new materials that are not listed in Ref. 16. In such cases a special ruling based upon substantiating tests performed by the material manufacturer or the user must be obtained from the procuring or certifying agency.

3-10 MATERIAL SELECTION

Minimum structural weight is of paramount importance in aerospace vehicles. As a result, the use of expensive materials and fabrication methods is often justifiable. Because of the larger number of materials to be considered, the aerospace-vehicle designer has a more difficult material-selection problem than the designer of a civil or machine structure.

The designer attempts to develop a structure of minimum weight (or in some cases of minimum volume) to carry a prescribed load over a given span. The process of selecting the proper combination of material and structural proportions for minimum weight is known as *optimum design*. In this section we shall give some elementary examples to illustrate the considerations involved in material selection for minimum weight. For simplicity, we shall consider problems in which there is only one free dimension to be chosen in the design. More complete treatments of optimum-design principles can be found in Refs. 18 to 20.

In determining the weight of a structure to meet a given design criterion we need:

1. A design criterion that establishes the allowable stress or deformation of the structure
2. An equation that relates the applied loads to the actual stresses or deformations
3. An equation for the weight of the structure in terms of the specific weight of the material and the prescribed and free geometric dimensions

By equating the allowable stress or deflection of step 1 to the actual stress or deflection of step 2, the free dimension can be determined in terms of the loading, the fixed dimensions, and a material property that depends upon the design criterion. Using this relationship, the free dimension can be eliminated from the weight equation of step 3. The result will be the weight of the structure in terms of a parameter involving the applied load and specified dimensions of the structure and a material-property parameter. The effects that the loads, dimensions, and materials have upon the weight are then readily apparent.

Example 3-1   Determine the weight of a thin-walled cylindrical shell of fixed radius R and length L which is subjected to a bending moment M. The material and wall thickness are to be chosen so that failure by elastic buckling does not occur on the compression side of the cylinder.

The allowable bending stress at which elastic buckling occurs in a thin-walled cylinder is approximately[21]

$$\sigma_{\text{allow}} = KE \frac{t}{R} \qquad (a)$$

where K is a constant that is often taken to be 0.39. The actual bending stress is given by $\sigma = Mc/I$, where M is the applied bending moment, c is the distance from the neutral axis to the most highly stressed fiber, and I is the moment of inertia of the cross section. Noting that $c = R$ and $I = \pi R^3 t$, we find

$$\sigma_{\text{actual}} = \frac{M}{\pi R^2 t} \qquad (b)$$

Letting $\sigma_{\text{allow}} = \sigma_{\text{actual}}$ and solving for the free dimension t, we obtain $t = (M/\pi KRE)^{1/2}$. The weight of the cylinder is

$$W = 2\pi RLt\omega \qquad (c)$$

where ω is the specific weight of the material. Substituting for t, we find

$$W = 2 \left(\frac{\pi}{K} \right)^{1/2} (MRL^2)^{1/2} \frac{\omega}{E^{1/2}} \qquad (d)$$

where $(MRL^2)^{1/2}$ is the load-geometry parameter and $\omega/E^{1/2}$ is the material-property parameter.

If M, R, and L are prescribed, the ratio of the weights for two different materials is

$$\frac{W_1}{W_2} = \frac{(E^{1/2}/\omega)_2}{(E^{1/2}/\omega)_1} \qquad (e)$$

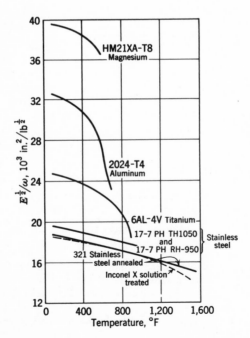

Fig. 3-17 $E^{1/2}/\omega$ as a function of temperature (Ref. 22).

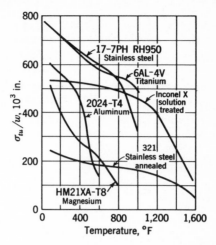

Fig. 3-18 σ_{tu}/ω as a function of temperature (Ref. 22).

We see from Eqs. (d) and (e) that for minimum weight, a high value of $E^{1/2}/\omega$ is desired. This parameter is shown in Fig. 3-17 as a function of temperature for a variety of materials. It is seen that material selection is strongly dependent upon temperature.

Example 3-2 Determine the weight of the cylindrical surface of a thin-walled circular cylinder of given radius R and length L subjected to an internal pressure p. (Vessels of this type occur in pressurized fuel tanks, hydraulic accumulators, rocket motor cases, jet and ramjet engines, etc.)

Assuming that rupture occurs when the maximum principal stress reaches the ultimate tensile stress σ_{tu}, we find $\sigma_{allow} = \sigma_{tu}$. The actual principal stress is given by the hoop-stress equation $\sigma_{actual} = pR/t$. Equating the allowable and actual stresses gives $t = pR/\sigma_{tu}$ for the free dimension. Substituting this into Eq. (c) of Example 3-1 gives

$$W = 2\pi(pR^2L)\frac{\omega}{\sigma_{tu}}$$

The ratio of the weights of two cylinders with the same values of p, R, and L but of different materials will then be

$$\frac{W_1}{W_2} = \frac{(\sigma_{tu}/\omega)_2}{(\sigma_{tu}/\omega)_1}$$

The lightest structure will be the one whose material has the highest value of σ_{tu}/ω. This parameter is shown as a function of temperature for several materials in Fig. 3-18. Reference 22 derives the material-weight parameters for additional types of structures and loadings and gives curves (similar to Figs. 3-17 and 3-18) for these parameters as a function of temperature.

3-11 THREE–DIMENSIONAL LINEARLY ELASTIC STRESS–STRAIN RELATIONSHIPS

It has been seen in Secs. 3-2 to 3-9 that the stress-strain-temperature-time relationships of materials are complicated functions, even for a uniaxial state of stress.   In three-dimensional cases completely general mathematical relationships are not known and, indeed, would be too cumbersome for practical computations.   For this reason we shall restrict our discussion of two- and three-dimensional stress-strain laws to linearly elastic materials.   We have seen that this is an adequate representation of structural materials if the stresses do not exceed the elastic and proportional limits, the temperatures are not excessive, and the duration of loading is short when elevated temperatures occur.

The fact that the strains can be expressed as a linear function of the stresses leads to six relationships of the form

$$\epsilon_{xx} = c_{11}\sigma_{xx} + c_{12}\sigma_{yy} + c_{13}\sigma_{zz} + c_{14}\sigma_{xy} + c_{15}\sigma_{yz} + c_{16}\sigma_{zx}$$

$$\epsilon_{yy} = c_{21}\sigma_{xx} + c_{22}\sigma_{yy} + c_{23}\sigma_{zz} + c_{24}\sigma_{xy} + c_{25}\sigma_{yz} + c_{26}\sigma_{zx}$$

$$\cdots \cdots \cdots \cdots \cdots \cdots \cdots \cdots \cdots \cdots \cdots \cdots \cdots \cdots \cdots$$ $$(3\text{-}16)$$

$$\epsilon_{zz} = c_{61}\sigma_{xx} + c_{62}\sigma_{yy} + c_{63}\sigma_{zz} + c_{64}\sigma_{xy} + c_{65}\sigma_{yz} + c_{66}\sigma_{zx}$$

where the c_{ij}'s are material constants.   At first it would appear that there are 36 independent constants, but it can be shown that symmetry exists such that $c_{ij} = c_{ji}$ (Sec. 6-13).   The number of independent material constants is therefore only 21.   Equations (3-16) are often referred to as the *generalized Hooke's law equations* and apply to any nonisotropic linearly elastic material.   A *nonisotropic*, or *anisotropic*, medium is one having varying properties in different directions.

In most structural materials there are additional elastic symmetries which simplify the strain-stress relationships and reduce the number of material constants.   A material having three mutually perpendicular planes of elastic symmetry is referred to as *orthotropic*.   Examples of such materials are rolled sheet metals, woods, and honeycomb materials.   In these materials the shearing stresses produce no normal strains or shearing strains out of their own planes if the coordinate axes are taken parallel to the planes of symmetry.   The Hooke's law equations for such materials become

$$\epsilon_{xx} = c_{11}\sigma_{xx} + c_{12}\sigma_{yy} + c_{13}\sigma_{zz}$$

$$\epsilon_{yy} = c_{21}\sigma_{xx} + c_{22}\sigma_{yy} + c_{23}\sigma_{zz}$$

$$\epsilon_{zz} = c_{31}\sigma_{xx} + c_{32}\sigma_{yy} + c_{33}\sigma_{zz}$$ $$(3\text{-}17)$$

$$\epsilon_{xy} = c_{44}\sigma_{xy} \qquad \epsilon_{yz} = c_{55}\sigma_{yz} \qquad \epsilon_{zx} = c_{66}\sigma_{zx}$$

Considering the symmetry $c_{ij} = c_{ji}$, it is seen that there are nine independent constants for an orthotropic material.

In many cases the elastic properties are the same in all directions at each point in the body (although the properties may vary from point to point).   Such a material is called *isotropic*.   A material with the same properties at all points is said to be *homogeneous*.

Consider a small parallelepiped from a linearly elastic isotropic body subjected to a three-dimensional state of stress (Fig. 2-3) and further assume that the temperature of the element is *increased* by T. Since the body is linearly elastic, the principle of superposition may be used, and the final strains do not depend upon the order in which the stresses and temperatures are applied. Let us assume that σ_{xx} is applied first. From Eq. (3-1) this leads to a strain in the x direction equal to σ_{xx}/E. Now assume that σ_{yy} is applied. This will cause a strain in the y direction equal to σ_{yy}/E, which will induce an additional strain in the x direction. From Eq. (3-4) this added strain is $-\nu\sigma_{yy}/E$. Similarly the strain in the x direction which results from σ_{zz} is $-\nu\sigma_{zz}/E$. If the temperature is *increased* by an amount T, there will be a thermal strain in all directions equal to αT, where α is the *mean* coefficient of linear expansion for the temperature change T.

From the principle of superposition the total strain in the x direction will be the sum of the strains, or

$$\epsilon_{xx} = \frac{1}{E}\left[\sigma_{xx} - \nu(\sigma_{yy} + \sigma_{zz})\right] + \alpha T$$

The shear stresses induce only shearing strains, and no additional longitudinal strains are produced. From Eq. (3-5) the application of σ_{xy} leads to a shear strain $\epsilon_{xy} = \sigma_{xy}/G$. The remaining shearing stresses cause no additional shear strain in the xy plane, nor does the temperature change produce shearing strains, so that σ_{xy}/G is the total shearing strain. Repeating the above procedures in the other directions gives the following set of six strain-stress-temperature relationships:

$$\epsilon_{xx} = \frac{1}{E}\left[\sigma_{xx} - \nu(\sigma_{yy} + \sigma_{zz})\right] + \alpha T$$

$$\epsilon_{yy} = \frac{1}{E}\left[\sigma_{yy} - \nu(\sigma_{xx} + \sigma_{zz})\right] + \alpha T$$

$$\epsilon_{zz} = \frac{1}{E}\left[\sigma_{zz} - \nu(\sigma_{xx} + \sigma_{yy})\right] + \alpha T$$

$$\epsilon_{xy} = \frac{\sigma_{xy}}{G} \qquad \epsilon_{yz} = \frac{\sigma_{yz}}{G} \qquad \epsilon_{zx} = \frac{\sigma_{zx}}{G}$$

(3-18)

Equations (3-18) provide the required six strain-stress equations, which, together with Eqs. (2-8), (2-27), and the boundary conditions, give the relationships required to determine the six stress, six strain, and three displacement components in a linearly elastic isotropic body. It will be noted in the derivation of Eqs. (3-18) that the elastic properties were assumed to be the same in the x, y, and z directions, which is permissible only for an isotropic material.

For the case of plane stress Eqs. (3-18) simplify to

$$\epsilon_{xx} = \frac{1}{E}(\sigma_{xx} - \nu\sigma_{yy}) + \alpha T \qquad \epsilon_{yy} = \frac{1}{E}(\sigma_{yy} - \nu\sigma_{xx}) + \alpha T$$

$$\epsilon_{zz} = -\frac{\nu}{E}(\sigma_{xx} + \sigma_{yy}) + \alpha T \qquad \epsilon_{xy} = \frac{\sigma_{xy}}{G} \qquad \epsilon_{yz} = \epsilon_{zx} = 0 \tag{3-19}$$

In many cases, stress as a function of strain is desired instead of strain as a function of stress. For plane stress we find from Eqs. (3-19) that

$$\sigma_{xx} = \frac{E}{1 - \nu^2}(\epsilon_{xx} + \nu\epsilon_{yy}) - \frac{E\alpha T}{1 - \nu}$$

$$\sigma_{yy} = \frac{E}{1 - \nu^2}(\epsilon_{yy} + \nu\epsilon_{xx}) - \frac{E\alpha T}{1 - \nu} \tag{3-20}$$

$$\sigma_{xy} = G\epsilon_{xy} \qquad \sigma_{zz} = \sigma_{zy} = \sigma_{zz} = 0$$

Plane-strain equations corresponding to Eqs. (3-19) and (3-20) can also be written (see Prob. 3-2). For the general three-dimensional case the solution of Eqs. (3-18) gives

$$\sigma_{xx} = \lambda e + 2G\epsilon_{xx} - (3\lambda + 2G)\alpha T$$

$$\sigma_{yy} = \lambda e + 2G\epsilon_{yy} - (3\lambda + 2G)\alpha T$$

$$\sigma_{zz} = \lambda e + 2G\epsilon_{zz} - (3\lambda + 2G)\alpha T \tag{3-21}$$

$$\sigma_{xy} = G\epsilon_{xy} \qquad \sigma_{yz} = G\epsilon_{yz} \qquad \sigma_{zz} = G\epsilon_{zz}$$

where e is defined in Eq. (3-24) and the *Lamé constant* λ is

$$\lambda = \frac{\nu E}{(1 + \nu)(1 - 2\nu)} \tag{3-22}$$

In the preceding equations it appears that E, G, and ν are three independent material constants. However, we now show that these properties can be related by an equation, so that any two of the constants and α define the behavior of a linearly elastic isotropic material.

Consider an arbitrary body subjected to loads and a temperature change. Let us focus our attention on a typical point in the body and assume that the x, y, and z axes are taken in the directions of the principal strains ϵ_{11}, ϵ_{22}, and ϵ_{33} at the point. Noting that the shear strains are zero in the principal directions, we find from Eq. (2-26b) that the shear strain referred to an arbitrary pair of orthogonal x', y' axes is

$$\epsilon_{x'y'} = 2l_1l_2\epsilon_{11} + 2m_1m_2\epsilon_{22} + 2n_1n_2\epsilon_{33}$$

Using the first three of Eqs. (3-18), this equation becomes

$$\epsilon_{x'y'} = \frac{2}{E}\{[\sigma_{11} - \nu(\sigma_{22} + \sigma_{33})]l_1l_2 + [\sigma_{22} - \nu(\sigma_{11} + \sigma_{33})]m_1m_2$$

$$+ [\sigma_{33} - \nu(\sigma_{11} + \sigma_{22})]n_1n_2\} + 2\alpha T(l_1l_2 + m_1m_2 + n_1n_2)$$

After rearranging and noting that $l_1 l_2 + m_1 m_2 + n_1 n_2 = 0$, we find

$$\epsilon_{x'y'} = \frac{2(1 + \nu)}{E}(\sigma_{11}l_1 l_2 + \sigma_{22}m_1 m_2 + \sigma_{33}n_1 n_2)$$

The principal strain axes are also axes of principal stress, since the body is isotropic. Applying Eq. (2-13b) (with principal axes) to the last equation, we obtain

$$\epsilon_{x'y'} = \frac{2(1 + \nu)}{E}\sigma_{x'y'}$$

and it follows from the fourth of Eqs. (3-18) that

$$G = \frac{E}{2(1 + \nu)} \tag{3-23}$$

Most structural materials are not perfectly isotropic. However, the degree of anisotropy is small for polycrystalline materials when the stresses are below the elastic and proportional limits. In these cases it is usually sufficiently accurate to assume that the material behaves in an isotropic fashion. The mathematical simplicity that results from this assumption is apparent from comparing Eqs. (3-16) and (3-18).

As a body is strained, it may change in volume. Consider a small element of a body which initially has the dimensions dx, dy, and dz. Under the action of a three-dimensional state of stress the element will undergo a change in volume per unit volume e, equal to

$$e = \frac{(1 + \epsilon_{xx})\, dx(1 + \epsilon_{yy})\, dy(1 + \epsilon_{zz})\, dz - dx\, dy\, dz}{dx\, dy\, dz}$$

Expanding the right side and making use of the fact that the strains are small so that their products may be neglected in comparsion to the strains themselves, we find

$$e = \epsilon_{xx} + \epsilon_{yy} + \epsilon_{zz} \tag{3-24}$$

For small strains the shearing strains produce a change in shape of the element but do not change the volume [to the order of accuracy considered in Eq. (3-24)]. For a linearly elastic isotropic body we may substitute Eqs. (3-18) into the right-hand side of Eq. (3-24) and find

$$e = \frac{(1 - 2\nu)}{E}\theta + 3\alpha T \tag{3-25}$$

where

$$\theta = \sigma_{xx} + \sigma_{yy} + \sigma_{zz} \tag{3-26}$$

We can use Eq. (3-25) to obtain the range of values which ν may have. Consider a body at constant temperature subjected to a hydrostatic pressure p, so that $\sigma_{xx} = \sigma_{yy} = \sigma_{zz} = -p$. Then

$$e = -\frac{3(1 - 2\nu)p}{E} \tag{3-27}$$

The quantity $E/[3(1 - 2\nu)]$ is known as the *bulk modulus* of the material. Since the body cannot increase in volume under a pressure, we see from Eq. (3-27) that $\nu \le 0.5$. In addition, if we consider a uniaxial tensile straining of a body, the transverse dimensions of the body cannot increase, so that we find from Eq. (3-4) that $\nu \ge 0$. Combining the results of the foregoing considerations gives $0 \le \nu \le 0.5$ for an isotropic material.

As previously mentioned, most structural materials have a Poisson's ratio in the range from 0.25 to 0.33 below the elastic and proportional limits. It has also been indicated that above the proportional limit ν increases and approaches 0.5. It is seen from Eq. (3-27) that an isotropic material with a Poisson's ratio of 0.5 is incompressible, an assumption that is made in the theory of perfectly plastic bodies. Rubber and rubberlike materials, such as some types of solid propellants, have a value of ν close to 0.5. Although one tends intuitively to think of rubber as being very compressible, this is not the case; the change in shape is mistaken for a change in volume.

In the next chapter we return to our discussion of the stresses and strains in deformable bodies but for the specific case of a linearly elastic isotropic material, as described by the stress-strain-temperature relationships of Eqs. (3-18).

REFERENCES

1. Crandall, S. H., and N. C. Dahl (eds.): "An Introduction to the Mechanics of Solids," McGraw-Hill Book Company, New York, 1959.
2. Shanley, F. R.: "Strength of Materials," McGraw-Hill Book Company, New York, 1957.
3. Marin, J.: "Mechanical Behavior of Engineering Materials," Prentice-Hall, Inc., Englewood Cliffs, N.J., 1962.
4. "ASTM Standards, pt. 3, Methods of Testing Metals," American Society for Testing Materials, Philadelphia, 1962.
5. Ramberg, W., and W. R. Osgood: Description of Stress-Strain Curves by Three Parameters, *NACA Tech. Note* 902, July, 1943.
6. Hill, H. W.: Determination of Stress-Strain Relations from Offset Yield Strength Values, *NACA Tech. Note* 927, 1944.
7. Rivello, R. M.: Ramberg-Osgood and Hill Parameters of Aircraft Structural Materials at Elevated Temperatures, *Univ. Maryland Dept. Aeron. Eng. Aeron. Rept.* 60-1, March, 1960.
8. Cozzone, F. P., and M. A. Melcon: Nondimensional Buckling Curves: Their Development and Application, *J. Aeron. Sci.*, **13**(10): 511–517 (October, 1946).

9. Bruhn, E. F.: "Analysis and Design of Flight Vehicle Structures," Tri-state Offset Co., Cincinnati, Ohio, 1965.
10. Sokolnikoff, I. S.: "Mathematical Theory of Elasticity," 2d ed., McGraw-Hill Book Company, New York, 1956.
11. Alfrey, T.: "Mechanical Behavior of High Polymers," Interscience Publishers, Inc., New York, 1948.
12. Miner, M. A.: Cumulative Damage in Fatigue, *Trans. ASME*, **67**: A-159 (1945).
13. Lundberg, B., Fatigue Life of Airplane Structures: *J. Aeron. Sci.*, **22**(6): 896 (June, 1955).
14. Schijve, J., J. R. Heath-Smith, and E. R. Welbourne: "Current Aeronautical Fatigue Problems," Pergamon Press, New York, 1965.
15. Trapp, W. J., and D. M. Forney, Jr.: "Acoustical Fatigue of Aerospace Structures," Syracuse University Press, Syracuse, N.Y., 1965.
16. Metallic Materials and Elements for Flight Vehicle Structures, *Military Handbook* MIL-HDBK-5A, Feb. 8, 1966.
17. Moon, D. P., and W. S. Hyler: MIL-HDBK-5 Guidelines for Presentation of Data, *Wright-Patterson Air Force Base Air Force Mater. Lab. Tech. Rept.* AFML-TR-66-386, Ohio, 1967.
18. Gerard, G.: Optimum Structural Design Concepts for Aerospace Vehicles, *J. Spacecraft and Rockets*, **3**(1): 5–18 (January, 1966).
19. Shanley, F. R.: "Weight-Strength Analysis of Aircraft Structures," McGraw-Hill Book Company, New York, 1952.
20. Gerard, G.: "Minimum Weight Analysis of Compression Structures," New York University Press, New York, 1956.
21. Gerard, G., and H. Becker: Handbook of Structural Stability, pt. III: Buckling of Curved Plates and Shells, *NACA Tech. Note* 3783, August, 1957.
22. Caywood, W. C., and R. M. Rivello: Material Strengths under Missile Load Conditions, *Johns Hopkins Univ. Appl. Phys. Lab. Bumblebee Rept.* 270, June, 1957.

PROBLEMS

3-1. A ceramic radome is proof-tested for thermal shock by placing it in a wind tunnel which simulates flight conditions. Two strain gages are attached to the inner surface at the point of expected maximum stress with their axes perpendicular to each other. During the test the gages indicate strains of 1750 and 1745 μin./in., and a thermocouple at the location of the gages shows a temperature rise of 110°F. The ceramic properties [$E = 17.4 \times 10^6$ psi, $\nu = 0.24$, and $\alpha = 2.3 \times 10^{-6}$ in./(in.)(°F)] are constant over the temperature range. Determine the stresses in the directions of the gages. [*Ans.* 34,310 and 34,210 psi.]

3-2. Show that the first two of Eqs. (3-20) apply to plane strain if E, ν, and α are replaced by E', ν', and α', where $E' = E/(1 - \nu^2)$, $\nu' = \nu/(1 - \nu)$, and $\alpha' = (1 + \nu)\alpha$.

3-3. Derive Eqs. (3-12) and (3-13).

3-4. Derive Eqs. (3-21) by inverting Eqs. (3-18).

3-5. Determine the tangent modulus of 17-7 PH stainless steel at 800°F and a stress of 130,000 psi from Fig. 3-9.

3-6. Determine the Ramberg-Osgood parameters of 17-7 PH stainless steel at 800°F from Fig. 3-9.

3-7. A material with the fatigue properties shown in Fig. 3-14 is subjected to the following loading:

Number of cycles	Alternating stress, ksi	Mean stress, ksi
4×10^4	80	40
10^5	60	60
10^5	40	100

Is failure expected?   [*Ans.*   No; $\Sigma n/N = 0.867$.]

3-8. The critical buckling load of a long pin-ended column is given by $P = \pi^2 EI/L^2$, where I is the moment of inertia of the cross section and L is the column length.

(*a*) Determine the ratio of the weights of two columns of different materials if the columns are of rectangular cross section, as shown, and b, L, and P are fixed in the design.   [*Ans.* $W_1/W_2 = (E^{\frac{1}{3}}/\omega)_2/(E^{\frac{1}{3}}/\omega)_1$.]

(*b*) Which of the materials of Fig. 3-17 will give the lightest design at 400°F if the material specific weights in pounds per cubic inch are: HM21XA-T8 magnesium 0.065, 2024-T4 aluminum 0.10, 6A1-4V titanium 0.16, 17-7 PH stainless steel 0.276, 321 stainless steel 0.286, and Inconel-X 0.30.

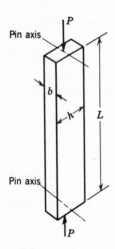

Fig. P3-8

3-9. Beryllium has a specific weight of 0.069 lb/in.³ and a room-temperature modulus of elasticity of 44×10^6 psi.   Does this indicate that it has potentialities for thin-walled cylinders subjected to bending?   Compare its weight with that of HM21XA-T8 magnesium at room temperature.

4

Introduction to the Theory of Elasticity

4-1 INTRODUCTION

It is well to review briefly the major results of Chaps. 2 and 3 at this point. For three-dimensional bodies a total of 15 unknown quantities have been introduced, namely, six stress, six strain, and three displacement functions. The physical requirements which these must satisfy are three equations of equilibrium, six strain-displacement relationships, and six stress-strain equations. In Chap. 3 it was found that three-dimensional stress-strain laws are mathematically complicated unless simplifying idealizations are made. For this reason it will be assumed in this chapter that the material is linearly elastic and isotropic. It will also be supposed that the displacements are small, so that linear strain-displacement equations (2-23) and the linear equilibrium equations (2-8a) to (2-8c) apply.

It is seen that there are 15 equations which the 15 unknown functions must satisfy at all points in the body. In addition, force and displacement boundary conditions for surface points have been given in Chap. 2. It therefore appears that sufficient conditions exist to obtain a solution. It can be proved that this is the case[1,2] and that a unique solution exists if the body is in stable equilibrium. It is convenient to utilize some of the equations to eliminate unknowns and reduce the problem to one in terms of displacements or stresses only. In this chapter the equations for the two- and three-dimensional displacement and stress formulations of the linear theory of elasticity will be given. More complete discussions of these equations and further examples of their application may be found in Refs. 1 to 9. For some simple problems these equations can be solved exactly. However, for bodies of arbitrary shape and loading, it is usually impossible to obtain exact solutions, and in these cases it is necessary to resort to numerical methods to obtain results. Approximate solutions can be determined by the method of finite difference (Chap. 5) or by the methods of weighted residuals.[10]

Alternate formulations of the elasticity problem, based upon work and energy principles, are given in Chap. 6. These principles lead to the Rayleigh-Ritz method, which can also be used to obtain approximate solutions for those cases in which the exact solutions are not known.

4-2 DISPLACEMENT FORMULATION

The elasticity problem can be formulated in terms of the displacements u, v, and w. The equilibrium equations (2-8) can be written in terms of strain by using the stress-strain equations (3-21). The strain-displacement equations (2-23) can then be used to express the equilibrium equations in terms of displacements. If this is done and it is assumed that the body is homogeneous, the equilibrium equations become

$$(\lambda + G)\frac{\partial e}{\partial x} + G\,\nabla^2 u - (3\lambda + 2G)\frac{\partial \alpha T}{\partial x} + X = 0$$

$$(\lambda + G)\frac{\partial e}{\partial y} + G\,\nabla^2 v - (3\lambda + 2G)\frac{\partial \alpha T}{\partial y} + Y = 0 \qquad (4\text{-}1)$$

$$(\lambda + G)\frac{\partial e}{\partial z} + G\,\nabla^2 w - (3\lambda + 2G)\frac{\partial \alpha T}{\partial z} + Z = 0$$

From Eqs. (3-24) and (2-23), e can be written

$$e = \frac{\partial u}{\partial x} + \frac{\partial v}{\partial y} + \frac{\partial w}{\partial z} \qquad (4\text{-}2)$$

The notation ∇^2 (*del squared*) in Eqs. (4-1) is the three-dimensional laplacian, or harmonic operator,

$$\nabla^2 = \frac{\partial^2}{\partial x^2} + \frac{\partial^2}{\partial y^2} + \frac{\partial^2}{\partial z^2} \qquad (4\text{-}3)$$

The boundary conditions for S_1, the portion of the surface upon which forces are prescribed, can be found by substituting Eqs. (3-21) into Eqs. (2-10) and using Eqs. (2-23). If this is done, it is found that the equations

$$\lambda e l + G\left(\frac{\partial u}{\partial x} l + \frac{\partial u}{\partial y} m + \frac{\partial u}{\partial z} n\right) + G\left(\frac{\partial u}{\partial x} l + \frac{\partial v}{\partial x} m + \frac{\partial w}{\partial x} n\right)$$
$$= \bar{X} + \frac{E\alpha T}{1 - 2\nu} l$$

$$\lambda e m + G\left(\frac{\partial v}{\partial x} l + \frac{\partial v}{\partial y} m + \frac{\partial v}{\partial z} n\right) + G\left(\frac{\partial u}{\partial y} l + \frac{\partial v}{\partial y} m + \frac{\partial w}{\partial y} n\right)$$
$$= \bar{Y} + \frac{E\alpha T}{1 - 2\nu} m \qquad (4\text{-}4)$$

$$\lambda e n + G\left(\frac{\partial w}{\partial x} l + \frac{\partial w}{\partial y} m + \frac{\partial w}{\partial z} n\right) + G\left(\frac{\partial u}{\partial z} l + \frac{\partial v}{\partial z} m + \frac{\partial w}{\partial z} n\right)$$
$$= \bar{Z} + \frac{E\alpha T}{1 - 2\nu} n$$

must be satisfied at all points on S_1, where $\bar{X}$, $\bar{Y}$, and $\bar{Z}$ are the prescribed surface forces per unit area. On S_2 the displacements must satisfy the equations

$$u = u_1(x,y,z) \qquad v = v_1(x,y,z) \qquad w = w_1(x,y,z) \qquad (4\text{-}5)$$

where u_1, v_1, and w_1 are prescribed functions of position on S_2. If both S_1 and S_2 exist, the boundary conditions are said to be *mixed*. The displacement formulation is most convenient in this case or when displacements are specified over the entire surface, i.e., when all of S belongs to S_2.

Equations (4-1), together with (4-4) and (4-5), constitute a boundary-value problem for the determination of u, v, and w. If a solution is found to these equations, the strains can then be computed from Eqs. (2-23) and the stresses from Eqs. (3-21).

It should be noted that it is unnecessary to use compatibility equations in the displacement formulation, as these equations were introduced to ensure that integration of the strains produced single-valued continuous displacements. Since u, v, and w are determined directly in the displacement formulation, it is assured that these functions are single-valued without recourse to the compatability equations.

4-3 STRESS FORMULATION

If the problem is to be reduced so that only the stresses appear as unknowns, it is necessary to write the compatibility equations (2-27) in terms of stress. Consider Eq. (2-27a) as an example; by using Eqs. (3-18) and (3-23) this can be written

$$2\frac{\partial^2}{\partial x\, \partial y}\left(\frac{1+\nu}{E}\sigma_{xy}\right) = \frac{\partial^2}{\partial x^2}\left\{\frac{1}{E}\left[(1+\nu)\sigma_{yy} - \nu\theta\right] + \alpha T\right\}$$
$$+ \frac{\partial^2}{\partial y^2}\left\{\frac{1}{E}\left[(1+\nu)\sigma_{xx} - \nu\theta\right] + \alpha T\right\}$$

If the body is homogeneous, E, ν, and α are constants, and the equation reduces to

$$2(1+\nu)\frac{\partial^2\sigma_{xy}}{\partial x\, \partial y} = (1+\nu)\left(\frac{\partial^2\sigma_{yy}}{\partial x^2} + \frac{\partial^2\sigma_{xx}}{\partial y^2}\right) - \nu\left(\frac{\partial^2\theta}{\partial x^2} + \frac{\partial^2\theta}{\partial y^2}\right)$$
$$+ E\left(\frac{\partial^2\alpha T}{\partial x^2} + \frac{\partial^2\alpha T}{\partial y^2}\right)$$

Finally by using Eqs. (2-8) we obtain

$$\nabla^2\sigma_{zz} + \frac{1}{1+\nu}\frac{\partial^2\theta}{\partial z^2} = -\frac{E}{1+\nu}\left(\frac{1+\nu}{1-\nu}\nabla^2\alpha T + \frac{\partial^2\alpha T}{\partial z^2}\right) - \frac{\nu}{1-\nu}$$
$$\left(\frac{\partial X}{\partial x} + \frac{\partial Y}{\partial y} + \frac{\partial Z}{\partial z}\right) - 2\frac{\partial Z}{\partial z} \qquad (4\text{-}6a)$$

In a similar manner the remaining compatibility equations (2-27b) to (2-27f) can be written

$$\nabla^2\sigma_{yy} + \frac{1}{1+\nu}\frac{\partial^2\theta}{\partial y^2} = -\frac{E}{1+\nu}\left(\frac{1+\nu}{1-\nu}\nabla^2\alpha T + \frac{\partial^2\alpha T}{\partial y^2}\right) - \frac{\nu}{1-\nu}$$
$$\left(\frac{\partial X}{\partial x} + \frac{\partial Y}{\partial y} + \frac{\partial Z}{\partial z}\right) - 2\frac{\partial Y}{\partial y} \tag{4-6b}$$

$$\nabla^2\sigma_{xx} + \frac{1}{1+\nu}\frac{\partial^2\theta}{\partial x^2} = -\frac{E}{1+\nu}\left(\frac{1+\nu}{1-\nu}\nabla^2\alpha T + \frac{\partial^2\alpha T}{\partial x^2}\right) - \frac{\nu}{1-\nu}$$
$$\left(\frac{\partial X}{\partial x} + \frac{\partial Y}{\partial y} + \frac{\partial Z}{\partial z}\right) - 2\frac{\partial X}{\partial x} \tag{4-6c}$$

$$\nabla^2\sigma_{xy} + \frac{1}{1+\nu}\frac{\partial^2\theta}{\partial x\,\partial y} = -\frac{E}{1+\nu}\frac{\partial^2\alpha T}{\partial x\,\partial y} - \frac{1}{1+\nu}\left(\frac{\partial X}{\partial y} + \frac{\partial Y}{\partial x}\right) \tag{4-6d}$$

$$\nabla^2\sigma_{yz} + \frac{1}{1+\nu}\frac{\partial^2\theta}{\partial y\,\partial z} = -\frac{E}{1+\nu}\frac{\partial^2\alpha T}{\partial y\,\partial z} - \frac{1}{1+\nu}\left(\frac{\partial Y}{\partial z} + \frac{\partial Z}{\partial y}\right) \tag{4-6e}$$

$$\nabla^2\sigma_{zx} + \frac{1}{1+\nu}\frac{\partial^2\theta}{\partial z\,\partial x} = -\frac{E}{1+\nu}\frac{\partial^2\alpha T}{\partial z\,\partial x} - \frac{1}{1+\nu}\left(\frac{\partial Z}{\partial x} + \frac{\partial X}{\partial z}\right) \tag{4-6f}$$

The problem reduces to finding the six stresses which satisfy the three equilibrium equations (2-8) and the six compatibility equations (4-6) at all interior points of the body. The stress formulation is not convenient for problems with displacement or mixed boundary conditions. For problems with force boundary conditions the stresses must also satisfy Eqs. (2-10) at all points on the surface.

It can be shown that a unique solution exists to Eqs. (2-8), (4-6), and (2-10) if the body is simply connected and has no dislocations. For a discussion of the additional compatibility conditions which must be satisfied if the body is multiply connected or has dislocations see Refs. 2 and 7.

After the stresses are found, the strains can be determined from the stress-strain equations (3-18). The strains can then be expressed in terms of the displacements by means of the strain-displacement equations (2-23). These equations must then be integrated to obtain the displacements.

4-4 TWO-DIMENSIONAL PROBLEMS

In many cases certain of the stress or displacement components are zero or negligibly small compared to the remaining ones, and the equations reduce to a two-dimensional problem. This is the situation for the bodies shown in Fig. 4-1. Figure 4-1a shows a body with a uniform thickness which is small compared to the other dimensions. In this case the body is in a state of *plane stress* if:

1. The upper and lower plane surfaces are free of surface forces.
2. On the cylindrical surface, $\bar{Z} = 0$ and $\bar{X}$ and $\bar{Y}$ are independent of z.
3. Within the body, $Z = 0$ and X, Y, and T are independent of z.

In this instance, displacement functions u and v must be found which satisfy the two-dimensional equations of equilibrium written in terms of displacements. These are found by substituting the expressions for ϵ_{xx}, ϵ_{yy}, and ϵ_{xy} from Eqs. (2-23) into Eqs. (3-20), which in turn are substituted in the second and third of Eqs. (2-11). This gives

$$\frac{E}{2(1-\nu)} \frac{\partial}{\partial x}\left(\frac{\partial u}{\partial x} + \frac{\partial v}{\partial y}\right) + \frac{E}{2(1+\nu)} \nabla^2 u = \frac{E}{1-\nu} \frac{\partial \alpha T}{\partial x} - X$$

$$\frac{E}{2(1-\nu)} \frac{\partial}{\partial y}\left(\frac{\partial u}{\partial x} + \frac{\partial v}{\partial y}\right) + \frac{E}{2(1+\nu)} \nabla^2 v = \frac{E}{1-\nu} \frac{\partial \alpha T}{\partial y} - Y$$

$$(4\text{-}7)$$

where ∇^2 is the 2-dimensional laplacian operator.

By a similar process the boundary conditions for S_1, the part of the cylindrical surface upon which forces $\bar{X}$ and $\bar{Y}$ are prescribed, are found from Eqs. (2-12) to be

$$\frac{E}{1+\nu}\left[\frac{1}{1-\nu}\left(\frac{\partial u}{\partial x} + \nu\frac{\partial v}{\partial y}\right)l + \frac{1}{2}\left(\frac{\partial v}{\partial x} + \frac{\partial u}{\partial y}\right)m\right] = \bar{X} + \frac{E\alpha T}{1-\nu}l$$

$$\frac{E}{1+\nu}\left[\frac{1}{1-\nu}\left(\frac{\partial v}{\partial y} + \nu\frac{\partial u}{\partial x}\right)m + \frac{1}{2}\left(\frac{\partial v}{\partial x} + \frac{\partial u}{\partial y}\right)l\right] = \bar{Y} + \frac{E\alpha T}{1-\nu}m$$

$$(4\text{-}8)$$

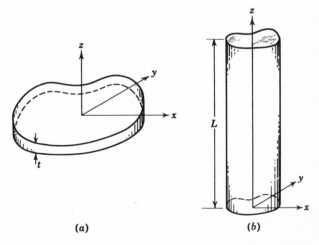

(a)   (b)

Fig. 4-1 Geometric shapes for two-dimensional problems. (a) Shape of body for plane stress; (b) shape of body for plane strain.

On the surface S_2, where displacements are prescribed,

$$u = u_1(x,y) \qquad v = v_1(x,y) \tag{4-9}$$

where u_1 and v_1 are prescribed function of position of S_2.

The plane-stress problem can also be formulated in terms of stresses. In this case the nonvanishing stresses σ_{xx}, σ_{yy}, and σ_{xy} must satisfy the equilibrium equations (2-11). The compatibility equation (2-27a) can be written in terms of stress by using the plane-stress Hooke's law equations (3-19). If use is made of Eqs. (2-11), the compatibility equation can finally be written

$$\nabla^2(\sigma_{xx} + \sigma_{yy}) = -E \, \nabla^2 \alpha T - (1 + \nu)\left(\frac{\partial X}{\partial x} + \frac{\partial Y}{\partial y}\right) \tag{4-10}$$

where ∇^2 is the two-dimensional laplacian operator. In addition to satisfying Eqs. (2-11) and (4-10), the stresses must satisfy Eqs. (2-12) at all points on the cylindrical surface if the boundary conditions are of the force type.

Another type of two-dimensional problem is shown in Fig. 4-1b, where the length of the cylindrical body is long compared to the other dimensions. A state of plane strain [defined by $w = 0$, $u = u(x,y)$, and $v = v(x,y)$] will exist in the body if:

1. The ends of the cylinder are restrained so that $w = 0$.
2. On the cylindrical surface, $\bar{Z} = 0$ and $\bar{X}$ and $\bar{Y}$ are independent of z.
3. Within the body, $Z = 0$ and X, Y, and T are independent of z.

For this case the stresses σ_{xx}, σ_{yy}, and σ_{xy} must satisfy Eqs. (2-11) at all interior points and Eqs. (2-12) on the cylindrical surface. The stress-strain relationships of Eqs. (3-20) apply to plane strain if E, ν, and α are replaced by E' ν', and α' as defined in Prob. 3-2. As a result the plane-strain compatability equation can be found from Eq. (4-10) by placing primes on all the material-property symbols in that equation.

4-5 STRESS–FUNCTION FORMULATION

In the stress formulation of the elasticity problem we found that for the three-dimensional case a total of six stress components must be found which simultaneously satisfy Eqs. (2-8) and (4-6) at all interior points and Eqs. (2-10) at all surface points of the body (assuming forces are prescribed over the entire surface). The problem would be greatly simplified if the six stresses could be related to fewer functions defined in such a manner that they identically satisfy the equilibrium conditions. Such functions are known as *stress functions*. Consider for simplicity the two-dimensional cases of plane stress and plane strain (see Ref. 1 for a discussion of the three-dimensional problem). The stresses σ_{xx}, σ_{yy}, and σ_{xy} in these cases must satisfy the equations of equilibrium (2-11). Let us assume that the body forces are conservative, so that their components are derivable from a *potential function* $V(x,y)$ (Sec. 6-2) as follows:

$$X = -\frac{\partial V}{\partial x} \qquad Y = -\frac{\partial V}{\partial y} \tag{4-11}$$

The second and third of Eqs. (2-11) then become

$$\frac{\partial \sigma_{xx}}{\partial x} + \frac{\partial \sigma_{xy}}{\partial y} - \frac{\partial V}{\partial x} = 0 \qquad \frac{\partial \sigma_{yx}}{\partial x} + \frac{\partial \sigma_{yy}}{\partial y} - \frac{\partial V}{\partial y} = 0 \tag{4-12}$$

The English mathematician Airy suggested a stress function $\varphi(x,y)$ defined in the following manner, which assures that Eqs. (4-12) are always satisfied:

$$\sigma_{xx} = \frac{\partial^2 \varphi}{\partial y^2} + V \qquad \sigma_{yy} = \frac{\partial^2 \varphi}{\partial x^2} + V \qquad \sigma_{xy} = -\frac{\partial^2 \varphi}{\partial x\,\partial y} \tag{4-13}$$

This is easily verified by substituting Eqs. (4-13) into Eqs. (4-12). Since φ automatically satisfies the equilibrium conditions, it is only necessary to find a stress function which fulfills the compatibility and boundary conditions.

For the case of plane stress we substitute Eqs. (4-11) and (4-13) into Eq. (4-10) and find the compatibility equation in terms of φ to be

$$\nabla^4 \varphi = -E\,\nabla^2 \alpha T - (1 - \nu)\,\nabla^2 V \tag{4-14}$$

where ∇^4 is the *biharmonic operator*, defined as

$$\nabla^4 = \left(\frac{\partial^2}{\partial x^2} + \frac{\partial^2}{\partial y^2}\right)^2 = \frac{\partial^4}{\partial x^4} + 2\frac{\partial^4}{\partial x^2\,\partial y^2} + \frac{\partial^4}{\partial y^4} \tag{4-15}$$

We have seen in Sec. 4-4 that the compatibility equation in terms of stress for plane strain differs from that of plane stress only by the addition of primes to the material-properties terms. The equation for plane strain which corresponds to Eq. (4-14) is therefore

$$\nabla^4 \varphi = -E'\,\nabla^2 \alpha' T - (1 - \nu')\,\nabla^2 V \tag{4-16}$$

It is noted from Eqs. (4-14) and (4-16) that if $\nabla^2 \alpha T = \nabla^2 V = 0$, the compatibility equations for plane stress and plane strain become identical and are

$$\nabla^4 \varphi = 0 \tag{4-17}$$

This is the case when body forces and α are constant and the temperatures are those of a two-dimensional steady-state heat-transfer problem with no heat addition or loss at interior points.

The boundary conditions for both plane stress and plane strain can be found by substituting Eqs. (4-13) into Eqs. (2-12), which gives

$$\frac{\partial^2 \varphi}{\partial y^2} l - \frac{\partial^2 \varphi}{\partial x\, \partial y} m = \bar{X} - Vl \qquad -\frac{\partial^2 \varphi}{\partial x\, \partial y} l + \frac{\partial^2 \varphi}{\partial x^2} m = \bar{Y} - Vm \quad (4\text{-}18)$$

The plane-stress and plane-strain problems for simply connected bodies are thus reduced to finding a function $\varphi(x,y)$ which satisfies Eq. (4-14) or (4-16) at all interior points of a two-dimensional domain D and Eqs. (4-18) at all points on the boundary curve C of the domain. If a function φ can be found which satisfies these conditions, the stresses can be computed from Eqs. (4-13). If deflections are desired, the strains are obtained from Eqs. (3-19) for plane stress. For plane strain $\epsilon_{zx} = \epsilon_{zy} = \epsilon_{zz} = 0$, while ϵ_{xx} and ϵ_{yy} can be found from Eqs. (3-19) if primes are placed upon E, ν, and α. The equation for ϵ_{xy} remains the same as for plane stress. The strains are then substituted into Eqs. (2-23), which must be integrated to obtain the deflections.

It is often convenient to express Eqs. (4-18) in terms of φ on the boundary and its derivative normal to the boundary.[2] Referring to Fig. 4-2, we see that

$$l = \frac{dy}{ds} \qquad m = -\frac{dx}{ds} \qquad\qquad (4\text{-}19)$$

where s is a coordinate measured along the boundary in the counter-

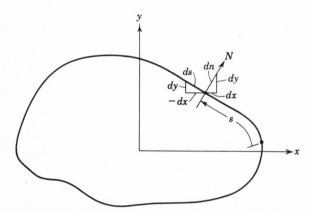

Fig. 4-2 Geometry for direction cosines of the normal and tangent to the boundary.

clockwise direction. Equations (4-18) can then be written

$$\frac{\partial}{\partial y}\left(\frac{\partial \varphi}{\partial y}\frac{dy}{ds} + \frac{\partial \varphi}{\partial x}\frac{dx}{ds}\right) = \frac{d}{ds}\frac{\partial \varphi}{\partial y} = \bar{X} - Vl$$

$$-\frac{\partial}{\partial x}\left(\frac{\partial \varphi}{\partial y}\frac{dy}{ds} + \frac{\partial \varphi}{\partial x}\frac{dx}{ds}\right) = -\frac{d}{ds}\frac{\partial \varphi}{\partial x} = \bar{Y} - Vm$$

Integrating these equations along the boundary, we obtain

$$\frac{\partial \varphi}{\partial y} = \int_0^s (\bar{X} - Vl)\, ds + C_1 = A + C_1$$

$$\frac{\partial \varphi}{\partial x} = -\int_0^s (\bar{Y} - Vm)\, ds + C_2 = -B + C_2$$

(4-20)

where A and B are functions of s given by

$$A = \int_0^s (\bar{X} - Vl)\, ds \qquad B = \int_0^s (\bar{Y} - Vm)\, ds \qquad (4\text{-}21)$$

and C_1 and C_2 are constants of integration. From Eqs. (4-20) and (4-19)

$$\frac{d\varphi}{ds} = \frac{\partial \varphi}{\partial x}\frac{dx}{ds} + \frac{\partial \varphi}{\partial y}\frac{dy}{ds} = Bm - C_2 m + Al + C_1 l$$

Integrating with respect to s, we find that φ on the boundary is given by

$$\varphi = \int_0^s (Al + Bm)\, ds + C_2 x + C_1 y + C_3$$

Equations (4-13) show that stresses depend only upon the second derivatives of φ; terms in φ which are constant or linear in x or y have no influence on the stresses. For this reason we may arbitrarily take $C_1 = C_2 = C_3 = 0$, so that

$$\varphi = \int_0^s (Al + Bm)\, ds \qquad (4\text{-}22)$$

The derivative of φ normal to the boundary is

$$\frac{\partial \varphi}{\partial n} = \frac{\partial \varphi}{\partial x}\frac{dx}{dn} + \frac{\partial \varphi}{\partial y}\frac{dy}{dn}$$

Noting from Fig. 4-2 that $dx/dn = l$ and $dy/dn = m$ and using Eqs. (4-20), we obtain

$$\frac{\partial \varphi}{\partial n} = -Bl + Am \qquad (4\text{-}23)$$

Since $\bar{X}$, $\bar{Y}$, and V are known, A and B can be computed from Eqs. (4-21), and φ and $\partial \varphi/\partial n$ can therefore be found at all points on the boundary from Eqs. (4-22) and (4-23). These two equations can be used as alternatives to Eqs. (4-18) for boundary conditions of φ.

In some problems, such as those with circular boundaries, it is convenient to use polar instead of cartesian coordinates. Equations (4-14) and (4-16) apply in this case if ∇^2 and ∇^4 are expressed in polar coordinates as

$$\nabla^2 = \frac{\partial^2}{\partial r^2} + \frac{1}{r}\frac{\partial}{\partial r} + \frac{1}{r^2}\frac{\partial^2}{\partial \theta^2} \tag{4-24}$$

and

$$\nabla^4 = \left(\frac{\partial^2}{\partial r^2} + \frac{1}{r}\frac{\partial}{\partial r} + \frac{1}{r^2}\frac{\partial^2}{\partial \theta^2}\right)\left(\frac{\partial^2}{\partial r^2} + \frac{1}{r}\frac{\partial}{\partial r} + \frac{1}{r^2}\frac{\partial^2}{\partial \theta^2}\right) \tag{4-25}$$

which are obtained from Eq. (4-15) by a transformation of coordinates. The conservative body forces R and Θ in this case are derivable from the potential function $V(r,\theta)$ by $R = -\partial V/\partial r$ and $\Theta = -\partial V/(r\,\partial\theta)$. The stress function for polar coordinates is defined by

$$\sigma_{rr} = \frac{1}{r}\frac{\partial\varphi}{\partial r} + \frac{1}{r^2}\frac{\partial^2\varphi}{\partial \theta^2} + V \qquad \sigma_{\theta\theta} = \frac{\partial^2\varphi}{\partial r^2} + V$$

$$\sigma_{r\theta} = \frac{1}{r^2}\frac{\partial\varphi}{\partial \theta} - \frac{1}{r}\frac{\partial^2\varphi}{\partial r\,\partial\theta} \tag{4-26}$$

Example 4-1 Determine the boundary conditions in terms of φ and $\partial\varphi/\partial n$ for the plane-stress problem with the parabolic edge loading shown in Fig. 4-3.

Since there are no body forces, $V = 0$, and Eqs. (4-21) become

$$A = \int_0^s \bar{X}\,ds \qquad B = \int_0^s \bar{Y}\,ds$$

Furthermore, $\bar{Y}$ is zero on all surfaces, so that $B = 0$ on the entire boundary. To evaluate A we arbitrarily take the origin of s at the point $(a,0)$. Measuring s in a counterclockwise direction around the boundary, we evaluate the line integral for A as follows:

$$A(s) = \int_0^s \bar{X}\,ds = \int_0^y cy^2\,dy = \frac{cy^3}{3} \qquad 0 \le s \le b$$

$$A(b) = \frac{cb^3}{3}$$

$$A(s) = A(b) - \int_a^x \bar{X}\,dx = \frac{cb^3}{3} - \int_a^x (0)\,dx = \frac{cb^3}{3} \qquad b \le s \le b + 2a$$

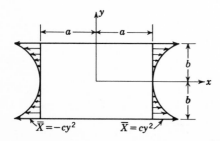

Fig. 4-3 Example 4-1.

$$A(b + 2a) = \frac{cb^3}{3}$$

$$A(s) = A(b + 2a) - \int_b^y \bar{X} \, dy$$

$$= \frac{cb^3}{3} - \int_b^y (-cy^2) \, dy = \frac{cy^3}{3} \qquad b + 2a \leq s \leq 3b + 2a$$

$$A(3b + 2a) = -\frac{cb^3}{3}$$

$$A(s) = A(3b + 2a) + \int_{-a}^x \bar{X} \, dx$$

$$= -\frac{cb^3}{3} + \int_{-a}^x (0) \, dx = -\frac{cb^3}{3} \qquad 3b + 2a \leq s \leq 3b + 4a$$

$$A(3b + 4a) = -\frac{cb^3}{3}$$

$$A(s) = A(3b + 4a) + \int_{-b}^y \bar{X} \, dy = -\frac{cb^3}{3} + \int_{-b}^y cy^2 \, dy = \frac{cy^3}{3}$$

$$3b + 4a \leq s \leq 4b + 4a$$

These values of A are shown in Fig. 4-4a.

Equation (4-23) reduces to $\partial\varphi/\partial n = Am$. Observing that $m = 0$ for the end faces, $m = 1$ for the top face, and $m = -1$ for the bottom face, we find the values of $\partial\varphi/\partial n$ given in Fig. 4-4b.

Next from Eq. (4-22) we find

$$\varphi = \int_0^s Al \, ds$$

from which we obtain

$$\varphi(s) = \int_0^y \left(\frac{cy^3}{3}\right)(1) \, dy = \frac{cy^4}{12} \qquad 0 \leq s \leq b$$

$$\varphi(b) = \frac{cb^4}{12}$$

$$\varphi(s) = \varphi(b) - \int_a^x Al \, dx$$

$$= \frac{cb^4}{12} - \int_a^x \left(\frac{cb^3}{3}\right)(0) \, dx = \frac{cb^4}{12} \qquad b \leq s \leq b + 2a$$

$$\varphi(b + 2a) = \frac{cb^4}{12}$$

$$\varphi(s) = \varphi(b + 2a) - \int_b^y Al \, dy$$

$$= \frac{cb^4}{12} - \int_b^y \left(\frac{cy^3}{3}\right)(-1) \, dy = \frac{cy^4}{12} \qquad b + 2a \leq s \leq 3b + 2a$$

$$\varphi(3b + 2a) = \frac{cb^4}{12}$$

$$\varphi(s) = \varphi(3b + 2a) + \int_{-a}^x Al \, dx$$

$$= \frac{cb^4}{12} + \int_{-a}^x \left(-\frac{cb^3}{3}\right)(0) \, dx = \frac{cb^4}{12} \qquad 3b + 2a \leq s \leq 3b + 4a$$

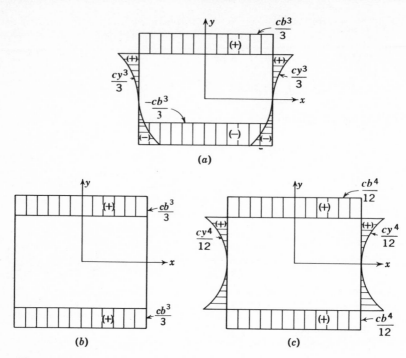

Fig. 4-4 Results of Example 4-1. (a) Distribution of A; (b) distribution of $\partial\varphi/\partial n$; (c) distribution of φ.

$$\varphi(3b + 4a) = \frac{cb^4}{12}$$

$$\varphi(s) = \varphi(3b + 4a) + \int_{-b}^{y} Al \, dy = \frac{cb^4}{12} + \int_{-b}^{y} \left(\frac{cy^3}{3}\right)(1) \, dy = \frac{cy^4}{12}$$

$$3b + 4a \le s \le 4b + 4a$$

The results are shown in Fig. 4-4c, which, with Fig. 4-4b constitutes the boundary conditions in terms of φ and $\partial\varphi/\partial n$. Approximate solutions for the stresses will be given in Examples 5-1 and 6-2.

4-6 THE INVERSE METHOD

In general, it is not possible to obtain a closed form solution to Eq. (4-14) or (4-16) and the boundary conditions (4-18) or (4-22) and (4-23) for an arbitrary distribution of body forces, surface forces, and temperatures acting upon a body of arbitrary shape. However, an *inverse method* can be used, which for plane stress without body forces is as follows:

1. A stress function $\varphi(x,y)$ is assumed and ∇^4 evaluated. In general it will be found that $\nabla^4\varphi = f(x,y)$. The assumed stress function is then the solution for problems in which $-E \, \nabla^2\alpha T = f(x,y)$.

2. An arbitrary boundary is chosen, and the surface loadings on the assumed boundary are evaluated by Eqs. (4-18), which in the absence of body forces become

$$\bar{X} = \frac{\partial^2 \varphi}{\partial y^2} l - \frac{\partial^2 \varphi}{\partial x\,\partial y} m \qquad \bar{Y} = -\frac{\partial^2 \varphi}{\partial x\,\partial y} l + \frac{\partial^2 \varphi}{\partial x^2} m$$

3. The stresses associated with the assumed stress function can be found from Eqs. (4-13), which reduce to

$$\sigma_{xx} = \frac{\partial^2 \varphi}{\partial y^2} \qquad \sigma_{yy} = \frac{\partial^2 \varphi}{\partial x^2} \qquad \sigma_{xy} = -\frac{\partial^2 \varphi}{\partial x\,\partial y}$$

These stresses are for the temperatures of step 1, the surface loadings of step 2, and the assumed body shape.

To illustrate the inverse method consider the case when $\nabla^2 T = V = 0$, for which Eq. (4-17) applies. Many functions satisfy this equation. For instance, some simple functions which identically satisfy Eq. (4-17) are the polynomials of the form $C_{ij}x^i y^j$, which are listed in Fig. 4-5. The stresses and surface forces associated with these functions and a rectangular boundary are also shown in this figure. The cases where $i = j = 0$ and where $i = 1, j = 0$ or $i = 0, j = 1$ have not been included in the figure because from Eqs. (4-13) there are no stresses associated with these functions. Higher-order polynomials do not identically satisfy Eq. (4-17), but it is often possible to combine them in a manner which does fulfill compatibility. Consider, for instance, the stress functions $C_{50}x^5$ and $C_{32}x^3 y^2$, which separately do not satisfy Eq. (4-17). The sum of these functions is

$$\varphi = C_{50}x^5 + C_{32}x^3 y^2$$

which substituted into Eq. (4-17) gives

$$120 C_{50}x + 24 C_{32}x = 0$$

It is seen that the function is admissible only if $C_{32} = -5C_{50}$, in which case

$$\varphi = C_{50}(x^5 - 5x^3 y^2)$$

While giving exact results, the inverse method has the obvious disadvantage of starting with a solution and proceeding to find the associated problem. In engineering we are faced with the reverse situation of finding the solution for a body of specific shape, loading, and temperature distribution. Nevertheless, useful solutions to some practical problems have been found by the method, and many of these can be found in books on the theory of elasticity.[1-9] Since the derived differential equations are linear, the principle of superposition applies. As a result, the solution for a body under a combination of loads can be

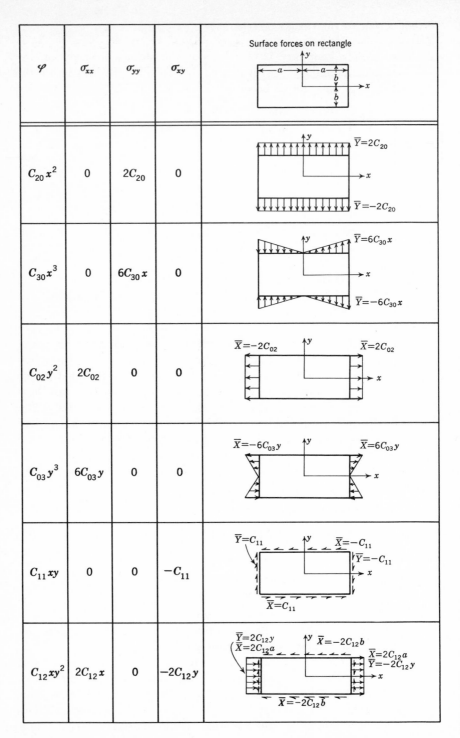

φ	σ_{xx}	σ_{yy}	σ_{xy}	Surface forces on rectangle
$C_{20}x^2$	0	$2C_{20}$	0	
$C_{30}x^3$	0	$6C_{30}x$	0	
$C_{02}y^2$	$2C_{02}$	0	0	
$C_{03}y^3$	$6C_{03}y$	0	0	
$C_{11}xy$	0	0	$-C_{11}$	
$C_{12}xy^2$	$2C_{12}x$	0	$-2C_{12}y$	

Fig. 4-5 Polynomial stress functions.

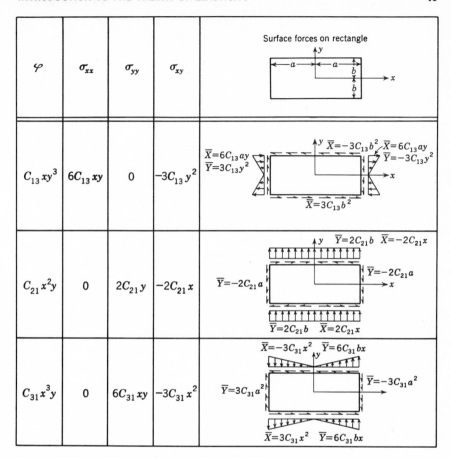

φ	σ_{xx}	σ_{yy}	σ_{xy}	
				Surface forces on rectangle
$C_{13}xy^3$	$6C_{13}xy$	0	$-3C_{13}y^2$	
$C_{21}x^2y$	0	$2C_{21}y$	$-2C_{21}x$	
$C_{31}x^3y$	0	$6C_{31}xy$	$-3C_{31}x^2$	

Fig. 4-5 Polynomial stress functions (concluded).

obtained by superimposing the solutions for each of the individual loadings.

Example 4-2 Show that the functions $C_{11}xy$ and $C_{13}xy^3$ can be combined to obtain the solution for a cantilevered beam of rectangular cross section loaded by a lateral force P (Fig. 4-6a). The stress function is

$$\varphi = C_{11}xy + C_{13}xy^3$$

and from Eqs. (4-13) the associated stresses are

$$\sigma_{xx} = 6C_{13}xy \qquad \sigma_{yy} = 0 \qquad \sigma_{xy} = -C_{11} - 3C_{13}y^2$$

On the boundaries $y = \pm h/2$, $\bar{X} = l = 0$, and the first of Eqs. (2-12) gives

$$\sigma_{xy}\Big|_{y=\pm h/2} = -C_{11} - 3C_{13}\left(\frac{h}{2}\right)^2 = 0$$

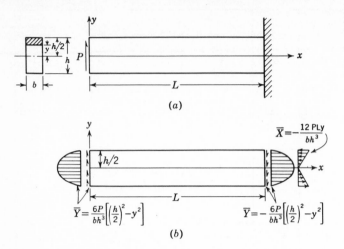

Fig. 4-6 Example 4-2. (a) Geometry of cantilever beam; (b) surface forces.

or $C_{11} = -3C_{13}(h/2)^2$, and therefore

$$\sigma_{xy} = 3C_{13}\left[\left(\frac{h}{2}\right)^2 - y^2\right]$$

On the boundary $x = 0$ the resultant of the $\bar{Y} = -\sigma_{xy}$ forces must equal P, which gives

$$P = \int_{-h/2}^{h/2} \bar{Y}b \, dy = -3C_{13}b \int_{-h/2}^{h/2}\left[\left(\frac{h}{2}\right)^2 - y^2\right] dy = -\frac{C_{13}bh^3}{2}$$

or $C_{13} = -2P/bh^3$. The stress function is then

$$\varphi = -\frac{2P}{bh^3}\left(xy^3 - \frac{3h^2xy}{4}\right)$$

and the stresses are

$$\sigma_{xx} = -\frac{12Pxy}{bh^3} = -\frac{My}{I} \qquad \sigma_{yy} = 0$$

$$\sigma_{xy} = -\frac{6P}{bh^3}\left[\left(\frac{h}{2}\right)^2 - y^2\right] = -\frac{VQ}{Ib}$$

where V and M are the shear force and bending moment at x, I is the moment of inertia of the cross section, and Q is the moment of the area between y and $h/2$ about the centroidal axis of the cross section. The surface forces associated with φ are shown in Fig. 4-6b.

The stresses are seen to be in agreement with the elementary theory of the bending of beams from strength of materials. However, for the solution to be exact at the ends, the tip force must be distributed in a parabolic fashion, and the reactions at the root must consist of a parabolic distribution of shearing forces and a linear distribution of normal forces. The problem has been solved as if forces were prescribed on all boundaries, when in reality the beam has mixed boundary conditions, and at $x = L$ the proper conditions are $u = v = 0$.

As a matter of fact, if deflections due to the above stresses are computed at $x = L$, it is found that the built-in end does not remain plane, and therefore the displacement boundary conditions are violated.[2] However, we shall see in Sec. 4-8 that the solution is good at a distance from the root greater than the height of the beam. It is for this reason that the strength-of-materials solution is applicable only to beams which are long relative to the cross-sectional dimensions and then only to points at distances from the ends that are greater than the largest cross-sectional dimension.

4-7 THE SEMI-INVERSE METHOD

We have seen that complicated simultaneous equations must be satisfied in a three-dimensional problem and the likelihood of obtaining an exact solution for a body of arbitrary shape and loading is remote. However, some problems have geometric shapes and loadings which permit simplifying assumptions that lead to solvable equations. A *semi-inverse method* was proposed by St. Venant to handle these problems. In this method conjectures are made about the stress or displacement components based upon physical intuition or experimental evidence (usually that certain of these are zero or negligible). It is necessary to restrict the shape and loading of the body to justify the assumptions, but some generality may still exist. It is necessary to show that the assumed functions, and those which remain to be determined, result in surface loadings which are coincident or equivalent (Sec. 4-8) to the actual surface forces. The equilibrium and compatibility equations are simplified by substituting the assumed stresses or deflections into them. It may then be possible to solve the simplified equations exactly, or if approximate methods must be used, the labor involved will be less than if the unsimplified equations were used.

St. Venant first applied the semi-inverse method to the twisting of a bar by end torques (Chap. 8) and to the development of a theory for beams subjected to lateral shearing forces.[1,2] Plane stress and plane strain may be thought of as being obtained from the three-dimensional equations by the semi-inverse method. For instance, for the case of plane strain, the displacement assumptions $u = u(x,y)$, $v = v(x,y)$, and $w = 0$ can be used to reduce the three-dimensional displacement formulation in Eqs. (4-1) and (4-4) to the two-dimensional formulation obtained by placing primes on the material constants in Eqs. (4-7) and (4-8). In order to effect these reductions it is necessary to introduce the restrictions for plane strain enumerated in Sec. 4-4.

4-8 ST. VENANT'S PRINCIPLE

In Example 4-2 the solution for a cantilevered beam with a lateral force at the free end was determined. However, it was found that for an exact solution the tip load must be parabolically distributed over the end. It was further noted that the surface forces on the supported end are not actually those associated with the displacement boundary conditions of a built-in end. One may then wonder whether the solution is of any value. To answer this question we shall consider an important and useful principle, stated by St. Venant:

> *If the loading distribution on a small section of the surface of an elastic body is replaced by another loading which has the same resultant force and moment as the original loading, then no appreciable changes will occur in the stresses in the body except in the region near the surface where the loading is altered.*

The region in which significant stress modification occurs will usually extend into the body for a distance equal to the greatest linear dimension of the portion of the surface on which the loading was altered. However, in the case of thin-walled stiffened shells and trusses the affected area may extend to several times the length of the surface over which the changes are made.

In Example 4-2, the stresses that were found would be expected to be very accurate at distances from the ends equal to, or greater than, the height of the beam. This is one reason why the strength-of-materials theory for beam stresses is applicable only to beams which have lengths that are long relative to their cross-sectional dimensions. For such beams, the regions of inaccuracy are then restricted to a small portion of the total length. More elaborate solutions are necessary if the stresses in the end regions are required.

REFERENCES

1. Love, A. E. H.: "A Treatise on the Mathematical Theory of Elasticity," 4th ed., Dover Publications, Inc., New York, 1927.
2. Timoshenko, S. P., and J. N. Goodier: "Theory of Elasticity," 2d ed., McGraw-Hill Book Company, New York, 1951.
3. Sechler, E. E.: "Elasticity in Engineering," John Wiley & Sons, Inc., New York, 1952.
4. Sokolnikoff, I. S.: "Mathematical Theory of Elasticity," 2d ed., McGraw-Hill Book Company, New York, 1956.
5. Green, A. E., and W. Zerna: "Theoretical Elasticity," Oxford University Press, New York, 1954.
6. Muskhelishvili, N. I.: "Some Basic Problems of the Theory of Elasticity," translation by J. R. M. Radok of 3d Russian ed., Ervin P. Noordhoff, Groningen, Netherlands, 1953.
7. Boley, B. A., and J. H. Weiner: "Theory of Thermal Stresses," John Wiley & Sons, Inc., New York, 1960.

8. Wang, C. T.: "Applied Elasticity," McGraw-Hill Book Company, New York, 1953.
9. Den Hartog, J. P.: "Advanced Strength of Materials," McGraw-Hill Book Company, New York, 1952.
10. Crandall, S. H.: "Engineering Analysis," McGraw-Hill Book Company, New York, 1956.

PROBLEMS

4-1. Verify the results in Fig. 4-5 for $C_{31}x^3y$.

4-2. (a) Show that

$$\varphi = -p\left\{x^2\left[\left(\frac{y}{h}\right)^3 - \frac{3}{4}\frac{y}{h} + \frac{1}{4}\right] - \frac{h^2}{5}\left[\left(\frac{y}{h}\right)^5 - \frac{1}{2}\left(\frac{y}{h}\right)^3\right]\right\}$$

is an admissable stress function for no body forces or temperature changes.

(b) Determine the stresses associated with φ.

(c) Evaluate the surface forces on the boundary that is shown.

(d) Find the resultant forces and moments on the boundaries $x = 0$ and L. Discuss the relationship of the problem to that of a beam of rectangular cross section, cantilevered at $x = L$, and loaded by a uniform pressure on its lower surface.

Fig. P4-2

4-3. A thin rectangular slab with boundaries at $x = \pm a$ and $y = \pm b$ is subjected to a tensile force per unit area of $C \cos(\pi y/2b)$ on the ends $x = \pm a$. Evaluate the boundary conditions for φ and $\partial\varphi/\partial n$ at the point $(a, b/2)$. [*Ans.* $\partial\varphi/\partial n = 0$; $\varphi = 0.1187b^2C$.]

4-4. Show that the two-dimensional equilibrium equations in polar coordinates given in Prob. 2-2 are identically satisfied by an Airy stress function $\varphi(r,\theta)$ defined by

$$\sigma_{rr} = \frac{1}{r}\frac{\partial\varphi}{\partial r} + \frac{1}{r^2}\frac{\partial^2\varphi}{\partial\theta^2} + V \qquad \sigma_{\theta\theta} = \frac{\partial^2\varphi}{\partial r^2} + V$$

$$\sigma_{r\theta} = \frac{1}{r^2}\frac{\partial\varphi}{\partial\theta} - \frac{1}{r}\frac{\partial^2\varphi}{\partial r\,\partial\theta}$$

when R and Θ are conservative body forces derivable from a potential function $V(r,\theta)$ by $R = -\partial V/\partial r$, $\Theta = -\partial V/(r\,\partial\theta)$.

4-5. Using Eqs. (4-14), (4-24), and (4-25), show that the compatibility equation for a circular disk rotating at a constant angular velocity Ω about its axis and subjected to a temperature change $T(r)$ is

$$\frac{d^4\varphi}{dr^4} + \frac{2}{r}\frac{d^3\varphi}{dr^3} - \frac{1}{r^2}\frac{d^2\varphi}{dr^2} + \frac{1}{r^3}\frac{d\varphi}{dr} = -E\left(\frac{d^2\alpha T}{dr^2} + \frac{1}{r}\frac{d\alpha T}{dr}\right) + 2(1-\nu)\frac{\omega\Omega^2}{g}$$

where ω is the specific weight of the disk and φ is the stress function defined in Prob. 4-4.

4-6. Derive the differential equations and boundary conditions for the displacement formulation of the plane-strain problem.

4-7. A thin circular disk of inside radius a and outside radius b is subjected to a temperature increase T which is a function of radius only. Use the results of Probs. 2-2 and 2-4 to show that the displacement formulation is given by the differential equation

$$\frac{d}{dr}\left(\frac{1}{r}\frac{dru}{dr}\right) = (1 + \nu)\frac{d\alpha T}{dr}$$

which by double integration gives

$$u = (1 + \nu)\frac{1}{r}\int \alpha Tr\,dr + C_1 r + \frac{C_2}{r}$$

where C_1 and C_2 are constants of integration to be determined from the radial surface loads at the inner and outer radii.

4-8. Show that the radial stress in Prob. 4-7 is

$$\sigma_{rr} = -\frac{E}{r^2}\int \alpha Tr\,dr + E\left[\frac{C_1}{1 - \nu} - \frac{C_2}{r^2(1 + \nu)}\right]$$

Use this result to evaluate C_1 and C_2 for a disk with no surface forces at $r = a$ and b.

5
Finite-difference Methods

5-1 INTRODUCTION

The differential equations and boundary conditions which apply to various structural problems are derived throughout this book. In some cases it is possible to solve these equations exactly and in closed form, but in many situations solutions to the differential equations are not known. An engineer is usually satisfied with an approximate answer, however, as the loading conditions and material properties are seldom known exactly. Fortunately techniques exist for obtaining approximate solutions. They reduce the continuum problem with an infinite number of degrees of freedom to an approximating system with a finite number of degrees of freedom. Thus the problem of solving a differential equation or system of differential equations is replaced by that of solving a set of algebraic equations. If the differential equations are linear, the approximating algebraic equations are also linear. The techniques for accomplishing this mathematical simplification have been known for many years. However, the calculations are lengthy, and it has only been in recent years, when high-speed digital computers have been available, that the methods have been practical.

In this chapter we consider a method in which the differential equations and boundary conditions are replaced by finite-difference equations. To do this, the derivatives in the differential equations and boundary conditions are replaced by their finite-difference approximations. The application of the method to eigenvalue and equilibrium boundary-value problems is described and illustrated with examples. Additional examples are given throughout the remainder of the book. Further information on the method can be found in Refs. 1 to 6.

Other procedures that are useful in obtaining approximate solutions are the Rayleigh-Ritz method, described in Chap. 6, the finite-element matrix methods introduced in Chap. 12, and the weighted-residual methods (collocation, subdomain, Galerkin, and least squares) given in

Refs. 1 and 3.   The Rayleigh-Ritz method is applied to the potential
energy or complementary potential energy of the body, while the weighted-
residual methods, like the finite-difference method, are applied to the
differential equations.   The Rayleigh-Ritz and weighted-residual methods are particularly useful when physical intuition is of assistance in
choosing functions to approximate the solution.   In these cases, accurate
solutions can often be obtained with relatively few degrees of freedom.
While requiring more degrees of freedom for an accurate solution, the
finite-difference method is usually simpler to apply when the differential
equation and boundary conditions are known.

5-2 FINITE–DIFFERENCE OPERATORS

In the finite-difference method we replace the continuous domain by a
pattern of discrete points within the domain, referred to as *grid* or *mesh
points*.   Instead of determining the unknown function at all points in
the domain, we find an approximation of the function at the grid points
only.   If the value of the function at intermediate points is required,
we obtain it by interpolation between the values at the grid points.
While difference approximations can be derived for unequally spaced
grid points, the relationships are cumbersome, and we shall restrict the
discussion to equally spaced points.

Let us first consider the one-dimensional case shown in Fig. 5-1,
where ψ is a function of the coordinate x.   We assume a lattice of points
a distance h apart and let the value of ψ at a typical point x_j be ψ_j.   By
using the Taylor's series expansion we can write the value of ψ at the
point $x_{j+1} = x_j + h$ in terms of ψ and its derivatives at x_j.   Thus

$$\psi_{j+1} = \psi(x_j + h) = \psi_j + h\left(\frac{d\psi}{dx}\right)_j + \frac{h^2}{2!}\left(\frac{d^2\psi}{dx^2}\right)_j + \frac{h^3}{3!}\left(\frac{d^3\psi}{dx^3}\right)_j + \cdots \quad (5\text{-}1)$$

Solving this equation for $(d\psi/dx)_j$, we find

$$\left(\frac{d\psi}{dx}\right)_j = \frac{\psi_{j+1} - \psi_j}{h} + O(h) \quad (5\text{-}2)$$

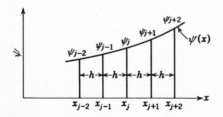

Fig. 5-1   One-dimensional finite-difference grid.

where $O(h)$ indicates that the sum of remaining terms is of the order of h; that is, all the remaining terms contain h as a factor. If we let h go to zero, we obtain one of the several possible definitions for the first derivative

$$\left(\frac{d\psi}{dx}\right)_j = \lim_{h \to 0} \frac{\psi_{j+1} - \psi_j}{h} \tag{5-3}$$

Taking the first term on the right side of Eq. (5-2) as an approximation of the derivative, we find

$$\left(\frac{d\psi}{dx}\right)_j = \frac{1}{h}\left(\psi_{j+1} - \psi_j\right) \tag{5-4}$$

It is seen that the sum of the neglected terms, or *truncation error*, is of order h. Equations (5-2) and (5-4) express the derivative at x_j in terms of the function at this point and at the forward point x_{j+1}. These equations are therefore referred to as *forward-difference formulas*.

The value of ψ at the point x_{j-1} can be found by replacing h by $-h$ in Eq. (5-1), to give

$$\psi_{j-1} = \psi_j - h\left(\frac{d\psi}{dx}\right)_j + \frac{h^2}{2!}\left(\frac{d^2\psi}{dx^2}\right)_j - \frac{h^3}{3!}\left(\frac{d^3\psi}{dx^3}\right)_j + \cdots \tag{5-5}$$

Solving this equation for $(d\psi/dx)_j$, we obtain

$$\left(\frac{d\psi}{dx}\right)_j = \frac{1}{h}\left(\psi_j - \psi_{j-1}\right) + O(h) \tag{5-6}$$

In this case the derivative at x_j is expressed in terms of ψ_j and ψ_{j-1}; for this reason the equation is referred to as a *backward-difference formula*.

Finally, an expression for the first derivative can be found by subtracting Eq. (5-5) from Eq. (5-1) and solving the resulting equation for $(d\psi/dx)_j$. If this is done, we obtain

$$\left(\frac{d\psi}{dx}\right)_j = \frac{1}{2h}\left(\psi_{j+1} - \psi_{j-1}\right) + O(h^2) \tag{5-7}$$

Since the derivative is expressed in terms of the function at both forward and backward points, Eq. (5-7) is referred to as a *central-difference formula*.

We see from Eqs. (5-2), (5-6), and (5-7) that there is no unique finite-difference expression for the derivative. Each of the equations gives the exact value of the derivative if all terms in the infinite series are retained or if h goes to zero. However, if we approximate the derivative by truncating the equations after the first terms, the approximations obtained from the three equations are not of equal accuracy. The advantage of using the central-difference form is that the terms which are neglected are $O(h^2)$ instead of $O(h)$, as in the forward and backward differences. Physically we see that the central-difference approximation gives the slope of the chord line from the point x_{j-1} to x_{j+1}. By using additional points it is possible to obtain forward- or backward-difference formulas in which the neglected terms are of the same order as in the central-difference formulas (see Prob. 5-1). These may be useful at boundary points. In the remaining explanation we shall consider only central-difference forms.

To obtain a difference expression for the second derivative we add Eqs. (5-1) and (5-5) and solve for $(d^2\psi/dx^2)_j$, which gives

$$\left(\frac{d^2\psi}{dx^2}\right)_j = \frac{1}{h^2}\,(\psi_{j-1} - 2\psi_j + \psi_{j+1}) + O(h^2) \tag{5-8}$$

Expressions for ψ_{j+2} and ψ_{j-2}, the values of ψ at x_{j+2} and x_{j-2}, are found by replacing h in Eqs. (5-1) and (5-5) by $2h$. By properly combining ψ_{j-2}, ψ_{j-1}, ψ_j, ψ_{j+1}, and ψ_{j+2} we can eliminate the first and second derivatives and solve for the third derivative. This gives

$$\left(\frac{d^3\psi}{dx^3}\right)_j = \frac{1}{2h^3}\,(-\psi_{j-2} + 2\psi_{j-1} - 2\psi_{j+1} + \psi_{j+2}) + O(h^2) \tag{5-9}$$

In a similar fashion the fourth derivative is found to be

$$\left(\frac{d^4\psi}{dx^4}\right)_j = \frac{1}{h^4}\,(\psi_{j-2} - 4\psi_{j-1} + 6\psi_j - 4\psi_{j+1} + \psi_{j+2}) + O(h^2) \tag{5-10}$$

In calculations it is convenient to use computational modules which give the coefficients by which the values of ψ at the grid points are multiplied to obtain the derivatives. These modules are shown in Fig. 5-2. Also shown in this figure are two modules[2] which are useful in numerical integration. The first is based upon the trapezoidal rule and may be used to integrate over an interval divided into any number of subintervals of length h. The second module, which is derived from Simpson's rule, is more accurate but may be applied only when the number of subintervals is even.

Next consider the case where ψ is a function of the two variables x and y, as shown in Fig. 5-3. In this case the preceding equations apply if the ordinary derivatives are replaced by the corresponding partial derivatives so that the modules shown in Fig. 5-4 are obtained. Mixed partial derivatives can occur in two-dimensional problems. The value of ψ at the point (x_{j+1}, y_{k+1}) is found from the Taylor's series expansion

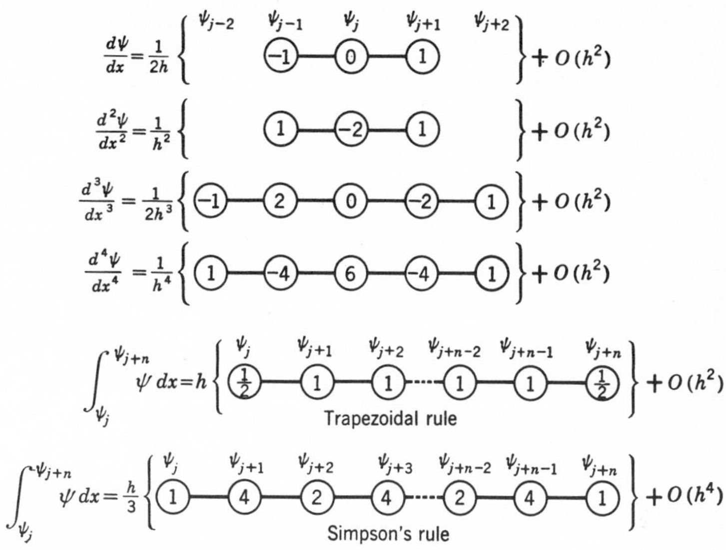

$$\frac{d\psi}{dx}=\frac{1}{2h}\left\{\begin{array}{cc} & \psi_{j-2} \quad \psi_{j-1} \quad \psi_j \quad \psi_{j+1} \quad \psi_{j+2} \\ & \boxed{-1}-\boxed{0}-\boxed{1} \end{array}\right\}+O(h^2)$$

$$\frac{d^2\psi}{dx^2}=\frac{1}{h^2}\left\{\boxed{1}-\boxed{-2}-\boxed{1}\right\}+O(h^2)$$

$$\frac{d^3\psi}{dx^3}=\frac{1}{2h^3}\left\{\boxed{-1}-\boxed{2}-\boxed{0}-\boxed{-2}-\boxed{1}\right\}+O(h^2)$$

$$\frac{d^4\psi}{dx^4}=\frac{1}{h^4}\left\{\boxed{1}-\boxed{-4}-\boxed{6}-\boxed{-4}-\boxed{1}\right\}+O(h^2)$$

$$\int_{\psi_j}^{\psi_{j+n}}\psi\,dx=h\left\{\begin{array}{c}\psi_j \quad \psi_{j+1} \quad \psi_{j+2} \quad \psi_{j+n-2} \quad \psi_{j+n-1} \quad \psi_{j+n} \\ \boxed{\tfrac{1}{2}}-\boxed{1}-\boxed{1}\cdots\boxed{1}-\boxed{1}-\boxed{\tfrac{1}{2}}\end{array}\right\}+O(h^2)$$

Trapezoidal rule

$$\int_{\psi_j}^{\psi_{j+n}}\psi\,dx=\frac{h}{3}\left\{\begin{array}{c}\psi_j \quad \psi_{j+1} \quad \psi_{j+2} \quad \psi_{j+3} \quad \psi_{j+n-2} \quad \psi_{j+n-1} \quad \psi_{j+n} \\ \boxed{1}-\boxed{4}-\boxed{2}-\boxed{4}\cdots\boxed{2}-\boxed{4}-\boxed{1}\end{array}\right\}+O(h^4)$$

Simpson's rule

Fig. 5-2 One-dimensional computational modules.

for two dimensions,

$$\psi_{j+1,k+1} = \psi_{j,k} + h\left(\frac{\partial\psi}{\partial x}\right)_{j,k} + h\left(\frac{\partial\psi}{\partial y}\right)_{j,k} + \frac{h^2}{2!}\left(\frac{\partial^2\psi}{\partial x^2}\right)_{j,k}$$
$$+ \frac{h^2}{2!}\left(\frac{\partial^2\psi}{\partial y^2}\right)_{j,k} + h^2\left(\frac{\partial^2\psi}{\partial x\,\partial y}\right)_{j,k} + \cdots \quad (5\text{-}11)$$

where on the right side of the equation ψ and its partial derivatives are evaluated at (x_j,y_k). The value of $\psi_{j-1,k+1}$ is found from this equation by replacing h in the x-derivative terms by $-h$. The expressions for

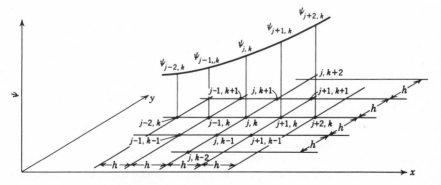

Fig. 5-3 Two-dimensional finite-difference grid.

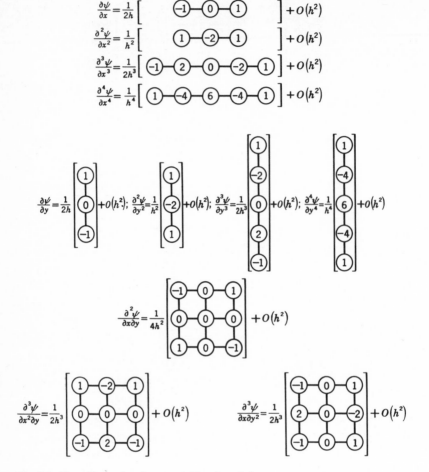

Fig. 5-4 Two-dimensional computational modules.

$\psi_{j-1,k-1}$ and $\psi_{j+1,k-1}$ are determined in a similar fashion. The reader can easily verify that the mixed second partial derivative is obtained by combining these expansions as follows:

$$\left(\frac{\partial^2 \psi}{\partial x\, \partial y}\right)_{j,k} = \frac{1}{4h^2}\left(\psi_{j-1,k-1} - \psi_{j-1,k+1} - \psi_{j+1,k-1} + \psi_{j+1,k+1}\right) + O(h^2) \quad (5\text{-}12)$$

Expressions for the operators ∇^2 and ∇^4 are derived by adding the modules for each of the derivatives in the operators. The modules for these operators and those for two-dimensional integration by the trapezoidal and Simpson's rules are shown in Fig. 5-5. In the modules of Figs. 5-4 and 5-5, the point (x_j, y_k) occupies the central position in all modules

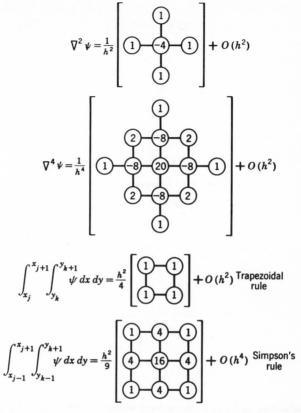

Fig. 5-5 Computational modules for two-dimensional operators for differentiation and integration.

except the one for integration by the trapezoidal rule, where it lies at the lower left corner.

Computational modules for different grid spacings in the x and y directions and for polar and skew coordinates may be found in Ref. 2.

5-3 APPLICATION TO EQUILIBRIUM BOUNDARY-VALUE PROBLEMS

In many physical problems the steady-state value of a function is sought at all points within a domain and on its boundary. Examples include steady subsonic fluid flow, steady-state heat transfer, and the stress or displacement analysis of stable elastic structures with prescribed static forces and temperatures. These seemingly unrelated physical phenomena have similar mathematical characteristics and may be grouped into a class known as *equilbrium boundary-value problems*. An excellent discussion of the mathematical properties of these problems and methods of solving them will be found in Ref. 1.   In this section we shall briefly mention some of their characteristics and see how approximate solutions can be obtained by means of the finite-difference method.

In general, the equilibrium boundary-value problem consists of determining a function ψ which satisfies the differential equation

$$L_{2m}(\psi) = f \tag{5-13}$$

at all interior points of a domain D and in addition satisfies the equations

$$B_i(\psi) = g_i \qquad i = 1, 2, \ldots, m \tag{5-14}$$

at all points on the boundary of the domain.[1]   In Eq. (5-13), L_{2m} represents a differential operator having a highest derivative of order $2m$. The derivatives in the operator will be ordinary or partial depending upon whether D is one- or multidimensional.   The term f is a known function of position at all points within D.   The notation B_i designates a differential operator with derivatives of order $\leq 2m - 1$, and g_i is a prescribed function of position at all points on the boundary of D.   There are m independent boundary conditions at each point on the boundary.   If $L_{2m}(\psi)$ and $B_i(\psi)$ are linear, that is, ψ and its derivatives appear linearly, the system is linear, and the principle of superposition is applicable.   In a structural problem the operators L_{2m} and B_i are determined by the elastic and geometric characteristics of the system, while f and g_i are related to the applied loads and temperatures.

To fix ideas, consider the Airy stress-function formulation of the plane-stress problem, which is developed in Sec. 4-5.   In this case we must find a function φ which satisfies Eq. (4-14) at all points within the body and Eqs. (4-22) and (4-23) on the boundary of the body.   We see that the unknown function $\psi = \varphi$ and the operator $L_{2m} = \nabla^4$, so that $2m = 4$.   The prescribed function of position in D is

$$f = -E \, \nabla^2 \alpha T - (1 - \nu) \, \nabla^2 V$$

As expected (because $m = 2$), there are two boundary conditions that must be satisfied at each boundary point.   We note that $B_1 = 1$ and $B_2 = \partial/\partial n$, while $g_1 = \int_0^s (Al + Bm) \, ds$ and $g_2 = -Bl + Am$.   The equations therefore describe an equilibrium boundary-value problem.

In general, Eqs. (5-13) and (5-14) cannot be solved exactly. Approximate solutions can be obtained by replacing the continuous domain by a grid of equally spaced points which cover the domain.   The unknowns then become the values of ψ at each of the grid points.   At interior points the differential equation (5-13) must be satisfied.   However, instead of using the differential operator L_{2m} we replace the derivatives in the operator by their finite-difference counterparts, neglecting the higher-order terms in the finite-difference forms. The resulting difference equation is applied at each grid point, giving a set of algebraic equations which approximate the differential equation (5-13) at the mesh points. The number of simultaneous algebraic equations is equal to the number of the interior grid points. For simplicity we shall consider only problems in which boundary points coincide with mesh points. For a discussion of the more general problem, in which the boundary does not pass through the mesh points, see Refs. 1 and 3. When the difference equation is applied to points next to the boundary, the equation will contain values of ψ at points that are on or outside of the boundary. Because of this there will be more unknowns than equations. To obtain additional equations the boundary conditions are expressed in finite-difference form and applied at each of the boundary mesh points. In doing so the number of equations will be increased to equal the number of unknown ψ_j. The result is a set of algebraic equations in terms of the unknown values of ψ at the grid points. If L_{2m} and B_i are linear operators, the algebraic equations will be linear and in the form

$$a_{11}\psi_1 + a_{12}\psi_2 + \cdots + a_{1n}\psi_n = b_1$$
$$a_{21}\psi_1 + a_{22}\psi_2 + \cdots + a_{2n}\psi_n = b_2$$
$$\cdots \cdots \cdots \cdots \cdots \cdots \cdots \cdots \cdots \cdots$$
$$a_{n1}\psi_1 + a_{n2}\psi_2 + \cdots + a_{nn}\psi_n = b_n$$

(5-15)

These equations may be written in matrix notation[1] as

$$\mathbf{a}\psi = \mathbf{b} \qquad (5\text{-}16)$$

where

$$\mathbf{a} = \begin{bmatrix} a_{11} & a_{12} & \cdots & a_{1n} \\ a_{21} & a_{22} & \cdots & a_{2n} \\ & \cdots \cdots \cdots & \\ a_{n1} & a_{n2} & \cdots & a_{nn} \end{bmatrix} \qquad \psi = \begin{Bmatrix} \psi_1 \\ \psi_2 \\ \cdot \\ \cdot \\ \cdot \\ \psi_n \end{Bmatrix} \qquad \mathbf{b} = \begin{Bmatrix} b_1 \\ b_2 \\ \cdot \\ \cdot \\ \cdot \\ b_n \end{Bmatrix} \qquad (5\text{-}17)$$

The computational modules involve ψ only at points in the neighborhood of the point at which it is applied; as a result many of the a_{ij} coefficients will usually be zero. If the points are numbered systematically, the nonzero elements of $\mathbf{a}$ will be clustered in a band about the main diagonal.

The solution of the simultaneous equations gives ψ at the mesh points, and, if needed, intermediate values can be obtained by interpolation. If functions which depend upon derivatives of ψ are also required, they can be found by applying the finite-difference forms for the derivatives to the values of ψ at the grid points.

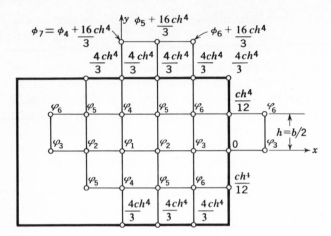

Fig. 5-6   Finite-difference grid for Example 5-1.

Example 5-1   Determine the stress distribution of the plane-stress problem shown in Fig. 4-3 when $a = 1.5b$.

The assumed difference grid is shown in Fig. 5-6; because of the symmetry of the body and loading, the stress function is symmetric about both axes. We take advantage of this in designating the unknowns by assigning identical numbers to corresponding points in each quadrant, as shown in Fig. 5-6. The value of φ at each mesh point on the boundary is known from Example 4-1 and is entered directly in Fig. 5-6.

There are no body forces or temperature changes, so that Eq. (4-17) must be satisfied at all interior points. To approximate this we apply the difference module for ∇^4 (Fig. 5-5) to φ at each of the interior mesh points and equate the result to zero. As an example, using 1 as a pivotal point, we obtain

$$\frac{1}{h^4}\,(20\varphi_1 - 16\varphi_2 + 2\varphi_3 - 16\varphi_4 + 8\varphi_5 + \tfrac{8}{3}ch^4) = 0$$

Points lying one mesh distance outside of the boundary become involved when we apply the ∇^4 module to a point next to the boundary. Rather than introduce additional unknowns, we use the $\partial\varphi/\partial n$ boundary condition to determine φ at each of these outside points in terms of φ at the corresponding point one mesh space inside of the boundary. Consider for instance the point $(0,b)$. From Example 4-1 we have $\partial\varphi/\partial n = cb^3/3$ at this point. Designating the outside mesh point above $(0,b)$ as 7, and using the $\partial/\partial y$ operator of Fig. 5-4, we obtain $(-\varphi_4 + \varphi_7)/2h = 8ch^3/3$, which gives $\varphi_7 = \varphi_4 + 16ch^4/3$. Similarly, at the point $(a,0)$, $\partial\varphi/\partial n = 0$, so that φ at the outside point to the right of $(a,0)$ is equal to φ_3. This method has been used to express φ at all exterior points in terms of φ at previously designated interior points, as shown in Fig. 5-6. The number of unknowns is therefore equal to the number of interior points. Applying the difference operator equivalent to the differential equation at each interior point gives as many equations as there are unknowns.

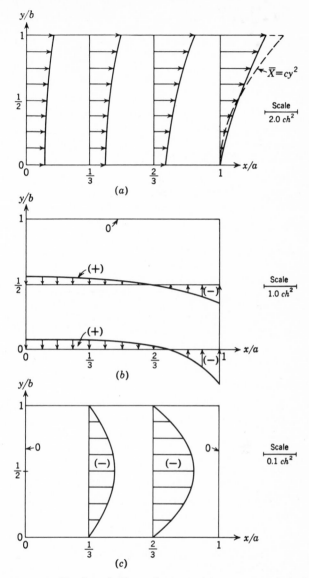

Fig. 5-7 Results of Example 5-1. (a) σ_{xx}, symmetrical about x and y axes; (b) σ_{yy}, symmetrical about x and y axes; (c) σ_{xy}, antisymmetrical about x and y axes.

These equations can be written in the matrix form

$$\begin{bmatrix} 20 & -16 & 2 & -16 & 8 & 0 \\ -8 & 21 & -8 & 4 & -16 & 4 \\ 1 & -8 & 21 & 0 & 4 & -16 \\ -8 & 4 & 0 & 22 & -16 & 2 \\ 2 & -8 & 2 & -8 & 23 & -8 \\ 0 & 2 & -8 & 1 & -8 & 23 \end{bmatrix} \begin{Bmatrix} \varphi_1 \\ \varphi_2 \\ \varphi_3 \\ \varphi_4 \\ \varphi_5 \\ \varphi_6 \end{Bmatrix} = -\frac{ch^4}{3} \begin{Bmatrix} 8 \\ 8 \\ 9 \\ 0 \\ \frac{1}{4} \\ -2 \end{Bmatrix}$$

Solving, we find

$$\pmb{\phi} = -ch^4 \{ 1.05647 \quad 0.91985 \quad 0.51064 \quad 0.51561 \quad 0.43539 \quad 0.19766 \}$$

where the column matrix has been written horizontally to conserve space.

The stresses are found by applying the operators of Fig. 5-4, which are equivalent to Eqs. (4-13), to each of the mesh points. As an example, to find $\sigma_{xx}(0,0)$ we use the approximate relationship

$$\sigma_{xx}(0,0) = \frac{1}{h^2} (\varphi_4 - 2\varphi_1 + \varphi_4) = 1.0817ch^2$$

The resulting stresses are shown in Fig. 5-7. It is seen that as the distance from the ends increases, σ_{xx} becomes more uniform and σ_{yy} and σ_{xy} tend toward zero, as expected from St. Venant's principle (Sec. 4-8). In the example a/b is so small that σ_{xx} never becomes uniform, and σ_{yy} and σ_{xy} do not actually become zero. It is interesting to note that because of the coarse grid, σ_{xx} at $x = \pm a$ is not in good agreement with the actual edge loads (shown by the dotted lines in Fig. 5-7).

5-4 APPLICATION TO EIGENVALUE PROBLEMS

Another class of physical problems frequently encountered is known as the *eigenvalue problem*. The prediction of buckling loads (Chaps. 14 and 15), natural frequencies of free vibrations (Chaps. 7, 8, and 13), and aeroelastic divergence (Chap. 8) are examples of this type of problem in the field of structures. In the typical eigenvalue problem we must determine a function ψ and an associated constant λ which satisfy the equation

$$L_{2m}(\psi) = \lambda M_{2n}(\psi) \tag{5-18}$$

at all points in a domain and the equations

$$B_i(\psi) = \lambda C_i(\psi) \qquad i = 1, 2, \cdots, m \tag{5-19}$$

at all points on the boundary of the domain.[1] In these equations L_{2m}, M_{2n}, B_i, and C_i are differential operators. The order of the highest derivative in L_{2m} and M_{2n} is $2m$ and $2n$, respectively, and $m > n$. There are m independent boundary conditions at each boundary point, and the operator B_i that appears in these conditions may contain derivatives of order $\leq 2m - 1$.

Equations (5-18) and (5-19) are homogeneous and therefore have the trivial solution $\psi = 0$. It is the nontrivial solutions of these equations that are of physical interest, however, and these occur only for specific values of λ known as the *characteristic* values, or *eigenvalues* (German for characteristic), of the system. Associated with each of these values of λ is a function ψ called the *eigenfunction* or *mode shape*. We note that if ψ is a solution to Eqs. (5-18) and (5-19), then ψ multiplied by a constant is also a solution. As a result, the absolute magnitude of the mode shape cannot be determined uniquely. In a continuum problem there are an infinite number of pairs of eigenvalues and associated eigenfunctions which satisfy the equations.

As an example of such a problem, it is shown in Sec. 14-2 that the bending deflections w of a buckled column with pinned ends must satisfy the differential equation

$$EI \frac{d^2w}{dx^2} = -Pw \qquad (5\text{-}20)$$

at all points in the interval $0 < x < L$, where L is the column length, and the boundary conditions $w(0) = 0$ and $w(L) = 0$. We note that the differential equation and boundary conditions are homogeneous and have the trivial solution $w = 0$. Nontrivial solutions are possible only for specific values of P. The lowest of these is the buckling load, because it is the smallest compressive force that can hold the column in the bent configuration $w \neq 0$. In Eq. (5-20) $L_{2m} = EI d^2/dx^2$, $\psi = w$, $\lambda = P$, and $M_{2n} = -1$ (a zero-order derivative). We observe that $2m = 2$, or $m = 1$, so that there is one boundary condition at each of the end points $x = 0$ and L. The boundary-condition operators are $B_1(0) = B_1(L) = 1$ and $C_1(0) = C_1(L) = 0$. In some cases it is possible to obtain exact solutions to problems of the form of Eqs. (5-18) and (5-19). The methods used in these cases are described in Chaps. 14 and 15.

As in equilibrium problems, to obtain an approximate solution we cover the domain by a grid of points and apply the finite-difference approximation of the differential equation at each of the interior points. The finite-difference approximations of the boundary conditions are imposed at each of the boundary points. If Eqs. (5-18) and (5-19) are linear, the approximating difference equations will result in a set of homogeneous linear algebraic equations of the form

$$
\begin{aligned}
a_{11}\psi_1 + a_{12}\psi_2 + \cdots + a_{1n}\psi_n &= \lambda(b_{11}\psi_1 + b_{12}\psi_2 + \cdots + b_{1n}\psi_n) \\
a_{21}\psi_1 + a_{22}\psi_2 + \cdots + a_{2n}\psi_n &= \lambda(b_{21}\psi_1 + b_{22}\psi_2 + \cdots + b_{2n}\psi_n) \\
&\cdots \cdots \cdots \cdots \cdots \cdots \cdots \cdots \cdots \cdots \\
a_{n1}\psi_1 + a_{n2}\psi_2 + \cdots + a_{nn}\psi_n &= \lambda(b_{n1}\psi_1 + b_{n2}\psi_2 + \cdots + b_{nn}\psi_n)
\end{aligned}
$$

$$(5\text{-}21)$$

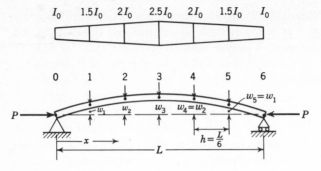

Fig. 5-8   Example 5-2.

By using matrix notation, these equations may be written

$$\mathbf{a}\psi = \lambda\mathbf{b}\psi \qquad\qquad (5\text{-}22)$$

where **a** and ψ are defined in Eq. (5-17) and

$$\mathbf{b} = \begin{bmatrix} b_{11} & b_{12} & \cdots & b_{1n} \\ b_{21} & b_{22} & \cdots & b_{2n} \\ \multicolumn{4}{c}{\cdots\cdots\cdots\cdots} \\ b_{n1} & b_{n2} & \cdots & b_{nn} \end{bmatrix} \qquad (5\text{-}23)$$

Methods for solving Eq. (5-21) or (5-22) are given in the next section.

Example 5-2 Find the matrices **a** and **b** for the finite-difference approximation of the buckling load and mode shape of the nonuniform column with pinned ends shown in Fig. 5-8.

The boundary condition $w(0) = 0$ gives $w_0 = 0$. The mode shape of the buckled column is symmetric about the center of the column; because of this, $w_5 = w_1$ and $w_4 = w_2$. As a result, there are only three unknown displacements for the grid shown, and $\mathbf{w} = \{w_1 \ w_2 \ w_3\}$. Using the symmetry condition is equivalent to restricting the domain to $0 < x < L/2$ and applying the boundary condition $(dw/dx)_3 = (1/2h)(w_4 - w_2) = 0$, from which $w_4 = w_2$.

By using the computational module for d^2/dx^2 from Fig. 5-2 we write Eq. (5-20) in the difference form

$$\frac{EI_i}{h^2}(w_{i-1} - 2w_i + w_{i+1}) = -Pw_i \qquad (a)$$

where $h = L/6$. Applying Eq. (a) to points 1, 2, and 3 and using the values of EI in Fig. 5-8 gives

$$-3w_1 + 1.5w_2 = -\frac{PL^2}{36EI_0}w_1$$

$$2w_1 - 4w_2 + 2w_3 = -\frac{PL^2}{36EI_0}w_2$$

$$5w_2 - 5w_3 = -\frac{PL^2}{36EI_0}w_3$$

These equations are of the form of Eqs. (5-22), where

$$a = \begin{bmatrix} -3 & 1.5 & 0 \\ 2 & -4 & 2 \\ 0 & 5 & -5 \end{bmatrix} \qquad b = \begin{bmatrix} 1 & 0 & 0 \\ 0 & 1 & 0 \\ 0 & 0 & 1 \end{bmatrix}$$

and $\lambda = -PL^2/36EI_0$.

5-5 SOLUTION OF MATRIX EIGENVALUE EQUATIONS

Equations (5-21) or (5-22) may be solved by direct or iterative methods. In the direct method we write Eqs. (5-21) as

$$(a_{11} - \lambda b_{11})\psi_1 + (a_{12} - \lambda b_{12})\psi_2 + \cdots + (a_{1n} - \lambda b_{1n})\psi_n = 0$$

$$(a_{21} - \lambda b_{21})\psi_1 + (a_{22} - \lambda b_{22})\psi_2 + \cdots + (a_{2n} - \lambda b_{2n})\psi_n = 0$$

$$\cdots\cdots\cdots\cdots\cdots\cdots\cdots\cdots\cdots\cdots\cdots\cdots\cdots \qquad (5\text{-}24)$$

$$(a_{n1} - \lambda b_{n1})\psi_1 + (a_{n2} - \lambda b_{n2})\psi_2 + \cdots + (a_{nn} - \lambda b_{nn})\psi_n = 0$$

or in matrix notation as $(a - \lambda b) = 0$. If we attempt to solve Eqs. (5-24) for ψ_j by Cramer's rule of determinants, we find

$$\psi_j = \frac{|\bar{a} - \lambda \bar{b}|}{|a - \lambda b|}$$

where $\bar{a}$ and $\bar{b}$ are obtained by placing zeros in the jth column of a and b, respectively. All the elements of the jth column of the determinant in the numerator are zero, and so the numerator is zero. A nonzero solution for ψ_j is possible only if the determinant $|a - \lambda b| = 0$, which gives the indeterminate form $\psi_j = 0/0$. The criterion for a nontrivial solution is therefore

$$|a - \lambda b| = 0 \qquad (5\text{-}25)$$

Expansion of the determinant of Eq. (5-25) gives an nth-order polynomial in λ, which is the *characteristic equation* of the system. The roots of this polynomial $(\lambda_1, \lambda_2, \ldots, \lambda_n)$ approximate the first n eigenvalues of Eqs. (5-18) and (5-19). If the mode shape associated with λ_j is also desired it is found by (1) discarding one of Eqs. (5-24), (2) substituting $\lambda = \lambda_j$ into the remaining equations, (3) transferring the term $(a_{ik} - \lambda_j b_{ik})\psi_k$ (k arbitrary) to the right side of each equation, and (4) solving these equations for the remaining $(n - 1)$ values of ψ_i in terms of ψ_k.

For simplicity of illustration, small values of n (usually 3) have been used in the examples throughout the book. In most problems it is necessary to use higher-order matrices to obtain acceptable accuracy. In these cases the expansion of the determinant in Eq. (5-25) and the determination of the roots of the resulting nth-order polynomial is tedious, and the direct method which has been described is not recommended. Equation (5-22) may be written in the form

$$\mathbf{c\psi} = \lambda\mathbf{\psi} \tag{5-26}$$

where $\mathbf{c} = \mathbf{b}^{-1}\mathbf{a}$. The matrix $\mathbf{b}$ is often diagonal, which greatly simplifies the inversion. The solution of the matrix equation (5-26) for the n pairs of λ_i and $\mathbf{\psi}_i$ is routine for the computer center that is certain to be a part of any large aerospace organization. The engineer need determine only the elements of $\mathbf{a}$ and $\mathbf{b}$; the remaining computational operations are carried out by the digital computer.

When $n > 3$ and computations must be made by hand or desk calculator, an iterative procedure[1] is more convenient than the direct method. The iteration method, which gives the mode shape as well as the eigenvalue, converges to the highest eigenvalue of the approximating matrix equation. In most problems it is the lowest eigenvalue that is of physical interest, in which case it is necessary to write Eq. (5-22) in the form

$$\mathbf{d\psi} = \bar{\lambda}\mathbf{\psi} \tag{5-27}$$

where $\bar{\lambda} = 1/\lambda$ and $\mathbf{d} = \mathbf{a}^{-1}\mathbf{b}$. Iteration then converges to the largest $\bar{\lambda}$, from which the smallest λ is computed.

The procedure is begun by assuming $\mathbf{\psi}_0$, an approximation of the mode shape. The assumption is arbitrary, but the solution is speeded by using physical intuition to make a reasonable first approximation. For convenience, one of the elements of $\mathbf{\psi}_0$, say ψ_{0_i}, is taken as unity. Premultiplying $\mathbf{\psi}_0$ by $\mathbf{d}$ gives $\mathbf{d\psi}_0 = \bar{\lambda}_1\mathbf{\psi}_1$, where $\bar{\lambda}_1$ and $\mathbf{\psi}_1$ are obtained by factoring the constant $\bar{\lambda}_1$ from the product $\mathbf{d\psi}_0$, so that the ψ_{1_i} element of $\mathbf{\psi}_1$ is reduced to unity. If $\mathbf{\psi}_0$ is a true eigenfunction of Eq. (5-27), $\mathbf{\psi}_1$ will equal $\mathbf{\psi}_0$, and $\bar{\lambda}_1$ is the true eigenvalue associated with $\mathbf{\psi}_0$. However, it is unlikely that the guess for $\mathbf{\psi}_0$ will be a true eigenfunction. Nevertheless, $\mathbf{\psi}_1$ is closer to the true mode shape than $\mathbf{\psi}_0$, and $\bar{\lambda}_1$ and $\mathbf{\psi}_1$ may be considered first approximations of the highest eigenvalue and eigenfunction of Eq. (5-27).

To obtain a second approximation we compute $\mathbf{d\psi}_1 = \bar{\lambda}_2\mathbf{\psi}_2$, where the element ψ_{2_i} has again been normalized to unity by factoring $\bar{\lambda}_2$ from the product $\mathbf{d\psi}_1$. The process is repeated according to the recurrence formula

$$\mathbf{d\psi}_{n-1} = \bar{\lambda}_n\mathbf{\psi}_n \tag{5-28}$$

until $\mathbf{\psi}_{n-1}$ and $\mathbf{\psi}_n$ agree to the desired number of significant figures. The lowest eigenvalue of Eq. (5-22) is then $\lambda = 1/\bar{\lambda}_n$, and the associated eigenfunction is $\mathbf{\psi}_n$. For hand computations, the method has the advantage that a computational error will not affect the final results but will merely increase the time to convergence.

The procedure for extending the iteration method to give eigenvalues in addition to the highest or lowest ones is given in Ref. 1, along with the proof of convergence and the exact conditions under which convergence is certain.   It should be noted, however, that while rather restrictive conditions are necessary to prove that convergence will exist, they are not always necessary for convergence.

Example 5-3   Obtain an approximate value for the buckling load and deflection shape of the nonuniform pin-ended column of Example 5-2.

The **b** matrix of Example 5-2 is a unit matrix, so that $\mathbf{d} = \mathbf{a}^{-1}\mathbf{b} = \mathbf{a}^{-1}$, which upon inverting the **a** matrix gives

$$\mathbf{d} = -\frac{1}{15}\begin{bmatrix} 10 & 7.5 & 3 \\ 10 & 15 & 6 \\ 10 & 15 & 9 \end{bmatrix}$$

Equation (5-27) then becomes

$$\begin{bmatrix} 10 & 7.5 & 3 \\ 10 & 15 & 6 \\ 10 & 15 & 9 \end{bmatrix} \begin{Bmatrix} w_1 \\ w_2 \\ w_3 \end{Bmatrix} = \bar{\lambda} \begin{Bmatrix} w_1 \\ w_2 \\ w_3 \end{Bmatrix} \qquad (a)$$

where

$$\bar{\lambda} = \frac{15 \times 36EI_0}{PL^2} \qquad (b)$$

A uniform column with pinned ends buckles into the deflection shape $\sin(\pi x/L)$, which should provide a reasonable first approximation for the non-uniform column.   Letting $x = L/6$, $L/3$, and $L/2$, we obtain

$$\mathbf{w_0} = \{0.500 \quad 0.866 \quad 1.000\}$$

Substituting this for **w** in the left side of Eq. (a) gives

$$\begin{bmatrix} 10 & 7.5 & 3 \\ 10 & 15 & 6 \\ 10 & 15 & 9 \end{bmatrix} \begin{Bmatrix} 0.500 \\ 0.866 \\ 1.000 \end{Bmatrix} = \begin{Bmatrix} 14.49 \\ 24.00 \\ 27.00 \end{Bmatrix} = 27.00 \begin{Bmatrix} 0.536 \\ 0.889 \\ 1.000 \end{Bmatrix}$$

giving $\bar{\lambda}_1 = 27.00$ for a first approximation of $\bar{\lambda}$.   We note, however, that

$$\mathbf{w_1} = \{0.536 \quad 0.889 \quad 1.000\} \neq \mathbf{w_0}$$

and so the assumed mode is not the true buckle shape.   For a second approximation we substitute $\mathbf{w_1}$ into the left side of Eq. (a) and obtain

$$\begin{bmatrix} 10 & 7.5 & 3 \\ 10 & 15 & 6 \\ 10 & 15 & 9 \end{bmatrix} \begin{Bmatrix} 0.536 \\ 0.889 \\ 1.000 \end{Bmatrix} = \begin{Bmatrix} 15.03 \\ 24.70 \\ 27.70 \end{Bmatrix} = 27.70 \begin{Bmatrix} 0.543 \\ 0.855 \\ 1.000 \end{Bmatrix}$$

giving $\bar{\lambda}_2 = 27.70$ and

$$\mathbf{w_2} = \{0.543 \quad 0.855 \quad 1.000\} \neq \mathbf{w_1}$$

Repeating the process, we find

$$\bar{\lambda}_3 \mathbf{w_3} = 27.24 \begin{Bmatrix} 0.545 \\ 0.890 \\ 1.000 \end{Bmatrix} \qquad \bar{\lambda}_4 \mathbf{w_4} = 27.78 \begin{Bmatrix} 0.545 \\ 0.891 \\ 1.000 \end{Bmatrix}$$

We note that $w_3 \approx w_4$, so that $\bar{\lambda}_4 = \bar{\lambda} = 27.78$. Substituting this result into Eq. (b) gives the buckling load $P = 19.45 EI_0/L^2$, which agrees with the exact solution of Eq. (5-20) obtained by using Bessel functions[7] to within 1.2 percent.

REFERENCES

1. Crandall, S. H.: "Engineering Analysis," McGraw-Hill Book Company, New York, 1956.
2. Salvadori, M. G., and M. L. Baron: "Numerical Methods in Engineering," Prentice-Hall, Inc., Englewood Cliffs, N.J., 1961.
3. Wang, C. T.: "Applied Elasticity," McGraw-Hill Book Company, New York, 1953.
4. Williams, D.: "Introduction to the Theory of Aircraft Structures," Edward Arnold (Publishers) Ltd., London, 1960.
5. Panov, D. J.: "Formulas for the Numerical Solution of Partial Differential Equations by the Method of Differences," Fredrick Ungar Publishing Co., New York, 1963.
6. Southwell, R. V.: "Relaxation Methods in Theoretical Physics," Oxford University Press, Fair Lawn, N.J., vol. 1, 1946, vol. 2, 1956.
7. Timoshenko, S. P., and J. M. Gere: "Theory of Elastic Stability," 2d ed., McGraw-Hill Book Company, New York, 1961.

PROBLEMS

5-1. Derive the following forward- and backward-difference equations for the first derivative:

$$\left(\frac{d\psi}{dx}\right)_j = -\frac{1}{2h}(3\psi_j - 4\psi_{j+1} + \psi_{j+2}) + O(h^2)$$

$$\left(\frac{d\psi}{dx}\right)_j = \frac{1}{2h}(3\psi_j - 4\psi_{j-1} + \psi_{j-2}) + O(h^2)$$

Note that these three-point difference formulas are $O(h^2)$.

5-2. Derive Eqs. (5-9) and (5-10).

5-3. An unrestrained thin square slab undergoes a temperature change $T = T_0(x^2 + y^2)$, where the origin for the coordinates is at the center of the slab.

(a) Observing symmetry, set up the simultaneous difference equations for φ using a mesh size of $h = a/4$, where a is the length of the sides. Write the equations in matrix notation.

(b) Write difference equations for the stresses at the center of the slab in terms of φ at the mesh points.

5-4. Obtain the buckling load of the column of Example 5-2 by the direct method and compare the result with that of Example 5-3.

5-5. The buckling load and bending displacements of a column with clamped ends must satisfy the equation (Sec. 14-2)

$$\frac{d^2}{dx^2}\left(EI\frac{d^2w}{dx^2}\right) = -P\frac{d^2w}{dx^2}$$

for $0 < x < L$, and the boundary conditions

$$w(0) = 0 \qquad \frac{dw}{dx}\bigg|_{x=0} = 0 \qquad w(L) = 0 \qquad \frac{dw}{dx}\bigg|_{x=L} = 0$$

Show that the problem is in the form of Eqs. (5-18) and (5-19) and is therefore an eigenvalue problem.

5-6. Using the equations of Prob. 5-5, write the finite-difference equations for the column of Fig. 5-8 if the ends are clamped instead of pinned. Use a mesh spacing $h = L/6$. Note that EI is a linear function of x in the interval $0 < x < L/2$ and that w at a fictitious mesh point at $x = -L/6$ is equal to w_1 as a result of the boundary condition $dw/dx = 0$ at $x = 0$.

5-7. Use the iteration method to determine the buckling load in Prob. 5-6. [*Ans.* $P = 43.81\ EI_0/L^2$]

6

Introduction to Work and Energy Principles

6-1 INTRODUCTION

The principles of analytic mechanics can be developed in two independent ways. *Vectorial mechanics* is based upon Newton's laws, which are formulated in terms of the vector quantities of force and momentum, while the alternate approach is developed from *energy* or *variational principles*, which are stated in terms of the scalar quantities of work and energy. Undergraduate courses in engineering mechanics usually emphasize the vector approach; however, the variational methods often result in significant simplifications in solving problems. An interesting account of the history of the development of the latter approach is given in Ref. 1. It is worth noting that some of the original concepts of variational mechanics date back to the works of Aristotle and Galileo.

The displacement and stress formulations of the governing equations for elastic bodies are developed in Chap. 4. These may be considered vector methods, as they are derived from the conditions that the components of the resultant force and moment vectors that act upon a differential element of the body must vanish and that the components of the displacement vector must result in a compatible displacement pattern. In this chapter we develop alternative methods for ensuring equilibrium and compatibility, which are formulated in terms of the scalar functions of work and energy. These methods have the following advantages:

1. They result in general principles which are independent of the structural configuration, loading, and coordinate system.
2. They permit freedom to choose a generalized coordinate system that is convenient to the particular problem.
3. They provide an alternate approach for deriving the differential equations and boundary conditions for elastic bodies which often avoids complicated geometric or mechanical reasoning.
4. They eliminate extraneous details in computing deformations and analyzing statically indeterminate structures.

5. They may be used to obtain approximate solutions for problems in which the exact solution cannot be determined.

The energy principles are based upon Bernoulli's principle of virtual work. This fundamental concept of statics and the principles of the stationary and minimum values of the total potential, which are derived from it, are reviewed in Secs. 6-1 to 6-6. The principle of the stationary value of the total potential is then shown to apply to elastic bodies, where it provides an alternate to the equations of equilibrium. The principle of virtual work is also used to derive the principle of the stationary value of the total complementary potential, which may be used as an alternate to compatibility conditions. The potential and complementary-potential principles are utilized in the Rayleigh-Ritz method to obtain approximate solutions to elasticity problems. The principle of virtual work is also used to develop the unit-load method, which is an especially convenient procedure for computing the deformations of complex structures.

The energy principles are very general and therefore appear to be rather abstract at first reading. The application of the principles will be clarified by numerous examples throughout the remainder of the book. The references at the end of this chapter are suggested for additional reading.

6-2 WORK AND ENERGY

We begin our discussion by considering the equilibrium of a particle. A *particle* is an element of matter of vanishingly small dimensions, so that forces that act upon it must be concurrent, and moment equilibrium need not be considered. Geometrically the particle is represented by a point. Assume that a particle located by the radius vector $\mathbf{r}$ is acted upon by a force $\mathbf{F}$, and that the particle undergoes an infinitesimal displacement $d\mathbf{r}$ (Fig. 6-1). The *work* done by the force on the particle during the displacement is defined as the scalar product of the force and displacement vectors, or

$$dW = \mathbf{F} \cdot d\mathbf{r} \tag{6-1}$$

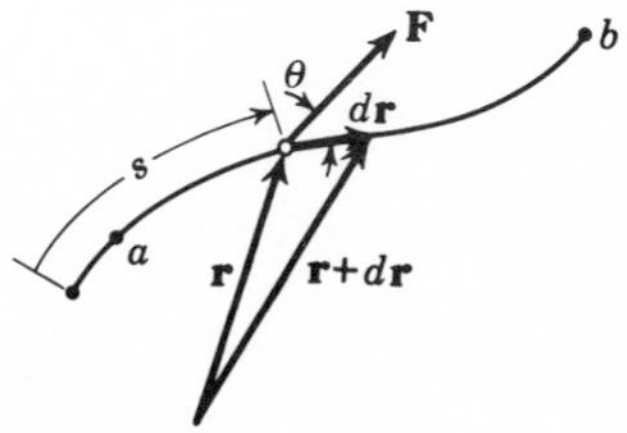

Fig. 6-1 Action of a force upon a particle.

By using the definition of the scalar product this may be written

$$dW = F \cos \theta \, dr$$

where θ is the angle between $\mathbf{F}$ and $d\mathbf{r}$ as shown in Fig. 6-1. As an alternative, vectors $\mathbf{F}$ and $d\mathbf{r}$ may be expressed as the vector sums of their components, or

$$\mathbf{F} = F_x\mathbf{i} + F_y\mathbf{j} + F_z\mathbf{k} \qquad d\mathbf{r} = dx\,\mathbf{i} + dy\,\mathbf{j} + dz\,\mathbf{k} \qquad (6\text{-}2)$$

where F_x, F_y, and F_z are the projections of $\mathbf{F}$ upon the x, y, and z axes and $\mathbf{i}$, $\mathbf{j}$, and $\mathbf{k}$ are unit vectors in the directions of these axes. The quantities dx, dy, and dz are the projections of $d\mathbf{r}$ on the x, y, and z axes. Substituting Eqs. (6-2) into (6-1) and making use of the properties of the scalar products of unit vectors gives

$$dW = F_x\,dx + F_y\,dy + F_z\,dz \qquad (6\text{-}3)$$

In the development of this equation it was assumed that $\mathbf{F}$ is constant during the infinitesimal displacement $d\mathbf{r}$, so that dW is a first-order differential of the work. The total work done on a particle as it moves from a point a to second point b (Fig. 6-1) is

$$W\Big|_a^b = \int_a^b \mathbf{F} \cdot d\mathbf{r} \qquad (6\text{-}4)$$

or, using Eqs. (6-2),

$$W\Big|_a^b = \int_a^b (F_x\,dx + F_y\,dy + F_z\,dz) \qquad (6\text{-}5)$$

To this point we have placed no restriction upon the nature of the forces. In many cases the right side of Eq. (6-3) is the negative of a perfect differential of a function of position only, $V(x,y,z)$. Under these circumstances we find from the chain rule for differentials that

$$dW = -dV = -\left(\frac{\partial V}{\partial x}\,dx + \frac{\partial V}{\partial y}\,dy + \frac{\partial V}{\partial z}\,dz\right) \qquad (6\text{-}6)$$

We see in Eqs. (6-5) and (6-6) that the components of $\mathbf{F}$ may be found from the relationships

$$F_x = -\frac{\partial V}{\partial x} \qquad F_y = -\frac{\partial V}{\partial y} \qquad F_z = -\frac{\partial V}{\partial z} \qquad (6\text{-}7)$$

if a function V exists which satisfies Eq. (6-6).

A force which is derivable from a function V by Eqs. (6-7) has interesting properties. In this case, where $dW = -dV$, Eq. (6-5) becomes

$$W\Big|_a^b = -\int_a^b dV$$

or upon integration

$$W \Big|_a^b = -(V_b - V_a) \tag{6-8}$$

where $V_a = V(x_a, y_a, z_a)$ and $V_b = V(x_b, y_b, z_b)$. We see that if the force on the particle is associated with a function V according to Eqs. (6-7), the work done on the particle as it moves from a to b depends only upon the coordinates of the end points and not upon the path taken between the points. We further observe that if the particle moves in any closed path from a back to a, the total work done by the force is zero. Since no net work is done as the particle moves in a closed path, energy is conserved, and forces which have the property of Eqs. (6-7) are therefore known as *conservative forces*. Forces associated with elastic deformations are conservative, as are those produced by gravitational and electrostatic fields. On the other hand the forces associated with friction and with plastic and creep deformations are nonconservative.

Energy is defined as the ability to do work, and energy possessed as a result of position in a force field is called *potential energy*. If $V_a > V_b$, we see from Eq. (6-8) that positive work is done on the particle in moving from a to b or that the particle at a has potential energy relative to the point b. The magnitude of this potential is determined by the function V, and for this reason it is called a *potential function*. From Eq. (6-8) $V_b = V_a - W \Big|_a^b$, and so the potential energy V_b is not absolutely defined, but it can be determined only relative to the potential energy at another point, in this case a. However, we see from Eqs. (6-7) that the force components are equal to the negative of the directional derivatives of V, and so the absolute value of V is not important. Because of this, the value of V may be arbitrarily defined as zero at some datum position a, and V at any other position b can be found from

$$V_b = -W \Big|_a^b \tag{6-9}$$

The potential energy at b is then equal to the negative of the work done by the force field on a particle as it moves from the datum configuration a to the point b.

6-3 VIRTUAL WORK AND EQUILIBRIUM

In the preceding section the work of a force on a particle during a real displacement was considered. In variational mechanics we imagine displacements to occur when in reality no such displacements actually exist. These fictitious motions are called *virtual displacements*, and the work performed during these imaginary displacements is referred to as *virtual work.*   The virtual displacements are assumed small so that there will be no significant changes in geometry and so that the forces may be assumed to remain constant during the displacements.   In addition, virtual displacements are usually defined as motions which do not violate the physical restraints of the system but which are otherwise arbitrary. This condition is imposed so that the reactive forces at rigid supports do no work and therefore will not enter the analysis.   In summary, a virtual displacement is *any* small displacement which does not violate the constraints of the system.   To differentiate virtual displacements from real displacements the notation $\delta \mathbf{r}$ is used instead of $d\mathbf{r}$.   Mathematically $\delta \mathbf{r}$ behaves the same as the differential quantity $d\mathbf{r}$.   The virtual work done by the force during a virtual displacement is then $\delta W = \mathbf{F} \cdot \delta \mathbf{r}$.   If the force is conservative, the first-order variation in the potential energy during the virtual displacement will be $\delta V = -\delta W$.

Consider now a particle which is acted upon by n forces $\mathbf{F}_i$.   During any virtual displacement the virtual work will be

$$\delta W = \sum_{i=1}^{n} \mathbf{F}_i \cdot \delta \mathbf{r} = \left(\sum_{i=1}^{n} \mathbf{F}_i \right) \cdot \delta \mathbf{r}$$

If the forces are in equilibrium, their vector sum must vanish, so that $\delta W = 0$ for *any* virtual displacement.   Alternatively, if it is known that $\delta W = 0$, then it follows that the vector sum of the forces is zero (for $\delta \mathbf{r}$ is arbitrary).   These facts may be stated as:

A necessary and sufficient condition for the equilibrium of a particle is that the virtual work of the forces on the particle is zero for any virtual displacement.

If the forces are conservative $\delta V = 0$, so that

A necessary and sufficient condition for the equilibrium of a particle acted upon by conservative forces is that the first-order variation of the potential energy of the forces is zero for any virtual displacement.

Next consider a system of N particles as shown in Fig. 6-2.   The forces acting upon each particle in the system may be divided into two

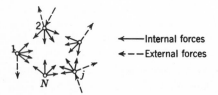

←——Internal forces
←– –External forces

Fig. 6-2 Equilibrium of a system of particles.

categories, those due to actions which are external to the system and those due to internal effects. The distinction is that the internal forces represent the actions of the particles upon each other and therefore occur as pairs of equal and opposite forces. Let δW_{e_j} and δW_{i_j} be the virtual work of the external and internal forces, respectively, on the jth particle. If the system is in equilibrium,

$$\sum_{j=1}^{N} (\delta W_{e_j} + \delta W_{i_j}) = 0$$

during any virtual displacement of the system, or

$$\delta W_e + \delta W_i = 0 \tag{6-10}$$

where

$$\delta W_e = \sum_{j=1}^{N} \delta W_{e_j} \qquad \delta W_i = \sum_{j=1}^{N} \delta W_{i_j}$$

A *continuum* is defined as a domain in which matter exists at every point. Since there are an infinite number of points in any one-, two-, or three-dimensional domain, we may think of the continuum as consisting of an infinite number of particles and apply Eq. (6-10), which may therefore be stated as:

A *necessary and sufficient condition for the equilibrium of a system of particles or a continuum system is that the work done by the external forces acting upon the system plus the work done by the internal forces of the system must vanish for any virtual displacement.*

This is a general statement of the *principle of virtual work.*

If the external and internal forces are conservative, Eq. (6-10) may be written as $\delta U + \delta V = 0$, where U is the potential energy of all the internal forces and V is the potential energy of all the external forces. This equation may also be written

$$\delta(U + V) = 0 \tag{6-11}$$

The quantity $U + V$ is referred to as the *total potential* of the system. Equation (6-11) may be stated as:

A *necessary and sufficient condition for the equilibrium of conservative forces acting on a system of particles or a continuum is that the first-order variation of the total potential must vanish for any virtual displacement.*

Since there is no first-order change in the total potential, this statement is known as the *principle of the stationary value of the total potential.*

6-4 COORDINATES AND DEGREES OF FREEDOM

The position of a particle is specified by giving its *coordinates*. These may be cartesian, polar, spherical, or any other set of *independent parameters* which *uniquely define* the position of the particle in space. A system of N particles in space may be located by giving the set of cartesian coordinates (x_j, y_j, z_j) for each of its N points.

It is often desirable to relate these $3N$ coordinates to another set of $3N$ independent quantities $(q_1, q_2, \ldots, q_{3N})$ by relationships of the general form

$$x_1 = x_1(q_1, q_2, \ldots, q_i, \ldots, q_{3N})$$
$$y_1 = y_1(q_1, q_2, \ldots, q_i, \ldots, q_{3N}) \qquad (6\text{-}12)$$
$$\ldots \ldots \ldots \ldots \ldots \ldots \ldots$$
$$z_N = z_N(q_1, q_2, \ldots, q_i, \ldots, q_{3N})$$

The quantities q_i are called *generalized coordinates*, and Eqs. (6-12) define the *coordinate transformation* from the generalized coordinates to the cartesian coordinates. It is emphasized that the q_i are not restricted to coordinates in the ordinary sense, but they may be any independent quantities from which the cartesian coordinates can be found. Great simplification is often possible by the proper choice of generalized coordinates.

Associated with the set of generalized displacements is a set of *generalized forces* defined in such a manner that the sum of the products of the generalized forces and the corresponding variation of the generalized displacements is equal to the virtual work. Considering the system of N particles, each under the action of a resultant force with components F_{x_j}, F_{y_j}, F_{z_j}, the virtual work for any virtual displacement of the particles will be

$$\delta W = \sum_{j=1}^{N} (F_{x_j}\, \delta x_j + F_{y_j}\, \delta y_j + F_{z_j}\, \delta z_j) \qquad (6\text{-}13)$$

If we express x_j, y_j, and z_j in terms of the generalized coordinates by Eq. (6-12), then the first-order variation in x_j resulting from virtual changes in the generalized coordinates will be

$$\delta x_j = \sum_{i=1}^{3N} \frac{\partial x_j}{\partial q_i}\, \delta q_i$$

and similar expressions can be written for δy_j and δz_j. Substituting these into Eq. (6-13) gives

$$\delta W = \sum_{i=1}^{3N} \left[\sum_{j=1}^{N} \left(F_{x_j} \frac{\partial x_j}{\partial q_i} + F_{y_j} \frac{\partial y_j}{\partial q_i} + F_{z_j} \frac{\partial z_j}{\partial q_i} \right) \right] \delta q_i$$

The fact that the product of the square bracket in this equation with δq_i has the units of work suggests defining the former quantity as the *generalized force* Q_i associated with the generalized displacement q_i, or

$$\delta W = \sum_{i=1}^{3N} Q_i \, \delta q_i \tag{6-14}$$

where

$$Q_i = \sum_{j=1}^{N} \left(F_{x_j} \frac{\partial x_j}{\partial q_i} + F_{y_j} \frac{\partial y_j}{\partial q_i} + F_{z_j} \frac{\partial z_j}{\partial q_i} \right) \tag{6-15}$$

The number of *degrees of freedom* of a system is the number of independent coordinates required to define uniquely the configuration of the system; this is the same in all coordinate systems; for a particle: in space it is three, in a plane it is two, and on a line it is one. A continuum has an infinite number of degrees of freedom because there are infinitely many points in a continuum.

In the preceding section the conditions for equilibrium of various systems were considered, and it was stated that the virtual work or first-order variation of the potential energy must vanish for *any* virtual displacement. To demonstrate that these quantities are zero for any virtual displacement it is only necessary to show that they are zero for an arbitrary variation of each of the coordinates of the system, since any virtual displacement can be taken as a linear combination of a variation of each of the coordinates. As a result, the number of independent variations is then always equal to the number of degrees of freedom. For a system with r degrees of freedom $U = U(q_1, q_2, \ldots, q_i, \ldots, q_r)$ and $V = V(q_1, q_2, \ldots, q_i, \ldots, q_r)$ and Eq. (6-11) becomes

$$\delta(U + V) = \sum_{i=1}^{r} \frac{\partial}{\partial q_i} (U + V) \, \delta q_i = 0 \tag{6-16}$$

Since the δq_i are independent and arbitrary, it follows that

$$\frac{\partial}{\partial q_i} (U + V) = 0 \qquad i = 1, 2, \ldots, r \tag{6-17}$$

6-5 STABILITY

We have seen that a necessary and sufficient condition for equilibrium of a conservative system is that the first-order change in the total potential energy of the system be zero for any virtual displacement. While this guarantees equilibrium, it provides no information on the nature of the stability, which may be of considerable importance. If a system is disturbed from its equilibrium configuration, forces are developed which may:

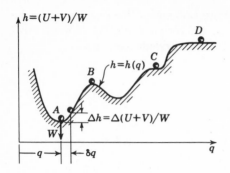

Fig. 6-3 Example of types of stability.

(1) return it to its original position, (2) hold it in the displaced position, or (3) move it away from its equilibrium position. These conditions are respectively known as *stable, neutral,* and *unstable equilibrium.*

As a simple illustration, consider the case of a particle of weight W which is constrained to slide along a frictionless plane curve as shown in Fig. 6-3. The height of the curve above the datum plane is given by h, and the equation of the curve is $h = h(q)$. Since only a single particle is involved, $U = 0$, and the only potential energy is that which is due to position in the gravitational field. Assuming that $V = 0$ at $h = 0$, we find from Eq. (6-9) that $V = Wh$. The particle has only a single degree of freedom, since its position is defined by the single coordinate q, and Eq. (6-17) gives $W(dh/dq) = 0$, or $dh/dq = 0$. This condition is satisfied at any point on the line where there is a horizontal tangent, such as at points A, B, C, or D. At each of these locations the first-order variation in $U + V$ is zero, and equilibrium of the particles will exist, but the type of equilibrium is different at each point. At A the equilibrium is stable, and it is seen that this condition is associated with a minimum in the potential. The equilibrium at B is unstable, a situation which prevails when the potential is a maximum. A mini-max in $U + V$ occurs at C, since V increases if q is increased and decreases if q decreases. The equilibrium is seen to be unstable at a mini-max point. At D the potential energy is a constant, and neutral stability exists.

The *total change* in potential $\Delta(U + V)$ that is associated with an arbitrary displacement δq from an equilibrium position (Fig. 6-3) can be determined from the Taylor's series expansion

$$\Delta(U + V) = \frac{d(U + V)}{dq} \delta q + \frac{1}{2!} \frac{d^2(U + V)}{dq^2} \delta q^2$$
$$+ \frac{1}{3!} \frac{d^3(U + V)}{dq^3} \delta q^3 + \cdots$$

where the derivatives are evaluated at the equilibrium point. But

$d(U + V)/dq = 0$ at an equilibrium point, so that

$$\Delta(U + V) = \frac{1}{2!} \frac{d^2(U + V)}{dq^2} \delta q^2 + \frac{1}{3!} \frac{d^3(U + V)}{dq^3} \delta q^3 + \cdots$$

For small δq the sign of the lowest nonvanishing derivative in this equation establishes the sign of $\Delta(U + V)$ and therefore the type of stability for the single-degree-of-freedom system. For stable equilibrium $\Delta(U + V) > 0$, which occurs when the lowest nonzero derivative is even and positive. On the other hand, if this derivative is (1) odd or (2) even and less than zero, $U + V$ is respectively a mini-max or a maximum, and the equilibrium is unstable. If all derivatives vanish, $\Delta(U + V) = 0$, and the equilibrium is neutral.

The statements that $U + V$ is (1) a minimum for stable equilibrium, (2) a maximum or mini-max for unstable equilibrium, and (3) a constant for neutral stability generalize to any number of degrees of freedom. However, the conditions which must be imposed on the derivatives of $U + V$ to assure these conditions become very complicated for multiple-degree-of-freedom systems.

6-6 SMALL DISPLACEMENTS OF A CONSERVATIVE SYSTEM

In many problems the internal forces are conservative, and under the action of these internal forces the system assumes an unloaded equilibrium configuration. Consider the case of a system with r degrees of freedom, and let the coordinates $q_1, q_2, \ldots, q_r$ be measured from the equilibrium position, which will be taken as the datum configuration for computing potential energies. The internal energy in the region of this position can be obtained by the Taylor's series expansion

$$U(q_1, q_2, \ldots, q_r) = U(0, 0, \ldots, 0) + \sum_{i=1}^{r} \frac{\partial U}{\partial q_i} q_i$$

$$+ \frac{1}{2} \sum_{i=1}^{r} \sum_{j=1}^{r} \frac{\partial^2 U}{\partial q_i\, \partial q_j} q_i q_j + \cdots \quad (6\text{-}18)$$

where the derivatives are evaluated at $q_1 = q_2 = \cdots = q_r = 0$. The term $U(0, 0, \ldots, 0) = 0$ as a result of our assumption of the datum configuration, while the second term vanishes because there are no external loads when the system is in the datum position, so that Eq. (6-17) reduces to $\partial U/\partial q_i = 0$ for $i = 1, 2, \ldots, r$.

If the displacements $(q_1, q_2, \ldots, q_r)$ are small enough, terms after the third will be negligible compared to the third term, and Eq. (6-18) will reduce to

$$U = \frac{1}{2} \sum_{i=1}^{r} \sum_{j=1}^{r} \frac{\partial^2 U}{\partial q_i \, \partial q_j} q_i q_j \tag{6-19}$$

Since the derivatives $\partial^2 U/(\partial q_i \, \partial q_j)$ are evaluated at

$$q_1 = q_2 = \cdot \cdot \cdot = q_r = 0$$

they are constants, and Eq. (6-19) may be written

$$U = \frac{1}{2} \sum_{i=1}^{r} \sum_{j=1}^{r} k_{ij} q_i q_j \tag{6-20}$$

where

$$k_{ij} = \frac{\partial^2 U}{\partial q_i \, \partial q_j} \tag{6-21}$$

The k_{ij} are known as the *stiffness influence coefficients* of the system, and it is seen from Eq. (6-21) that $k_{ij} = k_{ji}$.

Assume now that the system is subjected to a set of conservative external forces. Under the action of these forces the system will undergo displacements $q_1, q_2, \ldots, q_r$ to a new equilibrium configuration, and the potential of these forces will be a function of the displacements. From a Taylor's series expansion about the datum configuration we can write

$$V(q_1,q_2, \ldots ,q_r) = V(0,0, \ldots ,0) + \sum_{i=1}^{r} \frac{\partial V}{\partial q_i} q_i$$

$$+ \frac{1}{2} \sum_{i=1}^{r} \sum_{j=1}^{r} \frac{\partial^2 V}{\partial q_i \, \partial q_j} q_i q_j + \cdot \cdot \cdot$$

However, $V(0,0, \ldots ,0) = 0$ because it is the potential energy of the external forces in the datum configuration. For small displacements the terms beyond the second in the series will be negligible compared to the second, so that

$$V = \sum_{i=1}^{n} \frac{\partial V}{\partial q_i} q_i$$

where the derivatives $\partial V/\partial q_i$ are evaluated at $q_1 = q_2 = \cdot \cdot \cdot = q_r = 0$ and are therefore constants. The equation may then be written

$$V = - \sum_{i=1}^{r} Q_i q_i \tag{6-22}$$

where

$$Q_i = - \frac{\partial V}{\partial q_i} \qquad i = 1, 2, \ldots , r \tag{6-23}$$

are the generalized forces associated with the generalized coordinates. We note that Eq. (6-22) is in accordance with Eq. (6-9) in that V is the negative of the work done by the forces Q_i in moving through the displacements q_i when it is assumed that the forces remain constant.

The total potential in the loaded equilibrium configuration will then be

$$U + V = \frac{1}{2} \sum_{i=1}^{r} \sum_{j=1}^{r} k_{ij} q_i q_j - \sum_{i=1}^{r} Q_i q_i \qquad (6\text{-}24)$$

Since the loaded position is also an equilibrium position, Eq. (6-17) applies for any virtual displacements in which the q_i are varied by an amount δq_i. Substituting Eq. (6-24) into Eq. (6-17) gives

$$\sum_{j=1}^{r} k_{ij} q_j = Q_i \qquad i = 1, 2, \ldots, r \qquad (6\text{-}25)$$

Equations (6-25) are a set of r simultaneous linear algebraic equations which express the equilibrium that exists between the internal and external forces. These equations may be written in the matrix form

$$\mathbf{kq} = \mathbf{Q} \qquad (6\text{-}26)$$

where

$$\mathbf{k} = \begin{bmatrix} k_{11} & k_{12} & \cdots & k_{1r} \\ k_{21} & k_{22} & \cdots & k_{2r} \\ \cdot & \cdot & \cdot & \cdot & \cdot & \cdot & \cdot & \cdot & \cdot \\ k_{r1} & k_{r2} & \cdots & k_{rr} \end{bmatrix} \qquad \mathbf{q} = \begin{Bmatrix} q_1 \\ q_2 \\ \cdot \\ \cdot \\ \cdot \\ q_r \end{Bmatrix} \qquad \mathbf{Q} = \begin{Bmatrix} Q_1 \\ Q_2 \\ \cdot \\ \cdot \\ \cdot \\ Q_r \end{Bmatrix} \qquad (6\text{-}27)$$

Equations (6-25) can be solved simultaneously for the q_j to give equations of the form

$$\sum_{j=1}^{r} c_{ij} Q_j = q_i \qquad i = 1, 2, \ldots, r \qquad (6\text{-}28)$$

where the c_{ij} are known as *flexibility influence coefficients*. In matrix notation Eqs. (6-28) become

$$\mathbf{cQ} = \mathbf{q} \qquad (6\text{-}29)$$

where

$$\mathbf{c} = \begin{bmatrix} c_{11} & c_{12} & \cdots & c_{1r} \\ c_{21} & c_{22} & \cdots & c_{2r} \\ \cdot & \cdot & \cdot & \cdot & \cdot & \cdot & \cdot & \cdot \\ c_{r1} & c_{r2} & \cdots & c_{rr} \end{bmatrix} \qquad (6\text{-}30)$$

and it is seen from Eqs. (6-26) and (6-29) that

$$\mathbf{c} = \mathbf{k}^{-1} \qquad (6\text{-}31)$$

We note from Eq. (6-21) that the stiffness matrix is symmetric and therefore its inverse, the flexibility matrix, is also symmetric.

Since the set of equations (6-28) is linear, superposition applies, and a physical interpretation may be given to the c_{ij}. The coefficient c_{ij} is the increment of the displacement q_i due to a unit increment of the loading Q_j while all other loads are held constant. Likewise from Eqs. (6-25) the stiffness coefficient k_{ij} is the increment of the force Q_i which occurs as a result of a unit increment in the displacement q_j while all other displacements are held constant. In some cases where it is difficult to compute these system properties analytically it may be convenient to measure them experimentally.

Equation (6-20) may be expressed in matrix form as

$$U = \tfrac{1}{2}\mathbf{q'kq} \tag{6-32}$$

where the prime indicates the transposed matrix. The internal energy may also be expressed in terms of the generalized forces. Substituting Eqs. (6-26) and (6-29) into Eq. (6-32) and using the fact that because it is a symmetric matrix, $c' = c$, we find

$$U = \tfrac{1}{2}\mathbf{Q'cQ} \tag{6-33}$$

In summation notation this equation becomes

$$U = \frac{1}{2} \sum_{i=1}^{r} \sum_{j=1}^{r} c_{ij}Q_iQ_j \tag{6-34}$$

6-7 STRAIN ENERGY AND COMPLEMENTARY STRAIN ENERGY

To this point in the chapter we have developed principles of statics which are applicable to general systems of particles or a continuum. In the remaining sections of the chapter we shall apply these principles to the special case of the elastic body. In Sec. 6-3 it was shown that a necessary and sufficient condition for the equilibrium of a continuum system is that $\delta W_e = -\delta W_i$ for any virtual displacement. In this equation δW_e and δW_i are the virtual work of the external and internal forces of the system, respectively. Let us now consider the work of the internal forces if the continuum is an elastic solid. For simplicity we first investigate the case in which all stresses except σ_{xx} are zero. A differential element of a body subjected to such a stress is shown in Fig. 6-4. Let the stress acting upon this element be σ_{xx} at the time when the strain is ϵ_{xx}. From Fig. 6-4 it is seen that the work done by the external forces on the element during the increment of strain $d\epsilon_{xx}$ is $\sigma_{xx} \, d\epsilon_{xx} \, dx \, dy \, dz$. During this same increment of strain, the work done by the forces within the differential element will be the negative of that performed by the stresses acting upon it. The

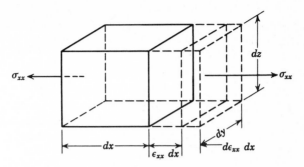

Fig. 6-4 Stresses and displacements of a differential element of a body with a uniaxial state of stress.

work of the internal forces as the strain increases from zero to its final value, ϵ_{xx_1}, is therefore

$$dW_i = - \left(\int_0^{\epsilon_{xx_1}} \sigma_{xx} \, d\epsilon_{xx} \right) dx \, dy \, dz$$

and the total internal work for the whole body is

$$W_i = - \int_V \left(\int_0^{\epsilon_{xx_1}} \sigma_{xx} \, d\epsilon_{xx} \right) dV \qquad (6\text{-}35)$$

where the integral extends over the entire volume of the body. The quantity within the parentheses is the negative of the internal work per unit volume, which is observed to be the area under a stress-strain diagram drawn for the stressing and temperature history of the element. Equation (6-35) applies for plastic and creep strains as well as elastic strains.

If the material is elastic, the internal work can be expressed in terms of a potential function. Stress-strain diagrams for two different load and temperature histories are shown in Fig. 6-5 for a nonlinearly elastic body. In Fig. 6-5a the temperature of the element is increased before stressing occurs, while in Fig. 6-5b the load is applied prior to the temperature change. In both cases it is assumed that the thermal distribution is such that no stresses arise as a result of the thermal strains. It is seen that the areas under the stress-strain curves $CABE$ are not equal, even though the final strains ϵ_{xx_1} are identical. Since the internal work depends upon the loading and temperature history and is not simply a function of the final displacements, the internal forces are nonconservative and cannot be obtained from a potential function. Equation (6-35) may be rewritten as follows without changing the value of W_i:

$$W_i = - \int_V \left(\int_0^{\epsilon_{xx_1}} \sigma_{xx} \, d\epsilon_{xx} - \int_0^{T_1} \sigma_{xx} \alpha \, dT + \int_0^{T_1} \sigma_{xx} \alpha \, dT \right) dV$$

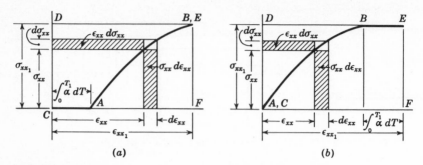

Fig. 6-5 Stress-strain diagrams for different load and temperature histories. (a) Temperature increase followed by stressing; (b) stressing followed by temperature increase.

where T_1 is the final value of the change in temperature. The first two terms within the parentheses are the areas under the curves AB of Fig. 6-5, which are seen to be the same in both cases. Since the work represented by these two integrals depends only upon the end values ϵ_{xx_1} and T_1 and not upon the sequence of stressing and heating, we may associate a potential function U_0 with them and write

$$W_i = - \int_V \left(U_0 + \int_0^{T_1} \sigma_{xx}\alpha \, dT \right) dV \tag{6-36}$$

where

$$U_0 = \int_0^{\epsilon_{xx_1}} \sigma_{xx} \, d\epsilon_{xx} - \int_0^{T_1} \sigma_{xx}\alpha \, dT \tag{6-37}$$

The quantity U_0 is the *elastic strain energy per unit volume*. Equation (6-36) may be written

$$W_i = -U - \int_V \left(\int_0^{T_1} \sigma_{xx}\alpha \, dT \right) dV \tag{6-38}$$

where

$$U = \int_V U_0 \, dV \tag{6-39}$$

is the total elastic strain energy of the body.

Another scalar quantity which proves useful is the area within the rectangles $CDEF$ of Fig. 6-5, which complements the areas under the curves AB. This area, defined as the *complementary energy per unit volume* U_0', is seen to be

$$U_0' = \int_0^{\sigma_{xx_1}} \left(\epsilon_{xx} - \int_0^T \alpha \, dT \right) d\sigma_{xx} + \sigma_{xx_1} \int_0^{T_1} \alpha \, dT \tag{6-40}$$

and the total complementary energy for the complete body is

$$U' = \int_V U'_0 \, dV \tag{6-41}$$

where the integration extends over the entire volume.

Equations (6-37) and (6-40) can be simplified if the material is linearly elastic and α is taken as the mean value of the coefficient of expansion for the range of the temperature change. If the material is linearly elastic, the curves AB of Fig. 6-5 become straight lines, and the area under AB is simply the area of a triangle. Thus

$$U_0 = \tfrac{1}{2}\sigma_{xx_1}(\epsilon_{xx_1} - \alpha T_1) \tag{6-42}$$

$$U'_0 = \tfrac{1}{2}\sigma_{xx_1}(\epsilon_{xx_1} + \alpha T_1) \tag{6-43}$$

It is seen that $U_0 = U'_0$ for a linearly elastic material if $T = 0$. The quantities U_0 and U'_0 may be written entirely in terms of stress or in terms of strain by using the uniaxial stress-strain law $\epsilon_{xx} = (\sigma_{xx} + \alpha ET)/E$. Introducing this into Eqs. (6-42) and (6-43), we find

$$U_0 = \frac{E}{2}(\epsilon_{xx_1} - \alpha T_1)^2 \tag{6-44}$$

$$U'_0 = \frac{\sigma_{xx_1}{}^2}{2E} + \sigma_{xx_1}\alpha T_1 \tag{6-45}$$

Consider now the case in which the differential element is subjected to a general three-dimensional state of stress. The strain energy per unit volume in this case can be found by summing terms similar to Eq. (6-37) for each of the stress components. If the coefficient of linear expansion of the material is isotropic, the temperature change does not produce a shearing strain, and so for the shearing stresses there are no terms corresponding to the second integral of Eq. (6-37). The strain energy per unit volume is then

$$U_0 = \int_0^{\epsilon_{xx_1}} \sigma_{xx} \, d\epsilon_{xx} + \int_0^{\epsilon_{yy_1}} \sigma_{yy} \, d\epsilon_{yy} + \int_0^{\epsilon_{zz_1}} \sigma_{zz} \, d\epsilon_{zz} + \int_0^{\epsilon_{xy_1}} \sigma_{xy} \, d\epsilon_{xy}$$
$$+ \int_0^{\epsilon_{yz_1}} \sigma_{yz} \, d\epsilon_{yz} + \int_0^{\epsilon_{zx_1}} \sigma_{zx} \, d\epsilon_{zx} - \int_0^{T_1} (\sigma_{xx} + \sigma_{yy} + \sigma_{zz})\alpha \, dT \tag{6-46}$$

and the complementary energy per unit volume is

$$U'_0 = \int_0^{\sigma_{xx_1}} \epsilon_{xx} \, d\sigma_{xx} + \int_0^{\sigma_{yy_1}} \epsilon_{yy} \, d\sigma_{yy} + \int_0^{\sigma_{zz_1}} \epsilon_{zz} \, d\sigma_{zz} + \int_0^{\sigma_{xy_1}} \epsilon_{xy} \, d\sigma_{xy}$$
$$+ \int_0^{\sigma_{yz_1}} \epsilon_{yz} \, d\sigma_{yz} + \int_0^{\sigma_{zx_1}} \epsilon_{zz} \, d\sigma_{zx} - \int_0^{\sigma_{xx_1}} \left(\int_0^T \alpha \, dT\right) d\sigma_{xx}$$
$$- \int_0^{\sigma_{yy_1}} \left(\int_0^T \alpha \, dT\right) d\sigma_{yy} - \int_0^{\sigma_{zz_1}} \left(\int_0^T \alpha \, dT\right) d\sigma_{zz}$$
$$+ (\sigma_{xx_1} + \sigma_{yy_1} + \sigma_{zz_1})\int_0^{T_1} \alpha \, dT \tag{6-47}$$

If the material is linearly elastic,

$$U_0 = \tfrac{1}{2}(\sigma_{xx}\epsilon_{xx} + \sigma_{yy}\epsilon_{yy} + \sigma_{zz}\epsilon_{zz} + \sigma_{xy}\epsilon_{xy} + \sigma_{yz}\epsilon_{yz} + \sigma_{zx}\epsilon_{zx})$$
$$- \frac{\alpha T}{2}(\sigma_{xx} + \sigma_{yy} + \sigma_{zz}) \tag{6-48}$$

$$U_0' = \tfrac{1}{2}(\sigma_{xx}\epsilon_{xx} + \sigma_{yy}\epsilon_{yy} + \sigma_{zz}\epsilon_{zz} + \sigma_{xy}\epsilon_{xy} + \sigma_{yz}\epsilon_{yz} + \sigma_{zx}\epsilon_{zx})$$
$$+ \frac{\alpha T}{2}(\sigma_{xx} + \sigma_{yy} + \sigma_{zz}) \tag{6-49}$$

where all stresses and strains are the final values but the subscript 1 has been dropped for simplicity.

For a body of a linearly elastic isotropic material the stresses and strains are related by Eqs. (3-18) or (3-21), and Eqs. (6-48) and (6-49) may be written

$$U_0 = \frac{\nu E e^2}{2(1 + \nu)(1 - 2\nu)} + \frac{E}{2(1 + \nu)}\,[\epsilon_{xx}{}^2 + \epsilon_{yy}{}^2 + \epsilon_{zz}{}^2$$
$$+ \tfrac{1}{2}(\epsilon_{xy}{}^2 + \epsilon_{yz}{}^2 + \epsilon_{zx}{}^2)] - \frac{\alpha E T e}{1 - 2\nu} + \frac{3E(\alpha T)^2}{2(1 - 2\nu)} \tag{6-50}$$

$$U_0' = \frac{1}{2E}\,[\sigma_{xx}{}^2 + \sigma_{yy}{}^2 + \sigma_{zz}{}^2 + 2(1 + \nu)(\sigma_{xy}{}^2 + \sigma_{yz}{}^2 + \sigma_{zx}{}^2)$$
$$- 2\nu(\sigma_{xx}\sigma_{yy} + \sigma_{yy}\sigma_{zz} + \sigma_{zz}\sigma_{xx})] + \alpha T(\sigma_{xx} + \sigma_{yy} + \sigma_{zz}) \tag{6-51}$$

We can find U_0 for plane stress by substituting Eqs. (3-20) into Eq. (6-48), to obtain

$$U_0 = \frac{E}{2(1 + \nu)}\left[\frac{1}{1 - \nu}(\epsilon_{xx} + \epsilon_{yy})^2 - 2\epsilon_{xx}\epsilon_{yy} + \tfrac{1}{2}\epsilon_{xy}{}^2\right]$$
$$- \frac{\alpha E T}{1 - \nu}(\epsilon_{xx} + \epsilon_{yy}) + \frac{E(\alpha T)^2}{1 - \nu} \tag{6-52}$$

In a similar manner, for plane stress the use of Eqs. (3-19) and (6-49) gives

$$U_0' = \frac{1}{2E}\,[(\sigma_{xx} + \sigma_{yy})^2 - 2(1 + \nu)(\sigma_{xx}\sigma_{yy} - \sigma_{xy}{}^2)] + \alpha T(\sigma_{xx} + \sigma_{yy}) \tag{6-53}$$

Equation (6-53) may be written in terms of the Airy stress function by using Eqs. (4-13), with the result

$$U_0' = \frac{1}{2E}\left\{\left(\frac{\partial^2\varphi}{\partial x^2} + \frac{\partial^2\varphi}{\partial y^2} + 2V\right)^2 - 2(1 + \nu)\left[\left(\frac{\partial^2\varphi}{\partial x^2} + V\right)\left(\frac{\partial^2\varphi}{\partial y^2} + V\right)\right.\right.$$
$$\left.\left. - \left(\frac{\partial^2\varphi}{\partial x\,\partial y}\right)^2\right]\right\} + \alpha T\left(\frac{\partial^2\varphi}{\partial x^2} + \frac{\partial^2\varphi}{\partial y^2} + 2V\right) \tag{6-54}$$

If $T = 0$, (6-52) to (6-54) apply to plane strain if primes are placed on the material-property symbols and the equations of Prob. 3-2 are used. Expressions for U and U' for the bending, extension, shearing, and twisting of beams and for the bending and extension of plates are given in subsequent chapters.

6-8 POTENTIAL AND COMPLEMENTARY POTENTIAL OF EXTERNAL FORCES

As a body deforms, the external forces which act upon it do work. For simplicity let us first consider the case in which the only loads are surface forces $\bar{X}$. As $\bar{X}$ increases from zero to its final value $\bar{X}_1$, the displacements in the x direction increase from zero to u_1, as shown in Fig. 6-6. The total work done by the surface forces is then

$$W_e = \int_S \left(\int_0^{u_1} \bar{X} \, du \right) dS$$

where the integral extends over the entire surface of the body. The quantity within the parentheses is the area under the force-displacement curve. We shall see that the area which complements the work under the curve AB is useful in the analysis of statically indeterminate bodies, and we define

$$W_e' = \int_S \left(\int_0^{\bar{X}_1} u \, d\bar{X} \right) dS$$

as the *complementary work* of the external forces. If the body is linear in its response to load, the curve AB becomes a straight line, and

$$W_e = \frac{1}{2} \int_S \bar{X}_1 u_1 \, dS = W_e'$$

When the external forces are conservative, we may associate a potential function V with them as follows:

$$V = - \int_S \bar{X}_1 u_1 \, dS$$

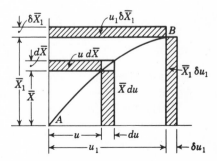

Fig. 6-6 Force-displacement diagram for surface forces.

We shall find that it is also useful to introduce a *complementary potential function V'*, defined as

$$V' = - \int_S u_1 \bar{X}_1 \, dS = V$$

We see from Fig. 6-6 that the virtual work associated with the virtual displacement δu_1 is

$$\delta W_e = \int_S \bar{X}_1 \, \delta u_1 \, dS = -\delta V$$

where δV is the first-order variation in V due to δu_1; that is, it is the change in V that occurs when u_1 is varied to $u_1 + \delta u_1$ while $\bar{X}_1$ is held constant. Similarly, for a variation in the forces $\delta \bar{X}_1$, we find from Fig. 6-6 that the first-order variation in W'_e is

$$\delta W'_e = \int_S u_1 \, \delta \bar{X}_1 \, dS = -\delta V'$$

where $\delta V'$ is the first-order variation of V' due to $\delta \bar{X}_1$. In this case u_1 is held constant during the variation $\delta \bar{X}_1$.

In general there may be surface forces with components $\bar{X}$, $\bar{Y}$, and $\bar{Z}$ and body forces having components X, Y, and Z. Under these circumstances

$$W_e = \int_S \left(\int_0^{u_1} \bar{X} \, du + \int_0^{v_1} \bar{Y} \, dv + \int_0^{w_1} \bar{Z} \, dw \right) dS$$
$$+ \int_V \left(\int_0^{u_1} X \, du + \int_0^{v_1} Y \, dv + \int_0^{w_1} Z \, dw \right) dV \quad (6\text{-}55)$$

$$W'_e = \int_S \left(\int_0^{\bar{X}_1} u \, d\bar{X} + \int_0^{\bar{Y}_1} v \, d\bar{Y} + \int_0^{\bar{Z}_1} w \, d\bar{Z} \right) dS$$
$$+ \int_V \left(\int_0^{X_1} u \, dX + \int_0^{Y_1} v \, dY + \int_0^{Z_1} w \, dZ \right) dV \quad (6\text{-}56)$$

The force-displacement curves are linear if the body is linearly elastic and the deformations are small; in this case

$$W_e = \frac{1}{2} \int_S (\bar{X}u + \bar{Y}v + \bar{Z}w) \, dS + \frac{1}{2} \int_V (Xu + Yv + Zw) \, dV = W'_e$$
$$(6\text{-}57)$$

where the subscripts 1 have been dropped for simplicity but it is understood that the forces and displacements are the final values.

If the forces are conservative,

$$V = - \int_S (\bar{X}u + \bar{Y}v + \bar{Z}w) \, dS - \int_V (Xu + Yv + Zw) \, dV = V' \quad (6\text{-}58)$$

and it is seen that

$$\delta W_e = \int_S (\bar{X}\,\delta u + \bar{Y}\,\delta v + \bar{Z}\,\delta w)\,dS$$
$$+ \int_V (X\,\delta u + Y\,\delta v + Z\,\delta w)\,dV = -\delta V \quad (6\text{-}59)$$

$$\delta W'_e = \int_S (u\,\delta\bar{X} + v\,\delta\bar{Y} + w\,\delta\bar{Z})\,dS$$
$$+ \int_V (u\,\delta X + v\,\delta Y + w\,\delta Z)\,dV = -\delta V' \quad (6\text{-}60)$$

These relationships will be used in the succeeding sections to develop the companion principles of the stationary values of the total potential and the total complementary potential.

6-9 THE PRINCIPLE OF THE STATIONARY VALUE OF THE TOTAL POTENTIAL

Let us now consider an isotropic elastic body which is in a state of equilibrium under the action of a set of surface forces, body forces, and a temperature change $T(x,y,z)$. A necessary and sufficient condition for the equilibrium of such a body is that $\delta W_e = -\delta W_i$ for any virtual displacements during which the forces and stresses are held constant. The virtual displacements are arbitrary on that portion of the surface S_1 on which forces are prescribed but are zero on the surface S_2 where displacements are given. The virtual work of the external forces is given by Eq. (6-59), where the first integral reduces to an integration over S_1 since the virtual displacements on S_2 are zero. If the forces are conservative, $\delta W_e = -\delta V$. The body is elastic, so that corresponding to Eq. (6-38) for the three-dimensional case we have

$$W_i = -U - \int_V \left[\int_0^{T_1} (\sigma_{xx} + \sigma_{yy} + \sigma_{zz})\alpha\,dT \right] dV$$

where U is obtained from Eqs. (6-39) and (6-46). We further assume that the temperature does not vary during the virtual displacement, and so from the last equation $\delta W_i = -\delta U$, where by using Eqs. (6-39) and (6-46)

$$\delta U = \int_V (\sigma_{xx}\,\delta\epsilon_{xx} + \sigma_{yy}\,\delta\epsilon_{yy} + \sigma_{zz}\,\delta\epsilon_{zz} + \sigma_{xy}\,\delta\epsilon_{xy}$$
$$+ \sigma_{yz}\,\delta\epsilon_{yz} + \sigma_{zx}\,\delta\epsilon_{zx})\,dV \quad (6\text{-}61)$$

The strain increments $\delta\epsilon_{xx}, \delta\epsilon_{yy}, \ldots, \delta\epsilon_{zz}$ are those associated with the virtual displacements, which for small displacement are found from Eqs. (2-23) to be

$$\delta\epsilon_{xx} = \frac{\partial\,\delta u}{\partial x} \qquad \delta\epsilon_{yy} = \frac{\partial\,\delta v}{\partial y} \qquad \delta\epsilon_{zz} = \frac{\partial\,\delta w}{\partial z}$$

$$\delta\epsilon_{xy} = \frac{\partial\,\delta v}{\partial x} + \frac{\partial\,\delta u}{\partial y} \qquad \delta\epsilon_{yz} = \frac{\partial\,\delta w}{\partial y} + \frac{\partial\,\delta v}{\partial z} \qquad \delta\epsilon_{zx} = \frac{\partial\,\delta u}{\partial z} + \frac{\partial\,\delta w}{\partial x} \qquad (6\text{-}62)$$

With $\delta W_e = -\delta V$ and $\delta W_i = -\delta U$, Eq. (6-10) gives $\delta U = -\delta V$, or finally

$$\delta(U + V) = 0 \qquad (6\text{-}63)$$

This equation expresses the principle of the stationary value of the total potential, which for an elastic body may be stated as:

> *Of all compatible states of deformation of an elastic body, the true deformations (those which satisfy equilibrium) are the ones for which the total potential has a stationary value for any virtual displacement.*

If the equilibrium is stable, the total potential $U + V$ is a minimum and we may say:

> *Among all compatible states of deformation of an elastic body in stable equilibrium, the true deformations are those for which the total potential is a minimum.*

This statement is known as the *principle of the minimum of the total potential*. In comparing approximate solutions obtained by energy methods the best approximate solution is the one having the smallest total potential.

6-10 THE PRINCIPLE OF THE STATIONARY VALUE OF THE TOTAL COMPLEMENTARY POTENTIAL

The equations of equilibrium are insufficient to determine the stresses in a statically indeterminate body, since there are an infinite number of stress distributions and reactive forces which satisfy equilibrium with the applied forces. Only one of these, the true one, also satisfies compatibility. By *applied loads* we mean those body and surface forces which are prescribed, and conversely *reactive forces* are defined as those forces which occur on surfaces where displacements are prescribed.

Consider the case of an isotropic elastic body which is restrained against rigid-body motion and is under the action of surface forces $\bar{X}$, $\bar{Y}$, and $\bar{Z}$, body forces, X, Y, and Z, and a temperature change $T(x,y,z)$. The body will come to an equilibrium state of stress σ_{xx}, σ_{yy}, . . . , σ_{zz} which satisfies compatibility. Let us consider variations in the stresses from this compatible state. We shall permit only stress variations $\delta\sigma_{xx}$, $\delta\sigma_{yy}$, . . . , $\delta\sigma_{zz}$ and reactive-force variations $\delta\bar{X}$, $\delta\bar{Y}$, $\delta\bar{Z}$ on S_2 which are self-equilibrating, so that both the true state and the varied state are in equilibrium with the applied forces. On S_1 we take $\delta\bar{X} = \delta\bar{Y} = \delta\bar{Z} = 0$, so that the prescribed surface forces are not altered. The set of stresses $\delta\sigma_{xx}$, etc., and the reactive forces $\delta\bar{X}$, $\delta\bar{Y}$, and $\delta\bar{Z}$ on S_2 form an equilibrium force system, and we may therefore apply the principle of virtual work. For virtual displacements we shall choose the *real displacements u, v, w,* so that

$$\delta W_e = \int_{S_2} (\delta \bar{X} \, u + \delta \bar{Y} \, v + \delta \bar{Z} \, w) \, dS \qquad (6\text{-}64)$$

The body-forces variations δX, δY, δZ and the surface-forces variations $\delta \bar{X}$, $\delta \bar{Y}$, $\delta \bar{Z}$ on S_1 are zero, and so from Eq. (6-60) we see that $\delta W_e = -\delta V'$. The virtual work of the internal forces is due to the stresses $\delta \sigma_{xx}$, $\delta \sigma_{yy}$, . . . , $\delta \sigma_{zz}$ moving through the real strains ϵ_{xx}, ϵ_{yy}, . . . , ϵ_{zz}. This gives

$$\delta W_i = -\int_{V} (\delta \sigma_{xx} \, \epsilon_{xx} + \delta \sigma_{yy} \, \epsilon_{yy} + \delta \sigma_{zz} \, \epsilon_{zz} + \delta \sigma_{xy} \, \epsilon_{xy} + \delta \sigma_{yz} \, \epsilon_{yz}$$
$$+ \, \delta \sigma_{zx} \, \epsilon_{zx}) \, dV \quad (6\text{-}65)$$

which from Eqs. (6-41) and (6-47) is found to be $-\delta U'$ if the temperatures are held constant during the virtual displacements. We then find from $\delta W_e = -\delta V'$, $\delta W_i = -\delta U'$, and Eq. (6-10) that

$$\delta(U' + V') = 0 \qquad (6\text{-}66)$$

where $U' + V'$ is the *total complementary potential* of the body.

Equation (6-66) is the *principle of the stationary value of the total complementary potential*, which may be stated as:

> *In a statically indeterminate elastic body there are an infinite number of stress distributions which together with the reactive forces are in a state of equilibrium with the applied forces. The true stresses and reactions (those which also satisfy compatibility) are the ones for which the total complementary potential has a stationary value for any self-equilibrating variation of the stresses and reactive forces.*

If the equilibrium is stable, $U + V$ is a minimum, and since $U + V = U' + V'$ for the equilibrium-stress and virtual-displacement systems which were used, the total complementary potential is also a minimum. We may then say:

> *Among all possible stress and reactive-force states which are in stable equilibrium with the applied forces on an elastic body, the true state is that one for which the total complementary potential has a minimum value.*

This statement is known as the *principle of the minimum total complementary potential*.

In most cases the prescribed deflections on S_2 are zero, and it is seen from Eq. (6-60) that $\delta V'$ is therefore zero. In such cases Eq. (6-66) reduces to

$$\delta U' = 0 \qquad (6\text{-}67)$$

This may be stated in the following principle:

> *Among all possible equilibrium stress states in an elastic body with no displacements at the supports, the true state is the one for which the complementary strain energy has a stationary value for any self-equilibrating variation of the stresses and reactions. If the equilibrium is stable, the complementary strain energy has a minimum value.*

If in addition, the body is linearly elastic and there is no temperature change, so that $U' = U$, we may say:

> *Among all possible equilibrium stress states in a linearly elastic body with no temperature changes or displacements at the supports, the true state is that which has the minimum elastic strain energy.*

This last statement is known as *Castigliano's theory of least work.*

6-11 DERIVATION OF EQUILIBRIUM AND COMPATIBILITY EQUATIONS BY VARIATIONAL METHODS

The principles described in Secs. 6-9 and 6-10 can be used to derive the governing differential equations of the theories of elasticity and structures. Deriving the equations in this manner is often simpler than the vector approach used in Chaps. 2 to 4, and the method has found widespread use in the development of theories for plates, shells, and sandwich structures. The vanishing of the first-order variation of the total potential assures equilibrium and may be used to derive equilibrium equations. In a similar manner the zero variation in the total complementary potential leads to compatibility equations. In both cases the appropriate number of boundary conditions are obtained along with the differential equation.

Euler was the first to systematically investigate the use of variational methods to derive the differential equations and boundary conditions for boundary-value problems, and equations derived in this manner are often referred to as *Euler differential equations.* In deriving the Euler equations certain of the boundary conditions automatically come out of the variational methods; they are known as the *natural boundary conditions.* On the other hand, some of the boundary conditions must be imposed in the variational method, and they are known as the *essential boundary conditions.* When the variational method is applied to the principle of the stationary value of the total potential the essential boundary conditions are the displacement boundary conditions, while the force (equilibrium) boundary conditions are the natural boundary conditions. In using the complementary-potential principle the essential boundary conditions are force conditions, and the natural boundary conditions are displacement conditions.

In the variational method we apply Eq. (6-63) for any virtual displacements or Eq. (6-66) for any self-equilibrating variation in stresses and reactions. The differential equations and boundary conditions are then obtained by integration by parts. The method is illustrated in Example 6-1. The procedure is also used in Example 7-9.

The following integration formulas[9] are helpful in single and double integration by parts in one-, two-, and three-dimensional domains.

$$\int_{x_1}^{x_2} a \frac{d\beta}{dx} \frac{d\gamma}{dx} \, dx = a\beta \frac{d\gamma}{dx} \Big|_{x_1}^{x_2} - \int_{x_1}^{x_2} \beta \frac{d}{dx} \left(a \frac{d\gamma}{dx} \right) dx \tag{6-68}$$

$$\iint_D a \left(\frac{\partial\beta}{\partial x} \frac{\partial\gamma}{\partial x} + \frac{\partial\beta}{\partial y} \frac{\partial\gamma}{\partial y} \right) dx \, dy = \oint a\beta \frac{\partial\gamma}{\partial n} \, ds$$
$$- \iint_D \beta \left[\frac{\partial}{\partial x} \left(a \frac{\partial\gamma}{\partial x} \right) + \frac{\partial}{\partial y} \left(a \frac{\partial\gamma}{\partial y} \right) \right] dx \, dy \tag{6-69}$$

$$\int_{x_1}^{x_2} a \frac{d^2\beta}{dx^2} \frac{d^2\gamma}{dx^2} \, dx = a \frac{d\beta}{dx} \frac{d^2\gamma}{dx^2} \Big|_{x_1}^{x_2} - \beta \frac{d}{dx} \left(a \frac{d^2\gamma}{dx^2} \right) \Big|_{x_1}^{x_2}$$
$$+ \int_{x_1}^{x_2} \beta \frac{d^2}{dx^2} \left(a \frac{d^2\gamma}{dx^2} \right) dx \tag{6-70}$$

$$\iint_D a \, \nabla^2\beta \, \nabla^2\gamma \, dx \, dy = \oint a \frac{\partial\beta}{\partial n} \nabla^2\gamma \, ds - \oint \beta \frac{\partial}{\partial n} (a \, \nabla^2\gamma) \, ds$$
$$+ \iint_D \beta \, \nabla^2(a \, \nabla^2\gamma) \, dx \, dy \tag{6-71}$$

$$\iint_D \left(\frac{\partial\beta}{\partial x} - \frac{\partial\gamma}{\partial y} \right) dx \, dy = \oint (\gamma \, dx + \beta \, dy) = \oint (-\gamma m + \beta l) \, ds \tag{6-72}$$

$$\int_V a \left(\frac{\partial\beta}{\partial x} \frac{\partial\gamma}{\partial x} + \frac{\partial\beta}{\partial y} \frac{\partial\gamma}{\partial y} + \frac{\partial\beta}{\partial z} \frac{\partial\gamma}{\partial z} \right) dV = \int_S a\beta \left(\frac{\partial\gamma}{\partial x} l + \frac{\partial\gamma}{\partial y} m + \frac{\partial\gamma}{\partial z} n \right) dS$$
$$- \int_V \beta \left[\frac{\partial}{\partial x} \left(a \frac{\partial\gamma}{\partial x} \right) + \frac{\partial}{\partial y} \left(a \frac{\partial\gamma}{\partial y} \right) + \frac{\partial}{\partial z} \left(a \frac{\partial\gamma}{\partial z} \right) \right] dV \tag{6-73}$$

In these equations a is a known function of the coordinates, while β and γ are arbitrary functions which possess the differentiability indicated by the equations. In the two-dimensional equations the direction ds is along the boundary in the counterclockwise direction, and dn is in the direction of the outward normal to the boundary. The direction cosines of the outward normal to the boundary are l, m, and n.

Example 6-1   Derive the equations of equilibrium (2-8) and the equilibrium boundary conditions (2-10) by the variational method.

Since equilibrium equations are desired, we apply Eq. (6-63). If we assume that the temperature is fixed during the virtual displacements, δU is given by Eq. (6-61), where the virtual strains are found from Eqs. (6-62). Therefore

$$\delta U = \int_V \left[\sigma_{xx} \frac{\partial \, \delta u}{\partial x} + \sigma_{yy} \frac{\partial \, \delta v}{\partial y} + \sigma_{zz} \frac{\partial \, \delta w}{\partial z} + \sigma_{xy} \left(\frac{\partial \, \delta v}{\partial x} + \frac{\partial \, \delta u}{\partial y} \right) \right.$$
$$\left. + \sigma_{yz} \left(\frac{\partial \, \delta w}{\partial y} + \frac{\partial \, \delta v}{\partial z} \right) + \sigma_{zx} \left(\frac{\partial \, \delta u}{\partial z} + \frac{\partial \, \delta w}{\partial x} \right) \right] dV \qquad (a)$$

This equation must be integrated by parts to obtain δu, δv, or δw as a factor of each term under the integral sign. This is most conveniently accomplished by applying Eq. (6-73). Consider the first term of Eq. (a), for example, and let $a = 1$, $\beta = \delta u$, $\partial \gamma / \partial x = \sigma_{xx}$, and $\partial \gamma / \partial y = \partial \gamma / \partial z = 0$. We find from Eq. (6-73) that

$$\int_V \sigma_{xx} \frac{\partial \, \delta u}{\partial x} \, dV = \int_S \delta u \, \sigma_{xx} l \, dS - \int_V \delta u \frac{\partial \sigma_{xx}}{\partial x} \, dV$$

The remaining terms can be integrated in a similar fashion to give

$$\delta U = \int_S [(\sigma_{xx} l + \sigma_{xy} m + \sigma_{xz} n) \, \delta u + (\sigma_{yx} l + \sigma_{yy} m + \sigma_{yz} n) \, \delta v$$
$$+ (\sigma_{zx} l + \sigma_{zy} m + \sigma_{zz} n) \, \delta w] \, dS - \int_V \left[\left(\frac{\partial \sigma_{xx}}{\partial x} + \frac{\partial \sigma_{xy}}{\partial y} + \frac{\partial \sigma_{xz}}{\partial z} \right) \delta u \right.$$
$$\left. + \left(\frac{\partial \sigma_{yx}}{\partial x} + \frac{\partial \sigma_{yy}}{\partial y} + \frac{\partial \sigma_{yz}}{\partial z} \right) \delta v + \left(\frac{\partial \sigma_{zx}}{\partial x} + \frac{\partial \sigma_{zy}}{\partial y} + \frac{\partial \sigma_{zz}}{\partial z} \right) \delta w \right] dV$$

Adding δV from Eq. (6-59) to this result and equating the sum to zero, we find

$$\int_S [(\sigma_{xx} l + \sigma_{xy} m + \sigma_{xz} n - \bar{X}) \, \delta u + (\sigma_{yx} l + \sigma_{yy} m + \sigma_{yz} n - \bar{Y}) \, \delta v$$
$$+ (\sigma_{zx} l + \sigma_{zy} m + \sigma_{zz} n - \bar{Z}) \, \delta w] \, dS - \int_V \left[\left(\frac{\partial \sigma_{xx}}{\partial x} + \frac{\partial \sigma_{xy}}{\partial y} + \frac{\partial \sigma_{xz}}{\partial z} + X \right) \delta u \right.$$
$$\left. + \left(\frac{\partial \sigma_{yx}}{\partial x} + \frac{\partial \sigma_{yy}}{\partial y} + \frac{\partial \sigma_{yz}}{\partial z} + Y \right) \delta v + \left(\frac{\partial \sigma_{zx}}{\partial x} + \frac{\partial \sigma_{zy}}{\partial y} + \frac{\partial \sigma_{zz}}{\partial z} + Z \right) \delta w \right] dV = 0$$

We now impose the essential displacement boundary conditions that

$$\delta u = \delta v = \delta w = 0$$

on S_2, so that the integral on S is reduced to an integral on S_1. Otherwise δu, δv, and δw are arbitrary, and for the integrals to vanish the quantities in the parentheses under the integral signs must vanish. This requires that Eqs. (2-10) (the natural boundary conditions) be satisfied at all points on S_1 and that Eqs. (2-8) (the Euler differential equations) be satisfied at all interior points of the body.

6-12   THE RAYLEIGH-RITZ METHOD

We shall now consider how the energy principles can be used to obtain approximate solutions to problems where the exact solutions to the governing differential equations are not known. The procedure, known as the *Rayleigh-Ritz method*, can be used with either the total-potential or the total-complementary-potential principle.   The method approximates the continuum system with its infinite degrees of freedom by a system with a finite number of degrees of freedom and reduces the problem to that of solving a set of simultaneous algebraic equations.   Consider, for example, the case of plane stress.   The strain energy in terms of strains for a linearly elastic isotropic material can be found by integrating Eq. (6-52) over the volume of the body.   This in turn can be expressed in terms of displacements by using Eqs. (2-23).   The potential of the external forces is already in terms of displacements in Eq. (6-58), which may be simplified to the two-dimensional case by dropping the Z and $\bar{Z}$ terms. The total potential will therefore be of the form

$$U + V = U(u,v) + V(u,v) \tag{6-74}$$

The displacements of the body are approximated by the series

$$u = u_0(x,y) + \sum_{j=1}^{m} a_j u_j(x,y) \qquad v = v_0(x,y) + \sum_{j=1}^{n} b_j v_j(x,y) \tag{6-75}$$

where u_0 and v_0 are assumed functions which satisfy the nonhomogeneous essential (displacement) boundary conditions on S_2 and where u_j and v_j are linearly independent assumed functions which vanish on the boundary S_2; that is, they satisfy the homogeneous essential boundary conditions. Except for this, and the fact that they must be differentiable, the assumed functions are arbitrary.   The a_j and b_j are coefficients which are to be determined.   For a given set of assumed functions the constants a_j and b_j determine the configuration of the deformed body and therefore form a set of generalized coordinates $q_1, q_2, \ldots, q_{m+n}$.   By substituting Eqs. (6-75) into Eq. (6-74) $U + V$ can be expressed in terms of the coefficients a_j and b_j.   Equation (6-17) then gives

$$\begin{aligned} \frac{\partial}{\partial a_i}(U + V) = 0 \qquad i = 1, 2, \ldots, m \\[2mm] \frac{\partial}{\partial b_i}(U + V) = 0 \qquad i = 1, 2, \ldots, n \end{aligned} \tag{6-76}$$

Equations (6-76) are a set of $m + n$ simultaneous linear algebraic equations which can be solved for the a_j and b_j coefficients.   Substitution of these into Eqs. (6-75) then gives an approximate solution for the deflections at any point (x,y).   Substituting Eqs. (6-75) into Eqs. (2-23) gives the strains, which when substituted into the stress-strain equations (3-20) result in the stresses.

     In principle, an exact solution is obtained if an infinite number of terms are taken in the series and the series is complete (in the sense that the series is capable of representing any continuous function within the domain of the body). In practice it is usually necessary to truncate the series at a finite number of terms. In this case accuracy is improved if intuition regarding the deflection shape is used in choosing the assumed functions.

To illustrate how the Rayleigh-Ritz method is used with the complementary potential, consider again the case of plane stress. If the material is linearly elastic and isotropic, U_0' in terms of the stress function is given by Eq. (6-54). This equation can be simplified by setting $\nu = 0$ when the body is simply connected and body forces are constant, so that $\nabla^2 V = 0$. This is possible in this case because ν does not appear in Eqs. (4-14) and (4-18), and as a result φ does not depend upon ν. With these assumptions and the body forces equal to zero, the integral of Eq. (6-54) over the volume simplifies to

$$U' = \frac{1}{2E} \iint_D \left[\left(\frac{\partial^2 \varphi}{\partial x^2} \right)^2 + \left(\frac{\partial^2 \varphi}{\partial y^2} \right)^2 + 2 \left(\frac{\partial^2 \varphi}{\partial x \, \partial y} \right)^2 \right] dx \, dy$$
$$+ \iint_D \alpha T \, \nabla^2 \varphi \, dx \, dy \quad (6\text{-}77)$$

The stress function can be approximated by the series

$$\varphi = \varphi_0(x,y) + \sum_{j=1}^{n} a_j \varphi_j(x,y) \quad (6\text{-}78)$$

where φ_0 is an assumed function which satisfies the nonhomogeneous essential (force) boundary conditions and the φ_j are assumed functions which produce no stresses on the boundary; i.e., they satisfy the homogeneous essential boundary conditions. Other than this, and the fact that they must be twice differentiable to substitute into Eq. (6-77), the functions are arbitrary; however, physical intuition is helpful in choosing them. The a_j coefficients are constants which are found by the Rayleigh-Ritz method.

The stress-function formulation is practical only when forces are prescribed over the entire boundary so that there are no surfaces in class S_2. In this case $\delta V' = 0$, and Eq. (6-66) reduces to $\delta U' = 0$. The complementary strain energy is expressed in terms of the a_j by substituting Eq. (6-78) into (6-77). Varying the a_j coefficients alters φ, but the stresses associated with the variation of φ are self-equilibrating, because all stress functions automatically satisfy equilibrium. It follows then that

$$\delta U' = \sum_{i=1}^{n} \frac{\partial U'}{\partial a_i} \delta a_i = 0$$

which for arbitrary variations δa_i gives

$$\frac{\partial U'}{\partial a_i} = 0 \qquad i = 1, 2, \ldots, n \qquad (6\text{-}79)$$

Substituting Eq. (6-77) into the last equation gives

$$\iint\limits_{D} \left(\frac{\partial^2 \varphi}{\partial x^2} \frac{\partial}{\partial a_i} \frac{\partial^2 \varphi}{\partial x^2} + \frac{\partial^2 \varphi}{\partial y^2} \frac{\partial}{\partial a_i} \frac{\partial^2 \varphi}{\partial y^2} + 2 \frac{\partial^2 \varphi}{\partial x \, \partial y} \frac{\partial}{\partial a_i} \frac{\partial^2 \varphi}{\partial x \, \partial y} \right) dx \, dy$$

$$= -\iint\limits_{D} E\alpha T \frac{\partial}{\partial a_i} \nabla^2 \varphi \, dx \, dy \qquad i = 1, 2, \ldots, n \qquad (6\text{-}80)$$

and by using Eq. (6-78) we find

$$\sum_{j=1}^{n} \left[\iint\limits_{D} \left(\frac{\partial^2 \varphi_i}{\partial x^2} \frac{\partial^2 \varphi_j}{\partial x^2} + \frac{\partial^2 \varphi_i}{\partial y^2} \frac{\partial^2 \varphi_j}{\partial y^2} + 2 \frac{\partial^2 \varphi_i}{\partial x \, \partial y} \frac{\partial^2 \varphi_j}{\partial x \, \partial y} \right) dx \, dy \right] a_j$$

$$= -\iint\limits_{D} \left(\frac{\partial^2 \varphi_0}{\partial x^2} \frac{\partial^2 \varphi_i}{\partial x^2} + \frac{\partial^2 \varphi_0}{\partial y^2} \frac{\partial^2 \varphi_i}{\partial y^2} + 2 \frac{\partial^2 \varphi_0}{\partial x \, \partial y} \frac{\partial^2 \varphi_i}{\partial x \, \partial y} \right) dx \, dy$$

$$- \iint\limits_{D} E\alpha T \left(\frac{\partial^2 \varphi_i}{\partial x^2} + \frac{\partial^2 \varphi_i}{\partial y^2} \right) dx \, dy \qquad i = 1, 2, \ldots, n \qquad (6\text{-}81)$$

Equations (6-81) are a set of simultaneous linear algebraic equations that can be solved for $a_1, a_2, \ldots, a_n$. Substituting these into Eq. (6-78) gives an approximate solution for φ, which, when substituted into Eqs. (4-13), gives the stresses.

The method is illustrated in the following example. Additional examples are given throughout the remainder of the book. While very accurate results may often be obtained by the Rayleigh-Ritz method with relatively few well-chosen terms in the series, the evaluation of the integrals of Eq. (6-81) can be tedious.

Example 6-2 Obtain an approximate solution to the plane-stress problem shown in Fig. 4-3 (Example 4-1).

The nonhomogeneous essential (force) boundary conditions are satisfied if we take $\varphi_0 = cy^4/12$. To assure that the φ_j give no stresses on the boundary we take $(x^2 - a^2)^2(y^2 - b^2)^2$ as a factor of each of these functions. We then assume

$$\varphi = \frac{cy^4}{12} + (x^2 - a^2)^2(y^2 - b^2)^2(a_1 + a_2 x^2 + a_3 y^2 + \cdots) \qquad (a)$$

Only even power terms have been included in the polynomial series, since for the loading considered the normal stresses will be symmetrical about the x and y axes. If the loading were antisymmetric with respect to an axis, only odd-order polynomial terms in that variable which is normal to the axis would be included. For a general asymmetric loading both odd and even terms should be contained in the series.

If the series is truncated after the first three terms, the following matrix

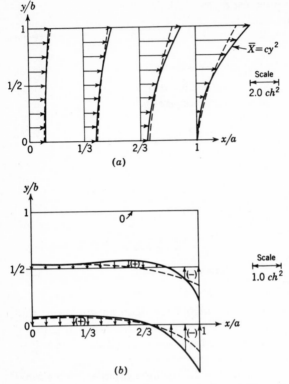

Fig. 6-7 Results of Rayleigh-Ritz solution of Example 6-2 and comparison with finite-difference solution of Example 5-1 (shown dashed). (a) σ_{xx}, symmetrical about x and y axes; (b) σ_{yy}, symmetrical about x and y axes.

equation for $a/b = 1.5$ is obtained by substituting the functions of Eq. (a) into Eq. (6-81):

$$\begin{bmatrix} 3.974 & 0.326 & 0.196 \\ 0.326 & 0.353 & 0.019 \\ 0.196 & 0.019 & 0.247 \end{bmatrix} \begin{Bmatrix} a_1 \\ a^2 a_2 \\ a^2 a_3 \end{Bmatrix} = -\frac{c}{a^4} \begin{Bmatrix} 0.300 \\ 0.043 \\ 0.019 \end{Bmatrix} \qquad (b)$$

Solving, we find $a_1 = -0.0700c/a^4$, $a_2 = -0.0557c/a^6$, and $a_3 = -0.0173c/a^6$. The stresses are obtained by substituting these results into Eq. (a), which in turn is substituted into Eqs. (4-13). The distributions of σ_{xx} and σ_{yy} are shown in Fig. 6-7. It is interesting to compare these with the finite-difference results of Example 5-1, which are shown dashed in Fig. 6-7. We see that the Rayleigh-Ritz method exactly predicts the σ_{xx} stresses at $x/a = \pm 1$ and is therefore more accurate in the region of the ends. Increased accuracy can be obtained by increasing the number of terms in the series.

6-13 THE RECIPROCAL THEOREMS OF BETTI AND MAXWELL

We now develop two useful theorems for linearly elastic bodies restrained by supports which do not displace. We assume that the body is first subjected to a set of applied body forces X_1, Y_1, Z_1 and surface forces $\bar{X}_1$, $\bar{Y}_1$, $\bar{Z}_1$. The displacements which occur as a result of these loads we designate as u_1, v_1, w_1. The work done by the applied forces during these deflections is

$$W_{e_{11}} = \frac{1}{2} \int_V (X_1 u_1 + Y_1 v_1 + Z_1 w_1)\, dV$$

$$+ \frac{1}{2} \int_{S_1} (\bar{X}_1 u_1 + \bar{Y}_1 v_1 + \bar{Z}_1 w_1)\, dS \quad (6\text{-}82)$$

Now let a second set of forces X_2, Y_2, Z_2 and $\bar{X}_2$, $\bar{Y}_2$, $\bar{Z}_2$ be applied which cause additional deflections u_2, v_2, w_2. The work of the second system of forces in moving through the displacements which it causes is

$$W_{e_{22}} = \frac{1}{2} \int_V (X_2 u_2 + Y_2 v_2 + Z_2 w_2)\, dV$$

$$+ \frac{1}{2} \int_{S_1} (\bar{X}_2 u_2 + \bar{Y}_2 v_2 + \bar{Z}_2 w_2)\, dS \quad (6\text{-}83)$$

An additional increment of work $W_{e_{12}}$ will be done by the first set of forces in moving through the displacements due to the second force system. Since the loads in the first system are constant during these additional displacements,

$$W_{e_{12}} = \int_V (X_1 u_2 + Y_1 v_2 + Z_1 w_2)\, dV + \int_{S_1} (\bar{X}_1 u_2 + \bar{Y}_1 v_2 + \bar{Z}_1 w_2)\, dS$$
$$(6\text{-}84)$$

The total work done by the application of both force systems is the sum of Eqs. (6-82) to (6-84), or

$$W_e = W_{e_{11}} + W_{e_{22}} + W_{e_{12}} \quad (6\text{-}85)$$

Now let us begin again and apply the forces in the reverse order. In this case the total work will be

$$W_e = W_{e_{22}} + W_{e_{11}} + W_{e_{21}} \quad (6\text{-}86)$$

where $W_{e_{11}}$ and $W_{e_{22}}$ are again given by Eqs. (6-82) and (6-83) and $W_{e_{21}}$ is the work done by force system 2 in moving through the incremental displacements associated with force system 1. It is easily seen that

$$W_{e_{21}} = \int_V (X_2 u_1 + Y_2 v_1 + Z_2 w_1)\, dV + \int_{S_1} (\bar{X}_2 u_1 + \bar{Y}_2 v_1 + \bar{Z}_2 w_1)\, dS$$
$$(6\text{-}87)$$

Since the body is linearly elastic and only small deflections are considered, the principle of superposition applies, and the total work of the applied forces must be independent of the order of loading. Therefore the values of W_e obtained from Eqs. (6-85) and (6-86) must be the same, and it follows that $W_{e_{12}} = W_{e_{21}}$. This is the reciprocal theorem of Betti, which may be stated:

If a linearly elastic body restrained by rigid supports is subjected to two force systems, the work done by the first system in moving through the displacements due to the second system is equal to the work performed by the second system in moving through the displacements resulting from the first system.

A special case of Betti's theorem is obtained when each of the systems consists of a single force. Let the two forces be designated by Q_i and Q_j, and let the work producing displacement at i due to a unit load at j be c_{ij}. Betti's theorem then gives $Q_i(c_{ij}Q_j) = Q_j(c_{ji}Q_i)$, which reduces to $c_{ij} = c_{ji}$. This is Maxwell's reciprocal theorem, which may be stated:

If a linearly elastic body with rigid supports is subjected to two unit forces, the deflection at (and in the direction of) the first force which is due to the second force is equal to the deflection at (and in the direction of) the second force which is due to the first force.

The forces and displacements in this theorem should be considered in their generalized senses. If Q_i is a force, for instance, and Q_j is a moment, then c_{ij} is the deflection at, and in the direction of, Q_i due to a unit moment at, and in the direction of, Q_j, and c_{ji} is the rotation at, and in the direction of, Q_j due to a unit force at, and in the direction of, Q_i.

From the manner in which they were defined the c_{ij} are seen to be flexibility influence coefficients, and Maxwell's reciprocal theorem confirms the results of Sec. 6-6 that the flexibility matrix is symmetric.

6-14 THE USE OF VIRTUAL WORK TO COMPUTE DEFLECTIONS

A very useful method for computing the deflections of structures subjected to loads and temperatures is based upon the principle of virtual work. It is variously known as the *unit-load, dummy-load, virtual-work, Maxwell-Mohr,* or *Mueller-Breslau method.* A general description of the procedure and a simple illustrative example will be given in this section. Additional examples will be found throughout the text, especially in Chap. 10, which is devoted to structural displacements. There the method is extended to include relative deflections and is also used to determine redundant reactions of structures which are supported in a statically indeterminate manner.

Consider the case of an elastic body supported against rigid-body motions which is subjected to a system of body and surface forces and to a temperature change $T(x,y,z)$. Let us assume that a generalized deflection q is desired at a particular point on the body in a certain direction. To determine this displacement we forget the real loads and temperatures for a moment and imagine that the body is loaded by a fictitious work producing generalized unit force $Q = 1$ applied to the body at the point where the deflection is to be computed and in the direction of the desired displacement. We assume that the unit load is reacted by forces at rigid supports so that a state of equilibrium exists. Since the reactive forces are applied at points of rigid support, they do no work during virtual displacements which do not violate the geometric constraints. In the case of a structure which is supported in a statically determinate manner, there is only one choice for the forces which react the unit load, and that is the reactions which would occur for the actual body supports. On the other hand, in a structure with redundant supports we may choose any convenient set of reactions which meet the requirements that (1) the reactive forces are in equilibrium with the unit load, and (2) they do no work during the virtual displacements. In such cases, the most convenient reactions will be those in which the redundant supports are assumed absent so that the remaining restraints are statically determinate.

We designate the stresses which result from the unit load and its reactions by $\bar{\sigma}_{xx}, \bar{\sigma}_{yy}, \ldots, \bar{\sigma}_{zx}$. It is easier to compute these stresses, of course, if we have taken statically determinate supports as noted. The unit load, its reactions, and the associated stresses form an equilibrium system of external and internal forces. As a result of this we may say that $\delta W_e = \delta U$ for any virtual displacements of the body. In this equation δW_e is the virtual work performed by the external forces during the virtual displacements, and since the reactions do no work, this is equal to the unit load multiplied by the virtual displacement at, and in the direction of, the unit load. The quantity δU is the variation of the strain energy resulting from the stresses $\bar{\sigma}_{xx}, \bar{\sigma}_{yy} \ldots, \bar{\sigma}_{zz}$ moving through the strains associated with the virtual displacements.

As virtual displacements we choose the *real* displacements which result from the actual applied forces and temperature changes. We designate the strains associated with the real displacements by $\epsilon_{xx}, \epsilon_{yy}, \ldots, \epsilon_{zx}$. In this case δW_e is equal to the product of the unit load and the real displacement at, and in the direction of, the unit load or is simply the desired displacement q. Observing this and using Eq. (6-61) to evaluate δU (where $\sigma_{xx}, \sigma_{yy}, \ldots, \sigma_{zz}$ are equal to $\bar{\sigma}_{xx}, \bar{\sigma}_{yy}, \ldots, \bar{\sigma}_{zx}$

and $\delta\epsilon_{xx}$, $\delta\epsilon_{yy}$, . . . , $\delta\epsilon_{zz}$ are ϵ_{xx}, ϵ_{yy}, . . . , ϵ_{zz}), we find from $\delta W_e = \delta U$ that

$$q = \int_V (\bar{\sigma}_{xx}\epsilon_{xx} + \bar{\sigma}_{yy}\epsilon_{yy} + \bar{\sigma}_{zz}\epsilon_{zz} + \bar{\sigma}_{xy}\epsilon_{xy} + \bar{\sigma}_{yz}\epsilon_{yz} + \bar{\sigma}_{zx}\epsilon_{zx}) \, dV \quad (6\text{-}88)$$

This equation is general and applies to a body with arbitrary shape and loads. For specific bodies, such as beams and bars, Eq. (6-88) can be simplified by expressing $\bar{\sigma}_{xx}$, $\bar{\sigma}_{yy}$, . . . , $\bar{\sigma}_{zz}$ in terms of the stress resultants in the unit-load system and ϵ_{xx}, ϵ_{yy}, . . . , ϵ_{zz} in terms of the stress resultants in the actual-load system. Simplified equations for these special cases are given in Chap. 10. We defined q in the preceding discussion as a generalized displacement and Q as a unit generalized force, so that its product with q has the units of work. If q is a rotation, for instance, then Q is a unit couple.

Example 6-3 Obtain an expression for δU for a beam that is loaded in bending about one of its principal axes and use the result to compute the slope at the center of a uniform cantilevered beam subjected to a lateral tip force P.

Let us designate the bending moment associated with the real loads on the beam by M and the moment that results from the unit-load system by $\bar{M}$. Using the strength-of-materials theory of bending, we assume that all stresses other than σ_{xx} are negligible (where the x axis coincides with the centroidal axis of the beam) and obtain $\sigma_{xx} = -Mz/I$. In this equation, z is the distance from the neutral axis of bending, and I is the moment of inertia of the beam cross section about the neutral axis. For the uniaxial-stress condition we find from Eq. (3-1) that the strain resulting from the real loads is

$$\epsilon_{xx} = -\frac{Mz}{EI} \quad (a)$$

The stress due to the unit-load system is

$$\bar{\sigma}_{xx} = -\frac{\bar{M}z}{I} \quad (b)$$

Substituting Eqs. (a) and (b) into the right side of Eq. (6-88) and neglecting all strains except ϵ_{xx}, we find

$$\delta U = \int_L \frac{M\bar{M}}{EI^2} \left(\int_A z^2 \, dA \right) dx \quad (c)$$

where the integrations extend over the cross-sectional area A and the length L of the beam. We note that the quantity in the parentheses is I, so that Eq. (c) reduces to

$$\delta U = \int_L \frac{M\bar{M}}{EI} \, dx \quad (6\text{-}89)$$

The real-load system for the cantilevered beam is shown in Fig. 6-8a. If the slope (rotation) at $x = L/2$ is desired, we must apply a unit moment at that point as shown in Fig. 6-8b and react it by a shear force and moment at the wall, where they will do no work. From Fig. 6-8, $M = Px$ for $0 < x < L$,

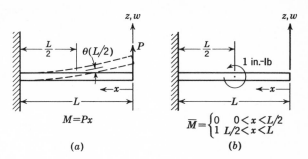

Fig. 6-8 Example 6-3. (a) Real-load system; (b) unit-load system.

while $\bar{M} = 0$ for $0 < x < L/2$, and $\bar{M} = 1$ for $L/2 < x < L$. Substituting these into Eq. (6-89), we find

$$\delta U = \int_{L/2}^{L} \frac{(Px)(1)}{EI} \, dx = \frac{3PL^2}{8EI}$$

The virtual work done by the unit moment during the real displacements is $\delta W_e = 1 \times \theta(L/2) = q$, where $\theta(L/2)$ is the rotation of the tangent to the beam axis at $x = L/2$. Equation (6-88) then gives $\theta(L/2) = 3PL^2/8EI$.

REFERENCES

1. Lanczos, C.: "The Variational Principles of Mechanics," University of Toronto Press, Toronto, 1957.
2. Hoff, N. J.: "The Analysis of Structures," John Wiley & Sons, Inc., New York, 1956.
3. Kármán, T. V., and M. A. Biot: "Mathematical Methods in Engineering," McGraw-Hill Book Company, New York, 1940.
4. Bisplinghoff, R. L., H. Ashley, and R. L. Halfman: "Aeroelasticity," Addison-Wesley Publishing Company, Inc., Reading, Mass., 1955.
5. Argyris, J. H., and S. Kelsey: "Energy Theorems and Structural Analysis," Butterworth & Co. (Publishers) Ltd., London, 1960.
6. Williams, D.: "An Introduction to the Theory of Aircraft Structures," Edward Arnold (Publishers) Ltd., London, 1960.
7. Van den Broek, J. A.: "The Elastic Energy Theory," John Wiley & Sons, Inc., New York, 1931.
8. Langhaar, H. L.: "Energy Methods in Applied Mechanics," John Wiley & Sons, Inc., New York, 1962.
9. Crandall, S. H.: "Engineering Analysis," McGraw-Hill Book Company, New York, 1956.

PROBLEMS

6-1. Verify the following in Example 6-2:

(a) The element of the first row and second column of the square matrix in Eq. (b).

(b) The first element in the column matrix on the right side of Eq. (b).

(c) The value of σ_{xx} at $x/a = \frac{2}{3}$, $y/b = 1$ in Fig. 6-7a.

6-2. Write a suitable Rayleigh-Ritz series solution for the plane-stress problems that are shown.

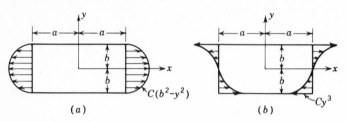

Fig. P6-2

6-3. The slab of Example 6-2 is subjected to a temperature change $T = Cy^2$ instead of the edge loading. Determine the matrix equation that replaces Eq. (b) in Example 6-2.

6-4. A cantilevered beam is subjected to a concentrated moment M_0 at its center. Determine the tip deflection by the unit-load method. Using your results and those of Example 6-3, show that the displacements for the two load systems satisfy Betti's reciprocal theorem.

6-5. A uniform cantilevered beam is loaded by lateral forces Q_1 and Q_2 at its center and tip, respectively. Use the unit-load method to obtain the flexibility matrix c which relates the forces Q_1 and Q_2 to the lateral displacements q_1 and q_2 at their points of application. Verify that c is symmetric. Find the displacements from Eq. (6-29) when $Q_1 = Q_2 = P$.

6-6. Invert c in Prob. 6-5 to obtain k and verify that it is symmetric. Use Eq. (6-26) to find Q_1 and Q_2 for $q_1 = 1$ and $q_2 = 0$.

7
Bending and Extension of Beams

7-1 INTRODUCTION

It was shown in Chap. 4 that the theory of elasticity leads to partial differential equations and that solutions to these equations are known only for a relatively few special cases. Approximate solutions can be found by the finite-difference and Rayleigh-Ritz methods; however, these require lengthy computations which are frequently unwarranted, and approximate theories based upon simplifying assumptions are preferable in these circumstances. In introducing simplifying assumptions it is necessary to make restrictions on the shape and loading of the body. Furthermore, experimental evidence or a strong basis for intuitive judgment is required to introduce realistic approximations. In some cases these lead to theories which prove to be exact when checked by the theory of elasticity, but in most instances the solutions do not satisfy all the equilibrium, compatibility, stress-strain, and boundary-condition equations. Even so, the resulting theory may be accurate enough for engineering purposes. Theories of this type belong to the fields of *applied elasticity* or *strength of materials*. The theories for beams, bars, cables, plates, and membranes given in this and subsequent chapters fall into this category. We begin our discussion of these approximate theories with the theory of bending and extension of beams, which we shall treat with considerable generality.

7-2 STRESS RESULTANTS

The term *beam* is used to describe a member that is capable of resisting bending moments and whose length is large compared to its cross-sectional dimensions. In beam theory it is convenient to work with force and moment resultants of the stresses that act upon cross-sections of the beam. Consider the beam shown in Fig. 7-1a, where the x axis is in the direction of the length of the beam. The stresses on a differential area dA at point B with coordinates (x,y,z) are σ_{xx}, σ_{xy}, and σ_{xz}. The resultant of these

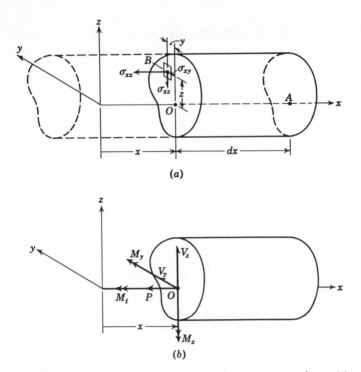

Fig. 7-1 Stresses and stress resultants on a beam cross section. (*a*) Stresses at an arbitrary point *B* in the cross section; (*b*) stress resultants on the cross section.

stresses on the cross-section consist of a force and a moment. The force may be resolved into an axial component P and shear components V_y and V_z, which are parallel to the y and z axes, respectively. The moment may be divided into a torsional moment M_t and bending moments M_y and M_z, which are about axes that are parallel to the y and z axes, respectively. The positive directions that are assumed for these *stress resultants* are shown in Fig. 7-1*b*, where the moment vectors are indicated by double arrowheads. It is seen from Fig. 7-1 that

$$P = \int_A \sigma_{xx}\, dA \qquad V_y = -\int_A \sigma_{xy}\, dA \qquad V_z = -\int_A \sigma_{xz}\, dA \quad (7\text{-}1)$$

$$M_t = \int_A (\sigma_{xz}y - \sigma_{xy}z)\, dA \tag{7-2}$$

$$M_y = -\int_A \sigma_{xx}z\, dA \qquad M_z = -\int_A \sigma_{xx}y\, dA \tag{7-3}$$

In statically determinate beams the stress resultants can be found from the applied forces by equations of equilibrium alone. On the other hand, it is necessary to consider deformations in computing the stress resultants in indeterminate beams. An equation for the stresses that result from P, M_y, and M_z is derived in the next section, while expressions for the stresses associated with M_t, V_y, and V_z are developed in Chaps. 8 and 9.

7-3 STRESSES DUE TO EXTENSION AND BENDING

Consider a nonhomogeneous beam subjected to axial forces, bending loads, and a temperature change $T(x,y,z)$, which give rise to stress resultants P, M_y, and M_z (restrictions on the form of T will be given in Sec. 7-5). Instead of determining the stresses which satisfy the compatibility equations of the theory of elasticity, we assume the displacements in accordance with the *Bernoulli-Euler theory of bending*. In this theory, frequently called the *engineers' theory of bending*, it is assumed that cross-sectional planes of the beam remain plane and normal to the axis of the beam as it deforms. This assumption makes it possible to determine the deflections of any point in the beam in terms of the deflections of points on the axis of the beam.

Consider a point O with coordinates $(x,0,0)$ and a point B with coordinates (x,y,z) which lie in a cross section at a distance x from the origin (Fig. 7-1a). The displacement of B parallel to the x axis can be written

$$u_B = u(x) - \theta_y(x)z - \theta_z(x)y \tag{7-4}$$

where $u(x)$ is the displacement of O in the direction of the x axis as a result of extension of the beam and $\theta_y(x)$ and $\theta_z(x)$ are the angles of rotation of the cross-sectional plane about axes through O which are parallel to the y and z axes, respectively.

The longitudinal strain at B is found by substituting Eq. (7-4) into the first of Eqs. (2-23), which gives

$$\epsilon_{xx} = C_1 + C_2 y + C_3 z \tag{7-5}$$

where

$$C_1 = \frac{du(x)}{dx} \qquad C_2 = -\frac{d\theta_z(x)}{dx} \qquad C_3 = -\frac{d\theta_y(x)}{dx} \tag{7-6}$$

For a given cross section, C_1, C_2, and C_3 are constants, and we see from Eq. (7-5) that the Bernoulli-Euler displacement assumptions are equivalent to postulating a linear strain distribution over the cross section. To determine the stresses associated with the strains of Eq. (7-5) we must adopt a stress-strain law. Again, in keeping with the Bernoulli-Euler theory, we assume that all stress components other than σ_{xx} are negligibly small compared to σ_{xx}. The stress is then given by the uniaxial stress-strain relationship

$$\sigma_{xx} = E(\epsilon_{xx} - \alpha T) \tag{7-7}$$

where E and αT are evaluated at B. Substituting Eq. (7-5) into Eq. (7-7), we find

$$\sigma_{xx} = E(C_1 + C_2 y + C_3 z - \alpha T) \tag{7-8}$$

To determine C_1, C_2, and C_3 we substitute Eq. (7-8) into the first of Eqs. (7-1) and into Eqs. (7-3). After dividing by an arbitrary *reference modulus* E_1 (for reasons which will become apparent), we obtain

$$A^*C_1 + \bar{y}^*A^*C_2 + \bar{z}^*A^*C_3 = \frac{P^*}{E_1}$$

$$\bar{z}^*A^*C_1 + I_{yz}^*C_2 + I_{yy}^*C_3 = -\frac{M_y^*}{E_1} \tag{7-9}$$

$$\bar{y}^*A^*C_1 + I_{zz}^*C_2 + I_{yz}^*C_3 = -\frac{M_z^*}{E_1}$$

where the *modulus-weighted section properties* are defined by

$$A^* = \int_A dA^* \tag{7-10}$$

$$\bar{y}^* = \frac{1}{A^*} \int_A y \, dA^* \qquad \bar{z}^* = \frac{1}{A^*} \int_A z \, dA^* \tag{7-11}$$

$$I_{yy}^* = \int_A z^2 \, dA^* \qquad I_{zz}^* = \int_A y^2 \, dA^* \qquad I_{yz}^* = \int_A yz \, dA^* \tag{7-12}$$

and the *effective stress resultants* are

$$P^* = P + P_T \qquad M_y^* = M_y + M_{yT} \qquad M_z^* = M_z + M_{zT} \tag{7-13}$$

The integrals extend over the cross-sectional area A, and dA^* is a modulus-weighted differential area defined by

$$dA^* = \frac{E}{E_1} dA \tag{7-14}$$

The thermal equivalents of stress resultants in Eqs. (7-13) are defined as

$$P_T = \int_A E\alpha T \, dA \qquad M_{yT} = -\int_A zE\alpha T \, dA \qquad M_{zT} = -\int_A yE\alpha T \, dA \tag{7-15}$$

We note that if the beam is homogeneous and we choose $E_1 = E$, the modulus-weighted section properties reduce to the familiar geometric section properties of the cross section, and the asterisks can be dropped in Eqs. (7-10) to (7-12).

The solution of the simultaneous equations (7-9) is simplified if we choose the y and z axes so they have the property that $\bar{y}^* = \bar{z}^* = 0$.

These axes are found by choosing an arbitrary set of axes y_0 and z_0 as shown in Fig. 7-2 and letting $\bar{y}_0^*$ and $\bar{z}_0^*$ be the distances from these to the y and z axes. Then by applying the first of Eqs. (7-11) we find

$$\bar{y}^* A^* = \int_A y \, dA^* = \int_A (y_0 - \bar{y}_0^*) \, dA^* = \int_A y_0 \, dA^* - \bar{y}_0^* A^* = 0$$

which gives

$$\bar{y}_0^* = \frac{1}{A^*} \int_A y_0 \, dA^* \tag{7-16a}$$

In a similar fashion

$$\bar{z}_0^* = \frac{1}{A^*} \int_A z_0 \, dA^* \tag{7-16b}$$

The coordinates $(\bar{y}_0^*, \bar{z}_0^*)$ locate a point which we define as the *modulus-weighted centroid* of the cross section. When the beam is homogeneous, Eqs. (7-16) reduce to the equations for the coordinates of the geometric centroid.

Taking the y and z axes through the modulus-weighted centroid so that $\bar{y}^* = \bar{z}^* = 0$, we find from Eqs. (7-9) that

$$C_1 = \frac{P^*}{E_1 A^*} \tag{7-17}$$

$$C_2 = -\frac{1}{E_1} \frac{M_z^* I_{yy}^* - M_y^* I_{yz}^*}{I_{yy}^* I_{zz}^* - (I_{yz}^*)^2} \qquad C_3 = -\frac{1}{E_1} \frac{M_y^* I_{zz}^* - M_z^* I_{yz}^*}{I_{yy}^* I_{zz}^* - (I_{yz}^*)^2} \tag{7-18}$$

Substituting these results into Eq. (7-8) gives

$$\sigma_{xx} = \frac{E}{E_1} \left[\frac{P^*}{A^*} - \frac{M_z^* I_{yy}^* - M_y^* I_{yz}^*}{I_{yy}^* I_{zz}^* - (I_{yz}^*)^2} \, y - \frac{M_y^* I_{zz}^* - M_z^* I_{yz}^*}{I_{yy}^* I_{zz}^* - (I_{yz}^*)^2} \, z - E_1 \alpha T \right] \tag{7-19}$$

Equation (7-19) is the general equation for the stresses in a non-homogeneous beam with axial and bending loads and a temperature change. The following special cases illustrate how Eq. (7-19) can be simplified when additional restrictions are imposed. If the beam is

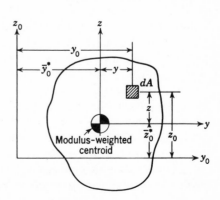

Fig. 7-2 Coordinate systems.

homogeneous and we take $E_1 = E$, the asterisks can be removed from the modulus-weighted section properties, and Eq. (7-19) becomes

$$\sigma_{xx} = \frac{P^*}{A} - \frac{M_z^* I_{yy} - M_y^* I_{yz}}{I_{yy}I_{zz} - I_{yz}^2} \, y - \frac{M_y^* I_{zz} - M_z^* I_{yz}}{I_{yy}I_{zz} - I_{yz}^2} \, z - E\alpha T \quad (7\text{-}20)$$

where A is the cross-sectional area, I_{yy} and I_{zz} are the moments of inertia about the y and z axes, and I_{yz} is the product of inertia for these same axes. The accuracy of this equation is discussed in Sec. 7-5. If there are no applied loads and no external restraints against thermal expansion, $P = M_y = M_z = 0$, and the stresses are entirely due to the thermal gradients. In this case Eq. (7-20) for the homogeneous beam becomes

$$\sigma_{xx} = \frac{P_T}{A} - \frac{M_{zT} I_{yy} - M_{yT} I_{yz}}{I_{yy}I_{zz} - I_{yz}^2} \, y - \frac{M_{yT} I_{zz} - M_{zT} I_{yz}}{I_{yy}I_{zz} - I_{yz}^2} \, z - E\alpha T \quad (7\text{-}21)$$

This equation has been derived and discussed in Ref. 1. On the other hand, if there are no temperature changes, Eq. (7-20) reduces to

$$\sigma_{xx} = \frac{P}{A} - \frac{M_z I_{yy} - M_y I_{yz}}{I_{yy}I_{zz} - I_{yz}^2} \, y - \frac{M_y I_{zz} - M_z I_{yz}}{I_{yy}I_{zz} - I_{yz}^2} \, z \quad (7\text{-}22)$$

If the y and z axes are chosen so that $I_{yz}^* = 0$, Eq. (7-19) becomes

$$\sigma_{xx} = \frac{E}{E_1} \left(\frac{P^*}{A^*} - \frac{M_z^* y}{I_{zz}^*} - \frac{M_y^* z}{I_{yy}^*} - \alpha E_1 T \right) \quad (7\text{-}23)$$

When the beam is homogeneous and $T = P = M_z = 0$, Eq. (7-23) simplifies to the familiar strength of materials equation

$$\sigma_{xx} = - \frac{M_y z}{I_{yy}} \quad (7\text{-}24)$$

Additional special cases can be readily found from the general equation (7-19) in a similar manner.

It is always possible to find a set of *principal axes* through the modulus-weighted centroid which have the property that $I_{yz}^* = 0$. The simplicity of Eq. (7-23) suggests that computations can be simplified if these axes are used. However, unless principal axis are obvious by inspection, this is not the case. In general the computation of the angle to the principal axes, the principal modulus-weighted moments of inertia, and the principal coordinates of points where stresses are to be determined consumes more time than using the equation for arbitrary modulus-weighted centroidal axes. From the last of Eqs. (7-12) we see that $I_{yz}^* = 0$ if either the y or z axis is an axis of geometric and elastic symmetry.

The application of Eq. (7-19) is illustrated in Example 7-1, which follows Sec. 7-6. References 2 and 3 contain numerous examples of the use of Eq. (7-22) when the axial force is zero; Ref. 3 also applies the special form of Eq. (7-19) which results when $P = T = 0$ to the bending of non-homogeneous beams.

7-4 MODULUS–WEIGHTED SECTION PROPERTIES

The modulus-weighted section properties are defined by integrals in Eqs. (7-10) to (7-12). These equations are inconvenient when E is not an analytic function of y and z or when the boundary of the cross section complicates the integrations. In these instances the integrals must be evaluated numerically. In some cases the trapezoidal or Simpson's rule approximations (Fig. 5-5) may be useful, while in other situations it may be more convenient to approximate the integrals by simple summations.

We noted in Chap. 3 that E is a function of temperature; therefore a beam with large thermal gradients will be elastically nonhomogeneous even if it is made from a single material. In some cases the beam may be a composite of different materials, so that E is piecewise constant. The cross section of a beam of this type is shown in Fig. 7-3, where within each subarea A_i the modulus is constant and equal to E_i. As a result, Eq. (7-10) becomes

$$A^* = \sum_{i=1}^{n} A_i^* \tag{7-25}$$

where n is the number of subareas and

$$A_i^* = \frac{E_i}{E_1} A_i \tag{7-26}$$

To determine the location of the modulus-weighted centroid we begin with arbitrary y_0, z_0 axes as shown in Fig. 7-3 and let $\bar{y}_i$ and $\bar{z}_i$ be

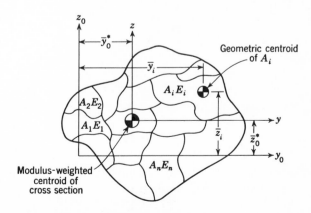

Fig. 7-3 Cross section of a nonhomogeneous beam.

the coordinates of the geometric centroid of A_i. From Eq. (7-16a) we find

$$\bar{y}_0^* = \frac{\sum\limits_{i=1}^{n} \dfrac{E_i}{E_1} \int_{A_i} y_0 \, dA}{\sum\limits_{i=1}^{n} \dfrac{E_i}{E_1} \int_{A_i} dA}$$

which reduces to

$$\bar{y}_0^* = \frac{1}{A^*} \sum_{i=1}^{n} \bar{y}_i A_i^* \tag{7-27a}$$

In a similar manner, from Eq. (7-16b) we find

$$\bar{z}_0^* = \frac{1}{A^*} \sum_{i=1}^{n} \bar{z}_i A_i^* \tag{7-27b}$$

From the first of Eqs. (7-12) the weighted moment of inertia about the y_0 axis is

$$I_{y_0 y_0}^* = \int_A z_0^2 \, dA^* = \sum_{i=1}^{n} \frac{E_i}{E_1} \int_{A_i} z_0^2 \, dA = \sum_{i=1}^{n} \frac{E_i}{E_1} I_{y_0 y_0_i}$$

where $I_{y_0 y_0_i}$ is the moment of inertia of A_i about the y_0 axis. By using the transfer-of-axis equation for moments of inertia, $I_{y_0 y_0}^*$ becomes

$$I_{y_0 y_0}^* = \sum_{i=1}^{n} \frac{E_i}{E_1} (\bar{I}_{yy_i} + \bar{z}_i^2 A_i) \tag{7-28a}$$

where $\bar{I}_{yy_i}$ is the geometric moment of inertia of A_i about its own centroidal axis. In a similar fashion we find

$$I_{z_0 z_0}^* = \sum_{i=1}^{n} \frac{E_i}{E_1} (\bar{I}_{zz_i} + \bar{y}_i^2 A_i) \tag{7-28b}$$

$$I_{y_0 z_0}^* = \sum_{i=1}^{n} \frac{E_i}{E_1} (\bar{I}_{yz_i} + \bar{y}_i \bar{z}_i A_i) \tag{7-28c}$$

Noting that $z_0 = z + \bar{z}_0^*$, we may also write

$$I_{y_0 y_0}^* = \int_A (z + \bar{z}_0^*)^2 \, dA^* = I_{yy}^* + 2\bar{z}_0^* \bar{z}^* A^* + (\bar{z}_0^*)^2 A^*$$

Observing that $\bar{z}^* = 0$ (because the z axis passes through the modulus-weighted centroid), we find

$$I_{yy}^* = I_{y_0 y_0}^* - (\bar{z}_0^*)^2 A^* \tag{7-29a}$$

By following the same procedure we obtain

$$I^*_{zz} = I^*_{z_0 z_0} - (\bar{y}^*_0)^2 A^* \qquad (7\text{-}29b)$$

$$I^*_{yz} = I^*_{y_0 z_0} - \bar{y}^*_0 \bar{z}^*_0 A^* \qquad (7\text{-}29c)$$

The similarity of Eqs. (7-27) to (7-29) to the equations for the corresponding geometric section properties is immediately obvious, and it is seen that the equations reduce to the relationships for the geometric section properties when the beam is homogeneous and E_1 is taken equal to E. The reference modulus E_1 was introduced to make this simple reduction possible.

In computing I^*_{yy}, I^*_{zz}, and I^*_{yz} it is usually preferable to (1) find the weighted section properties for convenient y_0 and z_0 axes from Eqs. (7-28), (2) determine $\bar{y}^*_0$ and $\bar{z}^*_0$ from Eqs. (7-27), and (3) find the properties for the y and z axes from Eqs. (7-29). The procedure is illustrated in Example 7-1. If the location of the weighted centroid is known by inspection, the weighted properties are of course evaluated for the y and z axes directly. The same procedures and equations apply to homogeneous beams, in which case the asterisks can be dropped from the properties. The preceding equations can also be used for beams with temperature-dependent moduli. In this case the cross section is subdivided into small areas A_i, and E_i is taken equal to the modulus at the mean temperature of A_i.

7-5 ACCURACY OF THE BEAM–STRESS EQUATION

In general, Eq. (7-19) gives only an approximate solution for the longitudinal stresses in a beam. However, in some situations the results are exact, and in this section we shall determine the circumstances for which this is the case. To do this we shall determine the restrictions which must be imposed upon the beam-theory solution to have it satisfy the theory-of-elasticity equations. In our development of the theory of elasticity we assumed that the body was homogeneous, and so we shall limit our examination to Eq. (7-20), which is for the homogeneous beam.

Recalling that all stresses except σ_{xx} are assumed to be negligible, we find that Eq. (7-20) satisfies the equilibrium equations (2-8b) and (2-8c) if $Y = Z = 0$. The remaining equilibrium equation (2-8a) is satisfied if the cross section is constant, $dP/dx = dM^*_y/dx = dM^*_z/dx = X = 0$, and for a temperature-independent α if

$$\frac{1}{A} \int_A \frac{\partial T}{\partial x}\, dA = \frac{\partial T}{\partial x}$$

This condition is fulfilled if $T = T_0 x + T_1(y,z)$, where T_0 is a constant.

With these restrictions the compatibility equation (4-6c) reduces to $\partial^2 T_1/\partial y^2 + \partial^2 T_1/\partial z^2 = 0$, which requires that T_1 be plane harmonic in the cross-sectional coordinates. The remaining compatibility conditions for a singly connected cross section [Eqs. (4-6a), (4-6b), and (4-6d) to (4-6f)] provide no additional restrictions.

On the cylindrical surface of the beam the direction cosine $l = 0$, and it follows from Eqs. (2-10) that it is necessary that $\bar{X} = \bar{Y} = \bar{Z} = 0$ on this surface. On the end surfaces $m = n = 0$, so that Eqs. (2-10) require that $\bar{Y} = \bar{Z} = 0$ and that $\bar{X} = \sigma_{xx}$ on the right end while $\bar{X} = -\sigma_{xx}$ on the left end. This requires the distribution of the surface loads on the ends to be the same as the stress distribution computed from Eq. (7-20). It is seldom that these end conditions are satisfied, but from St. Venant's principle we see that this will cause significant errors only in the regions of the ends. In the end regions, however, the results may be seriously in error. For example, Eq. (7-20) predicts stresses at a free end of a beam with a thermal gradient, when in reality the end stresses are zero. However, in this case $P = M_y = M_z = 0$, and at distances from the ends greater than several cross-sectional dimensions it is unimportant whether $\bar{X} = 0$ or the resultants of $\bar{X}$ are zero. Since discrepancies often occur in the end conditions, the beam-theory solution is usually applicable only to long beams in the region away from the ends.

We see that the Bernoulli-Euler assumptions lead to an exact solution for a homogeneous linearly elastic uniform beam with a simply connected cross section if (1) the axial force and bending moments are constant, (2) T is plane harmonic in the y and z coordinates and linear in x, and (3) the external forces are applied only over the ends, where their distribution is the same as that of the internal stresses. The restriction that $T_1(y,z)$ be plane harmonic is an interesting one in that it is shown in texts on heat transfer that $\nabla^2 T = 0$ for steady-state temperatures in a homogeneous two-dimensional solid if no heat is added or removed at interior points.

That the solution may be incorrect for a beam with a multiply connected cross section can be seen by referring to Fig. 7-4a, where the steady-state temperature distribution through a thick pipe is shown. In this case the inner diameter tries to expand more than the outer diameter, but it is restrained by continuity from doing so. It is easy to see that tangential and radial stresses are developed in addition to the longitudinal stresses, and the uniaxial stress-strain law used in the development of Eq. (7-20) is incorrect. As another example where sizable errors occur, consider Fig. 7-4b, where the beam is nonhomogeneous, with one material in the center and a second material with a different coefficient of expansion on the outside. It is seen here that even a uniform temperature change will produce tangential and radial stresses.

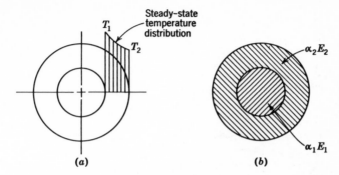

Fig. 7-4 Examples where beam theory produces significant errors. (*a*) Heated multiply connected beams; (*b*) heated nonhomogeneous beam.

Although we have shown that the results are exact only if the axial force and moment are constant along the length, the equations are sufficiently accurate for engineering calculations if these quantities change gradually. The theory is exact only for beams of uniform cross section; however, if the cross section of the beam tapers gradually, the errors are not significant. An additional discussion of tapered beams is given in Chap. 9. Gross errors should be expected, however, in the region of concentrated lateral forces or at cross-sectional discontinuities in the beam.

The equations of Sec. 7-3 are often used in situations where their accuracy is questionable to obtain approximate results for preliminary design purposes. In such cases the expense of a more accurate solution may not be justified because of the tentative nature of the work. However, a thorough understanding of the theory is required to interpret the results intelligently.

7-6 IDEALIZATION OF STIFFENED–SHELL STRUCTURES

The structure of a flight vehicle usually has a dual function: it transmits and resists the forces which are applied to the vehicle, and it acts as a cover which provides the aerodynamic shape and protects the contents of the vehicle from the environment. This combination of roles is fortunate since, from the standpoint of structural weight, the most efficient location for the structural material is at the outer surface of the vehicle. As a result, the structures of most flight vehicles are essentially thin shells. If these shells are not supported by stiffening members, they are referred to as *monocoque*. When the cross-sectional dimensions are large, the wall of a monocoque structure must be relatively thick to resist bending, com-

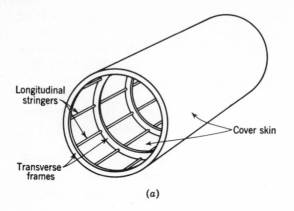

Longitudinal
stringers

Cover skin

Transverse
frames

(a)

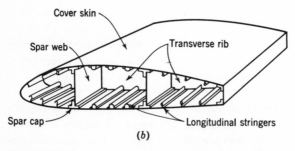

Cover skin

Spar web

Transverse rib

Spar cap

Longitudinal stringers

(b)

Fig. 7-5 Typical semimonocoque construction.   (a) Body
structures; (b) aerodynamic surface structures.

pressive, and torsional loads without buckling.   In such cases a more
efficient type of construction is one which contains stiffening members
that permit a thinner covering shell.   Stiffening members may also be
required to diffuse concentrated loads into the cover.   Constructions of
this type are called *semimonocoque*.   Typical examples of semimonocoque
body structures and aerodynamic surfaces are shown in Fig. 7-5.   While
at first glance these structures appear to differ considerably, functionally
there are similarities.   Both have thin-sheet coverings, longitudinal
stiffening members, and transverse supporting elements which play
similar structural roles.

In semimonocoque structures the *cover*, or *skin*, has the following
functions:

1. It transmits aerodynamic forces to the longitudinal and transverse
   supporting members by plate and membrane action (Chap. 13).
2. It develops shearing stresses which react the applied torsional moments
   (Chap. 8) and shear forces (Chap. 9).

3. It acts with the longitudinal members in resisting the applied bending and axial loads (Chaps. 7, 15, and 16).
4. It acts with the longitudinals in resisting the axial load and with the transverse members in reacting the hoop, or circumferential, load when the structure is pressurized.

In addition to these structural functions, it provides an aerodynamic surface and cover for the contents of the vehicle. *Spar webs* (Fig. 7-5*b*) play a role that is similar to function 2 of the skin.

The longitudinal members are known as *longitudinals, stringers,* or *stiffeners*. Longitudinals which have large cross-sectional areas are referred to as *longerons*. These members serve the following purposes:

1. They resist bending and axial loads along with the skin (Chap. 7).
2. They divide the skin into small panels and thereby increase its buckling and failing stresses (Chaps. 15 and 16).
3. They act with the skin in resisting axial loads caused by pressurization.

The *spar caps* in an aerodynamic surface perform functions 1 and 2.

The transverse members in body structures are called *frames, rings,* or if they cover all or most of the cross-sectional area, *bulkheads*. In aerodynamic surfaces they are referred to as *ribs*. These members are used to:

1. Maintain the cross-sectional shape.
2. Distribute concentrated loads into the structure and redistribute stresses around structural discontinuities (Chap. 9).
3. Establish the column length and provide end restraint for the longitudinals to increase their column buckling stress (Chap. 14).
4. Provide edge restraint for the skin panels and thereby increase the plate buckling stress of these elements (Chap. 16).
5. Act with the skin in resisting the circumferential loads due to pressurization.

The behavior of these structural elements is often idealized to simplify the analysis of the assembled component. The following assumptions are usually made:

1. The longitudinals carry only axial stresses.
2. The webs (skin and spar webs) carry only shearing stresses.
3. The axial stress is constant over the cross section of each of the longitudinals, and the shearing stress is uniform through the thickness of the webs.

4. The transverse frames and ribs are rigid within their own planes, so that the cross section is maintained unchanged during loading. However, they are assumed to possess no rigidity normal to their plane, so that they offer no restraint to warping deformations out of their plane.

When the cross-sectional dimensions of the longitudinals are very small compared to the cross-sectional dimensions of the assembly, assumptions 1 and 3 result in little error. The webs in an actual structure carry significant axial stresses as well as shearing stresses, and it is therefore necessary to use an analytical model of the structure which includes this load-carrying ability. This is done by combining the effective areas of the webs adjacent to a longitudinal with the area of the longitudinal into a *total effective area* of material which is capable of resisting bending moments and axial forces. A method for determining this effective area is given in Sec. 15-7. In the illustrative examples and problems on stiffened shells in this and succeeding chapters it may be assumed that this idealization has already been made and that areas given for the longitudinals are the total effective areas. The fact that the cross-sectional dimensions of most longitudinals are small compared with those of the stiffened-shell cross section makes it possible to assume without serious error that the area of the effective longitudinal is concentrated at a point on the midline of the skin where it joins the longitudinal. The locations of these idealized longitudinals will be indicated by small circles, as shown in Fig. 7-6b. In thin aerodynamic surfaces the depth of the longitudinals may not be small compared to the thickness of the cross section of the assembly, and a more elaborate idealized model of the structure may be required.

The fewer the number of longitudinals, the simpler the analysis, and in some cases several longitudinals may be lumped into a single effective longitudinal to shorten computations (Fig. 7-6). On the other hand, it is

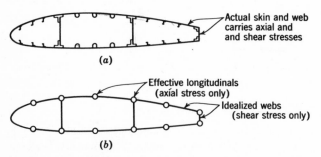

Fig. 7-6 Idealization of semimonocoque structure.   (a) Actual structure; (b) idealized structure.

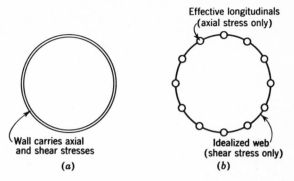

Fig. 7-7 Idealization of a monocoque shell. (*a*) Monocoque shell; (*b*) idealization.

sometimes convenient to idealize a monocoque shell into an idealized stiffened shell by lumping the shell wall area into idealized longitudinals, as shown in Fig. 7-7, and assuming that the skin between these longitudinals carries only shearing stresses. The simplification of an actual structure into an analytical model represents a compromise, since elaborate models which more nearly simulate the actual structure are usually difficult to analyze. A more complete discussion of the idealization of shell structures will be found in Ref. 4.

Once the idealization is made, the stresses in the longitudinals due to bending moments, axial load, and thermal gradients can be computed from the equations of this chapter if the structure is long compared to its cross-sectional dimensions and if there are no significant structural or loading discontinuities in the region where the stresses are computed. In many flight structures the cross section tapers; the effects of this taper upon the stresses are discussed in Chap. 9. When discontinuities or other conditions arise which violate the analytical assumptions made in the Bernoulli-Euler theory, it is necessary to analyze the stiffened shell as an indeterminate structure (Chaps. 11 and 12).

Example 7-1 The cross section shown in Fig. 7-8*a* is a simplified representation of the construction used in the aerodynamic surfaces of flight vehicles. The longitudinal stresses are desired when the beam is simultaneously subjected to the temperatures shown and to a bending moment about the horizontal axis of 10^6 in.-lb. It is assumed that the compression skin remains unbuckled and is therefore fully effective.

    To simplify the calculations we idealize the structure into longitudinals that carry only axial stress and webs that sustain only shear stress (the shear stresses will be computed in Example 9-4). In making the idealization we could divide the skin and spar webs into segments and lump the area of each segment at its centroid. A large number of segments would be required to

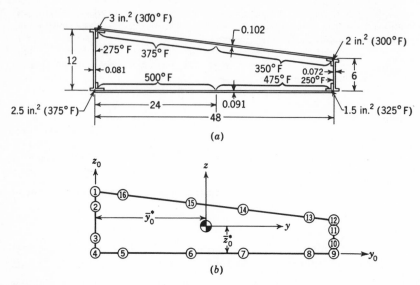

(a)

(b)

Fig. 7-8 Example 7-1. (a) Actual structure, 2024 aluminum alloy; (b) idealized structure.

obtain a good approximation of the moment of inertia with this method. A procedure which is often used on spar webs (when there are no thermal stresses) is to apply one-sixth of the web area with each of the spar-cap areas, thereby maintaining the same moment of inertia of the actual and idealized spar. This cannot be done in this problem, because in computing the thermal stresses we must preserve the area as well as the moment of inertia.

The spar-web idealization shown in Fig. 7-9 was chosen to keep the number of effective longitudinals to a minimum. The webs are lumped so that the idealized section has the same area, centroid, and moment of inertia as the

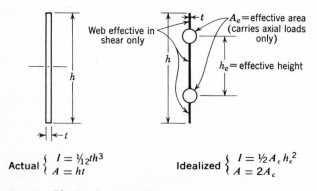

Actual $\begin{cases} I = \frac{1}{12}th^3 \\ A = ht \end{cases}$ Idealized $\begin{cases} I = \frac{1}{2}A_e h_e^2 \\ A = 2A_e \end{cases}$

Fig. 7-9 Idealization of a web.

original section. By equating the areas and moments of inertia of the actual and the idealized webs we find $A_e = th/2$ and $h_e = h/\sqrt{3}$. This same method was used to idealize the top and bottom skins. However, because the temperatures are not uniform along the full lengths of these members, each half of the skins was idealized into two lumped areas using the above formulas. The idealized cross section of the beam is shown in Fig. 7-8b; and the areas, coordinates, and temperatures of the idealized longitudinals are given in columns 2 to 4 of Table 7-1. (Circled numbers, e.g., ⑫, are used to represent column numbers.)

The modulus-weighted section properties of the structure are computed in ⑥ to ⑫. The room-temperature modulus of the beam material is taken as E_1. The ratio of the elevated temperature modulus to E_1 and the modulus-weighted area of each of the longitudinals are listed in ⑥ and ⑦. The first and second moments of the weighted areas with respect to the arbitrarily chosen y_0 and z_0 axes shown in Fig. 7-8b are given in ⑧ to ⑫. The weighted moments and products of inertia of the elements about their own centroidal axes are considered negligible compared to the transfer-of-axes terms. From the tabulation we find

$$A^* = \Sigma A_i^* = \Sigma ⑦ = 17.90 \text{ in.}^2$$

$$\bar{y}_0^* = \frac{1}{A^*}\Sigma \bar{y}_i A_i^* = \frac{1}{A^*}\Sigma ⑧ = \frac{373.9}{17.90} = 20.89 \text{ in.}$$

$$\bar{z}_0^* = \frac{1}{A^*}\Sigma \bar{z}_i A_i^* = \frac{1}{A^*}\Sigma ⑨ = \frac{93.4}{17.90} = 5.22 \text{ in.}$$

$$I_{yy}^* = I_{y_0 y_0}^* - (\bar{z}_0^*)^2 A^* = \Sigma ⑪ - (\bar{z}_0^*)^2 A^*$$
$$= 909.9 - 5.22^2 \times 17.90 = 422.2 \text{ in.}^4$$

$$I_{zz}^* = I_{z_0 z_0}^* - (\bar{y}_0^*)^2 A^* = \Sigma ⑩ - (\bar{y}_0^*)^2 A^*$$
$$= 14,846 - 20.89^2 \times 17.90 =: 7035 \text{ in.}^4$$

$$I_{yz}^* = I_{y_0 z_0}^* - \bar{y}_0^* \bar{z}_0^* A^* = \Sigma ⑫ - \bar{y}_0^* \bar{z}_0^* A^*$$
$$= 1483.4 - 20.89 \times 5.22 \times 17.90 = -468.6 \text{ in.}^4$$

The equivalent thermal loads are computed in ⑬ to ㉑. From these

$$P^* = P + P_T = P + \Sigma ⑲ = 0 + 694.1 \times 10^3 = 0.6941 \times 10^6 \text{ lb}$$
$$M_y^* = M_y + M_{yT} = M_y - \Sigma ㉑ = 10^6 + 440.8 \times 10^3 = 1.441 \times 10^6 \text{ in.-lb}$$
$$M_z^* = M_z + M_{zT} = M_z - \Sigma ⑳ = 0 + 103 \times 10^3 = 0.103 \times 10^6 \text{ in.-lb}$$

From Eqs. (7-17) and (7-18) we find

$$C_1 = \frac{0.6941 \times 10^6}{10.5 \times 10^6 \times 17.90} = 3.692 \times 10^{-3}$$

$$C_2 = -\frac{1}{10.5 \times 10^6} \frac{0.103 \times 10^6 \times 422.2 + 1.441 \times 10^6 \times 468.6}{422.2 \times 7035 - 468.6^2}$$
$$= -24.87 \times 10^{-6}$$

$$C_3 = -\frac{1}{10.5 \times 10^6} \frac{1.441 \times 10^6 \times 7035 + 0.103 \times 10^6 \times 468.6}{422.2 \times 7035 - 468.6^2}$$
$$= -352.4 \times 10^{-6}$$

The stresses are computed in ㉒ to ㉕ by using Eq. (7-8).

Table 7-1   Example 7-1

① Element i	② A_i, in.²	③ $\bar{y}_i$, in.	④ $\bar{z}_i$, in.	⑤ Temp., °F†	⑥ E_i/E_1‡	⑦ A_i^*, in.² (② × ⑥)	⑧ $\bar{y}_i A_i^*$, in.³ (③ × ⑦)	⑨ $\bar{z}_i A_i^*$, in.³ (④ × ⑦)	⑩ $\bar{y}_i^2 A_i^*$, in.⁴ (③ × ⑧)	⑪ $\bar{z}_i^2 A_i^*$, in.⁴ (④ × ⑨)	⑫ $\bar{y}_i \bar{z}_i A_i^*$, in.⁴ (③ × ⑨)	⑬ y_i, in. (③ − 20.89)	⑭ z_i, in. (④ − 5.22)	⑯ T_i, °F (⑤ − 70)
1	3.00	0.00	12.00	300	0.95	2.85	0.0	34.2	0	410.4	0.0	−20.89	6.78	230
2	0.49	0.00	9.47	275	0.96	0.47	0.0	4.5	0	42.6	0.0	−20.89	4.25	205
3	0.49	0.00	2.54	275	0.96	0.47	0.0	1.2	0	3.0	0.0	−20.89	−2.68	205
4	2.50	0.00	0.00	375	0.92	2.30	0.0	0.0	0	0.0	0.0	−20.89	−5.22	305
5	1.09	5.07	0.00	500	0.80	0.87	4.4	0.0	22	0.0	0.0	−15.82	−5.22	430
6	1.09	18.93	0.00	500	0.80	0.87	16.5	0.0	312	0.0	0.0	−1.96	−5.22	430
7	1.09	29.07	0.00	475	0.83	0.90	26.2	0.0	762	0.0	0.0	8.18	−5.22	405
8	1.09	42.93	0.00	475	0.83	0.90	38.6	0.0	1657	0.0	0.0	22.04	−5.22	405
9	1.50	48.00	0.00	325	0.94	1.41	67.7	0.0	3250	0.0	0.0	27.11	−5.22	255
10	0.22	48.00	1.27	250	0.97	0.21	10.1	0.3	485	0.4	14.4	27.11	−3.95	180
11	0.22	48.00	4.73	250	0.97	0.21	10.1	1.0	485	4.7	48.1	27.11	−0.49	180
12	2.00	48.00	6.00	300	0.95	1.90	91.2	11.4	4378	68.4	547.2	27.11	0.78	230
13	1.23	42.93	6.63	350	0.93	1.14	48.9	7.6	2099	50.4	326.3	22.04	1.41	280
14	1.23	29.07	8.37	350	0.93	1.14	33.1	9.5	962	79.5	276.2	8.18	3.15	280
15	1.23	18.93	9.63	375	0.92	1.13	21.4	10.9	405	105.0	206.3	−1.96	4.41	305
16	1.23	5.07	11.37	375	0.92	1.13	5.7	12.8	29	145.5	64.9	−15.82	6.15	305
Sum						17.90	373.9	93.4	14846	909.9	1483.4			

† Initial temperature is 70°F.
‡ E_i/E_1 from Fig. 3.2.3.14, MIL-HDBK-5 taking $E_1 = E_{\text{room temp}}$.

Table 7-1 Example 7-1 (continued)

① Element i	⑯ $(\alpha T)_i$, 10^{-6} in./in. $12.6 \times$ ⑮	⑰ E_i, 10^6 psi $10.5 \times$ ⑥	⑱ $(E\alpha T)_i$, ksi ⑯ $\times$ ⑰ $\times 10^{-3}$	⑲ $(E\alpha TA)_i$, 10^3 lb ② $\times$ ⑱	⑳ $(E\alpha TAy)_i$, 10^3 in.-lb ⑬ $\times$ ⑲	㉑ $(E\alpha TAz)_i$, 10^3 in.-lb ⑭ $\times$ ⑲	㉒ $C_2 y_i$, 10^{-6} in./in. $-24.87 \times$ ⑬	㉓ $C_3 z_i$, 10^{-6} in./in. $-352.4 \times$ ⑭	㉔ $\dfrac{\sigma_{zz}}{E_i}$, 10^{-6} in./in. §	㉕ σ_{zz}, ksi ⑰ $\times$ ㉔
1	2900	10.0	29.0	87.0	−1817	589.9	520	−2389	−1077	−10.8
2	2580	10.1	26.1	12.8	− 267	54.4	520	−1498	134	1.4
3	2580	10.1	26.1	12.8	− 267	− 34.3	520	944	2576	26.0
4	3840	9.7	37.2	93.0	−1943	−485.5	520	1840	2212	21.5
5	5420	8.4	45.5	49.6	− 785	−258.9	393	1840	505	4.2
6	5420	8.4	45.5	49.6	97	−258.9	49	1840	161	1.4
7	5100	8.7	44.3	48.3	395	−252.1	−203	1840	229	2.0
8	5100	8.7	44.3	48.3	1065	−252.1	−548	1840	116	1.0
9	3210	9.9	31.8	47.7	1293	−249.0	−674	1840	1648	16.3
10	2270	10.2	23.2	5.1	138	− 15.0	−674	1392	2140	21.8
11	2270	10.2	23.2	5.1	138	− 2.5	−674	173	921	9.4
12	2900	10.0	29.0	58.0	1572	45.2	−674	− 275	− 157	− 1.6
13	3530	9.8	34.6	42.6	939	60.1	−548	− 497	− 883	− 8.7
14	3530	9.8	34.6	42.6	348	134.2	−203	−1110	−1142	−11.2
15	3840	9.7	37.2	45.8	− 90	202.0	49	−1554	−1653	−16.0
16	3840	9.7	37.2	45.8	− 725	281.7	393	−2167	−1922	−18.6
Sum				694.1	− 103	−440.8				

§ $\sigma_{zz}/E_i = 3.692 \times 10^{-3} +$ ㉒ $+$ ㉓ $-$ ⑯.

7-7 EQUILIBRIUM EQUATIONS

The longitudinal stresses in a beam can be found from Eq. (7-19) when P, M_y, and M_z are known.   However, it is first necessary to obtain these stress resultants from the applied loads.   In this section we derive the equilibrium differential equations which relate the stress resultants and the loads per unit length that are applied to the beam.   The equations are useful in constructing shear, moment, axial-force, and torque curves and are the basis for the equilibrium equations in terms of displacements which are developed in Secs. 7-9 and 8-9.

In deriving the equilibrium equations in the theory of elasticity we ignored the geometric changes that accompany deformations.   This is usually permissible when the body is compact and the strains are elastic, but it is often necessary to consider the effects of deformation in writing the equilibrium equations for slender beams, and for generality we shall include these effects.   The stresses in a beam cause strains, and in the general case the beam undergoes stretching, bending, shearing, and twisting deformations.   We shall assume that the beam is slender, so that the shearing deformations are negligible compared to the bending deformations.

Let us focus our attention upon points O and A on the axis of the beam with coordinates of $(x,0,0)$ and $(x + dx, 0, 0)$, respectively, prior to deformation (Fig. 7-1a).   In studying the beam deformations it is convenient to refer to *local* sets of axes x_1, y_1, z_1 and x_2, y_2, z_2, which are parallel to the x, y, and z axes when the beam is undeformed and which have origins at O and A, respectively.   We assume that these axes are fixed in the beam, so that they displace and rotate with the beam when it stretches, bends, and twists.   As the beam is loaded and heated, the

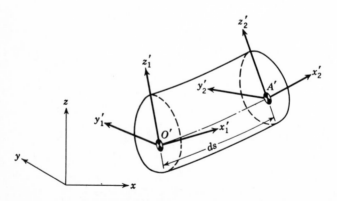

Fig. 7-10   Local axes of the deformed beam at O' and A'.

points O and A displace to O' and A', and the local axes translate and rotate to the new directions x_1', y_1', z_1' and x_2', y_2', z_2', as shown in Fig. 7-10. In doing so the local axes at O rotate through angles of φ, θ_y, θ_z relative to the original x_1, y_1, z_1 axes. Except for θ_y, these angles follow the right-hand sign convention. The angles φ, θ_y, and θ_z are the rotations of the cross-sectional plane through O due to twisting about the x_1 axis, bending about the y_1 axis, and bending about the z_1 axis, respectively.

In a similar fashion the x_2, y_2, z_2 axes undergo rotations of $\varphi + d\varphi$, $\theta_y + d\theta_y$, and $\theta_z + d\theta_z$ to x_2', y_2', z_2'. The differential angles $d\varphi$, $d\theta_y$, and $d\theta_z$ are infinitesimals; therefore their cosines are unity, and their sines are equal to the angles. As a result, we find the following direction cosines for the angles between the x_1', y_1', z_1', and x_2', y_2', z_2' axes:

	x_2'	y_2'	z_2'.
x_1'	1	$-d\theta_z$	$-d\theta_y$
y_1'	$d\theta_z$	1	$-d\varphi$
z_1'	$d\theta_y$	$d\varphi$	1

At any cross section we refer the stress resultants to the local axes that are fixed to the beam; and so the vectors associated with the force components P, V_y, V_z and the moment components M_t, M_y, M_z are parallel to the x_1', y_1', z_1' axes, as shown in Fig. 7-11. The stress resultants are functions of position along the length of the beam, and at A' they have the values $P + (dP/ds)\,ds$, $V_y + (dV_y/ds)\,ds$, etc., where ds is the length of the line segment $O'A'$. We assume that the beam is subjected to an applied force and moment per unit length and designate the force com-

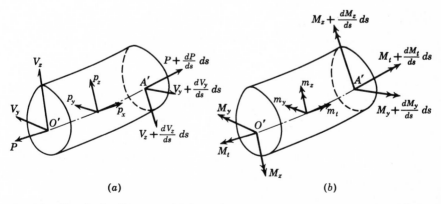

Fig. 7-11 Free-body diagrams of deformed-beam segment $O'A'$. (a) Applied forces per unit length and force stress resultants; (b) applied moments per unit length and moment stress resultants.

ponents by p_x, p_y, p_z and the moment components by m_t, m_y, m_z. These components are parallel to the local axes that are fixed to the beam and are defined positive when the vector components are in the directions of the positive axes (Fig. 7-11).

To determine the equilibrium equations we treat the beam segment $O'A'$ as a free body and require that the sums of the forces in the directions of the x_1', y_1', z_1' axes and the moments about these axes vanish. Summing forces in the x_1' direction in Fig. 7-11a and using the previously determined direction cosine for the angles between the x_2', y_2', z_2' and x_1', y_1', z_1' axes, we find

$$-P + \left(P + \frac{dP}{ds}\,ds\right) + \left(V_y + \frac{dV_y}{ds}\,ds\right) d\theta_z + \left(V_z + \frac{dV_z}{ds}\,ds\right) d\theta_y$$
$$+\ p_x\,ds = 0$$

Dividing by ds and taking the limit as ds approaches zero, we obtain

$$\frac{dP}{ds} + V_y\frac{d\theta_z}{ds} + V_z\frac{d\theta_y}{ds} + p_x = 0 \qquad (7\text{-}30)$$

The derivatives $d\theta_z/ds$ and $d\theta_y/ds$ are the curvatures of the projections of the deflected axis of the beam into the xy and xz planes, respectively. Therefore

$$\frac{d\theta_z}{ds} = \frac{1}{R_y} \qquad \frac{d\theta_y}{ds} = \frac{1}{R_z} \qquad (7\text{-}31)$$

where R_y and R_z are the radii of curvature of the projections in the xy and xz planes. By using Eqs. (7-31), Eq. (7-30) becomes

$$\frac{dP}{ds} + \frac{V_y}{R_y} + \frac{V_z}{R_z} + p_x = 0 \qquad (7\text{-}32a)$$

In a similar fashion, by summing forces in the y_1' and z_1' directions we find, respectively, that

$$-\frac{dV_y}{ds} + \frac{P}{R_y} + V_z\frac{d\varphi}{ds} + p_y = 0 \qquad (7\text{-}32b)$$

$$-\frac{dV_z}{ds} + \frac{P}{R_z} - V_y\frac{d\varphi}{ds} + p_z = 0 \qquad (7\text{-}32c)$$

Referring to Fig. 7-11b and summing moments about the x_1' axis, we obtain

$$-M_t + \left(M_t + \frac{dM_t}{ds}\,ds\right) + \left(M_y + \frac{dM_y}{ds}\,ds\right) d\theta_z - \left(M_z + \frac{dM_z}{ds}\,ds\right) d\theta_y$$
$$+\ m_t\,ds = 0$$

Dividing by ds, taking the limit as ds approaches zero, and using Eqs. (7-31) gives the result

$$\frac{dM_t}{ds} + \frac{M_y}{R_y} - \frac{M_z}{R_z} + m_t = 0 \qquad (7\text{-}32d)$$

By summing moments of the forces and moments in Fig. 7-11a and b about the y_1' axis we obtain

$$M_y - \left(M_y + \frac{dM_y}{ds} ds\right) + \left(V_z + \frac{dV_z}{ds} ds\right) ds - \left(M_z + \frac{dM_z}{ds} ds\right) d\varphi$$

$$+ \left(M_t + \frac{dM_t}{ds} ds\right) d\theta_z + m_y\, ds = 0$$

which upon dividing by ds and passing to the limit becomes

$$-\frac{dM_y}{ds} + V_z - M_z \frac{d\varphi}{ds} + \frac{M_t}{R_y} + m_y = 0 \qquad (7\text{-}32e)$$

In a similar manner, the summation of the moments about the z_1' axis gives

$$\frac{dM_z}{ds} - V_y - M_y \frac{d\varphi}{ds} + \frac{M_t}{R_z} + m_z = 0 \qquad (7\text{-}32f)$$

Equations (7-32) are the general forms of the equations of equilibrium of a beam with large deformations. The equations may be simplified when the displacements are small.

We shall use the notation $u(x)$, $v(x)$, $w(x)$ to designate the displacement components of the point O parallel to the x, y, and z axes. The functions v and w describe the shapes of the projections of the axis of the deformed beam in the xy and xz planes. From the calculus, the radii of curvature of the projections are related to v and w by

$$\frac{1}{R_y} = \frac{d^2v/dx^2}{[1 + (dv/dx)^2]^{3/2}} \qquad \frac{1}{R_z} = \frac{d^2w/dx^2}{[1 + (dw/dx)^2]^{3/2}} \qquad (7\text{-}33)$$

These formulas, which are nonlinear in v and w, lead to nonlinear equilibrium equations if they are substituted into Eqs. (7-32). If the deflections are small in the sense that $(dv/dx)^2 \ll 1$ and $(dw/dx)^2 \ll 1$, Eqs. (7-33) may be reduced to the approximations

$$\frac{1}{R_y} = \frac{d^2v}{dx^2} \qquad \frac{1}{R_z} = \frac{d^2w}{dx^2} \qquad (7\text{-}34)$$

Unless it is noted otherwise, we shall assume that the deflections are small enough to permit this simplification.

The shear forces in a beam are usually relatively small, and when the deflections are small, R_y and R_z are large and $d\varphi/ds$ is small. As a result, the products V_y/R_y, V_z/R_z, $V_y(d\varphi/ds)$, and $V_z(d\varphi/ds)$ in Eqs.

(7-32) are usually negligible. In addition, for small displacements the derivatives with respect to s may be replaced by derivatives with respect to x. Introducing these simplifications and using Eqs. (7-34), we obtain the following equations of equilibrium for small displacements:

$$\frac{dP}{dx} = -p_x \tag{7-35a}$$

$$\frac{dV_y}{dx} = p_y + P\frac{d^2v}{dx^2} \tag{7-35b}$$

$$\frac{dV_z}{dx} = p_z + P\frac{d^2w}{dx^2} \tag{7-35c}$$

$$\frac{dM_t}{dx} = -m_t - M_y\frac{d^2v}{dx^2} + M_z\frac{d^2w}{dx^2} \tag{7-35d}$$

$$\frac{dM_y}{dx} = V_z - M_z\frac{d\varphi}{dx} + M_t\frac{d^2v}{dx^2} + m_y \tag{7-35e}$$

$$\frac{dM_z}{dx} = V_y + M_y\frac{d\varphi}{dx} - M_t\frac{d^2w}{dx^2} - m_z \tag{7-35f}$$

The terms that involve displacements in Eqs. (7-35) can often be neglected. They are necessary, however, in problems involving buckling in compression, bending, or torsion, and they should also be used whenever the applied loads are not small compared to the buckling loads. When the loads are small compared to the buckling loads, the terms involving deflections can be neglected, and Eqs. (7-35) reduce to

$$\frac{dP}{dx} = -p_x \qquad \frac{dV_z}{dx} = p_z \qquad \frac{dV_y}{dx} = p_y$$

$$\frac{dM_t}{dx} = -m_t \qquad \frac{dM_y}{dx} = V_z + m_y \qquad \frac{dM_z}{dx} = V_y - m_z \tag{7-36}$$

These are the familiar beam-equilibrium equations derived by rigid-body statics in strength of materials.

It should be noted that in beam-theory equilibrium is enforced only at the macroscopic level of the stress resultants and not at the microscopic level of the stresses, as required in the theory of elasticity. In addition the loads per unit length p_x, p_y, p_z and m_t, m_y, m_z are applied at the beam axis, and no distinction is made between body forces and surface forces on the beam.

Some caution should be exercised in using Eqs. (7-35) or (7-36). The equations involve derivatives of the stress resultants and may be used only if the derivative of the stress resultant exists at the point where they are applied. The shear is discontinuous at the point of application of a concentrated lateral force, and as a result the equations involving the derivative of the shear cannot be applied at such a point. Likewise, care must be taken in integrating the equations if discontinuities exist in the integrand.

7-8 BEAM DEFLECTIONS

We now turn our attention to the deflections associated with P, M_y, M_z, and the temperature change. To obtain the deflections we again consider the points O and A on the axis of the beam which initially are a distance dx apart (Fig. 7-1). As the points deflect to O' and A', the length dx extends to $ds = [1 + \epsilon_{xx}(x,0,0)]\, dx$, where $\epsilon_{xx}(x,0,0)$ is the longitudinal strain at O. In practical cases $\epsilon_{xx}(x,0,0) \ll 1$, and we may say that $ds \approx dx$, in which case the second of Eqs. (7-31) becomes $1/R_z = d\theta_y/dx$. By using the last of Eqs. (7-6), (7-18), and (7-33) this relationship becomes

$$\frac{d^2w/dx^2}{[1 + (dw/dx)^2]^{3/2}} = \frac{1}{E_1} \frac{M_y^* I_{zz}^* - M_z^* I_{yz}^*}{I_{yy}^* I_{zz}^* - (I_{yz}^*)^2} \tag{7-37}$$

For small displacements, such that $(dw/dx)^2 \ll 1$, this differential equation may be linearized to

$$\frac{d^2w}{dx^2} = \frac{1}{E_1} \frac{M_y^* I_{zz}^* - M_z^* I_{yz}^*}{I_{yy}^* I_{zz}^* - (I_{yz}^*)^2} \tag{7-38a}$$

In a similar fashion we can show that

$$\frac{d^2v}{dx^2} = \frac{1}{E_1} \frac{M_z^* I_{yy}^* - M_y^* I_{yz}^*}{I_{yy}^* I_{zz}^* - (I_{yz}^*)^2} \tag{7-38b}$$

Special cases of Eqs. (7-38) are easily written. For a homogeneous beam we may take $E_1 = E$ and remove the asterisks from I_{yy}^*, I_{zz}^*, and I_{yz}^*. If there are no temperature changes, the asterisks can be dropped from M_y^* and M_z^*, or if $M_y = M_z = 0$, the terms reduce to M_{yT} and M_{zT}. We note from Eqs. (7-38) that a moment about one axis results in a lateral deflection in the direction of both axes if nonprincipal axes are used. If principal axes are used for a homogeneous beam with no temperature changes Eqs. (7-38) become

$$\frac{d^2w}{dx^2} = \frac{M_y}{EI_{yy}} \qquad \frac{d^2v}{dx^2} = \frac{M_z}{EI_{zz}} \tag{7-39}$$

which are the familar equations from strength of materials.

A differential equation which relates P^* to the longitudinal deflection u can also be written. From Eqs. (7-5) and (7-17) the strain at a point on the x axis is $\epsilon_{xx} = P^*/E_1 A^*$. This may be equated to ϵ_{xx} from Eqs. (2-22a); however, in doing so we observe that for a slender beam the lateral displacements v and w are usually much larger than the longitudinal displacement u. As a result we neglect $(du/dx)^2$ compared to

$(dv/dx)^2$ and $(dw/dx)^2$ and obtain

$$\frac{du}{dx} + \frac{1}{2}\left[\left(\frac{dv}{dx}\right)^2 + \left(\frac{dw}{dx}\right)^2\right] = \frac{P^*}{E_1 A^*} \tag{7-40}$$

where ordinary derivatives are used because u, v, and w are the deflection of points on the axis of the beam and are functions of x only. This non-linear equation reduces to the linear form

$$\frac{du}{dx} = \frac{P^*}{E_1 A^*} \tag{7-41}$$

when $(dv/dx)^2$ and $(dw/dx)^2$ are small compared to du/dx. In practical problems the nonlinear terms are important only when both ends of the beam are restrained against longitudinal motion so that the beam must stretch to deflect laterally. It will be shown in Example 7-2 that when both ends of the beam are longitudinally restrained, the lateral deflections must be small relative to the depth of the beam if the nonlinear terms are to be ignored.

When the beam is statically determinate, we can immediately find P^*, M_y^*, and M_z^* as functions of x from the applied loads and temperatures. The lateral deflections v and w can then be found by double integration of Eqs. (7-38), while u can be determined from a single integration of Eq. (7-41). The two constants of integration that arise from integrating Eq. (7-38a) or (7-38b) can be evaluated from two known conditions of slope and/or lateral displacement at the ends of the beam. The single constant of integration that results from integrating Eq. (7-41) is found from the longitudinal displacement that is specified at either end of the beam.

Equations (7-38) and (7-41) cannot be solved by direct integration when P, M_y, and M_z depend upon the displacements. In these cases we must determine the displacements from the differential equations that result from combining the equilibrium equations of Sec. 7-7 with the deformation equations derived in this section. These differential equations and associated boundary conditions are derived in the next section.

7-9 THE DIFFERENTIAL EQUATIONS OF BEAMS, BARS, AND CABLES

A straight member with a length which is large compared to its cross-sectional dimensions and which is subjected to loads and temperature changes that produce only extensional or twisting deformations is called a *bar*. A laterally loaded member that is long compared to its cross-sectional dimensions and which has a negligible bending rigidity is referred to as a *cable* or *string*. In this section we shall see that the differential equations and boundary conditions that govern the deformations of beams, bars, and cables can be derived by combining the equations of

Secs. 7-7 and 7-8.   To simplify the equations we shall use principal axes and assume that the deflections are small in the sense that $(v')^2 \ll 1$ and $(w')^2 \ll 1$, where primes are used to denote differentiation with respect to x.

From the last two of Eqs. (7-13) and Eqs. (7-38)

$$M_y = E_1 I_{yy}^* w'' - M_{yT} \qquad M_z = E_1 I_{zz}^* v'' - M_{zT} \qquad (7\text{-}42)$$

Placing these results into Eqs. (7-35e) and (7-35f) gives

$$V_y = (E_1 I_{zz}^* v'')' - V_{yT} - M_y \varphi' + M_t w'' + m_z \qquad (7\text{-}43a)$$

$$V_z = (E_1 I_{yy}^* w'')' - V_{zT} + M_z \varphi' - M_t v'' - m_y \qquad (7\text{-}43b)$$

where the thermal equivalents of shear forces are

$$V_{yT} = M_{zT}' \qquad (7\text{-}44a)$$

$$V_{zT} = M_{yT}' \qquad (7\text{-}44b)$$

Substituting Eqs. (7-43) into Eqs. (7-35b) and (7-35c) gives

$$(E_1 I_{zz}^* v'')'' - (M_y \varphi')' + (M_t w'')' - P v'' = p_y^* - m_z' \qquad (7\text{-}45)$$

$$(E_1 I_{yy}^* w'')'' + (M_z \varphi')' - (M_t v'')' - P w'' = p_z^* + m_y' \qquad (7\text{-}46)$$

where

$$p_y^* = p_y + p_{yT} \qquad p_z^* = p_z + p_{zT} \qquad (7\text{-}47)$$

$$p_{yT} = M_{zT}'' \qquad p_{zT} = M_{yT}'' \qquad (7\text{-}48)$$

From the first of Eqs. (7-13) and Eqs. (7-40) we find

$$P = E_1 A^* u' + \frac{E_1 A^*}{2} [(v')^2 + (w')^2] - P_T \qquad (7\text{-}49)$$

Substituting this result into Eq. (7-35a) gives

$$(E_1 A^* u')' + \tfrac{1}{2} \{ E_1 A^* [(v')^2 + (w')^2] \}' = -p_x^* \qquad (7\text{-}50)$$

where

$$p_x^* = p_x + p_{xT} \qquad p_{xT} = -P_T' \qquad (7\text{-}51)$$

Equations (7-45), (7-46), (7-50), and (8-74), which is derived in Sec. 8-9, are the equations which the displacements must satisfy to assure equilibrium of the deformed beam.   We note that in general these equations are coupled and must be solved simultaneously.   When the twisting moment is zero or negligible, the coupling occurs in the nonlinear terms of Eq. (7-50) and through P in Eqs. (7-45) and (7-46), which from Eq. (7-49) is seen to involve u, v, and w.   However, in practical problems the nonlinear terms are important only when longitudinal displacements are prevented at both ends of the beam and the beam must stretch to bend laterally. The nonlinear terms may be dropped from Eqs. (7-49) and (7-50) if either end of the beam is free to deflect longitudinally (the induced errors are on the order of terms that have already been neglected in the theory). The nonlinear terms may also be discarded if a large axial preload is applied to stretch the beam before the ends are fixed against axial motion. In this case P may be assumed constant and equal to the preload for small lateral deflections. It will be shown in Example 7-2 that the nonlinear terms may also be dropped for a beam with longitudinally restrained ends if the lateral displacements are small compared with the depth of the beam. Similar nonlinear effects exist in the bending and stretching of plates (Chap. 13) and shells.

In the linear cases, Eqs. (7-49) and (7-50) reduce to

$$P = E_1 A^* u' - P_T \tag{7-52}$$

$$(E_1 A^* u')' = -p_x^* \tag{7-53}$$

and the coupling of the equations for longitudinal and lateral displacements is removed.

Equations (7-45) and (7-46) are fourth-order differential equations, and each requires two boundary conditions at each of the end points of the beam. Consider the boundary conditions on w as an example; at each end either w or V_z in Eq. (7-43b) and either dw/dx or M_y in Eqs. (7-42) must be specified. Equation (7-53) [or Eq. (7-50)] is a second-order differential equation which requires one boundary condition at each end point. At each end either u or P in Eqs. (7-52) [or Eq. (7-49)] must be given.

We now consider several examples to show how the general equations can be simplified in a routine manner to give the differential equations and boundary conditions for special cases of beams, bars, and cables. The equations result in either equilibrium boundary-value or eigenvalue problems, and when only a single displacement function is involved, the equations are of the forms of Eqs. (5-13) and (5-14) or (5-18) and (5-19). Our intention here is to show how these equations are derived rather than how they are solved; however, references to solutions are given for the interested reader. When the equations are linear and have constant coefficients, the solution can be found by using the methods for nth-order linear ordinary differential equations with constant coefficients that are covered in texts on differential equations. Exact solutions are seldom possible when the coefficients of the differential equation are arbitrary functions of x. In these cases approximate solutions can be found by the finite-difference methods for equilibrium and eigenvalue problems given in Chap. 5.

The stresses can be found from Eqs. (7-23), (7-42), and (7-52) once the deflections are known. Combining these equations gives

$$\sigma_{xx} = E(u' - zw'' - yv'' - \alpha T) \tag{7-54}$$

Examples of equilibrium problems

Example 7-2 Determine the displacements of a uniform homogenous beam under the action of a distributed lateral loading $p_0 \sin (\pi x / L)$ when the ends of the beam at $x = 0$ and L are supported by pins that prevent longitudinal motion. The beam fits the end pins perfectly, so that $P = 0$ when the beam is straight. (This problem throws light on the limitations of the linear theory when the ends of the beam are fixed against longitudinal motion.)

For this special case Eq. (7-46) reduces to

$$EI_{yy}w^{\mathrm{iv}} - Pw'' = p_0 \sin \frac{\pi x}{L} \tag{a}$$

and the related boundary conditions for pinned ends are

$$w(0) = w''(0) = 0 \qquad w(L) = w''(L) = 0 \tag{b}$$

The displacements must also satisfy Eq. (7-50), which simplifies to

$$\left[EAu' + \frac{EA}{2} (w')^2 \right]' = 0 \tag{c}$$

and the associated boundary conditions

$$u(0) = 0 \qquad u(L) = 0 \tag{d}$$

From Eq. (7-49) we find

$$P = EAu' + \frac{EA}{2} (w')^2 \tag{e}$$

and so Eq. (c) indicates that P is constant along the beam.

We shall assume that the solution for the lateral displacements is

$$w = w_0 \sin \frac{\pi x}{L} \tag{f}$$

and show that this satisfies Eqs. (a) to (d). Substituting Eq. (f) into Eq. (e) and rearranging gives

$$u' = \frac{P}{EA} - \frac{w_0^2 \pi^2}{2L^2} \cos^2 \frac{\pi x}{L} \tag{g}$$

Integrating this equation between the limits of 0 and x and noting from Eqs. (d) that $u(0) = 0$, we obtain

$$u = \frac{Px}{EA} - \frac{w_0^2 \pi}{2L} \left(\frac{\pi x}{2L} + \frac{1}{4} \sin \frac{2\pi x}{L} \right) \tag{h}$$

Applying the boundary condition $u(L) = 0$ to Eq. (h) gives

$$P = \frac{\pi^2 EA w_0^2}{4L^2} \tag{i}$$

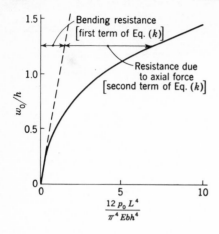

Fig. **7-12** Load-displacement curve for longitudinally restrained beam of Example 7-2 (Ref. 5).

We see that the axial load is proportional to the square of the lateral displacement. Substituting this result into Eq. (h) gives

$$u = -\frac{w_0{}^2\pi}{8L}\sin\frac{2\pi x}{L} \tag{j}$$

By substituting Eqs. (f) and (i) into Eq. (a) we find

$$EI_{yy}w_0 + \frac{EA}{4}w_0{}^3 = \frac{p_0 L^4}{\pi^4} \tag{k}$$

Equation (f) also satisfies Eqs. (b), so that Eqs. (f) [where w_0 is given by Eq. (k)] and (j) fulfill all the differential equations and boundary conditions and are therefore the solution to the problem.

We see from Eq. (k) that the deflection-load relationship for the beam is nonlinear. The first term on the left side of Eq. (k) is linear and is associated with the bending stiffness of the beam, while the second term is nonlinear and derives from the axial stiffness of the member as it behaves like a cable. Equation (k) is plotted in Fig. 7-12 for the case of a rectangular beam with a cross-sectional width b and height h. We observe that the linear term in Eq. (k) gives a good approximation for the displacements if $w_0/h < 0.3$. On the other hand, if $w_0/h > 2$, a reasonable approximation for the deflection can be found by using only the nonlinear term of Eq. (k). For intermediate cases both terms are required for an accurate solution.

We see that in beams with longitudinal end restraint, nonlinear terms in Eqs. (7-49) and (7-50) may be neglected when the lateral displacements are small compared to the depth of the beam.

The preceding solution was given by Donnell,[5] who has a more complete discussion of the problem and implications that carry over into plate and shell theory. In practice it is difficult to prevent longitudinal displacements at the ends because of the large axial forces that are developed. For example, from Eq. (i) a 15-in.-long aluminum beam with a 1-in. square cross-section would develop an axial force of roughly 100,000 lb when $w_0 = 1$ in. Few supports could react this force without sizable deformations. Furthermore the beam stresses greatly exceed the strength of any aluminum alloy.

Example 7·3   Derive the differential equation and boundary conditions for a uniform helicopter rotor blade in hovering flight. The root of the blade is hinged about a horizontal axis and a concentrated mass (due to a small ramjet which powers the rotor) is located at the tip (Fig. 7-13a).

Let the weight per unit length of the blade be $\bar{w}$ and the weight of the tip mass be W. The blade will deform under the actions of the aerodynamic lift and the weights and centrifugal forces of the blade and tip mass. A sketch of the applied forces on a differential length of the blade at x is shown in Fig. 7-13b. From this figure we find for small slopes that

$$p_x = \frac{\bar{w}}{g}\, \Omega^2 x \qquad p_z = p - \bar{w} - \frac{\bar{w}}{g}\, \Omega^2 x w' \tag{a}$$

where Ω is the angular velocity of rotation of the blade. In computing p_x we have neglected the component $\bar{w}w'$ of the weight, which is in the direction of the tangent to the blade, since it is small compared to the other forces.

Inserting p_x into Eq. (7-35a), integrating between the limits of x and L, and using the end condition $P(L) = WL\Omega^2/g$ [obtained from the free-body diagram of the tip mass in Fig. 7-13c by neglecting $Ww'(L)$ compared to the centrifugal-force component], we find

$$P = \frac{\Omega^2}{g} \left[WL + \frac{\bar{w}}{2}\, (L^2 - x^2) \right] \tag{b}$$

Substituting this result and the second of Eqs. (a) into Eq. (7-46) and assuming that the twisting moment on the blade is negligible, we find

$$EI_{yy}w^{iv} - \frac{\Omega^2}{g}\left[WL + \frac{\bar{w}}{2}\, (L^2 - x^2) \right]w'' + \frac{\Omega^2\bar{w}}{g}\, xw' = p - \bar{w} \tag{c}$$

Three of the boundary conditions are obviously

$$w(0) = 0 \qquad w''(0) = 0 \qquad w''(L) = 0 \tag{d}$$

The fourth boundary condition is obtained from the tip shear force, which from Fig. 7-13c is seen to be $V_z = W + WL\Omega^2 w'(L)/g$. Equation (7-43b) then gives

$$EI_{yy}w'''(L) - \frac{WL\Omega^2}{g}\, w'(L) = W \tag{e}$$

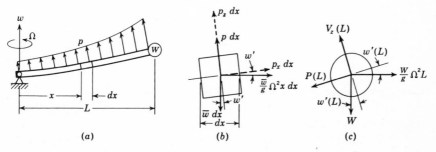

Fig. 7-13   Example 7-3. (a) Helicopter rotor; (b) differential element of rotor; (c) free-body diagram of tip jet.

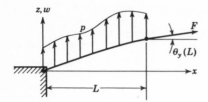

Fig. 7-14   Example 7-4.

Example 7-4   Determine the differential equation and boundary conditions for a cable that is loaded by a distributed lateral load p and a force F at $x = L$, as shown in Fig. 7-14.

The cable can resist loads only by the tensile forces that are developed along its length. This requires that $P(L) = F$ and $w'(L) = \theta_y(L)$, where $\theta_y(L)$ is the prescribed angle that F makes with the horizontal. The differential equation is found from Eq. (7-46) by setting $E_1 I_{yy}^* = M_z = M_t = m_y' = 0$, $P = F$, and $p_z^* = p$, which gives

$$w'' = -\frac{p}{F} \tag{a}$$

This second-order differential equation requires a single boundary condition at each end. These are

$$w(0) = 0 \qquad w'(L) = \theta_y(L) \tag{b}$$

Equation (a) is easily solved by double integration, and Eqs. (b) can be used to evaluate the constant of integration.

Example 7-5   Derive the differential equation and boundary conditions for a homogeneous bar of variable cross section that is subjected to a nonuniform axial load per unit length q and a compressive force F at $x = L$ while axial motion is prevented at $x = 0$. In addition to the loading there is a temperature change $T(x)$.

From the first of Eqs. (7-15) and Eqs. (7-51) and (7-53) we find for constant E that

$$(Au')' = -\frac{q}{E} + (\alpha T A)' \tag{a}$$

From the restraint condition at $x = 0$ and Eq. (7-52) applied at $x = L$ we find

$$u(0) = 0 \qquad u'(L) = -\frac{F}{EA(L)} + \alpha T(L) \tag{b}$$

Equation (a) can be solved by double integration, and the resulting constants of integration can be found from Eqs. (b).

Examples of eigenvalue problems   The following examples illustrate how the equilibrium equations of Sec. 7-9 can be used to derive the differential equations and boundary conditions encountered in column-buckling and beam-vibration problems. The equations for column buckling can be solved by the procedures described in Chap. 14. These same methods are also applicable to the vibration problem, since both are in the form of an eigenvalue problem.   Further information on the solution of the equations can be found in the references at the end of Chap. 14 for the buckling problem and Refs. 6 to 8 for the vibration problem.

In the examples that deal with buckling we consider the behavior of a column under the action of loads that induce no bending deformations when the column is perfectly straight.   For small values of the applied loads the straight column is in stable equilibrium, and if it is given a small lateral displacement and released, it will return to the straight configuration.   However, if the loads are made large enough, a state of neutral stability is reached in which the column remains in the disturbed position when it is released.   The smallest value of the loading that is able to hold the column in equilibrium in the slightly bent position is defined as the *buckling* or *critical load*, and the associated deflection curve is known as the *buckle* or *mode shape*.

Example 7-6 Derive the differential equation and boundary conditions associated with the homogeneous nonuniform column shown in Fig. 7-15a if the lines of action of the compressive force F and the distributed axial loading $q(x)$ remain horizontal as the column bends.

From Fig. 7-15b we see that for small displacements, $p_x = -q$ and $p_z = qw'$.   Substituting for p_x in the first of Eqs. (7-36), integrating between the limits of x and L, and using the condition $P(L) = -F$, we find

$$P = -F - \int_x^L q \, dx$$

Equation (7-46) then becomes

$$(EI_{yy}w'')'' + \left(F + \int_x^L q \, dx\right) w'' = qw' \tag{a}$$

Let us assume that q and F increase simultaneously so that we may write $q = F(q/F) = F f(x)$, where F and $f(x)$ establish respectively the magnitude and distribution of q.   Equation (a) may be written

$$(EI_{yy}w'')'' = -F\left[\left(1 + \int_x^L f \, dx\right) w'' - fw'\right] \tag{b}$$

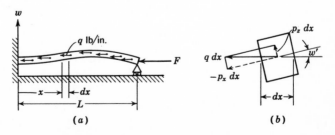

Fig. 7-15 Example 7-6.

The associated boundary conditions for the clamped-pinned ends are

$$w(0) = 0 \qquad w'(0) = 0 \qquad w(L) = 0 \qquad w''(L) = 0 \qquad (c)$$

Equations (b) and (c) are of the form of Eqs. (5-18) and (5-19), where $F = \lambda$ and $w = \psi$. The smallest value of F for which a nontrivial solution ($w \neq 0$) of Eqs. (b) and (c) exists is the buckling load.

Example 7-7 Derive the differential equations and boundary conditions for a column of circular cross section under the simultaneous action of a compressive thrust F and a torque M_t if the ends of column are supported by universal joints.

The bar may buckle as an Euler column under the action of F alone or into a space curve due to M_t alone, a form of instability observable in the kinking of a rubber band as it is twisted. An interaction occurs when the loads are applied simultaneously; i.e., the buckling loads are less for the combined loads than when they are applied separately. The torque couples Eqs. (7-45) and (7-46), and so the buckled shape will be a space curve with projections $v(x)$ and $w(x)$. For the circular cross section all centroidal axes are principal axes, and we let $I_{yy} = I_{zz} = I$. We measure displacements from the compressed and twisted configuration of the straight column in the prebuckled configuration. During buckling the beam bends, but no further twisting occurs, so that $\varphi' = \varphi'' = 0$ during buckling, and Eqs. (7-45) and (7-46) become

$$(EIv'')'' + M_t w''' = -Fv'' \qquad (EIw'')'' - M_t v''' = -Fw'' \qquad (a)$$

Each of these fourth-order differential equations requires two boundary conditions at each end point; these are

$$
\begin{array}{llll}
v(0) = 0 & v''(0) = 0 & v(L) = 0 & v''(L) = 0 \\
w(0) = 0 & w''(0) = 0 & w(L) = 0 & w''(L) = 0
\end{array} \qquad (b)
$$

The solution of this problem for a constant-cross-section shaft is given in Ref. 9.

Example 7-8 Derive the differential equations for free bending vibrations of a cantilever wing with a mass per unit length m and a tip tank of mass M_T (Fig. 7-16).

In vibration problems the displacements are functions of both x and time t, so that $w = w(x,t)$. We shall indicate derivatives with respect to x by primes and those with respect to t by dots. The only forces that act upon a freely vibrating beam are inertial. From d'Alembert's principle, the lateral force on a unit length of the beam is $p_z = -m\ddot{w}$, and the moment per unit

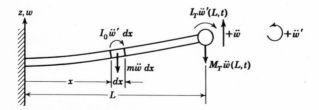

Fig. 7-16 Example 7-8.

length is $m_y = I_0 \ddot{\theta}_y = I_0 \ddot{w}'$, where I_0 is the mass moment of inertia per unit length of the beam. Substituting these results into Eq. (7-46), we find

$$(EI_{yy}w'')'' = -m\ddot{w} + (I_0\ddot{w}')' \tag{a}$$

The partial differential equation (a) can be reduced to two ordinary differential equations by the method of separation of variables. To do this we assume that the solution can be expressed in the form $w(x,t) = W(x)T(t)$, where W is a function of x only which describes the shape of the deflection curve and T is a function of t only which gives the time variation of the amplitude of the displacements. Substituting this expression into Eq. (a) gives

$$T(EI_{yy}W'')'' = -mW\ddot{T} + (I_0W')'\ddot{T}$$

which upon separation of variables gives

$$\frac{(EI_{yy}W'')''}{(I_0W')' - mW} = \frac{\ddot{T}}{T} = -\omega^2 \tag{b}$$

In this equation we have a function of x equal to a function of t for all values of x and t, which is possible only if both functions equal a separation constant which we designate $-\omega^2$. Equation (b) can be separated into two ordinary differential equations

$$\ddot{T} + \omega^2 T = 0 \tag{c}$$
$$(EI_{yy}W'')'' = \omega^2[mW - (I_0W')'] \tag{d}$$

In contrast to the differential equations derived in previous examples, which were boundary-value problems, Eq. (c) is an initial-value problem, which requires the specification of initial rather than boundary conditions. These come from the prescribed deflection and velocity distributions at $t = 0$. The solution of Eq. (c) is

$$T = C_1 \sin \omega t + C_2 \cos \omega t \tag{e}$$

and the initial conditions are

$$
\begin{aligned}
w(x,0) &= W(x)T(0) = C_2 W(x) = f(x) \\
\dot{w}(x,0) &= W(x)\dot{T}(0) = \omega C_1 W(x) = g(x)
\end{aligned} \tag{f}
$$

which can be used to evaluate C_1 and C_2 after W and ω are found.

Equation (d) is fourth-order and requires two boundary conditions at each end point. The following are obtained from the built-in end condition at $x = 0$ and the d'Alembert inertia force and moment of the tip tank at $x = L$:

$$
\begin{aligned}
w(0,t) = 0 \qquad w'(0,t) = 0 \qquad EI_{yy}(L)w''(L,t) &= -I_T\ddot{w}'(L,t) \\
[EI_{yy}(L)w''(L,t)]' &= M_T\ddot{w}(L,t)
\end{aligned} \tag{g}
$$

where I_T is the mass moment of inertia of the tip tank. Substituting

$$w = W(x)T(t),$$

noting that $T(t) \neq 0$ for all t, and recalling from Eq. (c) that $\ddot{T} = -\omega^2 T$, we obtain

$$
\begin{aligned}
W(0) = 0 \qquad W'(0) = 0 \qquad EI_{yy}(L)W''(L) &= \omega^2 I_T W'(L) \\
[EI_{yy}(L)W''(L)]' &= -\omega^2 M_T W(L)
\end{aligned} \tag{h}
$$

Equations (d) and (h) are of the form of Eqs. (5-18) and (5-19), where ω^2 is the eigenvalue λ and W is the eigenfunction ψ. Physically the values of ω that satisfy these equations are the *natural frequencies of vibration* of the system and the W's are the *natural mode shapes*. Solutions of Eq. (d) for uniform beams with I_0 negligible and with various end conditions are given in Ref. 7, which also contains approximate methods for nonuniform beams.

7-10 ENERGY EXPRESSIONS FOR BEAMS

Equations for the strain and complementary-strain energies of a beam can be derived from the expressions for arbitrary shaped bodies given in Chap. 6. Since all stress components except σ_{xx} are assumed to be negligible in the beam, Eqs. (6-44) and (6-45) apply. With these and Eqs. (6-39), (6-41), (7-7), (7-23), and (7-54) we find for principal axes that

$$U = \frac{1}{2} \int_L [E_1 A^*(u')^2 + E_1 I_{yy}^*(w'')^2 + E_1 I_{zz}^*(v'')^2$$

$$- 2(P_T u' + M_{yT} w'' + M_{zT} v'')]\, dx + \frac{1}{2} \iint_{LA} E(\alpha T)^2 \, dA \, dx \quad (7\text{-}55)$$

$$U' = \frac{1}{2} \int_L \left[\frac{(P^*)^2}{E_1 A^*} + \frac{(M_y^*)^2}{E_1 I_{yy}^*} + \frac{(M_z^*)^2}{E_1 I_{zz}^*} \right] dx - \frac{1}{2} \iint_{LA} E(\alpha T)^2 \, dA \, dx \quad (7\text{-}56)$$

In some cases the supports of a beam are so flexible that their elasticity must be considered in computing deformations and stresses. As an example, the longitudinal stiffeners in Fig. 7-5a are supported by the transverse frames, which provide an elastic restraint against lateral motion. A linearly elastic lateral support at $x = x_i$ produces a reaction $P_s = k_s w(x_i)$, which opposes the displacement $w(x_i)$. In this equation k_s, the *stiffness* or *spring constant* of the support, is the force required to deflect the support a unit distance. The energies of the support are

$$U = \int_0^{w(x_i)} P_s \, dw(x_i) = \tfrac{1}{2} k_s w^2(x_i) \quad (7\text{-}57)$$

$$U' = \int_0^{P_s} w(x_i) \, dP_s = \frac{P_s^{\,2}}{2k_s} \quad (7\text{-}58)$$

When there are many closely spaced supports, it is convenient to introduce the concept of an *elastic foundation* by considering the elastic restraints to be spread out so that the beam is continuously supported. Letting k be the force that must be applied to a unit length of the foundation to deflect it a unit distance, that is, $p_s = kw$, we find

$$U = \int_0^L \int_0^w p_s \, dw \, dx = \frac{1}{2} \int_0^L kw^2 \, dx \quad (7\text{-}59)$$

$$U' = \int_0^L \int_0^{p_s} w \, dp_s \, dx = \frac{1}{2} \int_0^L \frac{p_s^{\,2}}{k} \, dx \quad (7\text{-}60)$$

Elastic supports may also provide rotational restraint to the beam. In this case $M_r = k_r w'(x_i)$, where M_r is the moment imposed by the restraint, $w'(x_i)$ is the slope (rotation) at the support, and k_r is the moment required to rotate the support 1 rad. It follows that

$$U = \int_0^{w'(x_i)} M_r \, dw'(x_i) = \tfrac{1}{2}k_r[w'(x_i)]^2 \qquad (7\text{-}61)$$

$$U' = \int_0^{M_r} w'(x_i) \, dM_r = \frac{M_r^2}{2k_r} \qquad (7\text{-}62)$$

Equations (7-57) to (7-62) apply to restraints in the xz plane; similar expressions can be written for supports in the xy plane.

Specific equations can also be written for the potential and complementary-potential energies of conservative forces that are applied to the beam. Consider a concentrated force at $x = x_i$ with components F_x and F_z which are parallel to, and positive in the directions of, the x and z axes. Recalling that the potential is the negative of the work done by the force during the displacements, we find

$$V = -F_x u(x_i) - F_z w(x_i) = V' \qquad (7\text{-}63)$$

Equation (7-63) does not give the complete potential of F_x when the point of application of the force is free to move in the x direction as lateral bending occurs. Let $\Delta(x_i)$ be the additional displacement parallel to the x axis that develops at x_i as the beam bends (Fig. 7-17). The beam elongations are given by u, and no additional stretching occurs as a result of w unless both ends of the beam are axially restrained. When either end of the beam is free to move in the x direction, the developed length of the bent beam remains the same as the straight beam. The additional potential of F_x is $V = F_x \, \Delta(x_i)$. From Fig. 7-17

$$\Delta(x_i) = \int_0^{x_i} (ds - dx) = \int_0^{x_i} [(dx^2 + dw^2)^{1/2} - dx]$$

where s is a coordinate along the bent beam and it is assumed that motion

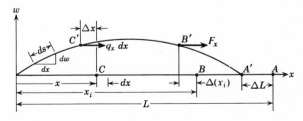

Fig. 7-17 Work producing displacements of horizontal forces during lateral deflections.

parallel to the x axis is prevented at $x = 0$. Factoring out dx and using the binomial expansion gives

$$\Delta(x_i) = \int_0^{x_i} \{[1 + (w')^2]^{1/2} - 1\}\, dx = \frac{1}{2} \int_0^{x_i} [(w')^2 + \cdots]\, dx$$

The higher-order terms in this equation contain fourth and higher powers of w'. We can neglect these terms compared to $(w')^2$ and get

$$V = \frac{F_x}{2} \int_0^{x_i} (w')^2\, dx = V' \tag{7-64}$$

which must be added to Eq. (7-63) to obtain the complete potential of the force.

Next consider a force per unit length of the beam with components q_x and q_z that are parallel to, and positive in the directions of, the x and z axes. The potentials of these forces are

$$V = - \int_L (q_x u + q_z w)\, dx = V' \tag{7-65}$$

where the integral extends over the length L of the beam. As with F_x, an additional term must be added to Eq. (7-65) to account for the work done by q_x during the w displacements. Referring to Fig. 7-17 and applying the same method that was used to derive Eq. (7-64), we find the additional potential to be

$$V = \frac{1}{2} \int_L q_x \left\{ \int_0^x [w'(\eta)]^2\, d\eta \right\} dx = V' \tag{7-66}$$

The potentials that are associated with a concentrated moment M_{y_i}, applied about an axis parallel to the y axis at x_i, are

$$V = M_{y_i} w'(x_i) = V' \tag{7-67}$$

In this equation M_{y_i} is positive when its vector obtained from the right-hand rule is in the direction of the positive y axis. Using the same sign convention, the potentials for an applied moment per unit length m_y are

$$V = \int_L m_y w'\, dx = V' \tag{7-68}$$

Equations that are similar to Eqs. (7-63) to (7-68) can be written for forces that produce bending in the xy plane.

The following examples illustrate how the preceding energy relationships can be applied to derive differential equations and boundary conditions by the variational method (Sec. 6-11) and to obtain approximate solutions by the Rayleigh-Ritz method (Sec. 6-12). An example of the application of the Rayleigh-Ritz method to column-buckling problems is given in Chap. 14.

Example 7-9 Derive the differential equations and boundary conditions for the beam-column on an elastic foundation shown in Fig. 7-18 (a *beam-column* is a beam that is subjected to compressive forces and loads which cause lateral bending). In addition to the loading shown, the beam is subjected to a temperature change $T(x,z)$.

The total potential is found from Eqs. (7-55) (with $v = 0$), (7-59), (7-63) to (7-65), and (7-67), giving

$$U + V = \frac{1}{2} \int_0^L [E_1 A^*(u')^2 + E_1 I_{yy}^*(w'')^2 - 2(P_T u' + M_{yT} w'')]\, dx$$

$$+ \frac{1}{2} \int_0^L \int_A E(\alpha T)^2\, dA\, dx + \frac{1}{2} \int_0^L kw^2\, dx - \int_0^L pw\, dx$$

$$+ F_{x_L} u(L) + F_{z_L} w(L) - \frac{F_{x_L}}{2} \int_0^L (w')^2\, dx - M_{y_L} w'(L) \quad (a)$$

As a result of the applied loads and temperatures, the beam will assume an equilibrium configuration described by $u(x)$ and $w(x)$. For the end restraints that are shown, the displacements must satisfy the essential boundary conditions

$$u(0) = 0 \qquad w(0) = 0 \qquad w'(0) = 0 \qquad (b)$$

For any virtual displacements δu and δw which do not violate the geometric constraints

$$\delta u(0) = 0 \qquad \delta w(0) = 0 \qquad \delta w'(0) = 0 \qquad (c)$$

and during which T does not change, $U + V$ will have a stationary value. Letting $\delta(U + V) = 0$, we find

$$\int_0^L (E_1 A^* u'\, \delta u' + E_1 I_{yy}^* w''\, \delta w'' - P_T \delta u' - M_{yT}\, \delta w'')\, dx$$

$$+ \int_0^L kw\, \delta w\, dx - \int_0^L p\, \delta w\, dx + F_{x_L}\, \delta u(L) + F_{z_L}\, \delta w(L)$$

$$- F_{x_L} \int_0^L w'\, \delta w'\, dx - M_{y_L}\, \delta w'(L) = 0 \quad (d)$$

Noting that $\delta u' = (\delta u)'$, $\delta w' = (\delta w)'$, and $\delta w'' = (\delta w)''$, we may rewrite Eq. (d) as

$$\int_0^L (E_1 A^* u' - P_T)(\delta u)'\, dx + \int_0^L (E_1 I_{yy}^* w'' - M_{yT})(\delta w)''\, dx$$

$$- F_{x_L} \int_0^L w'(\delta w)'\, dx + \int_0^L (kw - p)\, \delta w\, dx + F_{z_L}\, \delta w(L) - M_{y_L}[\delta w(L)]'$$

$$+ F_{x_L}\, \delta u(L) = 0 \quad (e)$$

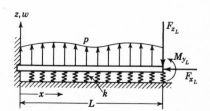

Fig. 7-18 Example 7-9.

Using Eq. (6-68) to integrate the first and third integrals by parts and Eq. (6-70) to integrate the second integral, we find after regrouping and taking note of Eqs. (c)

$$(E_1A^*u' - P_T + F_{x_L})_{x=L}\,\delta u(L) - \int_0^L (E_1A^*u' - P_T)'\,\delta u\,dx$$
$$+ (E_1I_{yy}^*w'' - M_{y_T} - M_{y_L})_{x=L}\,\delta w'(L)$$
$$- [(E_1I_{yy}^*w'' - M_{y_T})' - F_{z_L} + F_{x_L}w']_{x=L}\,\delta w(L)$$
$$+ \int_0^L [E_1I_{yy}^*w'' - M_{y_T})'' + F_{x_L}w'' + kw - p]\,\delta w\,dx = 0 \quad (f)$$

Observing that δu, δw, and $\delta w'$ are arbitrary at $x = L$ while δu and δw are arbitrary for $0 < x < L$, we obtain the Euler differential equations

$$(E_1A^*u')' = -p_{x_T} \qquad (E_1I_{yy}^*w'')'' + F_{x_L}w'' + kw = p + p_{z_T}$$

and the natural boundary conditions

$$E_1A^*(L)u'(L) = -F_{x_L} + P_T(L) \qquad E_1I_{yy}^*(L)w''(L) = M_{y_L} + M_{y_T}(L)$$
$$[E_1I_{yy}^*w'']'_{x=L} + F_{x_L}w'(L) = F_{z_L} + V_{z_T}(L)$$

which together with Eqs. (b) define the displacements. The notations of Eqs. (7-44b), (7-48b), and (7-51) have been used in writing the final equations. Solutions for uniform beams and beam-columns on uniform foundations are given in Ref. 10.

Example 7-10 Use the Rayleigh-Ritz method to reduce the problem of a laterally loaded beam on an elastic foundation to the solution of a set of simultaneous linear algebraic equations.

The total potential obtained from the appropriate terms of Eqs. (7-55), (7-59), and (7-65) is

$$U + V = \frac{1}{2}\int_0^L [E_1I_{yy}^*(w'')^2 + kw^2 - 2pw]\,dx \qquad (a)$$

where p is the lateral load per unit length. We approximate the deflections by the series

$$w(x) = w_0(x) + \sum_{j=1}^n a_jw_j(x) \qquad (b)$$

where the coefficients a_j are constants, the w_j are assumed functions that satisfy the homogeneous displacement boundary conditions, and w_0 is an assumed function that satisfies the nonhomogeneous-displacement boundary conditions. The coefficients a_j define the configuration of the beam and are therefore generalized coordinates. Applying Eq. (6-17) to Eq. (a), we find

$$\int_0^L \left(E_1I_{yy}^*w''\frac{\partial w''}{\partial a_i} + kw\frac{\partial w}{\partial a_i} - p\frac{\partial w}{\partial a_i}\right)dx = 0 \qquad i = 1, 2, \ldots, n \qquad (c)$$

Substituting Eq. (b) into Eq. (c) gives

$$\int_0^L \left[E_1I_{yy}^*\left(w_0'' + \sum_{j=1}^n a_jw_j''\right)w_i'' + k\left(w_0 + \sum_{j=1}^n a_jw_j\right)w_i - pw_i\right]dx = 0$$
$$i = 1, 2, \ldots, n$$

which after rearranging becomes

$$\sum_{j=1}^{n} \left[\int_0^L (E_1 I_{yy}^* w_i'' w_j'' + k w_i w_j)\, dx \right] a_j$$

$$= \int_0^L (p w_i - E_1 I_{yy}^* w_0'' w_i'' - k w_0 w_i)\, dx \qquad i = 1, 2, \ldots, n \quad (d)$$

The solution of the simultaneous equations obtained from Eq. (d) gives the coefficients a_j, which are substituted into Eq. (b) to obtain the deflections.

The choice of the functions for w_0 and w_j depends upon the displacement constraints at the ends of the beam. As an example, consider a beam with simply supported ends so that $w(0) = w(L) = 0$. In this case all of the displacement boundary conditions are homogeneous and $w_0 = 0$. A complete set of functions which satisfy the homogeneous-displacement boundary conditions is given by

$$w_j = \sin \frac{j\pi x}{L} \qquad (e)$$

Equation (d) then becomes

$$\sum_{j=1}^{n} \left[\int_0^L \left(\frac{i^2 j^2 \pi^4 E_1 I_{yy}^*}{L^4} + k \right) \sin \frac{i\pi x}{L} \sin \frac{j\pi x}{L}\, dx \right] a_j$$

$$= \int_0^L p \sin \frac{i\pi x}{L}\, dx \qquad i = 1, 2, \ldots, n \quad (f)$$

If $E_1 I_{yy}^*$ and k are constant, the simultaneous equations are uncoupled, and we find

$$a_i = \frac{2}{L[(i^4 \pi^4 E_1 I_{yy}^*/L^4) + k]} \int_0^L p \sin \frac{i\pi x}{L}\, dx \qquad (g)$$

If, in addition, p is constant, we obtain $a_i = 0$ when i is even and

$$a_i = \frac{4p}{i\pi[(i^4 \pi^4 E_1 I_{yy}^*/L^4) + k]}$$

when i is odd. Equation (b) then becomes

$$w = \frac{4p}{\pi} \sum_{j=1,3,5}^{n\ \text{odd}} \frac{\sin j\pi x/L}{j(j^4 \pi^4 E_1 I_{yy}^*/L^4 + k)} \qquad (h)$$

If $k = 0$, the first terms of the series gives $w(L/2) = 0.01307 p L^4 / E_1 I_{yy}^*$, which is within 0.4 percent of the exact solution for a simply supported beam with a uniform lateral load.

REFERENCES

1. Boley, B. A., and J. H. Weiner: "Theory of Thermal Stresses," John Wiley & Sons, Inc., New York, 1960.
2. Peery, David J.: "Aircraft Structures," McGraw-Hill Book Company, New York, 1950.
3. Steinbacher, F. R., and G. Gerard: "Aircraft Structural Mechanics," Pitman Publishing Corporation, New York, 1952.

4. Kuhn, P.: "Stresses in Aircraft and Shell Structures," McGraw-Hill Book Company, New York, 1956.

5. Donnell, L. H.: Shell Theory, *Proc. Fourth Midwest Conf. Solid Mechanics*, Austin, Texas, 1959.

6. Flugge, W.: "Handbook of Engineering Mechanics," McGraw-Hill Book Company, New York, 1962.

7. Bisplinghoff, R. L., H. Ashley, and R. L. Halfman: "Aeroelasticity," Addison-Wesley Publishing Company, Inc., Reading, Mass., 1955.

8. Timoshenko, S. P.: "Vibration Problems in Engineering," 3d ed., D. Van Nostrand Company, Inc., Princeton, N.J., 1955.

9. Timoshenko, S. P., and J. M. Gere: "Theory of Elastic Stability," 2d ed., McGraw-Hill Book Company, New York, 1961.

10. Hétenyi, M.: "Beams on an Elastic Foundation," University of Michigan Press, Ann Arbor, Mich., 1946.

PROBLEMS

7-1. Find the bending stress at point A if the Z section is subjected to the bending moment shown. [*Ans.* $\sigma_{zz} = 127,500$ psi.]

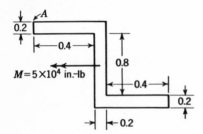

Fig. P7-1

7-2. The rectangular beam shown is subjected to a temperature change $T = 10z^2 + z^3$. Find the stress at points A and B if $\alpha = 10^{-5}$ in./(in.)(°F) and $E = 10^7$ psi. [*Ans.* $\sigma_{zz} = -21,650$ psi at A; $\sigma_{zz} = 8350$ psi at B.]

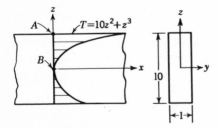

Fig. P7-2

7-3. The thick-skinned wing box that is shown is subjected to aerodynamic heating, which produces the temperature change indicated. Determine the stress in the bottom cover skin if the skins are 0.25 steel [$E = 30 \times 10^6$ psi, $\alpha = 10 \times 10^{-6}$ in./(in.)(°F)] and the spar webs are 0.125 aluminum [$E = 10 \times 10^6$ psi, $\alpha = 15 \times 10^{-6}$

in./(in.)(°F)]. All material is effective in carrying axial stress. [*Ans.* $\sigma_{xx} = -3300$ psi.]

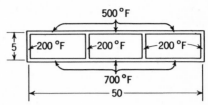

Fig. P7-3

7-4. A bimetallic strip is used in the thermostat which controls the cabin temperature in a manned space station. The element is of the size and materials shown. Find the clearance d which should be provided to have the contacts close for a temperature variation of $\pm 10°F$ if the strip is straight at 72°F. For invar $\alpha = 1.1 \times 10^{-6}$ in./(in.)(°F), $E = 21 \times 10^6$ psi; and for copper $\alpha = 9.5 \times 10^{-6}$ in./(in.)(°F), $E = 17 \times 10^6$ psi. [*Ans.* 0.0282 in.]

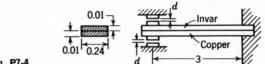

Fig. P7-4

7-5. The idealized fuselage section shown is subjected to a bending moment of 10^6 in.-lb about a horizontal axis. The effective areas of the longitudinals and their coordinates are given in the following table. Find the stresses in the longitudinals.

Longitudinal	A_i, in.^{2}	y_{0i}, in.	z_{0i}, in.
1	1.0	0.0	60.0
2	0.8	14.0	56.0
3	0.8	20.0	46.0
4	0.8	24.0	35.0
5	0.9	26.0	24.0
6	1.0	27.0	12.0
7	1.0	24.0	4.0
8	1.5	15.0	0.0
9	1.0	0:0	0.0

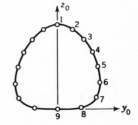

Fig. P7-5

7-6. A cantilevered thin-walled deployable boom, which forms a slit tube when it is unrolled, is used for gravity-gradient stabilization of a satellite. Solar radiation causes the circumferential temperature gradient shown. Derive the equation for the lateral tip deflection if the length of the boom is L. [*Ans.* $w(L) = \alpha T_2 L^2/2R$.]

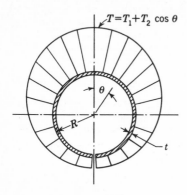

Fig. P7-6

7-7. A cantilevered Z-section beam with the cross section given in Prob. 7-1 is 10 in. long and is loaded by a vertical force of 300 lb at the tip. Determine the vertical and horizontal components of the tip deflection.

7-8. Derive the equation for the change in length of the helicopter rotor blade of Example 7-3.

7-9. A tapered shaft with a mass per unit length $m(x)$ rotates at an angular velocity Ω and is subjected to an axial thrust F, as shown. A bearing at the left end provides a clamped-end condition, while a self-aligning bearing at the right end gives a pin-ended support. If Ω or F is made large enough, the shaft can reach a state of neutral equilibrium in which a rotating bent position is possible. The x axis passes through the center of the bearings, and the z axis is in the plane of bending and rotates with the shaft. Assume that F is known and Ω for the bent configuration to be possible is to be determined. Show that Ω and w must satisfy $(EIw'')'' + Fw'' = \Omega^2 mw$, $w(0) = w'(0) = w(L) = w''(L) = 0$.

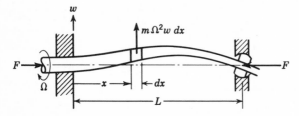

Fig. P7-9

7-10. A cable of mass per unit length m has both ends supported against lateral displacements and is subjected to a large initial tension F. It is given a small lateral displacement $f(x)$ and a velocity $g(x)$ at time zero.

(a) Write the partial differential equation, initial conditions, and boundary conditions which govern the ensuing free vibrations.

(b) Use the method of separation of variables to reduce the partial differential equation to two ordinary differential equations.

7-11. Derive the differential equations and boundary conditions of Example 7-9 from the equations of Sec. 7-9.

7-12. A variable-cross-section beam-column rests upon a nonuniform elastic foundation and is subjected to the loadings shown. Assume that I varies linearly from I_0 at the ends to $4I_0$ at the center and that k varies linearly from 0 at the ends to $3k_0$ at the center. Write finite-difference equations to obtain the displacements at $x = L/6$, $L/3$, and $L/2$, observing the symmetry about the center of the beam.

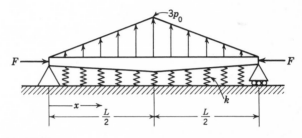

Fig. P7-12

7-13. Write the differential equation and boundary conditions of Example 7-3 in finite-difference form.

7-14. A pin-ended beam is supported by three equally spaced elastic supports having stiffnesses of k lb/in., as shown. It is loaded at the center by a force P. Use the Rayleigh-Ritz method with a two-term series to obtain an approximation of the center deflection.

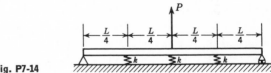

Fig. P7-14

8
The Torsion of Slender Bodies

8-1 INTRODUCTION

We have seen in Secs. 7-3 and 7-5 that a closed-form solution exists for the stresses in a slender beam with an arbitrary uniform cross section which is subjected to a constant bending moment and axial force and that this solution is a satisfactory approximation for a gradually changing cross section, bending moment, and axial force. We now consider the stresses and deformations associated with the torsional stress-resultant moment M_t (Sec. 7-2). We shall find that, unlike the case of bending and axial load, a closed-form solution for the torsional stresses does not exist for an arbitrary cross section and that a different solution must be determined for each cross-sectional shape. In this respect the torsion of a bar of circular cross section, which is treated in texts on strength of materials, is misleading because of its simplicity. In the circular bar it is hypothesized that the cross section remains plane during twisting and that diameters of the cross section remain straight lines. As a result of these assumptions the resultant shearing stress at any point in the plane of the cross section is normal to, and directly proportional to, the radius from the axis of twist to the point. We shall find that these assumptions lead to a result which agrees with the theory-of-elasticity solution for the circular section; however, for an arbitrary cross section, plane sections do not remain plane, and the direction of the resultant shearing stress and its distribution over the cross section are not known.

In this chapter we first consider the theory of torsion for bars of arbitrary uniform cross section. Examples of known solutions are given for specific cross-sectional shapes, and the finite-difference and Rayleigh-Ritz methods are applied to obtain approximate results where exact solutions are unknown. The simpler cases of thin-walled members with open and closed cross-sectional shapes are then considered. Finally the differential equation and boundary conditions that must be satisfied by the angle of twist φ (Sec. 7-7) are developed and used to derive the equations which govern several different torsional equilibrium and eigenvalue problems. Additional information on the torsion problem may be found in Refs. 1 to 4.

8-2 PRANDTL STRESS-FUNCTION FORMULATION

The torsion-theory equations are developed by the semi-inverse method (Sec. 4-7), in which assumptions about the displacement or stress components are made to simplify the differential equations. Naturally, these assumptions restrict the class of problems to which the solution is applicable. The semi-inverse method based upon assumed stresses is discussed in this section, and the method which uses displacement assumptions is treated in Sec. 8-4.

Consider a straight slender body with a singly connected constant cross section that is loaded by torsional moments at the ends $x = 0$ and L, as shown in Fig. 8-1 (the distribution of the forces which produce these moments is discussed later). It is assumed that the end $x = 0$ is restrained against rigid-body rotation and that both ends are free to warp; that is, u displacements at the ends are unrestrained. Let us hypothesize that for such a body

$$\sigma_{xx} = \sigma_{yy} = \sigma_{zz} = \sigma_{yz} = 0 \tag{8-1}$$

To verify this hypothesis we must show that the remaining stresses satisfy the equilibrium and compatibility equations at all interior points of the body and fulfill the equilibrium boundary conditions at all points on the surface.

Substituting Eqs. (8-1) into the equilibrium equations (2-8) and taking $X = Y = Z = 0$ gives

$$\frac{\partial \sigma_{xy}}{\partial y} + \frac{\partial \sigma_{xz}}{\partial z} = 0 \qquad \frac{\partial \sigma_{yx}}{\partial x} = 0 \qquad \frac{\partial \sigma_{zx}}{\partial x} = 0 \tag{8-2}$$

The last two of these equations require that σ_{xy} and σ_{xz} be functions of y

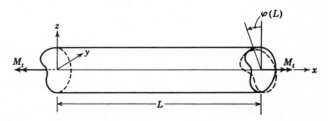

Fig. 8-1 Uniform bar with end torques.

and z only. The resultant shearing stress obtained from Eq. (2-5) is therefore constant at all points along the length of the body which have the same y and z coordinates. The first of Eqs. (8-2) may be identically satisfied by introducing the *Prandtl stress function*, defined by the equations

$$\frac{\partial \psi}{\partial z} = \sigma_{xy} \qquad \frac{\partial \psi}{\partial y} = -\sigma_{xz} \tag{8-3}$$

Turning to the compatibility equations, it is seen that Eqs. (4-6a) to (4-6c) and (4-6e) are identically satisfied, and Eqs. (4-6d) and (4-6f) reduce to $\nabla^2 \sigma_{xy} = 0$ and $\nabla^2 \sigma_{xz} = 0$. By using Eqs. (8-2) and (8-3) these equations may be simplified to

$$\left(\frac{\partial^2}{\partial y^2} + \frac{\partial^2}{\partial z^2}\right)\frac{\partial \psi}{\partial z} = \frac{\partial}{\partial z}\nabla^2 \psi = 0 \qquad -\left(\frac{\partial^2}{\partial y^2} + \frac{\partial^2}{\partial z^2}\right)\frac{\partial \psi}{\partial y} = -\frac{\partial}{\partial y}\nabla^2 \psi = 0$$
$$\tag{8-4}$$

where ∇^2 is the two-dimensional laplacian operator in the independent variables y and z. From Eqs. (8-4) it is seen that $\nabla^2 \psi$ is a constant, which we shall call F, so that ψ must satisfy the differential equation

$$\nabla^2 \psi = F \tag{8-5}$$

at all points within the body. It should be noted that Eq. (8-5), which is the compatibility equation in terms of the stress function, is in the form of Poisson's equation.

Next consider the boundary conditions. On the cylindrical surface there are no surface tractions, so that $\bar{X} = \bar{Y} = \bar{Z} = 0$. The direction cosine l is also zero; therefore the first of Eqs. (2-10) becomes

$$\sigma_{xy}m + \sigma_{xz}n = 0 \tag{8-6}$$

and the remaining equations are identically satisfied. Referring to Fig. 8-2, it is seen that

$$m = \frac{dz}{ds} \qquad n = -\frac{dy}{ds} \tag{8-7}$$

Substituting Eqs. (8-3) and (8-7) into (8-6) gives

$$\frac{\partial \psi}{\partial z}\frac{dz}{ds} + \frac{\partial \psi}{\partial y}\frac{dy}{ds} = 0$$

which by the chain rule of differentiation reduces to $\partial \psi/\partial s = 0$, and we observe that ψ is a constant on the cylindrical surface. Since from Eq. (8-3) the addition of an arbitrary constant to ψ does not affect the stresses, the constant may be conveniently taken as zero on the cylindrical surface for the singly connected body. The boundary condition on the cylindrical

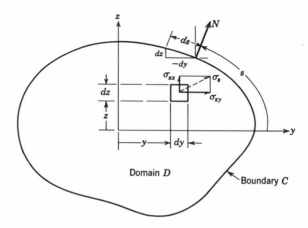

Fig. 8-2 Shearing stresses.

surface then becomes

$$\psi = 0 \qquad \text{on } C \tag{8-8}$$

where C is the boundary curve of the cross section.

Next consider the end surfaces of the body. On the end $x = L$ the direction cosines of the normal to the surface are $l = 1$ and $m = n = 0$, and so the boundary-condition equations (2-10) become $\bar{X} = 0$, $\bar{Y} = \sigma_{xy}$, and $\bar{Z} = \sigma_{xx}$. Recalling that σ_{xy} and σ_{xx} have already been found to be independent of x, we see that the surface forces on the end are distributed in the same manner as the stresses. From Eqs. (7-1) to (7-3) the resultants of the surface forces are found to be $P = M_y = M_z = 0$, and

$$V_y = - \iint\limits_A \sigma_{xy}\, dy\, dz \qquad V_z = - \iint\limits_A \sigma_{xx}\, dy\, dz$$

$$M_t = \iint\limits_A (\sigma_{xx}y - \sigma_{xy}z)\, dy\, dz$$

Using Eqs. (8-3), these may be written

$$V_y = - \iint\limits_A \frac{\partial \psi}{\partial z}\, dy\, dz \tag{8-9a}$$

$$V_z = + \iint\limits_A \frac{\partial \psi}{\partial y}\, dy\, dz \tag{8-9b}$$

$$M_t = - \iint\limits_A \left(\frac{\partial \psi}{\partial y} y + \frac{\partial \psi}{\partial z} z \right) dy\, dz \tag{8-10}$$

Replacing x, y, l, and m in Eq. (6-72) by y, z, m, and n, respectively, and applying the result to Eq. (8-9a) with $\gamma = \psi$ and $\beta = 0$, we find

$$V_y = - \oint \psi n \, ds$$

where the integral extends around C, the boundary curve of the end cross section. From Eq. (8-8) we see that $V_y = 0$, and in a similar manner $V_z = 0$. Equation (8-10) may be written

$$M_t = - \iint_A \left(\frac{\partial}{\partial y} \psi y + \frac{\partial}{\partial z} \psi z \right) dy \, dz + 2 \iint_A \psi \, dy \, dz$$

The first integral on the right of this equation is found to equal zero by using Eq. (6-72), modified as before to apply to the yz plane and with $\beta = -\psi y$ and $\gamma = \psi z$. As a result

$$M_t = 2 \iint_A \psi \, dy \, dz \tag{8-11}$$

Identical results are found for the end $x = L$, and it is seen that the only loading consists of torsional moments applied at the ends. The hypotheses of Eqs. (8-1) are therefore justified. If a stress function $\psi(y,z)$ can be found which vanishes on the boundary of the cross section and satisfies Eq. (8-5) at all interior points of the cross section, then it is an exact solution to the problem if the end torques are distributed in the same manner as the internal stress distribution which results from Eqs. (8-3). In practice, it is seldom that the end torques are applied in this manner; however, from St. Venant's principle this affects only the stresses in the end regions. At points in the body which are at distances from the ends greater than the largest cross-sectional dimension, the stresses will be in accordance with those which result from the solution of Eqs. (8-5) and (8-8).

The angle of rotation at $x = L$ relative to the end at $x = 0$ may be found by the unit-load method described in Sec. 6-14. We assume that the body is loaded by unit torques applied at the ends. If ψ is the stress function associated with the real torque M_t, then the stresses due to the unit torques are

$$\bar{\sigma}_{xy} = \frac{1}{M_t} \frac{\partial \psi}{\partial z} \qquad \bar{\sigma}_{xz} = - \frac{1}{M_t} \frac{\partial \psi}{\partial y} \tag{8-12}$$

The virtual strains are the real strains, which from Eqs. (8-3) are

$$\epsilon_{xy} = \frac{1}{G} \frac{\partial \psi}{\partial z} \qquad \epsilon_{xz} = - \frac{1}{G} \frac{\partial \psi}{\partial y} \tag{8-13}$$

The variation in strain energy is found by substituting Eqs. (8-12) and

(8-13) into Eq. (6-88).   The work done by the unit torques in moving
through the real displacements is $\varphi(L)$, the angle of twist at $x = L$.
Equating the work of the unit torque to the variation in strain energy
gives

$$\varphi(L) = \int_0^L \left\{ \frac{1}{M_t G} \iint_A \left[\left(\frac{\partial \psi}{\partial y} \right)^2 + \left(\frac{\partial \psi}{\partial z} \right)^2 \right] dy\, dz \right\} dx$$

or, since ψ, M_t, and G are constant with respect to the x integration,

$$\varphi(L) = \frac{L}{M_t G} \iint_A \left[\left(\frac{\partial \psi}{\partial y} \right)^2 + \left(\frac{\partial \psi}{\partial z} \right)^2 \right] dy\, dz \tag{8-14}$$

The integral in this equation is evaluated by Eq. (6-69) with x and y
replaced by y and z and with $a = 1$ while $\beta = \gamma = \psi$.   This gives

$$\iint_A \left[\left(\frac{\partial \psi}{\partial y} \right)^2 + \left(\frac{\partial \psi}{\partial z} \right)^2 \right] dy\, dz = \oint \psi \frac{\partial \psi}{\partial n} ds - \iint_A \psi \nabla^2 \psi \, dy\, dz$$

The first integral on the right vanishes as a result of Eq. (8-8), and the
second integral can be simplified by Eq. (8-5), so that

$$\iint_A \left[\left(\frac{\partial \psi}{\partial y} \right)^2 + \left(\frac{\partial \psi}{\partial z} \right)^2 \right] dy\, dz = -F \iint_A \psi \, dy\, dz \tag{8-15}$$

From Eq. (8-11) the integral on the right is equal to $M_t/2$; therefore Eq.
(8-14) may be written $\varphi(L) = -FL/2G$ or $F = -2G\theta$, where θ is defined
as the *angle of twist per unit length* of the bar.   Using this result, Eq. (8-5)
may be expressed as

$$\nabla^2 \psi = -2G\theta \tag{8-16}$$

It is convenient to introduce a *torsion constant J* defined by

$$J = \frac{M_t}{G\theta} \tag{8-17}$$

The product GJ is called the *torsional rigidity*, which from Eqs. (8-11),
(8-16), and (8-17) may be written

$$GJ = -\frac{4G}{\nabla^2 \psi} \iint_A \psi \, dy\, dz \tag{8-18}$$

Lines of constant ψ (Fig. 8-3), known as *lines of shearing stress*, play
an interesting role in the theory.   Using the chain rule for derivatives,
we find

$$\frac{\partial \psi}{\partial s} = \frac{\partial \psi}{\partial y} \frac{dy}{ds} + \frac{\partial \psi}{\partial z} \frac{dz}{ds} = 0$$

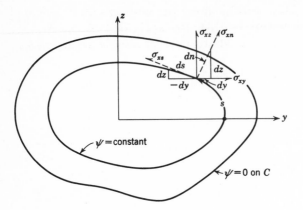

Fig. 8-3   Lines of shearing stress.

where s is measured along the line of constant ψ.   By using Eqs. (8-3) and (8-7) we may write this equation as

$$\frac{\partial \psi}{\partial s} = \sigma_{xz}n + \sigma_{xy}m = 0 \qquad (8\text{-}19)$$

where m and n are the directions cosines of the outward normal to the line of shearing stress.

From Fig. 8-3, the components of the shear stress in the directions that are normal to, and in the direction of, s are

$$\sigma_{xn} = \sigma_{xz}n + \sigma_{xy}m \qquad \sigma_{xs} = \sigma_{xz}m - \sigma_{xy}n \qquad (8\text{-}20)$$

By comparing Eq. (8-19) and the first of Eqs. (8-20) we see that $\sigma_{xn} = 0$, indicating that the resultant shearing stress at any point is tangent to the line of shearing stress through the point.   We have seen that ψ is constant on the boundary curve, and so the resultant shear stress at the surface is tangent to the boundary.   This is explained physically by the fact that σ_{xn} would require a companion stress σ_{ns} on the cylindrical surface, which is not possible since this surface is free of forces.

By using Eqs. (8-3) we may write the second of Eqs. (8-20) as $\sigma_{xs} = -(\partial \psi/\partial y)m - (\partial \psi/\partial z)n$.   Instead of expressing m and n in terms of ds by Eqs. (8-7), we may write them in terms of dn, a differential length normal to s.   Referring to Fig. 8-3, we see that $m = dy/dn$ and $n = dz/dn$, so that we may write

$$\sigma_{xs} = -\frac{\partial \psi}{\partial y}\frac{dy}{dn} - \frac{\partial \psi}{\partial z}\frac{dz}{dn} = -\frac{\partial \psi}{\partial n} \qquad (8\text{-}21)$$

Thus the resultant shearing stress at any point is tangent to the line of shearing stress through the point and is equal to the negative of the derivative of ψ in the direction normal to the line.

The theory which has been described is applicable to bars with singly connected cross sections; Refs. 3 and 4 give the additional equations that must be satisfied for multiply connected cross sections.

8-3 THE MEMBRANE ANALOGY

A useful analogy between ψ and the lateral deflection of a membrane was pointed out by Prandtl. A *membrane* is defined as a body whose thickness is small compared to its surface dimensions and which has negligible bending rigidity so that it can resist lateral loads only by internal forces which lie within its surface. A typical example of a membrane is a soap bubble. Consider a membrane that is initially stretched by a uniform biaxial tensile force per unit length N and which is restrained along its edge to conform to a closed curve C in the yz plane (Fig. 8-4a). Let the membrane be subjected to a lateral pressure q. It is assumed that the lateral displacements are small, so that no appreciable change in N occurs as the membrane deflects under the pressure. A free-body diagram of an element of the membrane is shown in Fig. 8-4b. For small deflections, the angles that the tangents to the surface make with the yz plane are small, and the sines of the angles can be replaced by the tangents. Summing forces in the x direction, we find

$$-N\frac{\partial u}{\partial y}\,dz + N\left(\frac{\partial u}{\partial y} + \frac{\partial^2 u}{\partial y^2}\,dy\right)dz - N\frac{\partial u}{\partial z}\,dy$$
$$+ N\left(\frac{\partial u}{\partial z} + \frac{\partial^2 u}{\partial z^2}\,dz\right)dy + q\,dy\,dz = 0$$

where u is the displacement parallel to the x axis. Canceling terms and dividing by $dy\,dz$ results in the differential equation of equilibrium

$$\nabla^2 u = -\frac{q}{N} \tag{8-22}$$

which must be satisfied at all points within the domain of the membrane D. The boundary condition for the deflections is

$$u = 0 \quad \text{on } C \tag{8-23}$$

By comparing Eqs. (8-22) and (8-23) with Eqs. (8-16) and (8-8) we see that an analogy exists between u and ψ when q is constant. If the boundary C is the same in both cases and $q/N = 2G\theta$, then

$$u(y,z) = \psi(y,z)$$

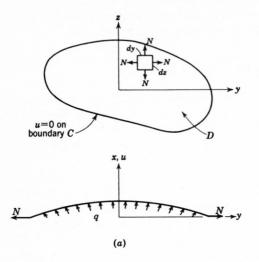

(a)

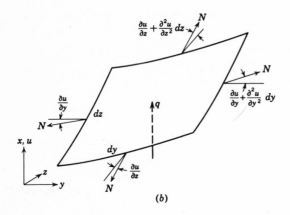

(b)

Fig. 8-4 Membrane analogy to torsion problem. (a) Membrane under the action of uniform biaxial tension and lateral pressure; (b) free-body diagram of an element of the membrane.

Contour lines of constant u correspond to lines of shearing stress. The maximum slope at any point on the membrane is $\partial u/\partial n$, where the direction n is normal to the contour line. From Eq. (8-21) we note that the resultant shear stress at any point is equal to the negative of the slope of the analogous membrane. The volume between the membrane and the

yz plane is

$$\text{Vol.} = \iint_A u \, dy \, dz$$

and it is seen from Eq. (8-11) that $M_t = 2 \text{ Vol.}$

As a result of the analogy, a solution that has been found for a membrane with a given boundary is also the solution for the torsion of a bar with a similar cross-sectional boundary. The membrane analogy provides a useful experimental method of solving the torsional problem for irregular boundaries for which solutions to Eqs. (8-16) and (8-8) are not known. Experimental techniques for accomplishing this are described in Ref. 4, which also discusses the membrane analogy for multiply connected cross sections.

Perhaps the greatest usefulness of the analogy is that it makes it easier to visualize the stress function. This is often helpful in obtaining approximate solutions by intuitively choosing a stress function guided by a visualization of the analogous membrane. Approximate values for the shear stresses and angle of twist can then be found from Eqs. (8-3) and (8-16). The analogy is also useful in visualizing the local stress conditions in a body subjected to torsion. For instance, from the membrane contour lines shown in Fig. 8-5 it is seen that the membrane slopes, and therefore the shear stresses at an external corner are zero, while at a sharp internal corner they are very large. We note that external corners do not add appreciably to the torsional rigidity and might well be eliminated to save weight. It is also seen that generous fillets are desirable at internal corners to reduce stress concentrations.

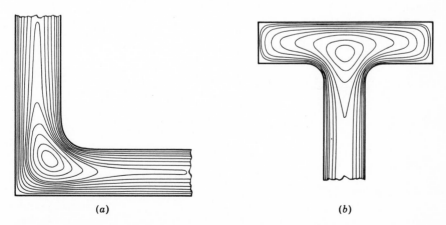

(a) (b)

Fig. 8-5 Membrane contour lines at external and internal corners (Ref. 7). (a) At L junction; (b) at T junction.

8-4 WARPING-FUNCTION FORMULATION

We noted in Sec. 8-2 that in the semi-inverse method we make simplifying assumptions regarding either the stresses or the deformations. The stress-assumption formulation was presented in Sec. 8-2, and in this section the displacement approach will be given. St. Venant made the hypothesis that when a bar is subjected to torsion, the cross sections rotate about the axis of twist; and even though the cross section may warp out of its original plane, its projection on the yz plane retains the initial shape. This is shown in Fig. 8-6, where the x axis has been chosen to coincide with the axis of twist. The rotation of a cross section at a distance x from the fixed end is θx, where, as before, θ is the constant angle of twist per unit length. Let an arbitrary point O in the cross section be located by either the cartesian coordinates (x,y,z) or the cylindrical coordinates (x,r,β). The v and w displacements of O are given by Fig. 8-6

$$v = -r\theta x \sin \beta = -zx\theta \qquad w = r\theta x \cos \beta = yx\theta \qquad (8\text{-}24)$$

It is reasonable to assume that the u displacement of O is directly proportional to θ and is the same at all cross sections, so that

$$u = \theta f(y,z) \qquad (8\text{-}25)$$

where $f(y,z)$ is an unknown function called the *warping function*.

Since we are dealing with displacements, it is not necessary to investigate compatibility, but it is required to show that the assumed displacements satisfy the equilibrium equations (4-1) and the force boundary conditions, Eqs. (4-4). Equations (8-24) and (8-25) identically satisfy

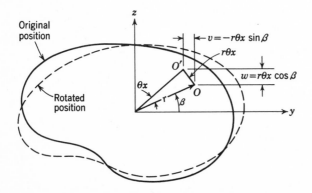

Fig. 8-6 Twisting deformations of a bar.

the last two of Eqs. (4-1), and the first equation reduces to

$$\nabla^2 f = 0 \tag{8-26}$$

where ∇^2 is the two-dimensional laplacian operator in y and z.

On the cylindrical surface the direction cosine $l = 0$, and so Eqs. (8-24) and (8-25) identically satisfy the last two of Eqs. (4-4), and the first equation reduces to

$$\left(\frac{\partial f}{\partial y} - z\right) m + \left(\frac{\partial f}{\partial z} + y\right) n = 0 \qquad \text{on } C \tag{8-27}$$

The direction cosines on the end $x = L$ are $l = 1$ and $m = n = 0$, so that upon substitution of Eqs. (8-24) and (8-25), Eqs. (4-4) become $\bar{X} = 0$, $\bar{Y} = G\theta(\partial f/\partial y - z)$, and $\bar{Z} = G\theta(\partial f/\partial z + y)$. The static resultants on the end are then $P = M_y = M_z = 0$, and

$$V_y = -G\theta \iint_A \left(\frac{\partial f}{\partial y} - z\right) dy\, dz \qquad V_z = -G\theta \iint_A \left(\frac{\partial f}{\partial z} + y\right) dy\, dz$$

$$M_t = G\theta \iint_A \left[\left(\frac{\partial f}{\partial z} + y\right) y - \left(\frac{\partial f}{\partial y} - z\right) z\right] dy\, dz \tag{8-28}$$

The equation for V_y can be written

$$V_y = -G\theta \iint_A \left\{\frac{\partial}{\partial y}\left[y\left(\frac{\partial f}{\partial y} - z\right)\right] + \frac{\partial}{\partial z}\left[y\left(\frac{\partial f}{\partial z} + y\right)\right]\right\} dy\, dz$$

$$+ G\theta \iint_A y \,\nabla^2 f \, dy\, dz$$

and from Eq. (8-26) the second integral is zero. The first integral can be evaluated with Eq. (6-69) by changing coordinates to y and z and letting $a = 1$, $\beta = y(\partial f/\partial y - z)$, and $\gamma = -y(\partial f/\partial z + y)$. This gives

$$V_y = -G\theta \oint y \left[\left(\frac{\partial f}{\partial z} + y\right) n + \left(\frac{\partial f}{\partial y} - z\right) m\right] ds = 0$$

as a result of Eq. (8-27). In a similar fashion $V_z = 0$, so that the hypothesized displacements of Eqs. (8-24) and (8-25) are correct for a uniform bar with end torques.

The problem reduces to finding the warping function $f(y,z)$ which satisfies Eqs. (8-26) and (8-27). The angle of twist for a given torque is found from the last of Eqs. (8-28). The strains are found by introducing Eqs. (8-24) and (8-25) into Eqs. (2-23), to give

$$\epsilon_{xx} = \epsilon_{yy} = \epsilon_{zz} = \epsilon_{yz} = 0 \qquad \epsilon_{xy} = \theta\left(\frac{\partial f}{\partial y} - z\right) \qquad \epsilon_{xz} = \theta\left(\frac{\partial f}{\partial z} + y\right) \tag{8-29}$$

Substituting these strains into Eqs. (3-21) gives

$$\sigma_{xx} = \sigma_{yy} = \sigma_{zz} = \sigma_{yz} = 0$$

and

$$
\begin{aligned}
\sigma_{xy} &= G\theta\left(\frac{\partial f}{\partial y} - z\right) = \frac{M_t}{J}\left(\frac{\partial f}{\partial y} - z\right) \\
\sigma_{xz} &= G\theta\left(\frac{\partial f}{\partial z} + y\right) = \frac{M_t}{J}\left(\frac{\partial f}{\partial z} + y\right)
\end{aligned}
\tag{8-30}
$$

We note that these results are consistent with the stress hypotheses given in Eq. (8-1). The stresses from Eqs. (8-3) and (8-30) must be identical if exact solutions are found for ψ and f, and so the following relationships exist between ψ and f.

$$\frac{\partial \psi}{\partial z} = G\theta\left(\frac{\partial f}{\partial y} - z\right) \qquad \frac{\partial \psi}{\partial y} = -G\theta\left(\frac{\partial f}{\partial z} + y\right) \tag{8-31}$$

From Eqs. (8-17) and the last of Eqs. (8-28) we find

$$J = \iint_A \left[\left(\frac{\partial f}{\partial z} + y\right)y - \left(\frac{\partial f}{\partial y} - z\right)z\right] dy\, dz \tag{8-32}$$

Since displacements rather than stresses are chosen as the unknowns in this section, it is not necessary to introduce integrability conditions on the strains to ensure that they result in single-valued (compatible) displacements. Therefore, unlike the stress-function formulation in Sec. 8-2, the warping-function theory is applicable to multiply connected cross sections without additional conditions. It is only necessary to apply Eq. (8-27) to both the external and internal boundary curves.

Example 8-1 Apply the stress-function formulation to derive the equations for the stresses and torsional rigidity of a bar of elliptical cross section (Fig. 8-7). Use the results to obtain the solution for a circular section.

The equation of the boundary is $(y/a)^2 + (z/b)^2 - 1 = 0$. Since ψ must vanish on the boundary, let us try the stress function

$$\psi = C_1\left[\left(\frac{y}{a}\right)^2 + \left(\frac{z}{b}\right)^2 - 1\right] \tag{a}$$

where C_1 is a constant. This satisfies Eq. (8-8) and upon substitution into Eq. (8-5) gives $\nabla^2\psi = 2C_1(a^2 + b^2)/a^2b^2$, which is a constant, and so the stress function has been chosen correctly. To evaluate C_1, we substitute ψ into Eq. (8-11) to get

$$M_t = 2C_1\left(\frac{1}{a^2}\iint_A y^2\, dy\, dz + \frac{1}{b^2}\iint_A z^2\, dy\, dz - \iint_A dy\, dz\right) \tag{b}$$

The first, second, and third integrals in this equation are observed to be I_{zz}, I_{yy}, and A of the section. These are given by $I_{zz} = \pi a^3 b/4$, $I_{yy} = \pi ab^3/4$, and

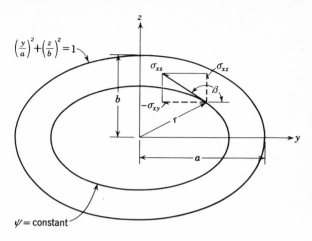

Fig. 8-7 Example 8-1, elliptical bar.

$A = \pi ab$, which when substituted into Eq. (b) yields $C_1 = -M_t/\pi ab$, so that
Eq. (a) becomes

$$\psi = -\frac{M_t}{\pi ab}\left(\frac{y^2}{a^2} + \frac{z^2}{b^2} - 1\right) \tag{c}$$

This equation shows that lines of shearing stress are ellipses with different
eccentricities than the boundary.

Substituting Eq. (c) into Eq. (8-16) produces the result

$$\theta = \frac{M_t(a^2 + b^2)}{G\pi a^3 b^3} \tag{d}$$

and from Eq. (8-17) it is seen that

$$J = \frac{\pi a^3 b^3}{a^2 + b^2} \tag{e}$$

Substituting Eq. (c) into Eqs. (8-3) gives

$$\sigma_{xy} = -\frac{2M_t z}{\pi ab^3} \qquad \sigma_{xz} = \frac{2M_t y}{\pi a^3 b} \tag{f}$$

The maximum resultant stresses occur at the ends of the minor axis.   Assuming
that $b < a$, the first of Eqs. (f) gives the maximum stress $\sigma_{s,\max} = \mp 2M_t/\pi ab^2$
for $z = \pm b$.   The resultant shearing stress at any point is found by substituting
Eqs. (f) into Eq. (2-5), giving

$$\sigma_{xs} = \frac{2M_t y}{\pi a^3 b}\left[1 + \left(\frac{z}{y}\right)^2 \left(\frac{a}{b}\right)^4\right]^{\frac{1}{2}} \tag{g}$$

The angle β that the resultant makes with the y axis is found from

$$\tan \beta = \frac{\sigma_{xz}}{\sigma_{xy}} = -\left(\frac{b}{a}\right)^2 \frac{y}{z} \tag{h}$$

Along any radial line y/z is a constant, and we see from Eqs. (g) and (h) that
σ_{xs} varies linearly and β is a constant along the line.

To obtain the solution for a circular cross section we let $a = b = R$, where R is the radius of the circle. In this case Eq. (d) reduces to

$$\theta = \frac{M_t}{GI_p} \tag{i}$$

where $I_p = \pi R^4/2$ is the polar moment of inertia of the circular section about the x axis. From Eq. (g), $\sigma_{xs} = 2M_t(y^2 + z^2)^{1/2}/\pi R^4$, which may be written

$$\sigma_{xs} = \frac{M_t r}{I_p} \tag{j}$$

where r is the radius to the point in the cross section with coordinates (y,z). The direction of the resultant shearing stress is found from Eq. (h) to be $\tan \beta = -y/z$, which is the negative reciprocal of the tangent of the angle that the radius vector makes with the y axis. This indicates that the resultant stress at any point is perpendicular to the radius vector to the point. Equations (i) and (j) are identical to the results for the circular section which are derived in elementary texts on the mechanics of materials. However, it should be noted that these results are not general but apply only to the circular section. In this respect the use of the symbol I_p is misleading, because it appears to imply that the equations may be used for other cross sections if the appropriate I_p is used.

Example 8-2   Show that circular cross section is solved by the warping function $f = 0$.
In this case Eq. (8-26) is satisfied, and Eq. (8-27) becomes $-zm + yn = 0$, which is satisfied because on the boundary of the circle $z = Rn$ and $y = Rm$. Therefore $f = 0$ is the correct solution, and from Eq. (8-25) we see that there is no warping of the cross section for the circular bar.
Substituting $f = 0$ into Eq. (8-32) yields

$$J = \iint\limits_A (y^2 + z^2)\, dy\, dz = I_p \tag{a}$$

which agrees with the result of Example 8-1. From Eqs. (8-30) we find $\sigma_{xy} = -M_t z/I_p$ and $\sigma_{xz} = M_t y/I_p$; this when substituted into Eq. (2-5) gives

$$\sigma_{xs} = \frac{M_t r}{I_p} \tag{b}$$

8-5 ANALYTICAL METHODS FOR APPROXIMATE SOLUTIONS

Exact solutions to the differential equations and boundary conditions of the torsion problem are known for relatively few cross-sectional shapes, and in general it is necessary to resort to approximate methods to obtain results. In this section we shall discuss the application of the Rayleigh-Ritz and finite-difference methods to such problems.

The Rayleigh-Ritz method may be used with either the total-potential or the total-complementary-potential principles (Sec. 6-12). By using both methods we can obtain upper and lower bounds for J. Considering the principle of the stationary value of the total potential first, we find from Eqs. (6-39), (6-50), and (8-29) that the strain energy per unit length of the member is

$$U = \frac{G\theta^2}{2} \iint\limits_{A} \left[\left(\frac{\partial f}{\partial y} - z \right)^2 + \left(\frac{\partial f}{\partial z} + y \right)^2 \right] dy \, dz \qquad (8\text{-}33)$$

The torques M_t that act upon a unit length do work in moving through the twist per unit length θ. The potential of these torques is $V = -M_t\theta$, and the total potential per unit length is therefore

$$U + V = \frac{G\theta^2}{2} \iint\limits_{A} \left[\left(\frac{\partial f}{\partial y} - z \right)^2 + \left(\frac{\partial f}{\partial z} + y \right)^2 \right] dy \, dz - M_t\theta \qquad (8\text{-}34)$$

The warping function may be approximated by the series

$$f = \sum_{j=1}^{n} c_j f_j(y,z) \qquad (8\text{-}35)$$

where the coefficients c_j are constants and the f_j are assumed functions. For an assumed set of functions f_j, the warping function is defined by the c_j, and we see from Eqs. (8-24) and (8-25) that this establishes the displacements for a given θ. The assumed displacements must satisfy displacement boundary conditions; however, the only boundary condition in the warping-function formulation is the equilibrium boundary condition of Eq. (8-27). As a result, the assumed functions f_j do not have to satisfy any boundary conditions, and it is necessary only that they be independent functions and that the series be capable of representing any warping function f in the region of the cross section. However, it is helpful to use physical judgment in choosing the f_j when only a few terms are used in the series.

The c_j constitute a set of generalized coordinates since they define the configuration of the twisted bar; and they may be determined from Eqs. (6-17) and (8-34), which give

$$\iint\limits_{A} \left[\left(\frac{\partial f}{\partial y} - z \right) \frac{\partial}{\partial c_i} \frac{\partial f}{\partial y} + \left(\frac{\partial f}{\partial z} + y \right) \frac{\partial}{\partial c_i} \frac{\partial f}{\partial z} \right] dy \, dz = 0 \qquad i = 1, 2, \ldots, n$$

By substituting Eq. (8-35) into this equation we find

$$\sum_{j=1}^{n} \left[\iint\limits_{A} \left(\frac{\partial f_i}{\partial y} \frac{\partial f_j}{\partial y} + \frac{\partial f_i}{\partial z} \frac{\partial f_j}{\partial z} \right) dy \, dz \right] c_j$$

$$= \iint\limits_{A} \left(z \frac{\partial f_i}{\partial y} - y \frac{\partial f_i}{\partial z} \right) dy \, dz \qquad i = 1, 2, \ldots, n \qquad (8\text{-}36)$$

Solution of this set of simultaneous linear algebraic equations [which are of the form of Eqs. (6-25) or (6-26)] yields the constants c_j, which when substituted into Eq. (8-35) give an approximate solution for f. Substitution of Eq. (8-35) into Eqs. (8-32) and (8-30) then gives J and the shear-stress components. Since Eq. (8-35) puts restraints on the warping deformations, the method underestimates deformations and therefore produces an upper bound to J.

Consider now the Rayleigh-Ritz method as applied to total complementary potential. The M_t-versus-θ curve is linear, and therefore $V' = V = -M_t\theta$, or from Eq. (8-11)

$$V' = -2\theta \iint_A \psi \, dy \, dz$$

An equation for U' is found from Eqs. (6-41), (6-51), (8-1), and (8-3). The total complementary potential per unit length is found to be

$$U' + V' = \frac{1}{2G} \iint_A \left[\left(\frac{\partial \psi}{\partial y} \right)^2 + \left(\frac{\partial \psi}{\partial z} \right)^2 \right] dy \, dz - 2\theta \iint_A \psi \, dy \, dz \quad (8\text{-}37)$$

The stress function can be approximated by the series solution

$$\psi(y,z) = \sum_{j=1}^{n} c_j \psi_j(y,z) \quad (8\text{-}38)$$

where the constants c_j are to be determined and the functions ψ_j are assumed. When the Rayleigh-Ritz method is applied to the complementary potential, the essential boundary conditions are the force boundary conditions. As a result the ψ_j must satisfy Eq. (8-8); i.e., they must vanish on C. The membrane analogy is helpful in choosing reasonable functions for ψ_j, especially when only a few terms are used in the series.

We recall that the stress function was defined so that equilibrium is always satisfied; as a result the variations of the stress that are associated with varying the coefficients c_j are self-equilibrating, and we may apply Eq. (6-66). The constants c_j are found by requiring $U' + V'$ to have a stationary value with respect to arbitrary variations in the c_j, so that $\partial(U' + V')/\partial c_i = 0$ for $i = 1, 2, \ldots, n$. Substituting Eq. (8-37) into this equation, we find

$$\frac{1}{G} \iint_A \left(\frac{\partial \psi}{\partial y} \frac{\partial}{\partial c_i} \frac{\partial \psi}{\partial y} + \frac{\partial \psi}{\partial z} \frac{\partial}{\partial c_i} \frac{\partial \psi}{\partial z} \right) dy \, dz - 2\theta \iint_A \frac{\partial \psi}{\partial c_i} dy \, dz \quad i = 1, 2, \ldots, n$$

and by using Eq. (8-38) we obtain

$$\sum_{j=1}^{n} \left[\frac{1}{G} \iint_A \left(\frac{\partial \psi_i}{\partial y} \frac{\partial \psi_j}{\partial y} + \frac{\partial \psi_i}{\partial z} \frac{\partial \psi_j}{\partial z} \right) dy \, dz \right] c_j = 2\theta \iint_A \psi_i \, dy \, dz$$

$$i = 1, 2, \ldots, n \quad (8\text{-}39)$$

The solution of this set of equations gives the constants c_j, and substitution into Eq. (8-38) gives an approximate solution for ψ. Substitution of Eq. (8-38) into Eqs. (8-3) and (8-18) then gives approximate results for the stresses and J. Equation (8-38) imposes restraints on the stresses, which leads to a lower bound for J.

A variation of the Rayleigh-Ritz method which reduces the two-dimensional problem to a set of ordinary differential equations is also useful in the torsion problem. This technique was used by Argyris and Kelsey[6] to obtain bounds to J for members with thin cross sections.

The finite-difference method can be used to obtain approximate values of ψ or f at a grid of points in the cross section. Both formulations follow the procedure for equilibrium boundary-value problems outlined in Sec. 5-3 that reduces the problem to a set of linear algebraic equations in which the unknowns are the values of the function at the mesh points. Since the boundary condition for ψ is simpler than that for f, the method will be illustrated for the stress-function formulation. To simplify the calculations, values of ψ for the torque $M_t = J$ associated with $\theta = 1/G$ will be determined. Since the system is linear, the results can be multiplied by M_t/J to obtain the solution for other torques. Equation (8-16) reduces to $\nabla^2 \psi = -2$ for $\theta = 1/G$ and becomes

$$\psi_{j-1,k} + \psi_{j,k-1} - 4\psi_{j,k} + \psi_{j+1,k} + \psi_{j,k+1} = -2h^2 \qquad (8\text{-}40)$$

when the ∇^2 module in Fig. 5-5 is used.

For simplicity we shall consider only problems in which the boundary passes through grid points; techniques for the more general case are given in Refs. 2 and 5. From Eq. (8-8), $\psi = 0$ at all boundary mesh points, and the number of unknown values of ψ is equal to the number of interior mesh points. Sufficient equations to determine these values of ψ are obtained by applying Eq. (8-40) to each of the interior mesh points. As noted earlier, $M_t = J$ for $\theta = 1/G$, so that J can be found from Eq. (8-11) by using one of the two-dimensional integration modules of Fig. 5-5.

The maximum stress occurs on the boundary and is found from Eq. (8-21). Forward- or backward-difference formulas are required to evaluate the derivatives in this formula because ψ is known only at boundary and interior grid points. The difference formulas of Prob. 5-1 can be used for this purpose. The stresses at interior points are found by applying the central-difference formulas for the first derivatives in Fig. 5-4 to Eqs. (8-3) and using Eq. (2-5).

Example 8-3 Use the Rayleigh-Ritz method to obtain the solution for the rectangular cross section shown in Fig. 8-8, where $a \geq b$.

We note from the membrane analogy that ψ should be symmetric about both axes. A series solution which satisfies this requirement and the essential

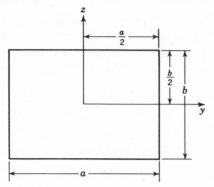

Fig. 8-8   Example 8-3, rectangular bar.

boundary condition of Eq. (8-8) is given by

$$\psi = \sum_{m=1,3,5}^{\infty} \sum_{n=1,3,5}^{\infty} a_{mn} \cos \frac{m\pi y}{a} \cos \frac{n\pi z}{b} \qquad (a)$$

where the constants a_{mn} correspond to the c_j coefficients in Eq. (8-38) and the cosine terms are the functions ψ_j. Taking $c_i = a_{rs}$, Eq. (8-39) gives

$$\sum_{m=1,3,5}^{\infty} \sum_{n=1,3,5}^{\infty} \left[\frac{1}{G} \int_{-b/2}^{b/2} \int_{-a/2}^{a/2} \left(\frac{\pi^2 mr}{a^2} \sin \frac{m\pi y}{a} \cos \frac{n\pi z}{b} \sin \frac{r\pi y}{a} \cos \frac{s\pi z}{b} \right. \right.$$

$$\left. \left. + \frac{\pi^2 ns}{b^2} \cos \frac{m\pi y}{a} \sin \frac{n\pi z}{b} \cos \frac{r\pi y}{a} \sin \frac{s\pi z}{b} \right) dy\, dz \right] a_{mn}$$

$$= 2\theta \int_{-b/2}^{b/2} \int_{-a/2}^{a/2} \cos \frac{r\pi y}{a} \cos \frac{s\pi z}{b} dy\, dz \qquad \begin{array}{l} r = 1,\, 3,\, 5,\, \ldots \\ s = 1,\, 3,\, 5,\, \ldots \end{array}$$

The integral on the left vanishes unless $m = r$ and $n = s$, in which case the equation reduces to

$$\frac{\pi^2 ab}{4G} \left[\left(\frac{r}{a} \right)^2 + \left(\frac{s}{b} \right)^2 \right] a_{rs} = \frac{8\theta ab}{\pi^2 rs} (-1)^{(r+s)/2-1}$$

so that

$$a_{rs} = \frac{32 G \theta b^2 (-1)^{(r+s)/2-1}}{\pi^4 rs [(b/a)^2 r^2 + s^2]}$$

Substituting this result into Eq. (a), we find

$$\psi = \frac{32 G \theta b^2}{\pi^4} \sum_{m=1,3,5}^{\infty} \sum_{n=1,3,5}^{\infty} \frac{(-1)^{(m+n)/2-1}}{mn[(b/a)^2 m^2 + n^2]} \cos \frac{m\pi y}{a} \cos \frac{n\pi z}{b} \qquad (8-41)$$

By inserting Eq. (8-41) into Eq. (8-11) we obtain

$$M_t = \frac{256 G \theta ab^3}{\pi^6} \sum_{m=1,3,5}^{\infty} \sum_{n=1,3,5}^{\infty} \frac{1}{m^2 n^2 [(b/a)^2 m^2 + n^2]}$$

so that from Eq. (8-17)

$$J = \beta a b^3 \tag{8-42}$$

where

$$\beta = \frac{256}{\pi^6} \sum_{m=1,3,5}^{\infty} \sum_{n=1,3,5}^{\infty} \frac{1}{m^2 n^2 [(b/a)^2 m^2 + n^2]} \tag{b}$$

The series for β converges rapidly and is plotted in Fig. 8-9 as a function of a/b. We note from this figure that β approaches $\frac{1}{3}$ as a/b becomes large.

From the membrane analogy we note that the maximum membrane slope occurs at the centers of the long edges. The maximum shear stresses occur at the same points and are found by substituting Eq. (8-41) into the first of Eqs. (8-3) and letting $y = 0$, $z = b/2$, and $\theta = M_t/GJ$. This gives

$$\sigma_{xs,\text{max}} = \frac{M_t}{\alpha a b^2} \tag{8-43}$$

where

$$\frac{1}{\alpha} = \frac{32}{\pi^3 \beta} \sum_{m=1,3,5}^{\infty} \sum_{n=1,3,5}^{\infty} \frac{(-1)^{(m+2n-1)/2}}{m[(b/a)^2 m^2 + n^2]} \tag{c}$$

The parameter α is plotted as a function of a/b in Fig. 8-9, where we observe that α approaches $\frac{1}{3}$ as a/b becomes large. The infinite series of Eqs. (8-42) and (8-43) are exact solutions since the series are complete.

Example 8-4 Determine J and the maximum shear stress for a square cross section by the finite-difference method.

From the membrane analogy we note that ψ is doubly symmetric and take advantage of this fact to reduce the number of unknowns by numbering the mesh points as shown in Fig. 8-10. Noting that $\psi = 0$ at the boundary mesh points, we find the following equation by successively applying Eq. (8-40) to points 1 to 3:

$$\begin{bmatrix} -4 & 2 & 1 \\ 2 & -4 & 0 \\ 4 & 0 & -4 \end{bmatrix} \begin{Bmatrix} \psi_1 \\ \psi_2 \\ \psi_3 \end{Bmatrix} = -\frac{a^2}{8} \begin{Bmatrix} 1 \\ 1 \\ 1 \end{Bmatrix}$$

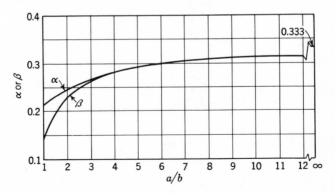

Fig. 8-9 Constants for rectangular bars.

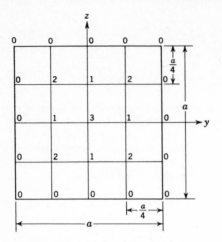

Fig. 8-10 Example 8-4, difference mesh for square bar.

Solving this equation, we find

$$\{\psi_1 \quad \psi_2 \quad \psi_3\} = a^2\{0.1094 \quad 0.0860 \quad 0.1406\}$$

In Eq. (8-40) we assumed that $\theta = 1/G$, so that $J = M_t$, which can be found from Eq. (8-11). Applying the Simpson's rule module from Fig. 5-5 to each quadrant of the square, we find

$$J = \frac{2 \times 4 \times a^2}{9 \times 16}(4\psi_1 + 16\psi_2 + \psi_3 + 4\psi_1) = 0.1328a^4$$

The exact solution obtained from Eq. (8-42) and Fig. 8-9 with $a/b = 1$ is $J = 0.1406a^4$, so that the finite-difference method with the coarse grid of Fig. 8-10 underestimates J by 5.6 percent.

From the membrane analogy, the maximum shear stress occurs at the center of the sides. Using Eq. (8-21) and the second equation in Prob. 5-1 (with y replacing x), we find for the point $(a/2, 0)$

$$\sigma_{zs,\text{max}} = -\frac{4}{2a}(0 - 4 \times 0.1094a^2 + 0.1406a^2) = 0.594a$$

for $M_t = J$. For an arbitrary M_t

$$\sigma_{zs,\text{max}} = 0.594a\frac{M_t}{J} = \frac{0.594aM_t}{0.1328a^4} = \frac{4.473M_t}{a^3}$$

which is 7 percent below the exact solution of $4.808M_t/a^3$ obtained from Eq. (8-43) and Fig. 8-9 with $a/b = 1$.

8-6 THIN-WALLED OPEN SECTIONS

Many structural members such as channel, I, and T sections are thin-walled. Exact solutions for these members are not known, but approximate methods exist for their analysis. These methods are based upon the

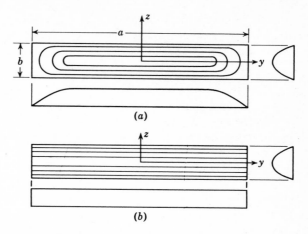

Fig. 8-11 Membrane contours for a slender rectangle. (*a*) Actual; (*b*) approximation.

equations for the slender rectangle shown in Fig. 8-11. By using the membrane analogy to ψ we see that except for small regions at the ends, the membrane surface has essentially the same shape at all points along its length. If we ignore the end regions and assume that ψ is independent of y, Eq. (8-16) reduces to $d^2\psi/dz^2 = -2G\theta$. Integrating twice and evaluating the constants of integration from the boundary condition $\psi = 0$ for $z = \pm b/2$, we find

$$\psi = -G\theta \left[z^2 - \left(\frac{b}{2}\right)^2 \right] \qquad (8\text{-}44)$$

We note that ψ has a parabolic variation in the z direction. Since ψ does not vanish on the short sides, it is not the true solution, which is given by the more complicated expression of Eq. (8-41). However, this is of little importance if the rectangle is slender, for the volume under the actual membrane (which is proportional to J) will differ only slightly from that beneath the assumed surface. Furthermore, the maximum stress occurs at the center of the long sides, where the slope of the membrane is

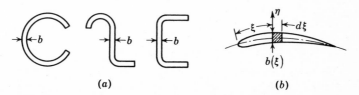

Fig. 8-12 (*a*) Thin bent-up sections; (*b*) variable-thickness sections.

essentially unaffected by the fact that the actual membrane is held down along the short edges. The torsion constant found by substituting Eq. (8-44) into Eq. (8-18) is

$$J = \frac{ab^3}{3} \tag{8-45}$$

Except in the vicinity of the ends, the shearing stress on the long sides is obtained by substituting Eq. (8-44) into the first of Eqs. (8-3) and using Eqs. (8-17) and (8-45), which gives

$$\sigma_{xy,\max} = \frac{3M_t}{ab^2} \tag{8-46}$$

Comparing Eqs. (8-45) and (8-46) with (8-42) and (8-43) and utilizing Fig. 8-9, we see that the results of the approximate and exact solutions coincide as a/b approaches infinity and that for $a/b > 10$ the approximate solution is in error by less than 6.5 percent.

From the membrane analogy we see that the solution for the narrow rectangle may be used as an approximation for sections formed from flat sheet (such as those in Fig. 8-12a) if the corner bends have generous radii so that large stress concentrations do not occur at reentrant corners. In such cases the dimension a is taken as the developed length of the cross section, and b is the thickness of the material from which the member is formed. The method can be modified to include cross sections which have a variable thickness, such as that shown in Fig. 8-12b. For these sections the thickness b is a function of the coordinate ξ measured along the midline of the section. At any point along the length it is assumed that the analogous membrane has the same parabolic shape and height as a thin rectangle with b equal to the local thickness, so that

$$\psi = -G\theta \left[\eta^2 - \left(\frac{b}{2}\right)^2 \right] \tag{8-47}$$

where η is a coordinate measured normal to ξ. The torque obtained from Eq. (8-11) is then

$$M_t = -2G\theta \int_0^a \int_{-b/2}^{b/2} \left[\eta^2 - \left(\frac{b}{2}\right)^2 \right] d\eta \, d\xi = \frac{G\theta}{3} \int_0^a b^3 \, d\xi$$

Substituting this result into Eq. (8-17), we find

$$J = \frac{1}{3} \int_0^a b^3 \, d\xi \tag{8-48}$$

The maximum stress occurs at the point on the boundary where the slope of the analogous membrane is the greatest, which is located at the point of maximum thickness. From Eq. (8-21) (with $n = \eta$) and Eq. (8-47) we find for $\eta = \pm b_{max}/2$ that $\sigma_{zs,max} = G\theta b_{max}$.   Using Eqs. (8-17) and (8-48), this becomes

$$\sigma_{zs,max} = \frac{3M_t b_{max}}{\displaystyle\int_0^a b^3 \, d\xi} \tag{8-49}$$

A more accurate method of analysis for sections of this type is described in Ref. 6.

Members with angle, channel, I, T, and similar cross sections are often produced by extrusion or rolling processes.   In these cases it is common practice to use fillets at reentrant corners as shown in Fig. 8-13 to reduce stress concentrations.   These fillets may increase the torsional rigidity significantly, so that the sections cannot be accurately analyzed by the method discussed for members that are bent up from flat sheet. This may be seen in Fig. 8-5, which shows the experimentally determined membrane contour lines of the flange and web junction areas of T and angle sections.   A bulge occurs at the intersection of the rectangles that make up the cross section, which increases the volume under the membrane and results in a greater torsional rigidity.   Trayer and March[7] developed a semiempirical method for these sections which is based upon membrane tests and the torsional theory for the rectangle.   In this method the volume under the membrane is taken as the sum of the volumes of the rectangular elements of the cross section and the bulges which occur at the junction points.

The rectangular elements in this method are divided into three classes, as shown in Fig. 8-13a to c.   In case 1, illustrated in Fig. 8-13a,

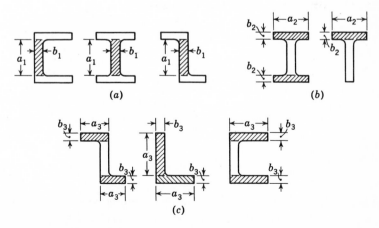

Fig. 8-13   Rolled and extruded sections.   (a) Webs; (b) flanges with center juncture; (c) flanges with edge juncture.

the rectangle is a web member. For such an element the analogous membrane is not held down on the short sides of the rectangle where it joins the flanges. The membrane shape therefore corresponds to that of a slender rectangle, and the torsion-constant contribution of this element, obtained from Eq. (8-45), is

$$J_1 = \tfrac{1}{3}a_1 b_1{}^3 \tag{8-50}$$

For rectangular elements similar to those illustrated in Fig. 8-13b, the membrane shapes correspond to rectangles of the same proportions. We then find from Eq. (8-42) that

$$J_2 = \beta a_2 b_2{}^3 \tag{8-51}$$

where β is obtained from Fig. 8-9 with $a/b = a_2/b_2$. The membranes for rectangular segments similar to those shown in Fig. 8-13c are held down on three sides but not on the side which is attached to the adjoining rectangle. For this reason its contribution to the torsion constant is taken as

$$J_3 = \beta a_3 b_3{}^3 \tag{8-52}$$

where β is found from Fig. 8-9 with $a/b = 2a_3/b_3$.

At juncture points the bulges in the membrane increase the volume under the membrane, as shown in Fig. 8-5. Based upon membrane measurements, Ref. 7 recommends the equation

$$J_4 = \alpha D^4 \tag{8-53}$$

for the torsion-constant contribution of this effect. In this equation α is a coefficient from Fig. 8-14, and D is the diameter of the largest circle which can be inscribed in the junction, including the fillets. The torsion contant for the entire section is then given by

$$J = \sum_{i=1}^{n_1} J_{1_i} + \sum_{i=1}^{n_2} J_{2_i} + \sum_{i=1}^{n_3} J_{3_i} + \sum_{i=1}^{n_4} J_{4_i} \tag{8-54}$$

where n_1 to n_4 are the number of elements in cases 1 to 4, respectively.

The maximum shearing stresses in members similar to those shown in Fig. 8-13 usually occur in the fillets, and therefore they cannot be computed from Eq. (8-49). A method for computing the stresses in these areas is given in Ref. 8.

8-7 THIN-WALLED CLOSED SECTIONS

The torsional rigidities of thin-walled open sections are relatively low, and the shearing stresses in these members are high. The most efficient method of increasing torsional rigidity and decreasing stresses is to use a closed rather than an open section. Except for the circular cross section,

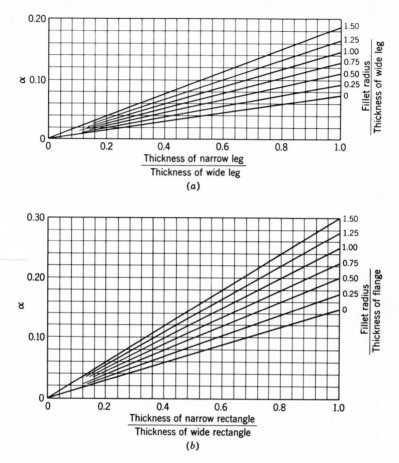

Fig. 8-14 Values of α for rolled and extruded sections (Ref. 7).   (a) L junctions; (b) T junctions.

thick-walled hollow sections are more difficult to analyze than solid sections.   However, when the walls of the hollow section are thin, the solution is greatly simplified.   We have seen that in general the direction and distribution of the torsional shear stresses are unknown at the outset of the analysis.   In a hollow section the resultant shearing stress must be in the direction of the tangents to the boundary at the inner and outer boundaries of the wall.   Since these directions are very nearly parallel in a thin-walled section, we may assume that the resultant shearing stresses throughout the wall are in the direction of the tangent to a median line drawn through the middle of the wall thickness.   Furthermore, except in the region of sharp corners, it is reasonable to assume that there is no

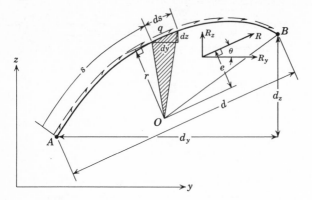

Fig. 8-15   Resultant of a constant shear flow.

appreciable variation in the magnitude of the stress through the thickness of a thin wall.

The fact that the stresses are uniform through the wall thickness makes it possible to deal with the product of the shearing stress and the wall thickness instead of the stress without any ambiguity.   For reasons which will become apparent we define this product as the *shear flow q*; thus

$$q = \sigma_{xs}t \tag{8-55}$$

where t is the wall thickness.   We see that the units of q are pounds per unit length of the wall.

Before deriving the torsion equations for closed thin-walled sections, let us consider some useful relations for a thin wall of arbitrary shape which is acted upon by a constant shear flow.   Referring to Fig. 8-15, the force on a differential length of the wall ds is $q\,ds$, and so the z component of the resultant force of the shear flow between A and B is

$$R_z = \int_A^B \frac{dz}{ds} q\,ds$$

Since q is constant, this reduces to

$$R_z = qd_z \tag{8-56}$$

where d_z is the z component of the distance from A to B.   In a similar fashion the y component is

$$R_y = qd_y \tag{8-57}$$

The resultant force is then $R = q(d_y{}^2 + d_z{}^2)^{1/2}$, which reduces to

$$R = qd \tag{8-58}$$

where d is the length of the straight line drawn from A to B. The angle which R makes with the y axis is $\theta = \tan^{-1}(R_z/R_y) = \tan^{-1}(d_z/d_y)$, which is the same angle that the straight line from A to B makes with the y axis.

Referring to Fig. 8-15, the resultant torque of q about an arbitrary moment center O is

$$M_t = \int_A^B rq\,ds$$

where r is the perpendicular distance from O to the line segment ds. Since q is constant, M_t becomes

$$M_t = q \int_A^B r\,ds = 2q \int_A^B dA \qquad (8\text{-}59)$$

where $dA = (r\,ds)/2$ is the area of the shaded triangular element and the integral is the area enclosed by the curved line AB and the radial lines OA and OB. Designating this area by A, we find

$$M_t = 2Aq \qquad (8\text{-}60)$$

The resultant force R must produce the same moment as q, so that $M_t = Re$, where e is the perpendicular distance from O to the line of action of R. Substituting Eq. (8-60) into this equation, we find

$$e = \frac{2A}{d} \qquad (8\text{-}61)$$

In using Eqs. (8-60) and (8-61) we must observe caution with sign conventions and the appropriate area for A. In this regard, it must be noted that the sign of M_t may not follow automatically from the sign of q but should be established by statics from a sketch of the problem. This is especially true when the lines OA or OB cross the curved line AB, as shown in Fig. 8-16a. If we consider counterclockwise torques to be positive, we find from Eq. (8-59) that

$$M_t = 2q \left(\int_A^C dA - \int_C^D dA + \int_D^B dA \right)$$

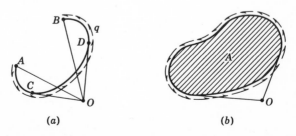

(a)   (b)

Fig. 8-16 Moment of a constant shear flow. (*a*) Open curve; (*b*) closed curve.

so that in using Eqs. (8-60) and (8-61) we should take

$$A = A_{OAC} - A_{ODC} + A_{ODB}$$

If the line along which q acts is closed, as shown in Fig. 8-16b, then from Eq. (8-58) $R = 0$, and the resultant is simply a couple given by Eq. (8-60), where A is the area bounded by the closed curve.

Now consider a thin-walled shell structure of arbitrary constant cross section, as shown in Fig. 8-17a. The area bounded by the outer wall is subdivided into an arbitrary number of cells, which are separated by thin webs. It is assumed that cross-sectional changes during twisting are prevented by transverse stiffening members (ribs or frames), which are considered to be rigid within their planes but perfectly flexible with regard to deformations normal to their planes. The structure is loaded by torques applied to the ends in the same manner that the shear stresses within the structure are distributed. It is further assumed that there is no warping restraint at the end cross sections.

With a constant torque and no warping restraint no longitudinal stresses are developed, and the only internal forces are the shearing stresses which result in the shear flows q. Let us consider the equilibrium of a portion of one of the walls between two points A and B, which lie

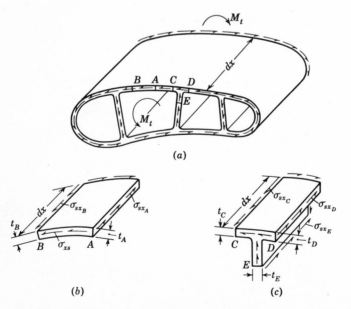

Fig. 8-17 (a) Torsion of a multicell structure; (b) free body of wall section; (c) free body of wall junction.

between wall junction points as shown in Fig. 8-17b. From the free-body diagram of this element we obtain the following equation for force equilibrium in the x direction:

$$\sigma_{xs_A} t_A - \sigma_{xs_B} t_B = 0$$

where t_A and t_B are the wall thicknesses at points A and B, respectively. By using Eq. (8-55) we may write this equation as $q_A = q_B$, which proves that though the wall thickness and shearing stress may vary, the shear flow in any wall is constant between the points where it joins with other walls.

Next consider the equilibrium of the shearing stresses at a point where an interior wall joins an exterior wall, as shown in Fig. 8-17c. At such a point the equilibrium of forces in the x direction leads to

$$\sigma_{xs_C} t_C - \sigma_{xs_D} t_D - \sigma_{xs_E} t_E = 0$$

which, by Eq. (8-55), becomes $q_E = q_C - q_D$. We see from this equation that the shear flow in any interior wall may be expressed in terms of the shear flows in the adjoining exterior walls. In general, when any number of walls intersect, the sum of the shear flows into the junction must equal the sum of the shear flows out of the junction for equilibrium to exist. An analogy then exists between the torsion problem and the flow of a fluid through a network of channels corresponding to the walls of the cross section. The flow rate must be constant at all points along each channel, and the sum of the flow rates into a junction must be equal to the sum of the flow rates out. Therefore the term *shear flow* is used.

Because the shear flow in any wall is constant between points where it joins other walls and the shear flow in any interior wall is equal to the difference of the shear flows in the adjoining exterior walls, we may select the unknowns in a thin-walled multicell structure as shown in Fig. 8-18. The shear flow around the ith cell is designated q_i, and so the shear flows in the outer walls of this cell are equal to q_i, and the shear flow in the web between cells i and $i + 1$ is $q_i - q_{i+1}$. A structure with n cells will therefore have n unknown shear flows. The shear flows must be in equilibrium with the applied torque. Using Eq. (8-60) to determine the resisting

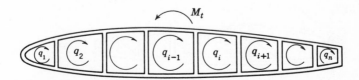

Fig. 8-18 Reacting shear flows in a multicell structure.

torque of each of the shear flows q_i, we find that for equilibrium

$$M_t = 2 \sum_{i=1}^{n} q_i A_i \qquad (8\text{-}62)$$

where A_i is the area enclosed by the midline of the ith cell. We note from this equation that if the cross section consists of but a single cell, the problem is statically determinate, and the shear flow is

$$q = \frac{M_t}{2A} \qquad (8\text{-}63)$$

and, from Eq. (8-55), the shear stress is given by

$$\sigma_{zs} = \frac{M_t}{2At} \qquad (8\text{-}64)$$

where t is the wall thickness at the point where the stress is determined. We note that this equation was derived from equilibrium considerations alone, and the results are therefore applicable to inelastic as well as elastic deformations.

For n cells there are an equal number of unknown shear flows and only the single equation of equilibrium (8-62), so that the problem is statically indeterminate, and the additional $n - 1$ relationships required for a solution must come from continuity of deformations. These equations are obtained from the compatibility condition that all cells must rotate through the same angle of twist per unit length θ if the cross sectional shape is maintained by ribs or frames which are rigid in their own planes. To derive these expressions we require an equation for θ of a thin-walled structure with an arbitrary shear-flow distribution. This equation will be derived by the unit-load method described in Sec. (6-14).

Consider a unit length of the multicell structure shown in Fig. 8-19 that has a shear-flow distribution q. The angle of twist at the rear face is

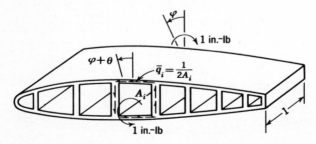

Fig. 8-19 Unit-load system for a multicell structure.

φ, and that of the front face is $\varphi + \theta$; these deformations and q may be the result of an applied torque, lateral shear force, or a thermal gradient. From Eqs. (8-55) and (3-5), the real strains that are associated with q are $\epsilon_{zs} = q/Gt$. To determine θ we assume a pair of unit torques to be applied to the structure, as shown in Fig. 8-19, and to maintain equilibrium with these torques we imagine a shear-flow distribution consisting of a constant shear flow $\bar{q}_i = 1/2A_i$ around the ith cell while all other shear flows are zero. From Eq. (8-55) the only nonzero stress in the unit-load system is then $\bar{\sigma}_{zs} = 1/2A_i t$ around the ith cell. Taking the real displacements and strains as the virtual displacements and strains and noting that the virtual work of the unit torque on the back face is $-\varphi$ while that of the unit torque on the front face is $\varphi + \theta$, we find $\delta W_e = \theta$. Equation (6-88) then becomes

$$\theta = \oint_i \bar{\sigma}_{zs} \epsilon_{zs} t \, ds = \oint_i \frac{1}{2A_i t} \frac{q}{Gt} t \, ds$$

where the line integral extends around the wall of the ith cell and A_i is the area of this cell. This equation may be written

$$\theta = \frac{1}{2A_i G_1} \oint_i \frac{q \, ds}{t^*} \tag{8-65}$$

where G_1 is an arbitrary reference modulus and t^* is a modulus-weighted thickness defined by

$$t^* = \frac{G}{G_1} t \tag{8-66}$$

When the cross section is homogeneous, the subscript on G and the asterisk on t may be dropped in Eq. (8-65).

In evaluating the line integral in Eq. (8-65) we take the positive directions of q and ds in the same direction as the unit torque on the front face. As a result, θ (the work of the unit torque) is positive when it is in the direction of the unit torque on the front face.

As noted, Eq. (8-65) is not restricted to torsion problems but also applies when q is due to a lateral shear force or a temperature gradient. When only a torque is applied, q is constant in each wall segment, and Eq. (8-65) may be written

$$\theta = \frac{1}{2A_i G_1} \left(q_i \oint_i \frac{ds}{t^*} - q_{i-1} \int_{\substack{\text{web} \\ i-1,i}} \frac{ds}{t^*} - q_{i+1} \int_{\substack{\text{web} \\ i,i+1}} \frac{ds}{t^*} \right) \tag{8-67}$$

where q_i is the constant shear flow around the ith cell and q_{i-1} and q_{i+1} are the shear flows around the $(i-1)$th and $(i+1)$th cells, respectively. Equation (8-67) may be applied to each of the n cells, and by equating θ

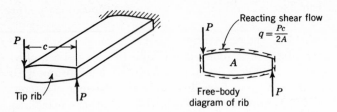

Fig. 8-20   Use of shear flows to determine rib loads.

for the first and ith cells we can write $n - 1$ equations of the form

$$\frac{1}{A_1}\left(q_1 \oint_1 \frac{ds}{t^*} - q_2 \int_{\substack{\text{web} \\ 1,2}} \frac{ds}{t^*}\right)$$

$$= \frac{1}{A_i}\left(q_i \oint_i \frac{ds}{t^*} - q_{i-1} \int_{\substack{\text{web} \\ i-1,i}} \frac{ds}{t^*} - q_{i+1} \int_{\substack{\text{web} \\ i,i+1}} \frac{ds}{t^*}\right) \quad (8\text{-}68)$$

by letting i run from 2 to n.   The simultaneous solution of these equations with Eq. (8-62) gives the n unknown values of q_i.   The value of θ can then be found by evaluating Eq. (8-67) for any one of the cells.   The torsional rigidity $G_1 J^*$ of the nonhomogeneous section can then be found from the defining equation

$$G_1 J^* = \frac{M_t}{\theta} \quad (8\text{-}69)$$

For a single-cell structure, we find from Eqs. (8-63) and (8-67) that

$$\theta = \frac{M_t}{4 A^2 G_1} \oint \frac{ds}{t^*} \quad (8\text{-}70)$$

which, for the homogeneous case, gives

$$J = \frac{4 A^2}{\oint \dfrac{ds}{t}} \quad (8\text{-}71)$$

The theory for thin-walled closed sections was developed by Bredt, and Eqs. (8-62) and (8-65) or those developed from them for the single cell are referred to as *Bredt's equations*.

The shear flows are used in the design of skins and interior webs, ribs and frames, and fasteners at skin splices, skin-web junctions, and joints where the ribs or frames meet the skins or webs.   The shearing stresses in the skins and webs are required to determine whether these members buckle and fail in diagonal tension.   The shear buckling of rectangular panels is discussed in Sec. 15-4.   Concentrated loads are introduced into shell structures by applying them to ribs or frames which distribute the loads into the skins and webs.   An example of this is shown in Fig. 8-20, where the couple on the end rib is maintained in equilibrium by the reacting shear flows in the skins and webs.   These reacting shear flows together with the applied couple permit a free-body diagram of the rib to be drawn, and this in turn can be used to determine the stresses in the rib. Problems of this type are discussed in Sec. 9-7.

The forces on the fasteners of a joint can be determined once the shear flow that is transmitted through the joint is known.   The shear flow gives the force that is transmitted across a unit length of the joint, so that

$$P_f = \frac{q}{n} \tag{8-72}$$

where P_f is the shear force per fastener and n is the number of fasteners per unit length of the joint.

Example 8-5   The cross section in Fig. 8-21 is the aft section of the fuselage of a jet airplane.   The lower surface, which is exposed to the jet exhaust, is made of stainless steel ($G = 11.5 \times 10^6$ psi), and the rest of the skin is 7075-T6 aluminum ($G = 3.9 \times 10^6$ psi).   The skin splice at the junction of the steel and aluminum consists of two rows of rivets with spacings of $\frac{3}{4}$ in.   The section is subjected to a torque of 100,000 in.-lb.   Determine (a) the shearing stresses in the aluminum and steel, (b) the angle of twist which occurs over a 50-in. length of the structure, and (c) the force per rivet in the splice.

The enclosed area is $A = \pi \times 30^2 = 2830$ in.^{2}, so that the stresses obtained from Eq. (8-64) are

$$\sigma_{xs,\text{alum}} = \frac{100,000}{2 \times 2830 \times 0.051} = 346 \text{ psi}$$

$$\sigma_{xs,\text{steel}} = \frac{100,000}{2 \times 2830 \times 0.064} = 276 \text{ psi}$$

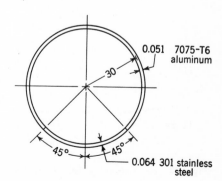

Fig. 8-21   Example 8-5.

The value of θ is found from Eq. (8-70) where (since t^* is piecewise constant) we may express the line integral as $\Sigma \, \Delta s_i / t_i^*$. Noting that

$$\Delta s_{\text{alum}} = \tfrac{3}{4}(60\pi) = 141.5 \text{ in.} \qquad \Delta s_{\text{steel}} = \tfrac{1}{4}(60\pi) = 47.2 \text{ in.}$$

and taking $G_1 = 3.9 \times 10^6$ psi, so that

$$t_{\text{alum}}^* = 0.051 \text{ in.} \qquad t_{\text{steel}}^* = 0.064 \frac{11.5}{3.9} = 0.1888 \text{ in.}$$

we find from Eq. (8-70)

$$\theta = \frac{100,000}{4 \times 2830^2 \times 3.9 \times 10^6} \left(\frac{141.5}{0.051} + \frac{47.2}{0.1888} \right) = 2.42 \times 10^{-6} \text{ rad/in.}$$

The total angle of twist for the 50-in. length is $\varphi = \theta L = 2.42 \times 10^{-6} \times 50 = 1.21 \times 10^{-4}$ rad.

From Eq. (8-63), $q = 100,000/2 \times 2830 = 17.66$ lb/in. The number of rivets per inch is $n = 2 \times \tfrac{4}{3} = 2.67$, so that from Eq. (8-72) $P_f = 17.66/2.67 = 6.61$ lb per rivet.

Example 8-6 The three-cell homogeneous wing section shown in Fig. 8-22a is subjected to a torque of 100,000 in.-lb. Determine the shear-flow distribution and J.

From Eq. (8-62) we find

$$2 \times 70q_1 + 2 \times 165q_2 + 2 \times 100q_3 = 100,000 \tag{a}$$

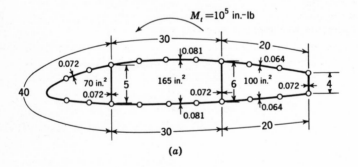

(a)

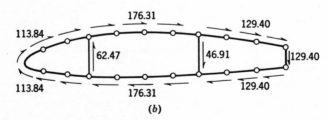

(b)

Fig. 8-22 Example 8-6. (a) Geometry and loading; (b) results.

By applying Eq. (8-68) for $i = 2$ and $i = 3$ we respectively obtain

$$\frac{1}{70}\left[q_1\left(\frac{40}{0.072} + \frac{5}{0.072}\right) - q_2\frac{5}{0.072}\right]$$

$$= \frac{1}{165}\left[q_2\left(\frac{30}{0.081} + \frac{5}{0.072} + \frac{30}{0.081} + \frac{6}{0.072}\right) - q_1\frac{5}{0.072} - q_3\frac{6}{0.072}\right] \quad (b)$$

$$\frac{1}{70}\left[q_1\left(\frac{40}{0.072} + \frac{5}{0.072}\right) - q_2\frac{5}{0.072}\right]$$

$$= \frac{1}{100}\left[q_3\left(\frac{20}{0.064} + \frac{6}{0.072} + \frac{20}{0.064} + \frac{4}{0.072}\right) - q_2\frac{6}{0.072}\right] \quad (c)$$

Solving Eqs. (a) to (c) simultaneously, we obtain

$$q_1 = 113.84 \text{ lb/in.} \qquad q_2 = 176.31 \text{ lb/in.} \qquad q_3 = 129.40 \text{ lb/in.}$$

and the shear-flow distribution shown in Fig. 8-22b. By applying Eq. (8-67) to the first cell we find

$$\theta = \frac{1}{2 \times 70G}\left[113.84\left(\frac{40}{0.072} + \frac{5}{0.072}\right) - 176.31\frac{5}{0.072}\right] = \frac{420.76}{G} \quad \text{rad/in.}$$

and from Eq. (8-17) we obtain $J = 100,000/420.76 = 237.7$ in.[4].

8-8 ACCURACY OF TORSION THEORY

Two formulations of the torsion problem, one in terms of ψ and the other in terms of f, have been given in Secs. 8-2 and 8-4. These methods lead to exact results (within the limits of the theory of elasticity, i.e., for small deflections and stresses below the elastic and proportional limits) if solutions can be found to the differential equations and boundary conditions. For the solution to be exact the bar must (1) be of uniform cross section, (2) have no warping restraint at any of its cross sections, and (3) be loaded by torques applied to the ends by surface forces which are distributed in the same manner as the computed internal stresses. These same restrictions apply to the Bredt theory for thin-walled closed sections. In practical applications these assumptions may be violated; in some cases the induced errors may be serious, while in other instances they may be negligible.

Michell has derived the differential equation which leads to an exact solution for the bar with a variable circular cross section (see Ref. 4); no simple theory exists for other cross sections. It may be stated, however, that in general abrupt changes in cross section invalidate the theories given in this chapter, but gradual changes produce only small errors. As a result of St. Venant's principle the theories which have been discussed may be used in structures with discontinuities but only at distances from the changes in cross section which are several times the largest cross-sectional dimension.

If the warping of the ends is unrestrained and the end torques are distributed in the same manner as the computed internal stresses, the solution is correct for any length torsion member.   If either of these conditions is violated, the length of the structure becomes a factor.   If free warping of the ends is prevented, longitudinal normal stresses are developed which induce a redistribution of the shearing stresses.   If the only loading is a torque, the longitudinal stresses must be self-equilibrating, and the actual distribution of shearing stresses must have the same static resultant as that developed from the theory.   Again as a result of St. Venant's principle, the theoretical solution will be satisfactory at distances from the ends that are greater than several times the largest cross-sectional dimension.   Warping restraint usually exists in the root sections of wing and tail surfaces, so that the Bredt theory is not applicable in these regions, which must be analyzed by the methods for statically indeterminate structures given in Chaps. 11 and 12.   If the surface has a large aspect ratio, the methods may be used at several chord lengths from the root.

It has been mentioned that the theories are exact only when the torques are applied to the ends of the member.   When a concentrated twisting moment is applied at an interior cross section, the torque is different on either side of the moment, and the warping displacements would be different if the structure were not constrained to be continuous at the location of the moment.   This warping discontinuity is not possible since it violates compatability, and longitudinal stresses are therefore developed in this region.   When twisting moments are distributed over the length of the structure and the torque varies gradually, the resulting errors are small.   A more complete discussion of torsion in shell structures with restrained warping is given in Ref. 9.

8-9 DIFFERENTIAL EQUATIONS FOR VARIABLE TORQUE

We noted in the preceding section that the theory which neglects the axial stresses due to restrained warping is sufficiently accurate when the cross section and torque vary gradually.   Furthermore, the warping restraint at the ends and the manner in which end torques are distributed have little influence upon the torsional deflections when the member is slender. In this section we consider the deformations of slender torsion members in which the cross section and/or torque vary gradually.

The torque M_t must satisfy the equilibrium equation (7-35d).   The angle of twist per unit length $\theta = d\varphi/dx$ varies when M_t or G_1J^* are functions of x.   In this case we find from Eq. (8-69) that

$$M_t = G_1J^* \frac{d\varphi}{dx} \tag{8-73}$$

Substituting this result into Eq. (7-35d), we find†

$$(G_1 J^* \varphi')' = -m_t - M_y v'' + M_z w'' \tag{8-74}$$

We note that in general the equilibrium equations (7-45), (7-46), and (8-74) are coupled. However, the bending moments in Eq. (8-74) are usually small compared to their critical values which cause buckling, in which case we may neglect the M_y and M_z terms and obtain

$$(G_1 J^* \varphi')' = -m_t \tag{8-75}$$

Equation (8-75) is a second-order differential equation and requires a single boundary condition at each end point. The boundary condition at each end is obtained from either the prescribed value of φ or from Eq. (8-73) if torque is specified. If the torsion member is statically determinate, $m_t(x)$ is known and φ may be found by double integration of Eq. (8-75). In other cases φ is determined by solving the differential Eqs. (8-75) [or (8-74)] with the boundary conditions. The equations that result in special cases may be of the form of an equilibrium boundary-value problem (Sec. 5-3) or an eigenvalue problem (Sec. 5-4). When the coefficients of the equations are functions of x, it may be necessary to use numerical procedures such as the finite-difference method (Chap. 5) to obtain solutions.

Energy methods are also useful when M_t or $G_1 J^*$ are variable. Consider a differential length dx of the member at a point x where the torque is M_t. The work that is done by M_t on the face at x is $-M_t \varphi/2$, while the work done by M_t on the face at $x + dx$ is $M_t (\varphi + \varphi' \, dx)/2$. The strain energy in the length dx is the net work done by M_t, so that $dU = M_t \varphi'/2$. The total strain energy obtained by integrating over the length of the member and using Eq. (8-73) is

$$U = \frac{1}{2} \int_L G_1 J^* (\varphi')^2 \, dx \tag{8-76}$$

and the complementary strain energy is

$$U' = \frac{1}{2} \int_L \frac{M_t^2}{G_1 J^*} \, dx \tag{8-77}$$

† Equation (8-73) must be modified when there are σ_{xx} stresses due to axial forces or temperature gradients.[10,11] The fibers of the twisted beam form an angle $r\varphi'$ with the axis of twist, where r is the radial distance from the axis of twist to the fiber. As a result, the σ_{xx} stresses (which are aligned with the fibers) have components $\sigma_{xx} r \varphi'$ that give rise to a torsional moment $\varphi' \int \sigma_{xx} r^2 \, dA$ about the axis of twist. In these cases, $G_1 J^*$ should be replaced by the effective torsional rigidity $G_1 J^* + \int \sigma_{xx} r^2 \, dA$ in Eqs. (8-73) to (8-75).

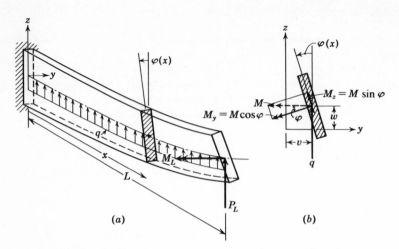

Fig. 8-23 Example 8-7.

The potentials of M_x, a conservative concentrated moment about the x axis at x_i, are

$$V = -M_x\varphi(x_i) = V' \tag{8-78}$$

and for a distributed moment per unit length m_x are

$$V = -\int_L m_x\varphi\, dx = V' \tag{8-79}$$

Example 8-7 Derive the differential equation and boundary conditions that govern the lateral buckling of a cantilevered beam of narrow rectangular cross section loaded by a uniformly distributed vertical load q, a vertical tip load P_L, and a tip moment about the horizontal axis M_L (Fig. 8-23).

A beam with a narrow rectangular cross section that is subjected to bending loads in the vertical plane will deflect only vertically when the loads are small. However, if the loads are made large enough, a state of neutral stability is reached in which the beam can bend laterally in the horizontal direction and twist as shown in Fig. 8-23. This type of buckling is easily demonstrated by bending a thin strip of paper about an axis which is normal to its larger cross-sectional dimension.

A section of the buckled beam at x is shown in Fig. 8-23b. In this figure M is the moment of the applied forces, while M_y and M_z are components of M about the principal axes that are fixed in the beam cross section, as assumed in the derivation of Eq. (8-74). For small φ, $M_y = M$, and $M_z = M\varphi$; and so for a homogeneous beam Eqs. (7-39) give $v'' = M\varphi/EI_{zz}$ and $w'' = M/EI_{yy}$. If q is applied at the centroidal axis of the beam, as shown in Fig. 8-23b, $m_t = 0$, and Eq. (8-74) becomes

$$(GJ\varphi')' = -\frac{M^2\varphi}{EI_{zz}} + \frac{M^2\varphi}{EI_{yy}} \tag{a}$$

Factoring $M^2\varphi/EI_{zz}$ from the right of this equation and noting that for the narrow rectangle $I_{zz}/I_{yy} \ll 1$, we find

$$(GJ\varphi')' = -\frac{M^2\varphi}{EI_{zz}} \tag{b}$$

Equation (b) is general and applies to the lateral buckling for any end conditions and loading for which $m_t = 0$. For the special case under consideration

$$M = M_L + P_L(L - x) + \tfrac{1}{2}q(L - x)^2 \tag{c}$$

Let us assume that the loads increase simultaneously and that the ratios $A_1 = M_L/P_L$ and $A_2 = q/P_L$ are constant during loading. Substituting Eq. (c) into Eq. (b), we find

$$(GJ\varphi')' = -P_L{}^2 \left[A_1 + L - x + \frac{A_2}{2}(L - x)^2 \right]^2 \frac{\varphi}{EI_{zz}} \tag{d}$$

The boundary conditions obtained from the conditions of zero twist at the root and zero torque at the tip are

$$\varphi(0) = 0 \qquad \varphi'(L) = 0 \tag{e}$$

Equations (d) and (e) define an eigenvalue problem. The lowest of the eigenvalues $P_L{}^2$ establishes the buckling load, and the associated eigenfunction φ gives the buckle shape.

Solutions for the lateral buckling loads of rectangular beams with various loads and end conditions are given in Ref. 12, which also contains solutions for I sections, where it is necessary to extend the theory to include warping restraint. However, if the I section is long and deep and the flanges are narrow, the theory in this example may be used as a reasonable approximation.

Example 8-8   Derive the differential equations and boundary conditions that govern free torsional vibrations of an unswept tapered wing with a tip tank.

For simplicity we assume that the centers of gravity of the wing sections and the tip tank coincide with the elastic axis of the wing, so that the torsional vibrations are uncoupled from the bending vibrations of Example 7-8. The *elastic axis*, which coincides with the *axis of twist*, is a line joining the shear centers of the cross sections (Sec. 9-4) of the wing. A lateral force on the elastic axis produces only bending with no twisting, while a torque causes only twisting about the elastic axis, and there is no lateral deflection along the elastic axis.

The angle of twist of a vibrating wing is a function of x and t. During free vibrations the only loading on a unit length of the wing is the d'Alembert inertial moment $m_t = -I_x\ddot{\varphi}$, where I_x is the mass moment of inertia per unit length of the wing about the elastic axis. Substituting m_t into Eq. (8-75), we find

$$(G_1J^*\varphi')' = I_x\ddot{\varphi} \tag{a}$$

The initial conditions are

$$\varphi(x,0) = f(x) \qquad \dot{\varphi}(x,0) = g(x) \tag{b}$$

in which $f(x)$ and $g(x)$ are prescribed functions; and the boundary conditions are

$$\varphi(0,t) = 0 \qquad G_1J^*\varphi'(L,t) = -I_T\ddot{\varphi}(L,t) \tag{c}$$

where I_T is the mass moment of inertia of the tip tank about the elastic axis.

The second of these boundary conditions is obtained from Eq. (8-73) with M_t equal to the d'Alembert inertial moment imposed by the tip tank.

We employ the method of separation of variables (Example 7-8) by writing

$$\varphi(x,t) = \Phi(x)T(t) \tag{d}$$

Substituting Eq. (d) into Eq. (a) and separating variables, we find the ordinary differential equations

$$\ddot{T} + \omega^2 T = 0 \tag{e}$$

$$(G_1 J^* \Phi')' = -\omega^2 I_x \Phi \tag{f}$$

The solution of Eq. (e) is $T = C_1 \sin \omega t + C_2 \cos \omega t$, which with Eq. (d) reduces Eqs. (b) to

$$C_2 \Phi(x) = f(x) \qquad C_1 \omega \Phi(x) = g(x) \tag{g}$$

From Eqs. (d) and (e), the boundary conditions of Eq. (c) become

$$\Phi(0) = 0 \qquad G_1 J^* \Phi'(L) = \omega^2 I_T \Phi(L) \tag{h}$$

Equations (f) and (h) are in the form of an eigenvalue problem in which the natural mode shape Φ is the eigenfunction and the square of the natural frequency of free vibration ω^2 is the eigenvalue. Methods for solving for Φ and ω are given in Ref. 13. The solution for $I_T = 0$ and constant $G_1 J^*$ and I_x is given in Prob. 8-12.

Example 8-9 Derive the differential equations and boundary conditions that govern the aeroelastic torsional deformations of a slender unswept wing (1) at a rigid body angle of attack α_R when the dynamic pressure of the flow is $q < q_D$ (where q_D is the dynamic pressure for torsional divergence), and (2) at $\alpha_R = 0$ when $q = q_D$.

For simplicity we assume that the elastic axis is a straight line perpendicular to the root chord. The forces on a unit length of the wing at a distance x from the root are shown in Fig. 8-24. The resultant aerodynamic force and moment may be referred to any point in the cross section; however, if they are referred to the *aerodynamic center*, the aerodynamic moment does not change with angle of attack. Assuming that α_R and φ are small, the moment per unit length about the elastic axis is

$$m_t = Ne + M_{AC} - n_z wd \tag{a}$$

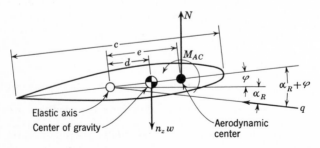

Fig. 8-24 Example 8-9.

where N is the aerodynamic normal force, M_{AC} is the aerodynamic moment about the aerodynamic center, n_z is the normal load factor (1 for level flight), and w is the weight per unit length of the wing. The lengths e and d are distances from the elastic axis to the aerodynamic center and the center of gravity of the wing section. The aerodynamic force and moment are proportional to q, and N is proportional to α, the total angle of attack of the airfoil section. Equation (a) may be written

$$m_t = q \frac{dC_n}{d\alpha} \alpha ce + qC_{m_{AC}}c^2 - n_z wd \qquad (b)$$

where $dC_n/d\alpha$ is the slope of the linear portion of the normal-force-coefficient curve, $C_{m_{AC}}$ is the moment coefficient referred to the aerodynamic center, and c is the chord length.

The bending moments on a wing are small compared to the critical values for lateral buckling (Example 8-7), and we may use Eq. (8-75). Substituting Eq. (b) into this equation and noting that $\alpha = \alpha_R + \varphi$, where φ is the angle of twist of the wing at x, we find

$$(G_1 J^* \varphi')' + q \frac{dC_n}{d\alpha} ce\varphi = -q \frac{dC_n}{d\alpha} ce\alpha_R - qC_{m_{AC}}c^2 + n_z wd \qquad (c)$$

The boundary conditions are

$$\varphi(0) = 0 \qquad \varphi'(L) = 0 \qquad (d)$$

where the second equation is obtained from Eq. (8-73). Equations (c) and (d) are of the form of an equilibrium boundary-value problem, the solution of which gives the twist distribution of the wing. When φ is known, the normal force distribution is found from $N = q(dC_n/d\alpha)c(\alpha_R + \varphi)$, and the torsional moment is determined from Eq. (8-73).

We have assumed that q in Eq. (c) is prescribed. The first term on the left of Eq. (c) is the elastic restoring moment, and the second term is the aerodynamic moment due to twist. Both of these terms are linearly related to the twist of the wing, but when the aerodynamic center is forward of the elastic axis (e positive), the moments are in opposing directions. For a given twist the elastic restoring moment is constant, while the aerodynamic moment increases with q. If q becomes large enough, a state of neutral stability is reached in which the increase in aerodynamic torque due to an increment of twist is equal to the increase in the elastic restoring torque associated with the same increment of twist. At this point the wing is unable to resist additional torques due to α_R and the inertial loads, i.e., the torques on the right of Eq. (c). The dynamic pressure for the condition of neutral stability is known as the *divergence dynamic pressure* q_D. Even if α_R and n_z are zero, the wing will be torsionally unstable and fail if it is given a small twisting disturbance when $q > q_D$.

To determine q_D we set $q = q_D$ and $\alpha_R = n_z = 0$ in Eq. (c) and obtain

$$(G_1 J^* \varphi')' = -q_D \frac{dC_n}{d\alpha} ce\varphi \qquad (e)$$

This equation and the boundary conditions of Eqs. (d) define an eigenvalue problem in which q_D is the lowest eigenvalue of the system and φ is the mode shape or eigenfunction. The pressure q_D may be found by the same methods that are given for computing the buckling loads of columns in Chap. 14. The solution of Eqs. (d) and (e) for constant G_1J^*, $dC_n/d\alpha$, c, and e is given in Problem 8-13. Methods for solving the problem when the coefficients are variable are given in Ref. 13. It was pointed out in a footnote that thermal stresses effect the torsional rigidity. The influence of these stresses upon q_D is discussed in Ref. 11.

REFERENCES

1. Timoshenko, S. F., and J. N. Goodier: "Theory of Elasticity," 2d ed., McGraw-Hill Book Company, New York, 1951.
2. Wang, C. T.: "Applied Elasticity," McGraw-Hill Book Company, New York, 1953.
3. Sokolnikoff, I. S.: "Mathematical Theory of Elasticity," 2d ed., McGraw-Hill Book Company, 1956.
4. Hetényi, M.: "Handbook of Experimental Stress Analysis," John Wiley & Sons, Inc., New York, 1950.
5. Crandall, S. H.: "Engineering Analysis," McGraw-Hill Book Company, New York, 1956.
6. Argyris, J. H., and S. Kelsey: "Energy Theorems and Structural Analysis," Butterworth & Co. (Publishers) Ltd., London, 1960.
7. Trayer, G. W., and H. W. March: "The Torsion of Members Having Sections Common in Aircraft Construction," *NACA Rept.* 334, 1929.
8. Beadle, C. W., and H. D. Conway: Stress Concentrations in Twisted Structural Beams, *J. Appl. Mech.*, **30**(1): 138–141 (March, 1963).
9. Kuhn, P.: "Stresses in Aircraft and Shell Structures," McGraw-Hill Book Company, New York, 1956.
10. van der Neut, A.: Buckling Caused by Thermal Stresses, in "High Temperature Effects in Aircraft Structures," Pergamon Press, New York, 1958.
11. Bisplinghoff, R. L., and J. Dugundji: Influence of Aerodynamic Heating on Aeroelastic Phenomena, in "High Temperature Effects in Aircraft Structures," Pergamon Press, New York, 1958.
12. Timoshenko, S. P., and J. M. Gere: "Theory of Elastic Stability," 2d ed., McGraw-Hill Book Company, New York, 1961.
13. Bisplinghoff, R. L., H. Ashley, and R. L. Halfman: "Aeroelasticity," Addison-Wesley Publishing Company, Inc., Reading, Mass., 1955.

PROBLEMS

8-1. Prove that the stress function

$$\psi = -G\theta\left[\tfrac{1}{2}(y^2 + z^2) - \frac{1}{2a}(y^3 - 3yz^2) - \frac{2a^2}{27}\right]$$

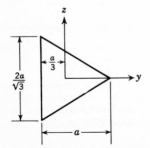

Fig. P8-1

is the solution for the equilateral triangle shown. At what points will the maximum shear stress occur? Determine its value in terms of θ. [*Ans,* $\sigma_{zs,\max} = G\theta a/2$ at the center of the sides.]

8-2. Show that

$$\psi = \left[y^2 - \left(\frac{a}{2}\right)^2 \right] \left[z^2 - \left(\frac{b}{2}\right)^2 \right] \sum_{m=0,2,4}^{\infty} \sum_{n=0,2,4}^{\infty} a_{mn} y^m z^n$$

is a satisfactory stress-function series to use in a Rayleigh-Ritz solution to the rectangular section shown in Fig. 8-8.

8-3. The warping-function series

$$f = (y^2 - z^2) \sum_{j=1,3,5}^{\infty} c_j y^i z^i$$

may be used with the Rayleigh-Ritz method to obtain a solution for the square section. Use the first two terms to obtain an approximate solution for J. Compare the result with that obtained from Eq. (8-42).

8-4. Obtain approximate equations for J and the maximum shear stress due to M_t for the thin modified double-wedge cross section shown by using Eqs. (8-48) and (8-49). [*Ans.* $J = ct^3/6$; $\sigma_{zs,\max} = 6M_t/ct^2$.]

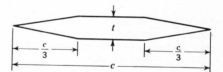

Fig. P8-4

8-5. Determine the torsion constant for the T section shown. [*Ans.* $J = 1.717 \times 10^{-2}$ in.^{4}.]

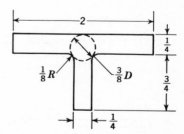

Fig. P8-5

8-6. Prove that $f = 0$ is the solution for a hollow circular cross section of inside radius a and outside radius b. Show that $J = I_p$ and $\sigma_{zs} = M_t r/I_p$, where I_p is the polar moment of inertia of the hollow section.

8-7. The slit and unslit tubes shown have the same radii and wall thicknesses and are subjected to the same torques. Determine the ratio of the angles of twist and the ratio of the shear stresses of the two sections. What does this indicate about the relative merits of open and closed sections?

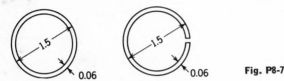

0.06   0.06   **Fig. P8-7**

8-8. Use the finite-difference method and the mesh shown to obtain an approximate solution for the torsion constant of the cruciform section.

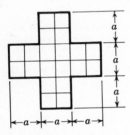

Fig. P8-8

8-9. The nonhomogeneous box beam shown is subjected to a torque of 10,000 in.-lb. Find the shear stress in each wall and the angle of twist per unit length. The cover skins are Inconel X ($G = 11.9 \times 10^6$ psi) and the spar webs are 6Al-4V titanium ($G = 6.2 \times 10^6$ psi).

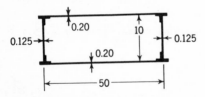

Fig. P8-9

8-10. The two-cell box beam shown is subjected to a torque of 10,000 in.-lb. Determine the shear flow in the center spar and the angle of twist per unit length if the material is 2024-T3 clad aluminum ($G = 4.0 \times 10^6$ psi). [*Ans.* $q = 2.66$ lb/in.; $\theta = 3.07 \times 10^{-3}$ deg/in.]

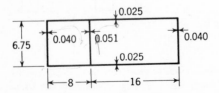

Fig. P8-10

8-11. A tapered cantilevered wing is acted upon by the pressure distribution shown. In addition a tip tank applies a torque M_T at the tip. Assume the elastic axis is at the midchord and that $GJ = GJ_T(1 + x/L)$. Determine the angle of twist as a function of x assuming that the pressure distribution is not appreciably changed by the twist.

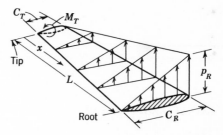

Fig. P8-11

8-12. Show that Eqs. (f) and (h) of Example 8-8 are satisfied by $\omega_n = \dfrac{n\pi}{2L}\left(\dfrac{G_1J^*}{I_x}\right)^{1/2}$ and $\Phi_n = A \sin (n\pi x/2L)$ for $n = 1, 3, 5, \ldots$ when G_1J^* and I_x are constant and $I_T = 0$.

8-13. Show that Eqs. (d) and (e) of Example 8-9 are satisfied by $q_D = \dfrac{\pi^2 G_1 J^*}{4L^2 ce}\bigg/\dfrac{dC_n}{d\alpha}$ and $\varphi = A \sin (\pi x/2L)$ when G_1J^*, c, e, and $dC_n/d\alpha$ are constant.

9

Stresses Due to Shear in Thin-walled Slender Beams

9-1 INTRODUCTION

The stress resultants in a slender beam were defined in Sec. 7-2, and equations for the stresses that are associated with the resultants M_y, M_z, and P were derived in Chap. 7, while those due to M_t were given in Chap. 8. In this chapter we complete the study of beams by determining the shear stresses that result from the shear forces V_y and V_z. The shear problem is similar to the torsion problem, since in both cases a closed-form solution for an arbitrary cross section is not known. The same difficulties arise in both problems, in that the directions and the distributions of the shear stresses are not generally known.

St. Venant used his semi-inverse method to simplify the three-dimensional equations of the theory of elasticity for the case of a prismatic bar with a lateral shear force. By introducing an appropriately defined stress function, the problem is reduced to solving a single second-order partial differential equation and its associated boundary condition.[1,2] For the most part, this equation has been solved only for the relatively few cross sections for which the torsion solution is known. A membrane analogy also exists for the shear problem, though it is more complicated than in the case of torsion.[1,2]

The shearing stresses due to lateral shear are seldom of concern in the design of slender beams of compact cross section, for the longitudinal stresses are ordinarily critical. As already noted, flight structures are usually fabricated with thin skins and webs supported by longitudinal and transverse stiffening members (Sec. 7-6). In these thin walled structures the shearing stresses are important, since they may produce buckling of the walls.

In this chapter we restrict ourselves to thin-walled structures, where fortunately the analysis is not so difficult as in compact sections. As in the Bredt theory of torsion (Sec. 8-7), in thin-walled structures subjected to shear we may assume that the shear stresses are uniformly distributed through the wall thickness and in the direction of the tangent to the midline of the wall.   As a result, we again find it convenient to use shear flows.   It will be shown that these shear flows may be used to determine the location of the shear center, design skin and web joints, and obtain the load distributions on ribs and frames.

9-2 OPEN SECTIONS

A differential length of a beam of arbitrary thin-walled uniform cross section is shown in Fig. 9-1a.   It is assumed that the beam is subjected to a shear force V and a temperature change $T(x,y,z)$.   The beam may be nonhomogeneous, but the elastic constants are considered to be functions of y and z only.   We assume that transverse ribs or frames, rigid in the y

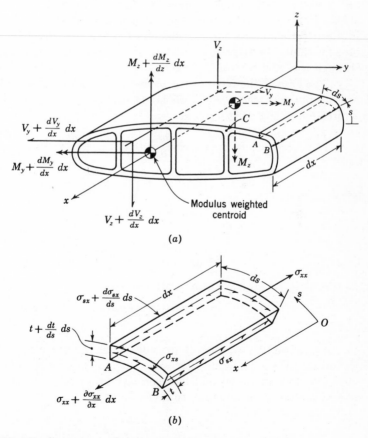

Fig. 9-1   Geometry and stress resultants on a thin-walled section.   (a) Thin-walled section with shear forces; (b) free-body diagram of an element of the wall.

and z directions but perfectly flexible in the x direction, maintain the cross-sectional shape but allow free warping out of the plane of the cross section.

We refer the analysis to axes which pass through the modulus-weighted centroid of the cross section and resolve V into components V_y and V_z that are parallel to the y and z axes, respectively. Since V_y and V_z are nonzero, the bending moments M_y and M_z are functions of x. The positive directions of the shear and moment components which are shown in Fig. 9-1a are consistent with those in Chap. 7.

Figure 9-1b shows a free-body diagram of a differential length of a typical wall of the structure in Fig. 9-1a. The coordinate s, which locates this element, is measured along the midline of the wall from an arbitrary point O in the wall. Since M_y, M_z, and T are functions of x, the σ_{xx} stresses that act upon the front and back faces of the element are different, and shearing stresses σ_{sx} are developed to maintain equilibrium. These shearing stresses, which are functions of s, are defined positive when σ_{xs} is in the direction of s on the face at $x + dx$ as shown. Since the wall thickness t is small, the stresses σ_{xx} and σ_{xs} at s may be assumed to be constant through the thickness of the wall.

The area of the faces of the element that are perpendicular to the x axis is approximately $[t + \frac{1}{2}(dt/ds)\,ds]\,ds$, and by summing forces in the x direction in Fig. 9-1b we find

$$-\sigma_{xx}\left(t + \frac{1}{2}\frac{dt}{ds}\,ds\right)ds + \left(\sigma_{xx} + \frac{\partial\sigma_{xx}}{\partial x}\,dx\right)\left(t + \frac{1}{2}\frac{dt}{ds}\,ds\right)ds$$

$$-\sigma_{sx}t\,dx + \left(\sigma_{sx} + \frac{d\sigma_{sx}}{ds}\,ds\right)\left(t + \frac{dt}{ds}\,ds\right)dx = 0$$

By canceling terms, dividing by $dx\,ds$, and taking the limit as ds approaches zero we find

$$\sigma_{xs}\frac{dt}{ds} + \frac{d\sigma_{xs}}{ds}\,t = -t\frac{\partial\sigma_{xx}}{\partial x}$$

Noting that $q = \sigma_{xs}t$, we may write this equation

$$\frac{dq}{ds} = -t\frac{\partial\sigma_{xx}}{\partial x} \tag{9-1}$$

Equation (9-1) applies at any point in the cross section except where the outer walls join the interior webs (such as point C in Fig. 9-1a). It may therefore be integrated between any two points O and s which lie within the same wall section, to give

$$q = q_0 - \int_0^s \frac{\partial\sigma_{xx}}{\partial x}\,t\,ds \tag{9-2}$$

where $q = q(s)$ and $q_0 = q(0)$.

To this point our only assumptions have been that σ_{xs} is uniform through the wall thickness and in the direction of the tangent to the midline of the wall. However, to proceed further we must know σ_{xx}. We assume that σ_{xx} is given by Eq. (7-19), but in doing so we must accept all the assumptions and limitations that apply to this equation. We note here a basic contradiction, for in the derivation of Eq. (7-19) it was assumed that M_y, M_z, and T did not vary with x. However, as noted in Sec. 7-5, the errors in Eq. (7-19) are small for a slender beam when M_y, M_z, and T vary gradually but are significant in the regions of large gradients in the shear or temperature.

Substituting Eq. (7-19) into Eq. (9-2), we find

$$q = q_0 - \int_0^s \frac{E}{E_1} \left[\frac{1}{A^*} (P_T)' - \frac{(M_z^*)' I_{yy}^* - (M_y^*)' I_{yz}^*}{I_{yy}^* I_{zz}^* - (I_{yz}^*)^2} y \right.$$
$$\left. - \frac{(M_y^*)' I_{zz}^* - (M_z^*)' I_{yz}^*}{I_{yy}^* I_{zz}^* - (I_{yz}^*)^2} z - E_1(\alpha T)' \right] t \, ds$$

where primes indicate differentiation with respect to x. In writing this equation it is assumed that P is constant, since no provision has been made for distributed axial forces in writing the equilibrium equation for the element AB in Fig. 9-1b. Using the second of Eqs. (7-51), we may write the last equation as

$$q = q_0 + \frac{A_s^*}{A^*} p_{xT} + C_2' Q_z^* + C_3' Q_y^* + \int_0^s E(\alpha T)' t \, ds \qquad (9\text{-}3)$$

where from Eqs. (7-18), the last two of Eqs. (7-36) (with $m_y = m_z = 0$), and (7-44) the symbols C_2' and C_3' are defined by

$$C_2' = \frac{V_y^* I_{yy}^* - V_z^* I_{yz}^*}{I_{yy}^* I_{zz}^* - (I_{yz}^*)^2} \qquad C_3' = \frac{V_z^* I_{zz}^* - V_y^* I_{yz}^*}{I_{yy}^* I_{zz}^* - (I_{yz}^*)^2} \qquad (9\text{-}4)$$

in which

$$V_y^* = V_y + V_{yT} \qquad V_z^* = V_z + V_{zT} \qquad (9\text{-}5)$$

In addition, the following notation is used in Eq. (9-3):

$$Q_y^* = \int_0^s z \, dA^* \qquad Q_z^* = \int_0^s y \, dA^* \qquad (9\text{-}6)$$

$$A_s^* = \int_0^s dA^* \qquad (9\text{-}7)$$

in which dA^* is given by Eq. (7-14). We note that A_s^* is the modulus-weighted area between O and s, and Q_y^* and Q_z^* are the moments of this weighted area about the y and z axes. When E is piecewise constant, we may write

$$Q_y^* = \Sigma \bar{z}_i A_i^* \qquad Q_z^* = \Sigma \bar{y}_i A_i^* \qquad (9\text{-}8)$$

where $\bar{y}_i$ and $\bar{z}_i$ are the centroidal coordinates of the ith area with constant modulus E_i, A_i^* is given by Eq. (7-26), and the sums extend over the wall material between O and s.   The special case of Eq. (9-3) in which the beam is homogeneous and $V_y = V_z = 0$ is derived and discussed in Ref. 3.

Next consider the equilibrium condition which the shear flows must satisfy at the intersection of several walls, such as at point C in Fig. 9-1a. A free-body diagram of a typical juncture of n walls is shown in Fig. 9-2, where for completeness we include the possibility of a longitudinal at the juncture.   It is assumed that the dimensions of the longitudinal are small compared to those of the beam cross section, so that the shear stresses in the longitudinal may be neglected, and the axial stresses may be considered constant over the area of the longitudinal.   If we designate the shear flows in the n intersecting walls by q_1, q_2, . . . , q_n (positive as shown), we find for equilibrium in the x direction that

$$\sum_{i=1}^{n} q_i = -P_L'$$ (9-9)

where P_L is the axial force in the longitudinal.   Noting that $P_L = \sigma_{xx}A_L$ and using Eq. (7-19), we may write Eq. (9-9) as

$$\sum_{i=1}^{n} q_i = \frac{A_L^*}{A^*} p_{xx} + C_2' Q_{z_L}^* + C_3' Q_{y_L}^* + E_L A_L (\alpha T)_L'$$ (9-10)

where the L subscripts refer to the longitudinal.   When there is no longitudinal at the intersection,

$$\sum_{i=1}^{n} q_i = 0$$ (9-11)

and the sum of the shear flows into the juncture equals the sum of the flows out.

If the beam is homogeneous, the asterisks may be dropped from the section-property terms, and if the temperatures do not change with x,

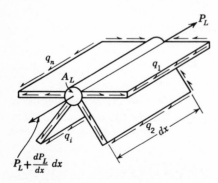

Fig. 9-2 Free-body diagram of wall junction.

they may be removed from V_y^* and V_z^*. If both of these simplifications apply and principal axes are used, Eqs. (9-3) and (9-10) reduce to

$$q = q_0 + \frac{V_y Q_z}{I_{zz}} + \frac{V_z Q_y}{I_{yy}} \tag{9-12}$$

$$\sum_{i=1}^{n} q_i = \frac{V_y Q_{z_L}}{I_{zz}} + \frac{V_z Q_{y_L}}{I_{yy}} \tag{9-13}$$

If Eq. (9-3) is to be used to determine the shear flow at s, it is necessary to know q_0. The manner in which q_0 is determined in beams with closed cells is given in Sec. 9-5. If the beam has an open cross section, the point O may be taken at a free edge so that $q_0 = 0$. When the cross section is open, V_y and V_z must pass through the shear center, or the section will twist, and the shear stresses will not be uniform through the wall thickness, as assumed in the deviation of Eqs. (9-3) and (9-10).

If a thin-walled homogeneous beam with an open section is loaded by a shear force V_z at the shear center and in the direction of the principal z axis, Eq. (9-12) reduces to

$$q = \frac{V_z Q_y}{I_{yy}} \tag{9-14}$$

and the shear stress is

$$\sigma_{xs} = \frac{V_z Q_y}{I_{yy} t} \tag{9-15}$$

This equation (which is usually derived in strength-of-materials texts)

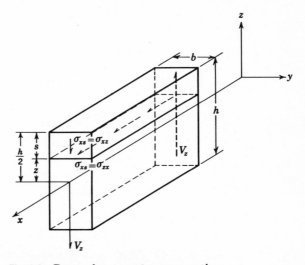

Fig. 9-3 Beam of rectangular cross section.

may be applied to the rectangular cross section shown in Fig. 9-3 when h/b is sufficiently large. For such a case $\sigma_{zs} = \sigma_{zz}$, and we see from Fig. 9-3 that $Q_y = b(h/2 - z)(h/2 + z)/2$, so that Eq. (9-15) becomes

$$\sigma_{xz} = \frac{6V_z}{bh^3}\left[\left(\frac{h}{2}\right)^2 - z^2\right]$$

indicating that the shear stresses are parabolically distributed through the depth of the beam. The theory-of-elasticity solution for the rectangle[1] shows that this equation is very accurate for $h/b > 2$, while for a square section it underestimates the maximum stress by 10 percent.

In the analysis of stiffened shells it is common practice to use the idealizations of Sec. 7-6, in which it is assumed that all the material that carries axial stress is concentrated into effective longitudinals and that the webs carry only shear stresses. In the development of Eq. (9-3) the section properties and thermal loads are obtained by integrating over the area which is carrying axial stress. In the idealized section this consists of the areas of the effective longitudinals and does not include the idealized webs, since their areas have already been lumped into the effective longitudinals. The quantities A_s^*, Q_y^*, Q_z^* and the integral of Eq. (9-3) are therefore constant between effective longitudinals and undergo step changes at the longitudinals.

The use of the preceding equations is illustrated in the following examples, the first of which deals with the actual cross section while the second uses the idealized section.

Example 9-1 Determine the shear-stress distribution in the circular section shown in Fig. 9-4 (this section is representative of monocoque-body structures or semi-monocoque bodies if t is the total effective thickness of skin and longitudinals obtained by spreading the area of the longitudinals uniformly around the circumference).

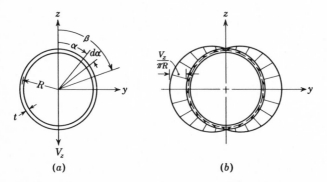

(a)   (b)

Fig. 9-4 Example 9-1. (a) Geometry and loading; (b) reacting shear flows.

The section and loading are symmetric about the z axis, so that q must also be symmetric about this axis. For this to be true the shear flow must vanish at $\beta = 0$ and π, and Eq. (9-14) applies if we measure s from the point where $\beta = 0$. Noting that $z = R \cos \alpha$ and $ds = R \, d\alpha$, we find from the first of Eqs. (9-6) that

$$Q_y = \int_0^\beta (R \cos \alpha)tR \, d\alpha = R^2 t \sin \beta$$

Observing that $I = \pi R^3 t$, we obtain from Eq. (9-14)

$$q(\beta) = \frac{V_z}{\pi R} \sin \beta \tag{9-16}$$

Example 9-2 A beam with the idealized cross section shown in Fig. 7-8b but with the skin cut between longitudinals 14 and 15 is subjected to a vertical shear force of 2000 lb at the shear center. In addition to the shear force there is a longitudinal thermal gradient $T_i' = -2.083 \times 10^{-3} T_i$ °F/in., where T_i' is the axial rate of temperature change in the ith effective longitudinal and T_i is the temperature in the ith longitudinal listed in column ⑯ of Table 7-1. Determine the shear-flow distribution. Circled numbers are used to refer to columns.

The shear flows are computed from Eq. (9-3), in which $q_0 = 0$ if we take the point O at the edge of the cut web between longitudinals 14 and 15. The sectional properties in this equation are given in Example 7-1. The values of Q_y^* and Q_z^* for each of the webs are calculated in ① to ⑧ of Table 9-1 by using Eq. (9-8) with the summation running in the counterclockwise direction from the cut web.

The computations for p_{xT}, V_{yT}, and V_{zT} are given in ⑨ to ⑬ of Table 9-1, from which

$$p_{xT} = -\sum_{i=1}^{16} E_i A_i (\alpha T)_i' = -\Sigma \, ⑪ = 1336.2 \text{ lb/in.}$$

$$V_y^* = V_y + V_{yT} = V_y - \sum_{i=1}^{16} E_i A_i \bar{y}_i (\alpha T)_i'$$

$$= V_y - \Sigma \, ⑬ = 0 - 3220 = -3220 \text{ lb}$$

$$V_z^* = V_z + V_{zT} = V_z - \sum_{i=1}^{16} E_i A_i \bar{z}_i (\alpha T)_i'$$

$$= V_z - \Sigma \, ⑫ = -2000 - 1024 = -3024 \text{ lb}$$

The coefficients C_2' and C_3' obtained from Eqs. (9-4) are

$$C_2' = \frac{-3220 \times 422.2 - 3024 \times 468.6}{422.2 \times 7035 - 468.6^2} = -1.009$$

$$C_3' = \frac{-3024 \times 7035 - 3220 \times 468.6}{422.2 \times 7035 - 468.6^2} = -8.284$$

The calculations for A_s^* and the integral in Eq. (9-3) are given in ⑭ and ⑯. The shear-flow computations using Eq. (9-3) are carried out in ⑮

Table 9-1 Computations for Example 9-2

① Longitudinal i	② A_i^*, in.²	③ $\bar{y}_i$, in.	④ $\bar{z}_i$, in.	⑤ $\bar{y}_i A_i^*$, in.³ ②×③	⑥ Q_z^*, in.³ Σ⑤	⑦ $\bar{z}_i A_i^*$, in.³ ②×④	⑧ Q_y^*, in.³ Σ⑦	⑨ $(\alpha T)_i'$, 10^{-6} in.⁻¹	⑩ $E_i A_i$, 10^6 lb 10.5×②	⑪ $E_i A_i (\alpha T)_i'$, lb/in. ⑨×⑩
15	1.13	− 1.96	4.41	− 2.2	0	4.98	0	− 8.00	11.9	− 95.2
16	1.13	−15.82	6.15	−17.9	− 2.2	6.95	4.98	− 8.00	11.9	− 95.2
1	2.85	−20.89	6.78	−59.5	− 20.1	19.32	11.93	− 6.04	30.0	−181.2
2	0.47	−20.89	4.25	− 9.8	− 79.6	2.00	31.25	− 5.37	4.9	− 26.3
3	0.47	−20.89	−2.68	− 9.8	− 89.4	− 1.26	33.25	− 5.37	4.9	− 26.3
4	2.30	−20.89	−5.22	−48.0	− 99.2	−12.01	31.99	− 8.00	24.3	−194.4
5	0.87	−15.82	−5.22	−13.8	−147.2	− 4.54	19.98	−11.29	9.2	−103.9
6	0.87	− 1.96	−5.22	− 1.7	−161.0	− 4.54	15.44	−11.29	9.2	−103.9
					−162.7		10.90			

7	0.90	8.18	−5.22	7.4	−155.3	−4.70	6.20	−10.62	9.5	−100.9
8	0.90	22.04	−5.22	19.8	−135.5	−4.70	1.50	−10.62	9.5	−100.9
9	1.41	27.11	−5.22	38.2	−97.3	−7.36	−5.86	−6.67	14.6	−97.3
10	0.21	27.11	−3.95	5.7	−91.6	−0.83	−6.69	−4.73	2.2	−10.4
11	0.21	27.11	−0.49	5.7	−85.9	−0.10	−6.79	−4.73	2.2	−10.4
12	1.90	27.11	0.78	51.5	−34.4	1.48	−5.31	−6.04	2.0	−12.1
13	1.14	22.04	1.41	25.1	−9.3	1.61	−3.70	−7.35	12.1	−88.9
14	1.14	8.18	3.15	9.3	0.0	3.59	−0.11	−7.35	12.1	−88.9
Sum										−1336.2

Table 9-1 Computations for Example 9-2 (continued)

① Longitudinal i	⑫ $E_i A_i \bar{z}_i (\alpha T)'_i$, lb ④ × ⑪	⑬ $E_i A_i \bar{y}_i (\alpha T)'_i$, lb ③ × ⑪	⑭ A_s^*, in.² Σ②	⑮ $p_{zr} A_s^*/A^*$, lb/in. $-74.65 \times$ ⑭	⑯ $\Sigma E_i A_i (\alpha T)'_i$, lb/in. Σ⑪	⑰ $C_2' Q_z^*$, lb/in. $1.009 \times$ ⑥	⑱ $C_3' Q_y^*$, lb/in. $8.284 \times$ ⑧	⑲ q, lb/in. ⑮ + ⑯ + ⑰ + ⑱
15	− 420	187	0	0	0	0	0	0
16	− 585	1506	1.13	84.3	− 95.2	2.2	− 41.3	− 50.0
1	−1228	3785	2.26	168.7	−190.4	20.3	− 98.8	−100.2
2	− 112	549	5.11	381.4	−371.6	80.3	−258.9	−168.8
3	70	549	5.58	416.5	−397.9	90.2	−275.4	−166.6
4	1015	4061	6.05	451.6	−424.2	100.1	−265.0	−137.5
5	545	1644	8.35	623.3	−618.6	148.5	−165.5	− 12.3
6	545	204	9.22	688.3	−722.5	162.4	−127.9	0.3
			10.09	753.2	−826.4	164.2	− 90.3	0.7

7	527	− 825	10.99	820.4	− 927.3	156.7	− 51.4	− 1.6
8	527	−2224	11.89	887.6	−1028.2	136.5	− 12.4	− 16.5
9	508	−2638	13.30	992.8	−1125.5	98.2	48.5	14.0
10	41	− 282	13.51	1008.5	−1135.9	92.4	55.4	20.4
11	5	− 282	13.72	1024.2	−1146.3	86.7	56.2	41.2
12	9	− 328	15.62	1166.0	−1158.4	34.7	44.0	86.3
13	− 125	−1959	16.76	1251.1	−1247.3	9.4	30.7	43.9
14	− 280	− 727	17.90	1336.2	−1336.2	0	0.9	0.9
Sum	1024	3220						

to ⑲. In passing across longitudinal 14 the shear flow should drop to zero;
the remainder of 0.9 lb/in. is attributable to the accumulation of round-off
errors. The distribution of the reacting shear flows is shown in Fig. 9-5.

9-3 FLUID–FLOW ANALOGY

In the analysis of aerodynamic surfaces and body structures it is common
practice to determine the axial stresses in the effective longitudinals at
each of the rib or frame locations. The axial forces in the longitudinals
may be found by multiplying these stresses by the effective areas. A
fluid-flow analogy which utilizes these axial forces often provides a more
convenient and general method for computing the shear flows in idealized
structures than Eq. (9-3) does.

A short section of an idealized stiffened-shell structure is shown in
Fig. 9-6a, in which we assume that the axial force at each end of the longitudinals is known from previous calculations. Let the axial forces in the
ith longitudinal due to applied loads and temperatures be P_i and $P_i + \Delta P_i$
at x and $x + \Delta x$, respectively. It is convenient to use a double-subscript
notation in identifying the shear flows, and we let q_{ij} be the shear flow
between the ith and jth longitudinals.

A free-body diagram of the ith longitudinal and the adjacent webs is
shown in Fig. 9-6b. The axial forces are defined positive when they are
tensile, and the shear flows are positive when they are in the positive s
direction on the face at $x + \Delta x$. Assuming that the shear flows are constant over Δx, we find for equilibrium in the x direction that

$$-P_i + (P_i + \Delta P_i) + (q_{i,i+1} - q_{i-1,i})\,\Delta x = 0$$

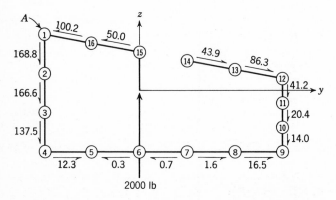

Fig. 9-5 Example 9-2, applied shear force and reacting shear
flows.

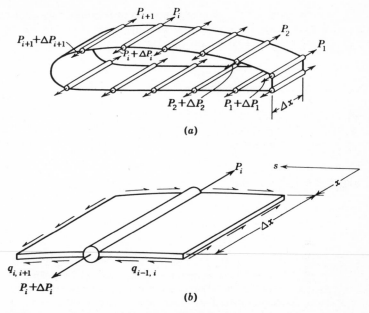

Fig. 9-6 (a) Forces in longitudinals and (b) free-body diagram of longitudinal and webs for the fluid-flow analogy.

so that

$$q_{i,i+1} = q_{i-1,i} - \frac{\Delta P_i}{\Delta x} \qquad (9\text{-}17)$$

We see that the shear flow can be found in any web if the shear flow in the preceding web and the axial forces at the ends of the intervening longitudinal are known.

We note from Eq. (9-17) and Fig. 9-6b that the shear flow leaving a longitudinal is $\Delta P_i/\Delta x$ less than the shear flow that enters the longitudinal. As a result, an analogy exists between the shear flow and a fluid flow along the wall in which fluid is lost at the ith longitudinal because of a sink of strength

$$\Delta q_i = -\frac{\Delta P_i}{\Delta x} \qquad (9\text{-}18)$$

if ΔP_i is negative, the flow in the web increases, and the longitudinal acts as a source of flow.

It should be noted that Eq. (9-17) does not depend upon the assumption that plane sections remain plane, as Eq. (9-3) does. It is therefore valid in the region of discontinuities in the cross section and/or shear force if these effects are accounted for in the end forces on the longitudinals.

If the shear flow in one of the webs of a cell (say $q_{i-1,i}$) is known, the shear flows around the cell may be found by successively applying Eq. (9-17) across each of the longitudinals.   As a result, $q_{m,m+1}$ may be written

$$q_{m,m+1} = q_{i-1,i} + \sum_{j=i}^{m} \Delta q_j \tag{9-19}$$

The application of this equation to closed single-cell or multicell structures is given in Sec. 9-5.   The problem is simplified in an open section because we may take $q_{i-1,i} = 0$ at the edge.   The shear flows may then be found by algebraically summing the source and sink flows of the longitudinals along the wall, so that

$$q_{m,m+1} = \sum_{j=i}^{m} \Delta q_j \tag{9-20}$$

Example 9-3   A beam has the idealized cross section shown in Fig. 9-7a.   The forces in the longitudinals at spanwise stations that are 10 in. apart are given in ② and ③ of Table 9-2.   Determine the shear-flow distribution when the webs between longitudinals 2 and 3 and 5 and 6 are cut as shown in Fig. 9-7b.

The calculations for ΔP_j and Δq_j from Eq. (9-18) are given in ④ and ⑤ of Table 9-2.   The values of Δq_j are also shown opposite each of the longitudinals

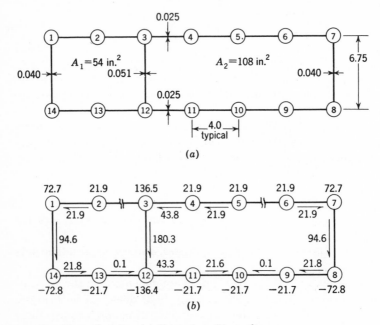

Fig. 9-7 Example 9-3. (a) Geometry; (b) results.

Table 9-2   Example 9-3

① Longitudinal j	② P_j, lb	③ $P_j + \Delta P_j$, lb	④ ΔP_j, lb ③ $-$ ②	⑤ Δq_j, lb/in. $\dfrac{-\;④}{10.0}$
1	-4241	-4968	$-\ 727$	72.7
2	-1279	-1498	$-\ 219$	21.9
3	-7955	-9320	-1365	136.5
4	-1279	-1498	$-\ 219$	21.9
5	-1279	-1498	$-\ 219$	21.9
6	-1279	-1498	$-\ 219$	21.9
7	-4241	-4968	$-\ 727$	72.7
8	4245	4973	728	$-\ 72.8$
9	1265	1482	217	$-\ 21.7$
10	1265	1482	217	$-\ 21.7$
11	1265	1482	217	$-\ 21.7$
12	7957	9321	1364	-136.4
13	1265	1482	217	$-\ 21.7$
14	4245	4973	728	$-\ 72.8$

in Fig. 9-7b.   The resulting shear flows in the figure are obtained by summing the Δq_j from longitudinals 2 to 12, 5 to 12, and 6 to 12.   At 12 the shear flows should satisfy the equation

$$q_{13,12} + q_{3,12} + q_{11,12} + \Delta q_{12} = 0$$

(where positive shear flows are toward 12).   Substituting the results of Fig. 9-7b into this equation, we find a closure error of 0.7 lb/in., attributable to round-off errors in computing the forces in the longitudinals.

9-4 SHEAR CENTER

The location of the elastic axis is important in the design of high-aspect-ratio aerodynamic surfaces (see Example 8-9).   This axis is the locus of points along the span of a beam at which a shear force produces bending deflections without twisting.

The elastic axis coincides with the axis of twist for torsional moments. This is proved with Betti's reciprocal theorem (Sec. 6-13) by considering a load system 1 consisting of a shear force V at the elastic axis and a system 2 composed of a torque M_t.   From the reciprocal theorem we find that $V\varphi_t d = M_t \varphi_v$, where d is the distance from the elastic axis to the axis of twist, while φ_t and φ_v are the angles of twist resulting from M_t and V. Noting that $\varphi_v = 0$, it follows that $d = 0$, and the two axes coincide.

The *shear center* is the point in the cross section of a *slender uniform* beam at which a shear produces no twisting. It is a property of the cross section. In an unswept high-aspect-ratio aerodynamic surface without discontinuities, the elastic axis is essentially a straight line which passes through the shear centers of the cross sections.

To determine the y coordinate of the shear center, we first compute the shear flows $q(V_z)$ due to V_z at the shear center. The method for calculating $q(V_z)$ for open sections was described in Secs. 9-2 and 9-3, and the procedure for closed sections is given in Sec. 9-5. The moment of V_z about an arbitrary moment center in the cross section is equal and opposite to that of the reacting shear flows, so that

$$V_z e_y + \int q(V_z) r \, ds = 0$$

where e_y is the y component of the distance from the moment center to the shear center and r is the perpendicular distance from the moment center to the tangent to the wall at s. The integral extends over all walls of the cross section. Solving for e_y and using the same procedure to determine e_z, the z component of the distance to the shear center, we find

$$e_y = -\frac{1}{V_z} \int q(V_z) r \, ds \qquad e_z = -\frac{1}{V_y} \int q(V_y) r \, ds \qquad (9\text{-}21)$$

where $q(V_y)$ is the shear flow due to a shear force V_y at the shear center.

In idealized structures the shear flow is constant between longitudinals, and Eq. (8-60) may be used to express the moment of the shear flows. In this case Eq. (9-21) becomes

$$e_y = -\frac{2}{V_z} \sum A_{ij} q_{ij}(V_z) \qquad e_z = -\frac{2}{V_y} \sum A_{ij} q_{ij}(V_y) \qquad (9\text{-}22)$$

where A_{ij} is the area swept by the radius vector from the moment center to a point in the web as it moves from the ith to the jth longitudinal. The summations extend over all webs of the cross section. The signs of the $2A_{ij} q_{ij}$ terms do not follow automatically from the sign of q_{ij} but should be determined from a free-body diagram using the sign convention for positive moments used in deriving Eq. (9-21).

The procedure for determining the location of the shear center is illustrated in Example 9-5. When the cross section has an axis of symmetry, the shear center must of course lie on this axis.

9-5 CLOSED SECTIONS

As stated earlier, Eqs. (9-3) and (9-19) apply to closed as well as to open sections. Either of these equations may be used to find the shear flow at any point in each cell in terms of the flow at an arbitrary chosen point in the cell.   For a section with n cells there are n unknown shear flows, one at the arbitrary chosen point in each cell.   These must be determined by additional conditions of equilibrium and compatibility.

Two methods for determining the shear flows will be described; in the *direct method* the flows are found without establishing the location of the shear center, while in the *shear-center method* the location of this point is obtained.   The latter method is longer when the position of the shear center is not obvious by inspection, but it is often necessary to determine the location of this point for aeroelastic analyses.

Consider a system with n cells in which we imagine a cut in each cell to reduce the structure to an open section (Fig. 9-8).   We designate the shear flows in the open section due to V_y, V_z, and T by $q^{(0)}$ and note that these flows may be computed by Eq. (9-3) (with $q_0 = 0$) or from Eq. (9-20).   The shear flows at the imaginary cuts are not zero, of course, and the actual shear flow at any point is

$$q = q^{(0)} + q^{(1)} \tag{9-23}$$

where $q^{(1)}$ is the flow due to the actual shear flows at the points where the structure is "cut."   If the cuts are made in the outer wall of each cell and the shear flow in the ith cell is designated as q_i, then $q^{(1)} = q_i$ in the outside walls of the ith cell, and $q^{(1)} = q_i - q_{i+1}$ in the interior web between cells i and $i + 1$.

The total shear flows must be in equilibrium with the applied loads. Equilibrium of forces in the y and z directions is satisfied by $q^{(0)}$, while equilibrium of forces in the x direction and moments about the y and z axes is assured by σ_{xx}.   The only remaining condition to be satisfied is moment equilibrium about an axis parallel to the x axis.

In the direct method we write the moment equation (Fig. 9-8)

$$M_t - V_y z_c + V_z y_c + 2 \sum A_{ij} q_{ij}^{(0)} + 2 \sum_{i=1}^{n} A_i q_i = 0 \tag{9-24}$$

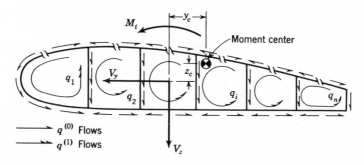

Fig. 9-8 Applied loads and reacting shear flows in a multicell structure.

where M_t is the applied torque, y_c and z_c are the distances from the moment center to V_z and V_y, and A_i is the area enclosed by the ith cell.

When the section forms a single cell, we find from Eq. (9-24) that

$$q_1 = \frac{1}{2A_1}\left(-M_t + V_y z_c - V_z y_c - 2\sum A_{ij}q_{ij}^{(0)}\right) \qquad (9\text{-}25)$$

When $n > 1$, we require $n - 1$ compatibility equations obtained from the condition that θ must be the same for all cells, since the transverse members are assumed rigid within their planes. Letting $\theta_1 = \theta_i$, we find from Fig. 9-8 and Eq. (8-65)

$$\frac{1}{A_1}\left(\oint_1 \frac{q^{(0)}\,ds}{t^*} + q_1 \oint_1 \frac{ds}{t^*} - q_2 \int_{\substack{\text{web}\\1,2}} \frac{ds}{t^*}\right)$$

$$= \frac{1}{A_i}\left(\oint_i \frac{q^{(0)}\,ds}{t^*} + q_i \oint_i \frac{ds}{t^*} - q_{i-1}\int_{\substack{\text{web}\\i-1,i}} \frac{ds}{t^*} - q_{i+1}\int_{\substack{\text{web}\\i,i+1}} \frac{ds}{t^*}\right) \qquad (9\text{-}26)$$

where $i = 2, 3, \ldots, n$. Solution of Eqs. (9-24) and (9-26) yield the n unknown q_i, which with Eq. (9-23) give the final shear flows.

In the shear-center method we assume that V_y and V_z act at the shear center and M_t is the torque about the shear center. The torque is unknown at the outset (because the location of the shear center is not known), so that Eq. (9-24) cannot be used as one of the equations to determine the q_i. Instead we separately find $q(V_y)$ and $q(V_z)$, the shear flows due to V_y and V_z at the shear center. Considering the $q(V_z)$ flows, we find from Eq. (8-65) with $\theta = 0$ that

$$q_{i-1}\int_{\substack{\text{web}\\i-1,i}} \frac{ds}{t^*} - q_i \oint_i \frac{ds}{t^*} + q_{i+1}\int_{\substack{\text{web}\\i,i+1}} \frac{ds}{t^*}$$

$$= \oint_i \frac{q^{(0)}\,ds}{t^*} \qquad i = 1, 2, \ldots, n \qquad (9\text{-}27)$$

where $q^{(0)}$ is the shear flow in the "cut" section due to V_z and q_i is the constant shear flow around the ith cell due to the shear flow at the cut. The q_i are found from Eq. (9-27), and with these the shear flow at any point is computed from $q(V_z) = q^{(0)} + q^{(1)}$, where $q^{(1)}$ is the flow due to the q_i around each cell. The moment of V_z about an arbitrary moment center must equal the moment of the shear flows about the moment center, so that in an idealized section

$$V_z e_y + 2\Sigma A_{ij}q_{ij}(V_z) = 0$$

where e_y is the y projection of the distance from the moment center to the shear center. The signs of the terms in this equation are found from a free-body diagram showing V_z and $q_{ij}(V_z)$. From this equation we find

$$e_y = -\frac{2}{V_z}\sum A_{ij}q_{ij}(V_z) \qquad (9\text{-}28)$$

The flow $q(V_y)$ and e_z, the z distance from the moment center to the shear center, are found in the same manner as $q(V_z)$ and e_y. With e_y and e_z known, M_t is found, and the shear flows $q(M_t)$ due to M_t are determined from Eqs. (8-62) and (8-68). The final flows are found from

$$q = q(V_y) + q(V_z) + q(M_t)$$

Example 9-4 Use the direct method to compute the shear flows in Example 9-2 when there is no cut between longitudinals 14 and 15 and the 2000-lb force is 10 in. to the right of the left spar web.

We imagine a cut between longitudinals 14 and 15 to determine $q^{(o)}$. These shear flows have been determined in Example 9-2 and are given in Fig. 9-5 and ② of Table 9-3. The shear flow q_1 is found from Eq. (9-25) with $M_t = V_y = 0$, $V_z = -2000$ lb, and $y_c = -10$ in. for the moment center at A in Fig. 9-5. The calculations for $2\Sigma A_{ij}q_{ij}^{(o)}$ are given in ③ and ④ of Table 9-3. Equation (9-25) gives

$$q_1 = \frac{1}{2 \times 432}(-2000 \times 10 + 4873) = -17.5 \text{ lb/in.}$$

where $+q_1$ is in the counterclockwise direction. The final shear flows are computed from Eq. (9-23) in ⑤ of Table 9-3. The applied shear force and the reacting shear flows are shown in Fig. 9-9.

Example 9-5 The longitudinal forces in ② and ③ of Table 9-2 (Example 9-3) are due to a vertical shear force of 2500 lb applied 3 in. to the right of the left spar web in Fig. 9-7. Determine the horizontal position of the shear center, the shear flows, and θ if the webs 2–3 and 5–6 are uncut.

The $q^{(o)}$ flows for cuts at webs 2–3 and 5–6 have been computed in Example 9-3 and are given in Fig. 9-7b. Let q_1 and q_2 be the shear flows in webs 2–3 and 5–6 when the 2500-lb force is applied at the shear center. These result in shear flows q_1 and q_2 around cells 1 and 2. Applying Eq. (9-27) to cells 1 and 2, we find

$$-q_1 \oint_1 \frac{ds}{t} + q_2 \int_{\substack{\text{web} \\ 1,2}} \frac{ds}{t} = \oint_1 \frac{q^{(o)} \, ds}{t} \qquad q_1 \int_{\substack{\text{web} \\ 1,2}} \frac{ds}{t} - q_2 \oint_2 \frac{ds}{t} = \oint_2 \frac{q^{(o)} \, ds}{t}$$

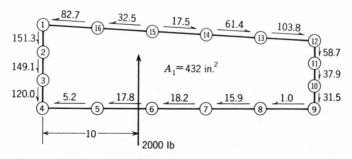

Fig. 9-9 Results of Example 9-4.

Table 9-3 Computations for Example 9-4

① Longitudinal, i	② $q_{ij}^{(o)}$, $lb/in.$	③ $2A_{ij}$, $in.^2$	④ $2A_{ij}q_{ij}^{(o)}$, $in.-lb$ ② $\times$ ③	⑤ q_{ij} $lb/in.$ ② $-$ 17.5
14				
	0	0	0	$-$ 17.5
15				
	50.0	0	0	32.5
16				
	100.2	0	0	82.7
1				
	168.8	0	0	151.3
2				
	166.6	0	0	149.1
3				
	137.5	0	0	120.0
4				
	12.3	60.8	748	$-$ 5.2
5				
	$-$ 0.3	166.3	$-$ 50	$-$ 17.8
6				
	$-$ 0.7	121.7	$-$ 85	$-$ 18.2
7				
	1.6	166.3	226	$-$ 15.9
8				
	16.5	60.8	1003	$-$ 1.0
9				
	$-$ 14.0	61.0	$-$ 854	$-$ 31.5
10				
	$-$ 20.0	166.1	$-$3388	$-$ 37.9
11				
	$-$ 41.2	61.0	$-$2513	$-$ 58.7
12				
	$-$ 86.3	0	0	$-$103.8
13				
	$-$ 43.9	0	0	$-$ 61.4
14				
Sum			$-$4873	

The integrals in these equations are evaluated in ② to ⑤ of Table 9-4. Substituting these into the above equation, we obtain

$$-941.2q_1 + 132.4q_2 = -896 \qquad 132.4q_1 - 1581.2q_2 = 21{,}792$$

which give $q_1 = -1.0$ lb/in. and $q_2 = -13.9$ lb/in. Superimposing these upon the $q^{(o)}$ flows of Fig. 9-7b, we obtain the results for $q(V_z)$ that are given in Fig. 9-10a.

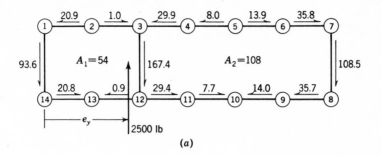

(a)

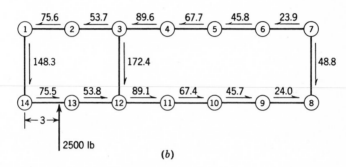

(b)

Fig. 9-10 Example 9-5. Results for 2500 lb at (a) shear center and (b) 3 in. aft of front spar.

With $q(V_z)$ known we may find e_y from Eq. (9-28). The summation in this equation is evaluated in ⑦ of Table 9-4 with the moment center at longitudinal 14. This gives $e_y = 26{,}290/2500 = 10.52$ in. The moment of the 2500-lb force which is 3 in. to the right of the left spar is then $M_t = 2500(10.52 - 3.00) = 18{,}800$ in.-lb. This torque causes additional shear flows around cells 1 and 2, which we designate by $q_1(M_t)$ and $q_2(M_t)$.

One equation for determining $q_1(M_t)$ and $q_2(M_t)$ is obtained from Eq. (8-62), which gives

$$2 \times 54 q_1(M_t) + 2 \times 108 q_2(M_t) = 18{,}800 \qquad (a)$$

A second equation is found from Eq. (8-68), where the line integrals are evaluated in ② and ③ of Table 9-4. This gives

$$\tfrac{1}{54}[941.2q_1(M_t) - 132.4q_2(M_t)] = \tfrac{1}{108}[1581.2q_2(M_t) - 132.4q_1(M_t)] \quad (b)$$

Solving Eqs. (a) and (b), we find $q_1(M_t) = 59.7$ lb/in. and $q_2(M_t) = 54.7$ lb/in. Superimposing these upon the shear flows of Fig. 9-10a gives the final shear flows shown in Fig. 9-10b.

The twist of the section is due solely to $q_1(M_t)$ and $q_2(M_t)$. Writing Eq. (8-67) for cell 1 with $q_1 = q_1(M_t)$ and $q_2 = q_2(M_t)$, we find for $G = 4 \times 10^6$ psi that

$$\theta = \frac{1}{2 \times 54 \times 4 \times 10^6}(54.7 \times 941.2 - 59.7 \times 132.4) = 1.009 \times 10^{-4} \text{ rad/in.}$$

Table 9-4 Example 9-5

①	②	③	④	⑤	⑥	⑦
Longitudinal i	Cell 1 $\dfrac{ds}{t}$	Cell 2 $\dfrac{ds}{t}$	Cell 1 $\dfrac{q^{(0)}\,ds}{t}$	Cell 2 $\dfrac{q^{(0)}\,ds}{t}$	$2A_{ij}$	$2A_{ij}q_{ij}(V_z)$
3						
	160.0		0		27.0	$-$ 27
2						
	160.0		3,504		27.0	564
1						
	168.8		15,968		0	0
14						
	160.0		3,488		0	0
13						
	160.0		16		0	0
12						
6						
		160.0		0	27.0	$-$ 375
5						
		160.0		3,504	27.0	216
4						
		160.0		7,008	27.0	807
3						
	132.4	132.4	$-23,872$	23,872	54.0	$-$ 9,040
12						
6						
		160.0		$-$ 3,504	27.0	$-$ 967
7						
		168.0		$-15,968$	162.0	$-17,469$
8						
		160.0		$-$ 3,488	0	0
9						
		160.0		-16	0	0
10						
		160.0		3,456	0	0
11						
		160.0		6,928	0	0
12						
Sum	941.2	1,581.2	$-$ 896	21,792		$-26,290$

9-6 EFFECTS OF TAPER

The theories for stresses in this chapter and in Chaps. 7 and 8 have assumed that the beam is of constant cross section. Aerodynamic surfaces and body structures are usually tapered, and we now consider modifications to the idealized stiffened-shell theories to account for the taper.

To introduce the concepts we first consider the simple case of an idealized tapered beam with two longitudinals and a web. A free-body diagram of a section of such a beam is shown in Fig. 9-11, where V and M are the shear and moment at x due to the applied loads. These loads are reacted by a shear force V_w in the web and by axial forces P_1 and P_2 in the longitudinals, as indicated.

The axial force P_i in longitudinal i may be resolved into components P_{x_i} and P_{z_i}, which are parallel to, and positive in the directions of, the x and z axes. In this simple case the internal forces are statically determinate. Summing moments about A and B, respectively, we find

$$P_{x_1} = -\frac{M}{h} \qquad P_{x_2} = \frac{M}{h} \qquad (9\text{-}29)$$

where h is the distance between the centroids of the longitudinals at x. We note from these equations that the horizontal components of the forces in the longitudinals are the same as those in an untapered beam of the same height h. The vertical components of the forces in the longitudinals are given by

$$P_{z_i} = P_{x_i} \frac{\Delta z_i}{\Delta x} \qquad (9\text{-}30)$$

where we note that Δz_i for longitudinal 2 is negative. The axial forces in

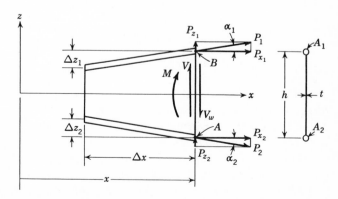

Fig. 9-11 Idealized tapered beam with two longitudinals.

the longitudinals are therefore

$$P_i = P_{x_i} \frac{(\Delta x^2 + \Delta z_i{}^2)^{1/2}}{\Delta x} \tag{9-31}$$

We observe that $\cos \alpha_i = \Delta x / (\Delta x^2 + \Delta z_i{}^2)^{1/2}$, where α_i is the angle that the ith longitudinal makes with the x axis. Substituting this result into Eq. (9-31) and using Eqs. (9-29), we find

$$P_1 = - \frac{M}{h \cos \alpha_1} \qquad P_2 = \frac{M}{h \cos \alpha_2} \tag{9-32}$$

The stresses in the longitudinals are found by dividing the forces by the areas, to give

$$\sigma_1 = - \frac{M}{h A_1 \cos \alpha_1} \qquad \sigma_2 = \frac{M}{h A_2 \cos \alpha_2} \tag{9-33}$$

We note that taper increases the stress in the longitudinals; however, when α_i is small, $\cos \alpha_i \approx 1$, and the effect is small.

Vertical equilibrium in Fig. 9-11 gives

$$V - V_w + P_{z_1} + P_{z_2} = 0$$

which by Eq. (9-30) may be written

$$V_w = V + \sum_{i=1}^{2} P_{x_i} \frac{\Delta z_i}{\Delta x} \tag{9-34}$$

Observing that $\tan \alpha_i = \Delta z_i / \Delta x$ and using Eqs. (9-29), we find

$$V_w = V - \frac{M}{h} (\tan \alpha_1 - \tan \alpha_2) \tag{9-35}$$

where α_2 is a negative angle. The shear flow in the web is $q = V_w/h$, and the shear stress is

$$\sigma_{xz} = \frac{V}{ht} - \frac{M}{h^2 t} (\tan \alpha_1 - \tan \alpha_2) \tag{9-36}$$

The first term in this equation is recognized as the shear flow in an untapered beam of height h, and the second term accounts for the portion of the applied shear that is reacted by the vertical components of the forces in the longitudinals. This term may increase or decrease σ_{xz} depending upon the relative signs of V, M, and α_i.

With the simple beam of Fig. 9-11 as a background we now consider the general case of a tapered idealized structure with an arbitrary number of longitudinals and cells. The structure is assumed to be subjected to shear forces V_y and V_z, a torque M_t, and a temperature distribution with equivalent shears V_{yT} and V_{zT}. A short length of the structure is shown in Fig. 9-12, where it should be noted that unlike the preceding cross-sectional figures in this chapter, Fig. 9-12a is a view in the direction of the positive x axis.

Proceeding in the same manner as for the beam with two longitudinals, we find the horizontal components of the forces in the longitudinals from

$$P_{x_i} = \sigma_{xx_i} A_i \qquad (9\text{-}37)$$

where σ_{xx_i} is the stress in the ith longitudinal due to M_y, M_z, and the temperature distribution. This stress is computed from Eqs. (7-19) by using the weighted cross-sectional properties at x. The vertical components of the forces in the longitudinals are given by Eq. (9-30), and in a similar fashion the y components in Fig. 9-12 are

$$P_{y_i} = P_{x_i} \frac{\Delta y_i}{\Delta x} \qquad (9\text{-}38)$$

The resultant axial force in the longitudinal is then

$$P_i = P_{x_i} \frac{(\Delta x^2 + \Delta y_i{}^2 + \Delta z_i{}^2)^{\frac{1}{2}}}{\Delta x} \qquad (9\text{-}39)$$

which is equivalent to Eq. (9-31) for the beam with two longitudinals.

The shear flows in the webs and the y and z components of the forces in the longitudinals must be in equilibrium with V_y, V_z, and M_t. Letting V_{w_y} and V_{w_z} be the horizontal and vertical components of the resultant of the web flows, we find for equilibrium in the y and z directions that

$$V_y - V_{w_y} + \sum_{i=1}^{m} P_{y_i} = 0 \qquad V_z - V_{w_z} + \sum_{i=1}^{m} P_{z_i} = 0$$

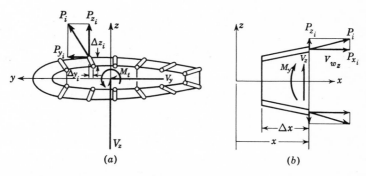

(a)

(b)

Fig. 9-12 Geometry and loading of idealized tapered beam.   (a) View in x direction; (b) view in y direction.

where m is the number of longitudinals. Using Eqs. (9-30) and (9-38), we find

$$V_{w_y} = V_y + \sum_{i=1}^{m} P_{x_i} \frac{\Delta y_i}{\Delta x} \qquad V_{w_z} = V_z + \sum_{i=1}^{m} P_{x_i} \frac{\Delta z_i}{\Delta x} \qquad (9\text{-}40)$$

which corresponds to Eq. (9-34) for the beam with two longitudinals.

The shear flows can be conveniently found by the direct method described in Sec. 9-5. In determining the $q_{ij}{}^{(0)}$ flows in the "cut" structure it is assumed that the beam is uniform with the cross-sectional properties at the point of analysis. The $q_{ij}{}^{(0)}$ flows can then be computed from Eq. (9-3) or the fluid-flow analogy. However, in this computation V_y^* and V_z^* are replaced by $V_{w_y}^*$ and $V_{w_z}^*$, where

$$V_{w_y}^* = V_{w_y} + V_{yT} \qquad V_{w_z}^* = V_{w_z} + V_{zT} \qquad (9\text{-}41)$$

The $q^{(1)}$ shear flows $(q_1, q_2, \ldots, q_n)$ are again found by the simultaneous solution of Eqs. (9-24) and (9-26). However, Eq. (9-24) must be modified to include the torsional moments produced by the forces P_{y_i} and P_{z_i}. Referring to Fig. 9-13, the torsional equilibrium equation becomes

$$M_t - V_y z_c + V_z y_c + 2 \sum A_{ij} q_{ij}{}^{(0)}$$
$$+ 2 \sum_{i=1}^{n} A_i q_i + \sum_{i=1}^{m} P_{y_i}(z_i - z_c) - \sum_{i=1}^{m} P_{z_i}(y_i - y_c) = 0 \qquad (9\text{-}42)$$

This equation may be simplified when the beam is conical, so that the lines of action of the forces in the longitudinals converge at a point. In such a case the sum of the last two terms is zero if the moment center is taken at the apex of the cone.

Example 9-6   The end cross sections of the segment of the tapered beam shown in Fig. 9-14 are 50 in. apart. The applied loads cause a bending moment $M_y =$

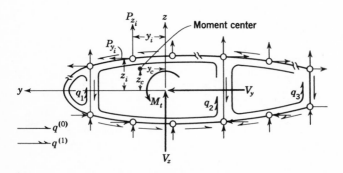

Fig. 9-13   Free-body diagram showing applied loads and reacting forces.

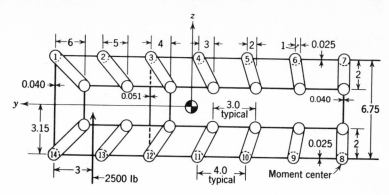

Fig. 9-14   Example 9-6, geometry.

14,563 in.-lb and a shear force $V_z = 2500$ lb (3 in. to the right of the left spar web) at the larger cross section.   At this cross section $I_{yy} = 81.6$ in.^{4}, and the centroidal location is as shown in Fig. 9-14.   Determine the forces in the longitudinals and the shear-flow distribution.

The areas of the idealized longitudinals are given in ② of Table 9-5, and the force components $P_{z_i} = -M_y z_i A_i/I_{yy}$ are listed in ③.   The values of P_{y_i}, P_{z_i}, and P_i are computed in ④ to ⑧.   From Eq. (9-40)

$$V_{w_y} = 0 - 2.4 \approx 0 \qquad V_{w_z} = 2500 - 861.2 = 1639 \text{ lb}$$

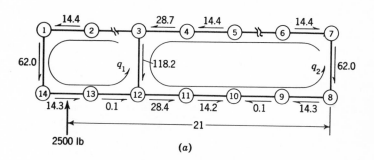

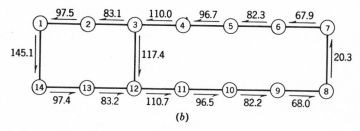

Fig. 9-15   Example 9-6, results.   (*a*) Shear flows $q^{(0)}$ and $q^{(1)}$; (*b*) resultant shear flows.

Table 9-5 Example 9-6

① Longitudinal i	② A_i, in.²	③ P_{zi}, lb	④ $\dfrac{\Delta y_i}{\Delta x}$	⑤ $\dfrac{\Delta z_i}{\Delta x}$	⑥ P_{yi}, lb ③×④	⑦ P_{zi}, lb ③×⑤	⑧ P_i, lb $(③^2 + ⑥^2 + ⑦^2)^{1/2}$	⑨ $z_i - z_c$, in.	⑩ $y_i - y_c$, in.	⑪ $P_{yi}(z_i - z_c)$, in.-lb ⑥×⑨	⑫ $P_{zi}(y_i - y_c)$, in.-lb ⑦×⑩
1	0.660	−4,241	0.12	0.02	− 508.9	− 84.8	−4,275	6.75	24	−3,435.1	− 2,035.2
2	0.199	−1,279	0.10	0.02	− 127.9	− 25.6	−1,287	6.75	20	− 863.3	− 512.0
3	1.238	−7,955	0.08	0.02	− 636.4	− 159.1	−7,987	6.75	16	−4,295.7	− 2,545.6
4	0.199	−1,279	0.06	0.02	− 76.7	− 25.6	−1,283	6.75	12	− 517.7	− 307.2
5	0.199	−1,279	0.04	0.02	− 51.2	− 25.6	−1,282	6.75	8	− 345.6	− 204.8
6	0.199	−1,279	0.02	0.02	− 25.6	− 25.6	−1,280	6.75	4	− 172.8	− 102.4
7	0.660	−4,241	0.00	0.02	− 0.0	− 84.8	−4,245	6.75	0	0.0	0.0
8	0.755	4,245	0.00	−0.02	0.0	− 84.8	4,249	0.00	0	0.0	0.0
9	0.225	1,265	0.02	−0.02	25.3	− 25.3	1,266	0.00	4	0.0	− 101.2
10	0.225	1,265	0.04	−0.02	50.6	− 25.3	1,268	0.00	8	0.0	− 202.4
11	0.225	1,265	0.06	−0.02	75.9	− 25.3	1,269	0.00	12	0.0	− 303.6
12	1.415	7,957	0.08	−0.02	636.6	− 159.1	7,989	0.00	16	0.0	− 2,545.6
13	0.225	1,265	0.10	−0.02	126.5	− 25.3	1,273	0.00	20	0.0	− 506.0
14	0.755	4,245	0.12	−0.02	509.4	− 84.9	4,279	0.00	24	0.0	− 2,037.6
Sum					− 2.4	−861.2				−9,630.2	−11,403.6

By assuming webs 2–3 and 5–6 to be "cut" the shear flows $q_{ij}^{(0)}$ for a shear force of 1639 lb are found by the method of Sec. 9-2, and the results are given in Fig. 9-15a. The shear flows q_1 and q_2 shown in this figure are found from Eqs. (9-26) and (9-42). For the moment center at longitudinal 8

$$2\Sigma A_{ij}q_{ij}^{(0)} = 4 \times 6.75(14.4 + 28.7 + 14.4 - 14.4) + 24 \times 6.75 \times 62.0$$
$$+ 16 \times 6.75 \times 118.2 = 23,974 \text{ in.-lb}$$

and using the sums of ⑪ and ⑫ of Table 9-5, Eq. (9-42) becomes

$$2 \times 54q_1 + 2 \times 108q_2 = 2500 \times 21 - 23,974 + 9630 - 11,404 = 26,752$$

From Eq. (9-26)

$$\tfrac{1}{54}(-585 + 941.2q_1 - 132.4q_2) = \tfrac{1}{108}(14,286 + 1581.2q_1 - 132.4q_2)$$

Solving the last two equations, we find $q_1 = 83.1$ lb/in. and $q_2 = 82.3$ lb/in. Superimposing the q_1 and q_2 flows upon the $q_{ij}^{(0)}$ flows, we obtain the results shown in Fig. 9-15b.

9-7 TRANSVERSE MEMBER LOADS

The role of transverse ribs and frames in semimonocoque structures is discussed in Sec. 7-6, where it is pointed out that one of their primary functions is to diffuse concentrated loads into the structure. In flight vehicles, concentrated forces occur at points where major structural components are joined or where propulsion systems, equipment, or payloads are mounted. It was shown in Sec. 8-7 that the reacting shear flow on a rib or frame is equal to the difference in the shear flows in the thin webs on either side of the transverse member.

The shear and torque undergo step changes at ribs or frames where concentrated loads are introduced. The reacting shear flows on these members can be found by the methods described in this chapter if V_y, V_z, and M_t are replaced by ΔV_y, ΔV_z, and ΔM_t, the increments in these stress resultants that are introduced at the rib or frame. While the process of taking the difference in the shear flows on either side of the transverse member is basically an exact procedure for determining the reacting shear flows, the method described in this chapter for determining the shear flows is only an approximation because cross-sectional warping restraint and rib and frame flexibility within their planes are neglected. In cases where these effects are believed to be important, it is necessary to analyze the structure by the methods for indeterminate structures given in Chaps. 11 and 12. A method for determining the internal forces that the applied loads and reacting shear flows cause in frames is given in Sec. 11-6.

Example 9-7 The fixed nose gear of a private airplane and the fuselage frame to which it is attached are shown in Fig. 9-16. A drift landing condition imposes vertical and horizontal design loads of 4000 and 1200 lb, as indicated in the figure.

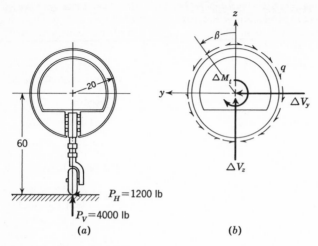

Fig. 9-16 Example 9-7.

The fuselage is monocoque and is without discontinuities at the frame location. Determine the reacting shear flows that the skin applies to the frame.

The shear center is at the center of the fuselage as result of symmetry. The incremental loads at the frame location are therefore $\Delta V_y = 1200$ lb, $\Delta V_z = 4000$ lb, and $\Delta M_t = 1200 \times 60 = 72{,}000$ in.-lb. From Eq. (8-63) the flow reacting ΔM_t is

$$q = \frac{\Delta M_t}{2A} = \frac{72{,}000}{2\pi \times 20^2} = 28.6 \text{ lb/in.} \qquad (a)$$

The reacting flow for ΔV_z from Eq. (9-16) is

$$q = \frac{\Delta V_z}{\pi R} \sin \beta = \frac{4000}{20\pi} \sin \beta = 63.7 \sin \beta \qquad (b)$$

and in a similar fashion for ΔV_y

$$q = \frac{\Delta V_y}{\pi R} \sin \left(\frac{\pi}{2} + \beta\right) = \frac{1200}{20\pi} \cos \beta = 19.1 \cos \beta \qquad (c)$$

Superimposing the results from Eqs. (a) to (c), we obtain the resultant reacting flow

$$q = 28.6 + 63.7 \sin \beta - 19.1 \cos \beta$$

which is positive in the direction of increasing β.

REFERENCES

1. Timoshenko, S. F., and J. N. Goodier: "Theory of Elasticity," 2d ed., McGraw-Hill Book Company, New York, 1951.
2. Hetényi, M.: "Handbook of Experimental Stress Analysis," John Wiley & Sons, Inc., New York, 1950.

3. Boley, B. A., and J. H. Weiner: "Theory of Thermal Stresses," John Wiley & Sons, Inc., New York, 1960.
4. Kuhn, P.: "Stresses in Aircraft and Shell Structures," McGraw-Hill Book Company, New York, 1956.
5. Wang, C. T.: "Applied Elasticity," McGraw-Hill Book Company, New York, 1953.

PROBLEMS

9-1. A built-up beam with the cross section shown is subjected to a vertical shear force of 10,000 lb.   The flange angles are steel ($E = 30 \times 10^6$ psi), and the web is aluminum ($E = 10 \times 10^6$ psi).

    (*a*) Determine the shear stress at the center of the web.

    (*b*) Determine the shear flow carried by the rivets joining the flange angles to the web.

9-2. The angle section shown is loaded by a vertical force of 1000 lb through the shear center.   Determine the shear stress at A.   [*Ans.* $\sigma_{zs} = 2340$ psi.]

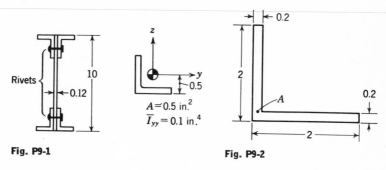

Fig. P9-1   **Fig. P9-2**

9-3. Determine the shear center of the idealized section shown.   The area of all longitudinals is 1 in.².   [*Ans.* $e_z = 5$ in.; $e_y = -4.33$ in. for the moment center at A.]

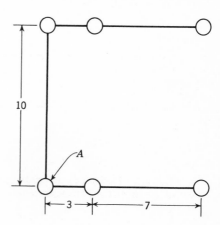

Fig. P9-3

9-4. Determine the shear-flow distribution in the idealized section shown. [*Ans.* $q = 70.7$ lb/in. in the vertical web.]

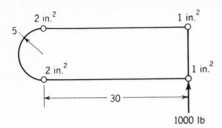

Fig. P9-4

9-5. (*a*) Determine the shear-flow distribution for a 1000-lb shear force at the shear center of the idealized section shown.

   (*b*) Compute the horizontal location of the shear center. [*Ans.* $e_y = 14.5$ in.]

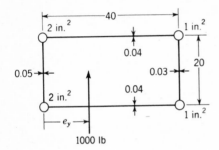

Fig. P9-5

9-6. An idealized section is subjected to a vertical shear force of 1000 lb, as shown. All longitudinals have an area of 1 in.², and all webs are 0.01 in. thick. Determine the shear-flow distribution.

9-7. Determine the horizontal location of the shear center of the section in Prob. 9-6. [*Ans.* $e_y = 16.8$ in.]

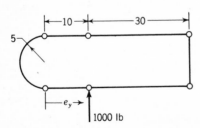

Fig. P9-6 and P9-7

9-8. Draw a free-body diagram of the rib at station 100 of the beam shown. The areas of the caps of the left spar are 2 in.², and areas of the caps of the right spar are 1 in.².

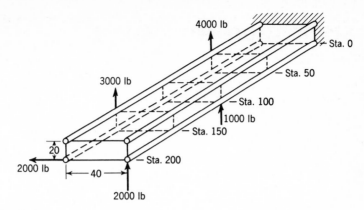

Fig. P9-8

9-9. (*a*) Determine the stresses in the upper and lower longitudinals at the center of the tapered beam shown.

(*b*) Compute the web shear stress on either side of the 1000-lb load. The areas of the upper and lower longitudinals are 1 and 1.5 in.², respectively, and the web is 0.05 in. thick. [*Ans.* $\sigma_1 = -2500$ psi; $\sigma_2 = 1675$ psi; $\sigma_{zz} = 500$ psi left of the load; $\sigma_{zz} = -1500$ psi right of the load.]

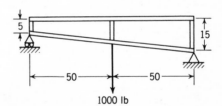

Fig. P9-9

9-10. Determine the shear flows and forces in the longitudinals at the large end of the tapered beam shown when the tip is subjected to a 5000-lb vertical shear force in the plane of symmetry. The areas of all longitudinals are 1 in.².

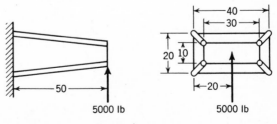

Fig. P9-10

10

Deflection Analysis of Structures

10-1 INTRODUCTION

The differential equations and boundary conditions that the bending and extensional deformations of beams must satisfy are derived in Chap. 7, and those which have to be satisfied by torsional displacements are developed in Chap. 8. When the loading per unit length of the beam does not depend upon the deflections, these equations can be solved by successive integrations. When this is not the case, the solution to the differential equation must be known to obtain an exact solution. Closed-form solutions are seldom possible when the differential equation does not have constant coefficients, in which case the finite-difference and Rayleigh-Ritz methods may be used. However, these methods are seldom practical for complex structures which consist of an assembly of members.

In this chapter we discuss energy methods which are suitable for the deflection analysis of complicated structures and introduce some of the notions of matrix methods, which are elaborated on in Chap. 12. As in the preceding chapters, we include the effects of thermal gradients and nonhomogeneity.

10-2 THE METHOD OF VIRTUAL WORK

The virtual-work, or unit-load, method was developed in Sec. 6-14 and the results given by the general equation (6-88). In this section we elaborate on the method and illustrate its use in beam problems. Explicit equations for δU of commonly encountered structural elements are given in the next section, which is followed by examples that show the application of the method to complex structures. References 1 to 4 are suggested for additional reading.

In Sec. 6-14 it was shown that in elastic bodies the deflection due to applied loads and temperatures can be found by applying an imaginary unit-load system to the body and giving the body a virtual displacement which is equal to the actual displacements. The unit-load system consists of a unit work producing generalized force $Q = 1$ (applied at, and in the direction of, the desired generalized displacement q) and its equilibrating reactions, which are applied in such a manner that they do no work during the real displacements.

In a statically determinate structure there is only one possibility for applying the reactions so that they do no work, namely, the way they would occur if the unit load were applied to the body with its actual supports. An exception to this statement occurs if the deflection or rotation of points other than reaction points is known to be zero from symmetry conditions. In an indeterminate structure there are choices for the unit-load reactions, and forethought will often simplify the solution of the problem. The reactions in this case need not be those that would occur in the structure with its actual supports but may be any set of reactions which are in equilibrium with $Q = 1$ and which do no work during the real displacements. It is usually advantageous to react the unit force in the simplest possible statically determinate manner.

During the virtual (real) displacement $\delta W_e = \delta U$, where

$$\delta W_e = 1 \times q$$

so that

$$q = \delta U \tag{10-1}$$

where

$$\delta U = \int_V \left(\bar{\sigma}_{xx}\epsilon_{xx} + \bar{\sigma}_{yy}\epsilon_{yy} + \bar{\sigma}_{zz}\epsilon_{zz} + \bar{\sigma}_{xy}\epsilon_{xy} + \bar{\sigma}_{yz}\epsilon_{yz} + \bar{\sigma}_{zx}\epsilon_{zx} \right) dV \tag{10-2}$$

In this equation $\bar{\sigma}_{xx}, \bar{\sigma}_{yy}, \ldots, \bar{\sigma}_{zz}$ is *any* set of stresses that is in equilibrium with the unit-load system, and $\epsilon_{xx}, \epsilon_{yy}, \ldots, \epsilon_{zz}$ is the set of actual strains that is due to the real loads and temperatures.

To fix ideas, consider a beam that is subjected to axial forces, bending moments, and thermal gradients. From slender-beam theory, all stresses except $\bar{\sigma}_{xx}$ are negligible, and Eq. (10-2) reduces to

$$\delta U = \int_L \left(\int_A \bar{\sigma}_{xx}\epsilon_{xx} \, dA \right) dx \tag{10-3}$$

where the integrals extend over the cross-sectional area and length of the beam. Let P^*, M_y^*, and M_z^* be the effective stress resultants due to the real loads and temperatures and $\bar{P}$, $\bar{M}_y$, and $\bar{M}_z$ be the stress resultants associated with the unit-load system. To simplify the equations we assume that y and z are principal axes. Then from Eqs. (7-5), (7-17), and (7-18)

$$\epsilon_{xx} = \frac{1}{E_1} \left(\frac{P^*}{A^*} - \frac{M_y^* z}{I_{yy}^*} - \frac{M_z^* y}{I_{zz}^*} \right) \tag{10-4}$$

and from Eq. (7-23)

$$\bar{\sigma}_{xx} = \frac{E}{E_1}\left(\frac{\bar{P}}{A^*} - \frac{\bar{M}_y z}{I_{yy}^*} - \frac{\bar{M}_z y}{I_{zz}^*} \right) \tag{10-5}$$

Substituting Eqs. (10-4) and (10-5) into (10-3), using Eqs. (7-10) to (7-12), and noting that $\bar{y}^* = \bar{z}^* = I_{yz}^* = 0$, we find

$$\delta U = \int_L \frac{P^* \bar{P}}{E_1 A^*} \, dx + \int_L \frac{M_y^* \bar{M}_y}{E_1 I_{yy}^*} \, dx + \int_L \frac{M_z^* \bar{M}_z}{E_1 I_{zz}^*} \, dx \tag{10-6}$$

The application of this equation to predicting thermal deflections of homogeneous beams is discussed in Ref. 4, where it is concluded that it is accurate enough for engineering calculations when the temperature variation in the x direction is smooth.

Several points about Eq. (10-6) are worth noting. To carry out the integrations P^*, $\bar{P}$, M_y^*, $\bar{M}_y$, M_z^*, and $\bar{M}_z$ must be expressed as functions of x. The origin for x is arbitrary; in fact, different origins may be used for different parts of the beam. However, for any segment the same origin *must* be used for both the real and the unit-load systems. The sign conventions for P^*, $\bar{P}$, etc., are also arbitrary and may be different in different parts of the beam, but the same conventions *must* be used in the real and unit-load systems. It is good practice to draw separate free-body diagrams of these two systems showing coordinates and sign conventions.

When $E_1 A^*$, $E_1 I_{yy}^*$, and $E_1 I_{zz}^*$ are constant, they may be factored through the integral signs in Eq. (10-6), otherwise they must be expressed as functions of x before the integrals can be evaluated. In some cases it may be necessary to evaluate the integrals numerically by using the trapezoidal or Simpson's rule (Fig. 5-2). If a negative result is found for δU, it indicates that $\delta W = q$ is negative; that is, q is in the direction opposite to that assumed for $Q = 1$. The use of Eqs. (10-1) and (10-6) is illustrated in the following examples.

Example 10-1 The uniform rectangular beam in Fig. 10-1a is subjected to the loading shown and to a temperature change given by $T = (T_L x/2L)(1 + 2z/h)$. Determine (a) the lateral deflection at $x = L/2$, (b) the slope at $x = 0$, (c) the axial displacement at $x = L/4$, and (d) the area under the deflection curve.

Substituting T into Eqs. (7-15), we find that $M_{z_T} = 0$ and

$$P_T = \int_{-h/2}^{h/2} \frac{\alpha E T_L x}{2L}\left(1 + \frac{2z}{h}\right) b \, dz = \frac{\alpha E T_L bhx}{2L}$$

$$M_{y_T} = -\int_{-h/2}^{h/2} \frac{\alpha E T_L x}{2L}\left(1 + \frac{2z}{h}\right) zb \, dz = -\frac{\alpha E T_L bh^2 x}{12L} \tag{a}$$

The real stress resultants are given in Fig. 10-1a, which, when substituted into

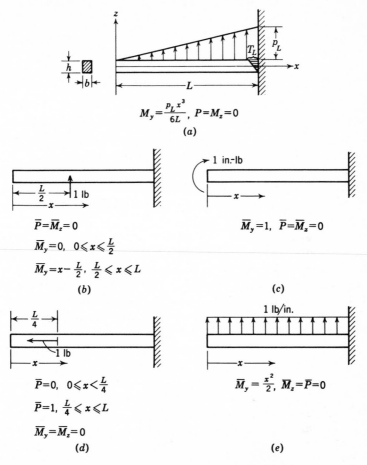

$$M_y = \frac{p_L x^3}{6L}, \quad P = M_z = 0$$

(a)

$\bar{P} = \bar{M}_z = 0$

$\bar{M}_y = 0, \quad 0 \leqslant x \leqslant \frac{L}{2}$

$\bar{M}_y = x - \frac{L}{2}, \quad \frac{L}{2} \leqslant x \leqslant L$

(b)

$\bar{M}_y = 1, \quad \bar{P} = \bar{M}_z = 0$

(c)

$\bar{P} = 0, \quad 0 \leqslant x < \frac{L}{4}$

$\bar{P} = 1, \quad \frac{L}{4} \leqslant x \leqslant L$

$\bar{M}_y = \bar{M}_z = 0$

(d)

$\bar{M}_y = \frac{x^2}{2}, \quad \bar{M}_z = \bar{P} = 0$

(e)

Fig. 10-1 Example 10-1. (a) Real loads and temperatures; (b) unit-load system for $w(L/2)$; (c) unit-load system for $\left.\dfrac{dw}{dx}\right|_{x=0}$; (d) unit-load system for $u(L/4)$; (e) unit-load system for S.

Eqs. (7-13) with Eqs. (a), give $M_z^* = 0$ and

$$P^* = \frac{\alpha E T_L b h x}{2L} \qquad M_y^* = \frac{p_L x^3}{6L} - \frac{\alpha E T_L b h^2 x}{12L} \tag{b}$$

(a) To find the lateral displacement at $x = L/2$ we apply the unit load in Fig. 10-1b, which is reacted by a shear force and moment at the wall. Using the equations for $\bar{P}$, $\bar{M}_y$, and $\bar{M}_z$ in this figure and $I_{yy} = bh^3/12$, we find from Eqs. (10-1) and (10-6) that

$$w\left(\frac{L}{2}\right) = \frac{1}{EI_{yy}} \int_{L/2}^{L} \left(\frac{p_L x^3}{6L} - \frac{\alpha E T_L b h^2 x}{12L}\right)\left(x - \frac{L}{2}\right) dx = \frac{49 p_L L^4}{3840 EI_{yy}} - \frac{5\alpha T_L L^2}{48h}$$

(b) To obtain the slope (rotation) at $x = 0$ we apply the unit moment shown in Fig. 10-1c. Using the stress resultants that are given in this figure, we find from Eqs. (10-1) and (10-6)

$$\frac{dw}{dx}\bigg|_{x=0} = \frac{1}{EI_{yy}} \int_0^L \left(\frac{p_L x^3}{6L} - \frac{\alpha ET_L bh^2 x}{12L} \right)(1)\, dx = \frac{p_L L^3}{24EI_{yy}} - \frac{\alpha T_L L}{2h}$$

(c) Using the unit-force and stress resultants in Fig. 10-1d, Eqs. (10-1) and (10-6) with $A = bh$ give

$$u\left(\frac{L}{4}\right) = \frac{1}{EA} \int_{L/4}^L \frac{\alpha ET_L bhx}{2L}(1)\, dx = \frac{15\alpha T_L L}{64}$$

(d) To determine the area S under the deflection curve we note that

$$S = \int_0^L (w)(1)\, dx$$

which is the work done by a 1 lb/in. uniformly distributed load during the real displacements. This unit-load system is shown in Fig. 10-1e, which with Eqs. (10-1) and (10-6) gives

$$S = \frac{1}{EI_{yy}} \int_0^L \left(\frac{p_L x^3}{6L} - \frac{\alpha ET_L bh^2 x}{12L} \right)\frac{x^2}{2}\, dx = \frac{p_L L^5}{72EI_{yy}} - \frac{\alpha T_L L^3}{96h}$$

Example 10-2   Determine the deflection and slope at point B of the uniform cantilever beam in Fig. 10-2a when it is loaded by a tip shear and moment. Use these results to obtain the values of the end forces and moments for given end displacements and rotations.

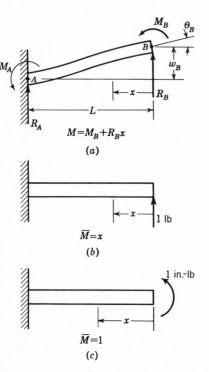

$M = M_B + R_B x$

(a)

$\overline{M} = x$

(b)

$\overline{M} = 1$

(c)

Fig. 10-2 Load systems, Example 10-2. (a) Real loads and displacements; (b) unit load for w_B; (c) unit load for θ_B.

Using the load systems of Fig. 10-2a and b, we find

$$w_B = \frac{1}{EI} \int_0^L (M_B + R_B x)(x) \, dx = \frac{L^2}{2EI} M_B + \frac{L^3}{3EI} R_B \qquad (a)$$

and with the load systems of Fig. 10-2a and c we obtain

$$\theta_B = \frac{1}{EI} \int_0^L (M_B + R_B x)(1) \, dx = \frac{L}{EI} M_B + \frac{L^2}{2EI} R_B \qquad (b)$$

where I is the moment of inertia about the y axis.

The simultaneous solution of Eqs. (a) and (b) leads to the following results for R_B and M_B when the displacements at B are w_B and θ_B and the displacement and rotation at A are zero:

$$R_B = \frac{12EI}{L^3} w_B - \frac{6EI}{L^2} \theta_B \qquad M_B = -\frac{6EI}{L^2} w_B + \frac{4EI}{L} \theta_B \qquad (c)$$

Equilibrium requires that $R_A = -R_B$ and $M_A = -M_B - R_B L$, which gives

$$R_A = -\frac{12EI}{L^3} w_B + \frac{6EI}{L^2} \theta_B \qquad M_A = -\frac{6EI}{L^2} w_B + \frac{2EI}{L} \theta_B \qquad (d)$$

In a similar fashion we find that for a given displacement and rotation at A, while displacement and rotation are prevented at B,

$$R_A = \frac{12EI}{L^3} w_A + \frac{6EI}{L^2} \theta_A \qquad M_A = \frac{6EI}{L^2} w_A + \frac{4EI}{L} \theta_A \qquad (e)$$

and

$$R_B = -\frac{12EI}{L^3} w_A - \frac{6EI}{L^2} \theta_A \qquad M_B = \frac{6EI}{L^2} w_A + \frac{2EI}{L} \theta_A \qquad (f)$$

Equations (c) to (f) are used in the determination of the stiffness matrices of beams in Sec. 10-5.

10-3 EQUATIONS FOR δU OF SIMPLE ELEMENTS

Complex structures may be subdivided into simple elements that behave like beams, bars, shear webs, etc. The total δU is then the sum of the values of δU for the individual elements. Expressions for δU of some commonly encountered structural elements are derived in this section.

Bending of beams The last two terms in Eq. (10-6) give δU for the bending deformations of beams. If bending occurs about only one of the principal axes, we obtain

$$\delta U = \int_L \frac{M^* \bar{M}}{E_1 I^*} \, dx \qquad (10\text{-}7)$$

The stresses and strains in curved beams are adequately predicted[5] by straight-beam theory when R/h, the ratio of the radius of the curved beam to its height, is greater than 10. In such cases we may rewrite Eq. (10-7) as

$$\delta U = \int_L \frac{M^* \bar{M}}{E_1 I^*} \, ds \qquad (10\text{-}8)$$

where s is a coordinate measured along the line which joins the weighted centroids of the cross sections of the curved beam and the integral extends along the curved length of the beam.

Extension of bars   The first term of Eq. (10-6) gives δU for the extensional deflections of straight bars.   Again, for curved bars with $R/h > 10$ we may write

$$\delta U = \int_L \frac{P^* \bar{P}}{E_1 A^*} \, ds \qquad (10\text{-}9)$$

In straight bars with pinned ends, P is constant if axial forces are applied only at the ends.   If $E_1 A^*$ and T are also constant along the length of the bar, Eq. (10-9) reduces to

$$\delta U = \frac{P^* \bar{P} L}{E_1 A^*} \qquad (10\text{-}10)$$

This equation is useful in analyzing pin-jointed trusses, for which the total δU is

$$\delta U = \sum_{i=1}^{n} \frac{P_i^* \bar{P}_i L_i}{E_1 A_i^*} \qquad (10\text{-}11)$$

where the subscript i refers to the ith member and the summation extends over all members of the truss.

In an idealized stiffened-shell structure the rate of change of force in the longitudinals is constant when it is assumed that the shear flows in the adjoining webs are constant [see Eq. (9-18)].   As a result, the force P in the longitudinal varies linearly between its ends.   If we further assume that T varies linearly along the length of the longitudinal and A is constant, then P_T and P^* will vary linearly.   Furthermore, the axial force $\bar{P}$ due to the unit-load system will also vary linearly, so that

$$P^* = P_A^* + (P_B^* - P_A^*)\frac{x}{L} \qquad \bar{P} = \bar{P}_A + (\bar{P}_B - \bar{P}_A)\frac{x}{L}$$

where the subscripts refer to the ends A and B of the longitudinal.   Substituting these expressions into Eq. (10-9) with $ds = dx$, we find after integration that

$$\delta U = \frac{L}{6EA}[\bar{P}_A(2P_A^* + P_B^*) + \bar{P}_B(P_A^* + 2P_B^*)] \qquad (10\text{-}12)$$

for a homogeneous longitudinal of constant cross section.

Shear deformation of thin-walled beams   No simple expression for δU exists for the shearing deformations of beams of arbitrary cross section.

For thin-walled beams we may write

$$\delta U = \int_V \bar{\sigma}_{xs}\epsilon_{xs}\, dV = \int_L \left(\int_s \frac{\bar{q}}{t}\frac{q}{Gt} t\, ds \right) dx$$

which reduces to

$$\delta U = \int_L \left(\int_s \frac{q\bar{q}}{Gt}\, ds \right) dx \qquad (10\text{-}13)$$

where the integral in the parentheses extends over all walls of the cross section. The shear flow q is due to the real loads and temperatures, and $\bar{q}$ is the flow due to the unit-load system. These can be computed by the methods given in Chap. 9.

In open sections with no thermal stresses we find from Eqs. (9-3) and (9-4) that

$$q = \frac{V_y Q_z^*}{I_{zz}^*} + \frac{V_z Q_y^*}{I_{yy}^*} \qquad \bar{q} = \frac{\bar{V}_y Q_z^*}{I_{zz}^*} + \frac{\bar{V}_z Q_y^*}{I_{yy}^*}$$

if principal axes are used. Substituting these results into Eq. (10-13) leads to

$$\delta U = \int_L \frac{K_{yy} V_z \bar{V}_z}{G_1 A^*}\, dx + \int_L \frac{K_{yz}(V_y \bar{V}_z + V_z \bar{V}_y)}{G_1 A^*}\, dx + \int_L \frac{K_{zz} V_y \bar{V}_y}{G_1 A^*}\, dx$$

$$(10\text{-}14)$$

where the K terms are defined by

$$K_{yy} = \frac{A^*}{(I_{yy}^*)^2} \int_s \frac{(Q_y^*)^2}{t^*}\, ds \qquad K_{zz} = \frac{A^*}{(I_{zz}^*)^2} \int_s \frac{(Q_z^*)^2}{t^*}\, ds$$

$$K_{yz} = \frac{A^*}{I_{yy}^* I_{zz}^*} \int_s \frac{Q_y^* Q_z^*}{t^*}\, ds \qquad (10\text{-}15)$$

where $t^* = tG/G_1$.

When one of the principal axes is an axis of symmetry, $K_{yz} = 0$, because the product $Q_y^* Q_z^*$ has an equal value but opposite sign at mirror-image points. Except for short beams with thin walls, δU for shear deformations is small compared to δU for bending deflections, and it is frequently neglected. In idealized stiffened-shell structures the shear flows in the webs are constant, and it is more convenient to compute δU for the shear deformations from Eq. (10-16) or (10-17), which follow, than to use Eq. (10-14).

Shear webs An idealized shear web which carries only shear stresses is shown in Fig. 10-3. The shear flow, thickness, and shear modulus are assumed constant in the s direction but may vary in the x direction. In

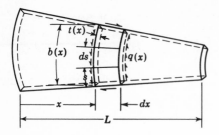

Fig. 10-3 Idealized shear web.

this case Eq. (10-13) becomes

$$\delta U = \int_L \frac{q\bar{q}b}{Gt}\, dx \tag{10-16}$$

When q, $\bar{q}$, G, t, and b are constant, this reduces to

$$\delta U = \frac{q\bar{q}bL}{Gt} \tag{10-17}$$

Torsional deformations   A differential length of a torsion member of arbitrary cross section is shown in Fig. 10-4.   If $\bar{M}_t$ is the torque at x that is due to the unit-load system and φ is the angle of twist at x that results from the real loads and temperatures, then the work that is done on the element by $\bar{M}_t$ in moving through the real displacement is

$$d\,\delta W_e = \bar{M}_t \frac{d\varphi}{dx}\, dx$$

Noting that $d\,\delta W_e = d\,\delta U$, integrating over the length, and using Eq. (8-73) gives

$$\delta U = \int_L \frac{M_t \bar{M}_t}{G_1 J^*}\, dx \tag{10-18}$$

where M_t is the torque due to the real loads and temperatures.

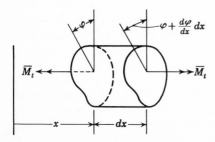

Fig. 10-4   Torsion member.

Elastic supports   Two types of elastic supports are schematically represented by the effective springs in Fig. 10-5.   If the spring force $\bar{P}$ in Fig. 10-5a is due to the unit-load system and d is the real displacement of the spring, then $\delta U = \delta W_e = \bar{P}d$.   If the support is linearly elastic, $d = P/k_s$, where P is the real force in the support and k_s is the equivalent spring constant of the support.   It follows that

$$\delta U = \frac{P\bar{P}}{k_s} \tag{10-19}$$

In a similar manner, for the elastic rotational restraint in Fig. 10-5b

$$\delta U = \frac{M\bar{M}}{k_r} \tag{10-20}$$

where M is the moment due to the real loads and temperatures, $\bar{M}$ is the moment due to the unit-load system, and k_r is the equivalent rotational spring constant of the support.

The following examples illustrate the use of the equations derived in this section.

Example 10-3   The rectangular cross-section semicircular frame in Fig. 10-6 is subjected to a temperature change, which varies linearly from T_1 at the inner radius to T_2 at the outer radius, and to the forces F_1 and F_2 shown.   Determine the horizontal displacement q at the right support.

The frame is slender, so that shear deformations can be neglected and only bending and extensional deformations need be considered.   In many cases the extensional displacements are also neglected; however, when there are temperature changes, these deformations may be significant.   We locate points on the frame by the coordinate θ and take advantage of symmetry by using the same coordinate for both halves of the frame.   From the free-body diagram of Fig. 10-6a

$$P = F_2 \sin \theta - \frac{F_1}{2} \cos \theta \qquad M = F_2R \sin \theta + \frac{F_1R}{2}(1 - \cos \theta)$$

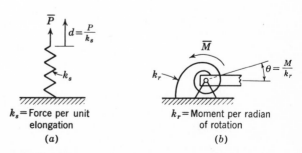

k_s = Force per unit
elongation

(a)

k_r = Moment per radian
of rotation

(b)

Fig. 10-5   Elastic supports.

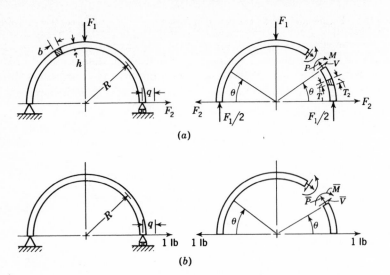

Fig. 10-6   Example 10-3.   (a) Real-load system; (b) unit-load system.

where the sign conventions for M and P are shown in the figure.

Measuring z from the centroidal axis of the cross section (positive in the direction in which M produces compressive stresses), we find

$$T = \frac{T_1 + T_2}{2} + \frac{(T_2 - T_1)z}{h}$$

Substituting this result into Eqs. (7-15) gives

$$P_T = \int_{-h/2}^{h/2} \alpha E \left[\frac{T_1 + T_2}{2} + \frac{(T_2 - T_1)z}{h} \right] b\, dz = \tfrac{1}{2} \alpha E (T_1 + T_2) A$$

$$M_T = - \int_{-h/2}^{h/2} \alpha E \left[\frac{T_1 + T_2}{2} + \frac{(T_2 - T_1)z}{h} \right] zb\, dz = - \frac{\alpha E (T_2 - T_1) I}{h}$$

where $A = bh$ and $I = bh^3/12$.   Adding these to P and M, we obtain

$$P^* = F_2 \sin \theta - \tfrac{1}{2} F_1 \cos \theta + \tfrac{1}{2} \alpha E A (T_1 + T_2)$$

$$M^* = F_2 R \sin \theta + \tfrac{1}{2} F_1 R (1 - \cos \theta) - \frac{\alpha E I (T_2 - T_1)}{h} \tag{a}$$

To obtain the horizontal displacement at the right support we use the unit-load system shown in Fig. 10-6b, from which

$$\bar{P} = \sin \theta \qquad \bar{M} = R \sin \theta \tag{b}$$

From Eqs. (10-8) and (10-9) the total δU is

$$\delta U = \frac{2R}{EI} \int_0^{\pi/2} M^* \bar{M} \, d\theta + \frac{2R}{EA} \int_0^{\pi/2} P^* \bar{P} \, d\theta$$

Substituting Eqs. (a) and (b) into this equation and using Eq. (10-1), we find

$$q = \frac{R^3}{2EI} (F_1 + \pi F_2) - \frac{R}{2EA} (F_1 - \pi F_2) + \alpha R \left[T_1 + T_2 - \frac{2R}{h} (T_2 - T_1) \right]$$

Example 10-4 Figure 10-7 shows the actuating system for the canard control surfaces of a missile. The system consists of a hydraulic actuator A with a push-pull rod B, a bell crank C, and torque shafts D and E that are supported by frictionless bearings. The compressibility of the fluid in the actuator is equivalent to a spring of stiffness k_A. Aerodynamic forces on the control surfaces apply a torque T_0 to the shafts D and E. Determine φ, the change in incidence angle of the control surfaces that results from deformations in the actuation system if it is assumed that the bearing and actuator supports are rigid (when this is not the case, the supports can be represented by springs of equivalent stiffnesses).

Free-body diagrams that show the forces due to the real loads are given in Fig. 10-8a. From these

$$P_A = \frac{2T_0}{L_C} \qquad M_A = M_{t_A} = 0$$

$$P_B = \frac{2T_0}{L_C} \qquad M_B = M_{t_B} = 0$$

$$M_C = \frac{2T_0}{L_C}x_C \qquad P_C = M_{t_C} = 0$$

$$M_D = \frac{T_0}{L_C}x_D \qquad M_{t_D} = T_0 \qquad P_D = 0$$

$$M_E = \frac{T_0}{L_C}x_E \qquad M_{t_E} = T_0 \qquad P_E = 0$$

To obtain the rotations at the roots of the control surfaces we apply a unit torque to the end of shaft D and react it at the bearings and actuator support, as shown in Fig. 10-8b. From the free-body diagrams in this figure

$$\bar{P}_A = \frac{1}{L_C} \qquad \bar{M}_A = \bar{M}_{t_A} = 0$$

$$\bar{P}_B = \frac{1}{L_C} \qquad \bar{M}_B = \bar{M}_{t_B} = 0$$

$$\bar{M}_C = \frac{x_C}{L_C} \qquad \bar{P}_C = \bar{M}_{t_C} = 0$$

$$\bar{M}_D = \frac{x_D}{2L_C} \qquad \bar{M}_{t_D} = 1 \qquad \bar{P}_D = 0$$

$$\bar{M}_E = \frac{x_E}{2L_C} \qquad \bar{M}_{t_E} = \bar{P}_E = 0$$

The total δU for the system (neglecting shear deformation) is

$$\delta U = \frac{P_A\bar{P}_A}{k_A} + \frac{P_B\bar{P}_B L_B}{E_B A_B} + \int_0^{L_C}\frac{M_C\bar{M}_C}{E_C I_C}\,dx_C + \int_0^{L_D}\frac{M_D\bar{M}_D}{E_D I_D}\,dx_D$$
$$+ \int_0^{L_E}\frac{M_E\bar{M}_E}{E_E I_E}\,dx_E + \int_0^{L_D}\frac{M_{t_D}\bar{M}_{t_D}}{G_D J_D}\,dx_D$$

Substituting the previously determined internal forces into this equation and noting that $\delta W_e = \varphi$, we find for $E_D I_D = E_E I_E$

$$\varphi = \frac{T_0}{L_C{}^2}\left(\frac{2}{k_A} + \frac{2L_B}{E_B A_B} + \frac{2L_C{}^3}{3E_C I_C} + \frac{L_D{}^3}{3E_D I_D} + \frac{L_C{}^2 L_D}{G_D J_D}\right)$$

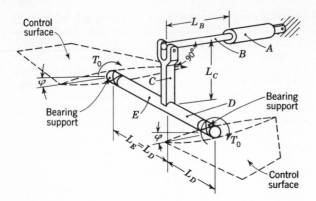

Fig. 10-7 Control-surface actuating mechanism, Example 10-4.

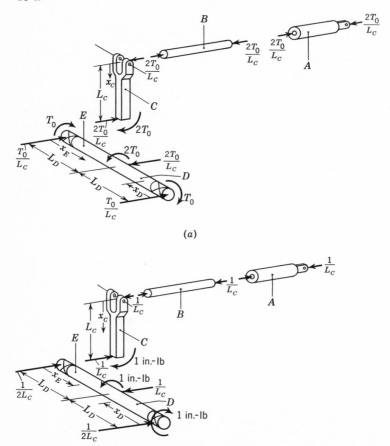

(a)

Fig. 10-8 Load systems for Example 10-4.   (a) Free-body diagrams of actual loads; (b) free-body diagram of unit load system.

Example 10-5 Determine the tip deflection of the idealized box beam shown in Fig. 10-9a, including the effects of shear deformations in the skins and spar webs. All longitudinals have the same area A; the thickness of the spar webs is t_s, and that of the cover skins is t_c. The root is rigid, while the tip rib is rigid in its own plane but is perfectly flexible normal to its plane. It is further assumed that the webs carry only shear flows and that these are constant within each panel.

The actual internal forces must be found before the deflections can be determined. A free-body diagram of the structure showing the root reactions is given in Fig. 10-9b. Symmetry, which has been used in designating the reactions, requires the shear flows in the center panel of the upper and lower surfaces to vanish. Vertical equilibrium gives $q_s = F/h$, and moment equilibrium about a horizontal axis through the lower longitudinals at the root gives

$$P_1 = \frac{FL}{h} - P_2 \tag{a}$$

The shear flow q_1 is determined from the free-body diagram of the center longitudinal in Fig. 10-10a, from which

$$q_1 = \frac{P_2}{L} \tag{b}$$

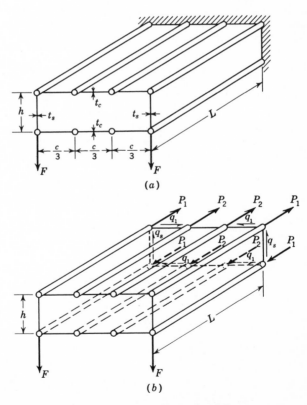

(a)

(b)

Fig. 10-9 Geometry and forces, Example 10-5.

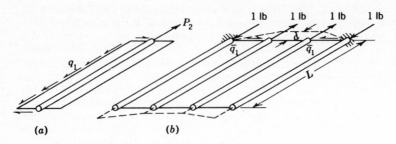

Fig. 10-10 Force systems in covers, Example 10-5.

We have determined P_1 and q_1 in terms of P_2 but do not have an equation relating P_2 to the applied loads F. The equilibrium conditions have been exhausted, and so the structure is statically indeterminate with one redundancy, which we take as P_2. We determine this force with the unit-load method by using the compatibility condition that the displacements of the center longitudinals must be zero at the root.

Imagine that the center longitudinals are cut at the wall and that the forces P_2 are applied rather than reactive forces. The displacements d at the cuts can be found by using the unit-load system shown in Fig. 10-10b. Noting that the forces in the longitudinals vary linearly when the shear flows in the webs are constant, we find from Eqs. (10-12) and (10-17) that

$$2d = \frac{2L}{6EA}[\bar{P}_{A_1}(2P_{A_1} + P_{B_1}) + \bar{P}_{B_1}(P_{A_1} + 2P_{B_1})]$$

$$+ \frac{2L}{6EA}[\bar{P}_{A_2}(2P_{A_2} + P_{B_2}) + \bar{P}_{B_2}(P_{A_2} + 2P_{B_2})] + \frac{2q_1\bar{q}_1cL}{3Gt_c} \qquad (c)$$

where the subscripts A and B refer to the tip and root ends of the longitudinals, respectively. From Eqs. (a) and (b) and Figs. 10-9 and 10-10 we obtain

$$P_{B_1} = \frac{FL}{h} - P_2 \qquad P_{B_2} = P_2 \qquad P_{A_1} = P_{A_2} = 0 \qquad q_1 = \frac{P_2}{L}$$

$$\bar{P}_{B_1} = -1, \qquad \bar{P}_{B_2} = 1, \qquad \bar{P}_{A_1} = \bar{P}_{A_2} = 0 \qquad \bar{q}_1 = \frac{1}{L}$$

Substituting these results into Eq. (c) and noting that $d = 0$ for root compatibility we obtain, after simplification,

$$P_2 = \frac{FL}{h}\frac{1}{2 + \beta} \qquad (d)$$

where $\beta = EAc/Gt_cL^2$. Substituting Eq. (d) into Eqs. (a) and (b), we find

$$P_1 = \frac{FL}{h}\frac{1 + \beta}{2 + \beta} \qquad (e)$$

$$q_1 = \frac{F}{h}\frac{1}{2 + \beta} \qquad (f)$$

The parameter β is seen to have an important bearing on the internal-force distribution. If $G = \infty$, then $\beta = 0$, and $P_1 = P_2 = FL/2h$; this agrees with simple-beam theory ($\sigma_{zz} = -Mz/I$), which assumes that plane sections remain plane. If, on the other hand, $G = 0$, then $\beta = \infty$, and $P_1 = FL/h$ while $P_2 = 0$, which indicates that all the load is resisted by the edge longitudinals. This phenomenon is known as *shear lag*, since the deformations and forces in the interior longitudinals lag behind those in the edge longitudinals as a result of the shear deformations in the cover skins.

To find the tip deflection we place a unit vertical force at the tip. This force does not have to be reacted as it would be in the actual structure but need only be reacted by forces that satisfy equilibrium and do no work during the real displacements. The simple system in Fig. 10-11, where all the load is resisted by the front spar, meets these requirements. For this system

$$\bar{P}_{B_1} = \frac{L}{h} \qquad \bar{q}_s = \frac{1}{h} \qquad \bar{P}_{A_1} = \bar{P}_{A_2} = \bar{P}_{B_2} = \bar{q}_1 = 0$$

From $\delta W_e = 1 \times d_t = \delta U$, the tip deflection d_t is

$$d_t = \frac{2L}{6EA}[\bar{P}_{A_1}(2P_{A_1} + P_{B_1}) + \bar{P}_{B_1}(P_{A_1} + 2P_{B_1})] + \frac{q_s\bar{q}_s hL}{Gt_s} \qquad (g)$$

where, from Eq. (e)

$$P_{B_1} = \frac{FL}{h}\frac{1 + \beta}{2 + \beta}$$

Substituting this into Eq. (g) and recalling that $q_s = F/h$, we find

$$d_t = \frac{FL^3}{3EAh^2} + \frac{FL^3}{3EAh^2}\frac{\beta}{2 + \beta} + \frac{FL}{Ght_s} \qquad (h)$$

Remembering that $\beta = 0$ when $G = \infty$, we observe from Eq. (h) that in this case $d_t = FL^3/3EAh^2$, which agrees with the beam-theory result for a cantilevered beam with a tip load of $2F$. The second term in Eq. (h) is the additional deflection that results from shear lag in the cover skins, and the third term is the shear deformation due to the shear strains in the spar webs.

10-4 RELATIVE DISPLACEMENTS

In the preceding sections absolute displacements were desired, and so the reactions were applied at fixed supports, where they would not contribute

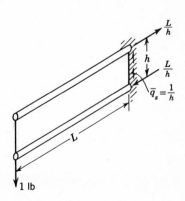

Fig. 10-11 Unit-load system for tip deflection, Example 10-5.

to δW_e. Sometimes relative rather than absolute displacements are required. In these cases it is often possible to apply the unit-load reactions in such a manner that they also do work and the total δW_e is equal to the desired relative displacement. The following cases illustrate the method; additional examples are given in Probs. 10-5 and 10-6.

Displacement at a point relative to a line through two other points   The points A, B, and C in Fig. 10-12 initially lie in a straight line. As a result of applied loads and temperature changes, the points undergo displacements d_A, d_B, and d_C (normal to AC), moving to the new positions A', B', C'. From the geometry of the figure we see that the deflection of B relative to $A'C'$ is

$$d_{B\,\mathrm{rel}\,AC} = d_B + d_A + \frac{d_C - d_A}{a + c}\,a = d_B + \frac{d_A c}{a + c} + \frac{d_C a}{a + c}$$

If we apply a unit force at B and react it by forces at A and C, the reactions are $c/(a + c)$ at A and $a/(a + c)$ at C. The virtual work of the unit-load system in moving through the real displacements is then (Fig. 10-12)

$$\delta W_e = 1 \times d_B + \frac{c}{a + c}\,d_A + \frac{a}{a + c}\,d_C = d_{B\,\mathrm{rel}\,AC}$$

Rotation of a line through two points relative to a line through two other points   In this case the relative rotation is

$$\varphi_{AB\,\mathrm{rel}\,CD} = \varphi_{AB} + \varphi_{CD}$$

as seen from Fig. 10-13. If a unit couple applied to the points A and B is reacted by a unit couple at the points C and D, as illustrated, we find

$$\delta W_e = 1 \times \varphi_{AB} + 1 \times \varphi_{CD} = \varphi_{AB\,\mathrm{rel}\,CD}$$

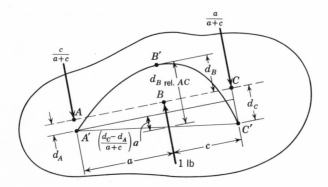

Fig. 10-12   Unit-load system for deflection at B relative to line AC.

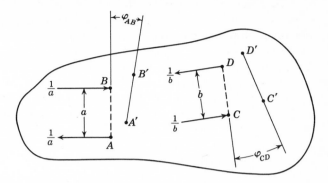

Fig. 10-13 Relative rotation of two lines.

Example 10-6 Determine the deflection at the center of the beam in Example 10-1 relative to a line from the tip to the root.

The unit load system for the relative displacement is shown in Fig. 10-14. The expression for M_y^* is given in Eq. (b) of Example 10-1. From these

$$
d_{B \text{ rel } AC} = \frac{1}{EI_{yy}} \left[\int_0^{L/2} \left(\frac{p_L x^3}{6L} - \frac{\alpha E T_L b h^2 x}{12L} \right) \left(-\frac{x}{2} \right) dx \right.
$$
$$
\left. + \int_{L/2}^L \left(\frac{p_L x^3}{6L} - \frac{\alpha E T_L b h^2 x}{12L} \right) \frac{x - L}{2} dx \right] = -\frac{3 p_L L^4}{64 E I_{yy}} + \frac{\alpha T_L L^3}{12 h^2}
$$

10-5 FLEXIBILITY AND STIFFNESS MATRICES

The unit-load method is useful for determining the deflections at a limited number of points in a structure. In designing flight vehicles it is usually necessary to determine the deflections at a large number of points in the structure for many different load conditions. As a result it is desirable to organize the calculations for machine computations so that they will not become burdensome. The digital computer is ideally suited for performing the operations of matrix algebra, and it is therefore convenient to formulate the calculations in matrix notation.

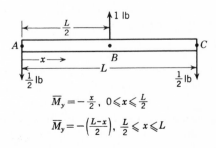

Fig. 10-14 Unit load system, Example 10-6.

$$
\overline{M}_y = -\frac{x}{2}, \quad 0 \leqslant x \leqslant \frac{L}{2}
$$
$$
\overline{M}_y = -\left(\frac{L-x}{2} \right), \quad \frac{L}{2} \leqslant x \leqslant L
$$

In this section we shall discuss the application of the flexibility and stiffness matrices (Sec. 6-6) to computing deflections. A more complete treatment of matrix methods, in which internal forces as well as displacements are determined, is given in Chap. 12. The following development is similar to that given by Argyris and Kelsey,[1] to which the reader is referred for further information.

Consider a linearly elastic body that is subjected to a set of applied forces R_1, R_2, . . . , R_n which gives rise to a set of displacements r_1, r_2, . . . , r_n at, and in the direction of, the applied forces. The terms force and displacement are used in the general sense. In addition to being elastic, the deformations are assumed small, so that the force-displacement relationships are linear, and the principle of superposition may be used. The displacements are therefore related to the forces by the linear equations

$$
\begin{aligned}
r_1 &= c_{11}R_1 + c_{12}R_2 + \cdots + c_{1n}R_n \\
r_2 &= c_{21}R_1 + c_{22}R_2 + \cdots + c_{2n}R_n \\
&\ \cdot\ \cdot\ \cdot\ \cdot\ \cdot\ \cdot\ \cdot\ \cdot\ \cdot\ \cdot\ \cdot\ \cdot\ \cdot\ \cdot\ \cdot\ \cdot \\
r_n &= c_{n1}R_1 + c_{n2}R_2 + \cdots + c_{nn}R_n
\end{aligned}
\tag{10-21}
$$

or in matrix notation by

$$
\mathbf{r} = \mathbf{cR} \tag{10-22}
$$

where the elements c_{ij} are the *flexibility influence coefficients* of the structure. The coefficient c_{ij} is the generalized displacement at (and in the direction of) the associated generalized force R_i, due to a unit generalized force at (and in the direction of) R_j. The jth column of $\mathbf{c}$ is a listing of the displacements r_i due to $R_j = 1$. We recall from Sec. 6-6 and Maxwell's reciprocal theorem (Sec. 6-13) that the flexibility matrix $\mathbf{c}$ is symmetric; that is, $c_{ij} = c_{ji}$. The elements c_{ii} are known as *direct flexibilities*, and the elements $c_{ij}(i \neq j)$ are called the *cross flexibilities*.

Equations (10-21) can be solved for the forces to give

$$
\begin{aligned}
R_1 &= k_{11}r_1 + k_{12}r_2 + \cdots + k_{1n}r_n \\
R_2 &= k_{21}r_1 + k_{22}r_2 + \cdots + k_{2n}r_n \\
&\ \cdot\ \cdot\ \cdot\ \cdot\ \cdot\ \cdot\ \cdot\ \cdot\ \cdot\ \cdot\ \cdot\ \cdot\ \cdot\ \cdot\ \cdot\ \cdot \\
R_n &= k_{n1}r_1 + k_{n2}r_2 + \cdots + k_{nn}r_n
\end{aligned}
\tag{10-23}
$$

or in matrix form

$$
\mathbf{R} = \mathbf{kr} \tag{10-24}
$$

where the k_{ij} coefficients are the *stiffness influence coefficients* of the structure. The coefficient k_{ij} is the generalized force at (and in the direction of) the associated generalized displacement r_i, due to a unit generalized displacement $r_j = 1$ while all other $r_k = 0$ $(k \neq j)$. The jth column of $\mathbf{k}$ is a listing of the forces that must be applied to produce a displacement $r_j = 1$

while all other $r_k = 0$ $(k \neq j)$. The elements k_{ii} are known as *direct stiffnesses*, and the elements k_{ij} $(i \neq j)$ are called *cross stiffnesses*.

When the structure is simple, c_{ij} can be found by integrating the differential equation that relates the displacements to the stress resultants, e.g., Eqs. (7-42) (with $M_{yr} = M_{zr} = 0$) for the bending of beams, Eq. (7-52) (with $P_T = 0$) for the extension of bars, and Eq. (8-73) for torsion members. The unit-load method is usually more convenient for computing flexibility influence coefficients, especially if the structure is complex. To fix ideas, consider the cantilevered beam in Fig. 10-15. The coefficient c_{ij} can be computed by using a real-load system with a 1-lb force at x_j and a unit-load system with 1 lb at x_i, as shown in the figure. Neglecting shear deformations, we find from Eq. (10-7) that

$$c_{ij} = \begin{cases} \int_0^{x_i} \dfrac{(x_i - x)(x_j - x)}{E_1 I^*} \, dx & x_i \le x_j \\ \int_0^{x_i} \dfrac{(x_i - x)(x_j - x)}{E_1 I^*} \, dx & x_i \ge x_j \end{cases} \qquad (10\text{-}25)$$

When the beam is uniform, Eqs. (10-25) reduce to

$$c_{ij} = \begin{cases} \dfrac{x_i{}^2}{6E_1 I^*} (3x_j - x_i) & x_i \le x_j \\ \dfrac{x_j{}^2}{6E_1 I^*} (3x_i - x_j) & x_i \ge x_j \end{cases} \qquad (10\text{-}26)$$

Applying Eqs. (10-26) to the uniform beam of Fig. 10-16 with

$$L_1 = L_2 = L_3 = L$$

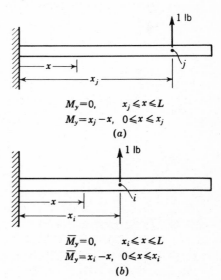

$M_y = 0, \qquad x_j \le x \le L$

$M_y = x_j - x, \quad 0 \le x \le x_j$

(a)

$\overline{M}_y = 0, \qquad x_i \le x \le L$

$\overline{M}_y = x_i - x, \quad 0 \le x \le x_i$

(b)

Fig. 10-15 (a) Real load system and (b) unit-load system for influence coefficients of a cantilevered beam.

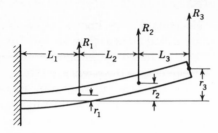

Fig. 10-16 Loading system and required displacements.

we find

$$c = \frac{L^3}{6EI} \begin{bmatrix} 2 & 5 & 8 \\ 5 & 16 & 28 \\ 8 & 28 & 54 \end{bmatrix}$$

Next let us examine the problems encountered in the direct computation of k for this same beam. To do this we consider the elements of the first column of k, which are the forces of a unit displacement at point 1 (Fig. 10-17). We note that it is considerably more difficult to compute these elements than it is to find the corresponding flexibilities; for to find the stiffness coefficients we must solve a statically indeterminate problem with two redundant supports.

To completely define the configuration of each segment of the beam we must specify both the lateral displacements and slopes at the ends of each segment. We now show that k is easier to determine if we consider all these deformations instead of just the lateral displacements. Consider the forces and displacements in Fig. 10-18 and assume that the stiffness matrix k relating R_1, R_2, R_3, to r_1, r_2, r_3 is desired. The complete sets of forces and displacements shown in Fig. 10-18 are related by

$$R = Kr$$

where R and r are column matrices with six elements and K is a 6×6 matrix. The matrix equation can be partitioned

$$\left\{ \begin{array}{c} R^f \\ \hline R^m \end{array} \right\} = \left[\begin{array}{c|c} k^{fd} & k^{fs} \\ \hline k^{md} & k^{ms} \end{array} \right] \left\{ \begin{array}{c} r^d \\ \hline r^s \end{array} \right\} \tag{10-27}$$

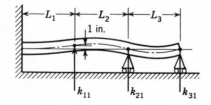

Fig. 10-17 Forces for unit displacement at point 1.

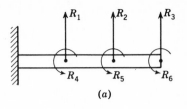

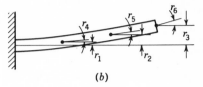

Fig. 10-18 (a) Force system and (b) displacement system for determining stiffness coefficients.

(b)

where $\mathbf{R}^f = \{R_1 \quad R_2 \quad R_3\}$ = column matrix of lateral forces

$\mathbf{R}^m = \{R_4 \quad R_5 \quad R_6\}$ = column matrix of moments

$$\mathbf{k}^{fd} = \begin{bmatrix} k_{11} & k_{12} & k_{13} \\ k_{21} & k_{22} & k_{23} \\ k_{31} & k_{32} & k_{33} \end{bmatrix}$$ = stiffness matrix of forces due to lateral displacements when the slopes are zero

$$\mathbf{k}^{fs} = \begin{bmatrix} k_{14} & k_{15} & k_{16} \\ k_{24} & k_{25} & k_{26} \\ k_{34} & k_{35} & k_{36} \end{bmatrix}$$ = stiffness matrix of forces due to slopes when the lateral displacements are zero

$$\mathbf{k}^{md} = \begin{bmatrix} k_{41} & k_{42} & k_{43} \\ k_{51} & k_{52} & k_{53} \\ k_{61} & k_{62} & k_{63} \end{bmatrix}$$ = stiffness matrix of moments due to lateral displacements when the slopes are zero

$$\mathbf{k}^{ms} = \begin{bmatrix} k_{44} & k_{45} & k_{46} \\ k_{54} & k_{55} & k_{56} \\ k_{64} & k_{65} & k_{66} \end{bmatrix}$$ = stiffness matrix of moments due to slopes when the lateral displacements are zero

$\mathbf{r}^d = \{r_1 \quad r_2 \quad r_3\}$ = column matrix of lateral displacements

$\mathbf{r}^s = \{r_4 \quad r_5 \quad r_6\}$ = column matrix of slopes

Equation (10-27) may be written as two equations

$$\mathbf{R}^f = \mathbf{k}^{fd}\mathbf{r}^d + \mathbf{k}^{fs}\mathbf{r}^s \qquad \mathbf{R}^m = (\mathbf{k}^{fs})'\mathbf{r}^d + \mathbf{k}^{ms}\mathbf{r}^s \qquad (10\text{-}28)$$

where we have replaced $\mathbf{k}^{md}$ by the transpose of $\mathbf{k}^{fs}$, since it follows from the symmetry of $\mathbf{K}$ that $\mathbf{k}^{md} = (\mathbf{k}^{fs})'$. In the actual load system of Fig. 10-16, $\mathbf{R}^m = \mathbf{0}$, so that the second of Eqs. (10-28) can be solved for $\mathbf{r}^s$ and the result substituted into the first equation to give

$$\mathbf{R}^f = \mathbf{k}\mathbf{r}^d \qquad (10\text{-}29)$$

where

$$\mathbf{k} = \mathbf{k}^{fd} - \mathbf{k}^{fs}(\mathbf{k}^{ms})^{-1}(\mathbf{k}^{fs})' \tag{10-30}$$

is the desired matrix that relates the forces to the lateral displacements.

Equation (10-29) can be solved for $\mathbf{r}^d$ for a given set of forces $\mathbf{R}^f$. When the deflections must be computed for many load cases, it is more convenient to compute the flexibility matrix from $\mathbf{c} = \mathbf{k}^{-1}$ and compute $\mathbf{r}^d$ from $\mathbf{r}^d = \mathbf{cR}^f$. The method can of course be generalized to any number of loads.

While it appears at first that this is a difficult method for computing $\mathbf{k}$ or $\mathbf{c}$, it involves relatively little work for the engineer, since the matrix operations of Eq. (10-30) and the inversion of $\mathbf{k}$ can be performed by a digital computer. The engineer must determine $\mathbf{K}$, but this is relatively simple when all possible deformation modes are considered. In the example cited we may write $\mathbf{K}$ immediately by using Eqs. (c) to (f) of Example 10-2 and summing the forces and moments from the spans that join at the displaced load point. The stiffness matrices are found to be

$$\mathbf{k}^{fd} = \begin{bmatrix} \alpha_1 + \alpha_2 & -\alpha_2 & 0 \\ & \alpha_1 + \alpha_2 & -\alpha_3 \\ \text{Sym.} & & \alpha_3 \end{bmatrix} \qquad \mathbf{k}^{fs} = \begin{bmatrix} \beta_2 - \beta_1 & \beta_2 & 0 \\ -\beta_2 & \beta_3 - \beta_2 & \beta_3 \\ 0 & -\beta_3 & -\beta_3 \end{bmatrix}$$

$$\mathbf{k}^{ms} = \begin{bmatrix} \gamma_1 + \gamma_2 & \dfrac{\gamma_2}{2} & 0 \\ & \gamma_2 + \gamma_3 & \dfrac{\gamma_3}{2} \\ \text{Sym.} & & \gamma_3 \end{bmatrix}$$

where $\alpha_i = 12(EI/L^3)_i$, $\beta_i = 6(EI/L^2)_i$, and $\gamma_i = 4(EI/L)_i$.

Computing $\mathbf{k}$ from Eq. (10-30) and inverting to find $\mathbf{c}$ is simpler than applying Eq. (10-25) when EI and L are different in the subspans of the beam. Furthermore, it is no more difficult to apply to indeterminate beams than it is to apply to determinate ones, which is not the case in computing $\mathbf{c}$ directly.

10-6 DISTRIBUTED LOADS AND WEIGHTING MATRICES

Structural loads are frequently distributed rather than concentrated, as assumed in Sec. 10-5. In these cases the structure may be divided into sections and the distributed load on each section may be replaced by its concentrated force and/or moment resultant. An alternate approach is to work with the distributed loading by using weighting or integrating matrices.[6,7]

Let the displacement of a beam at x due to a unit load at ξ be expressed by the function $C(x,\xi)$, known as the *flexibility influence function*. In a uniform cantilever beam, for example, this function is found from Eqs. (10-26) by letting $x_i = x$ and $x_j = \xi$, to give

$$C(x,\xi) = \begin{cases} \dfrac{x^2}{6E_1I^*}\,(3\xi - x) & x \le \xi \\[2mm] \dfrac{\xi^2}{6E_1I^*}\,(3x - \xi) & x \ge \xi \end{cases} \qquad (10\text{-}31)$$

If a distributed loading $p_z(\xi)$ acts over a differential length $d\xi$ at ξ, the displacement at x is $w(x) = C(x,\xi)p_z(\xi)\,d\xi$. Applying the principle of superposition, a distributed load over the length of the beam will give

$$w(x) = \int_0^L C(x,\xi)p_z(\xi)\,d\xi \qquad (10\text{-}32)$$

The function $C(x,\xi)$ is difficult to obtain for a beam of variable section, especially if it is indeterminate. In these cases it is easier to compute c_{ij} for specific grid points or obtain these flexibilities by tests. We can then determine the displacements at these points by using a *collocation method* in which Eq. (10-32) is satisfied at each of the points rather than at all values of x. Thus we require that

$$w(x_i) = \int_0^L C(x_i,\xi)p_z(\xi)\,d\xi \qquad i = 1, 2, \ldots, n \qquad (10\text{-}33)$$

where n is the number of points. The integrals from Eq. (10-33) can be evaluated numerically to give

$$w(x_i) = \sum_{j=1}^{n} c_{ij}\bar{W}_j p_{z_j} \qquad i = 1, 2, \ldots, n \qquad (10\text{-}34)$$

where $p_{z_j} = p_z(x_j)$ and the coefficients $\bar{W}_j$ are *weighting numbers* such as those employed by the trapezoidal or Simpson's rules (Fig. 5-2). Equation (10-34) may be written in the matrix form

$$\mathbf{w} = \mathbf{c}\bar{\mathbf{W}}\mathbf{p}_z \qquad (10\text{-}35)$$

where $\mathbf{c}$ is the flexibility matrix and

$$\mathbf{w} = \begin{Bmatrix} w_1 \\ w_2 \\ \cdot \\ \cdot \\ \cdot \\ w_n \end{Bmatrix} \qquad \bar{\mathbf{W}} = \begin{bmatrix} \bar{W}_1 & & & \\ & \bar{W}_2 & & \\ & & \cdot & \\ & & & \cdot \\ & & & & \bar{W}_n \end{bmatrix} \qquad \mathbf{p}_z = \begin{Bmatrix} p_{z_1} \\ p_{z_2} \\ \cdot \\ \cdot \\ \cdot \\ p_{z_n} \end{Bmatrix}$$

$$(10\text{-}36)$$

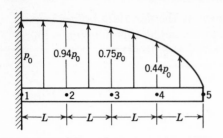

Fig. 10-19   Example 10-7.

Example 10-7   A uniform cantilever is subjected to the lateral loading in Fig. 10-19. Compute the deflections at the load points shown.

If point 1 is taken at the root, $c_{1j} = c_{i1} = 0$, and from Eq. (10-26) we find

$$c = \frac{L^3}{6EI} \begin{bmatrix} 0 & 0 & 0 & 0 & 0 \\ 0 & 2 & 5 & 8 & 11 \\ 0 & 5 & 16 & 28 & 40 \\ 0 & 8 & 28 & 54 & 81 \\ 0 & 11 & 40 & 81 & 128 \end{bmatrix}$$

If we use Simpson's rule (Fig. 5-2), the weighting matrix is

$$\bar{W} = \frac{L}{3} \begin{bmatrix} 1 & & & & \\ & 4 & & & \\ & & 2 & & \\ & & & 4 & \\ & & & & 1 \end{bmatrix}$$

The loading matrix is given by

$$p_z = p_0 \{1 \quad 0.94 \quad 0.75 \quad 0.44 \quad 0\}$$

Substituting c, $\bar{W}$, and p_z into Eq. (10-35) and carrying out the matrix multiplications gives

$$w = \frac{p_0 L^4}{18EI} \{0 \quad 29.1 \quad 92.1 \quad 167.1 \quad 243.9\}$$

REFERENCES

1. Argyris, J. H., and S. Kelsey: "Energy Theorems and Structural Analysis," Butterworth & Co. (Publishers) Ltd., London, 1960.
2. Langhaar, H. L.: "Energy Methods in Applied Mechanics," John Wiley & Sons, Inc., New York, 1960.
3. Hoff, N. J.: "The Analysis of Structures," John Wiley & Sons, Inc., New York, 1956.
4. Boley, B. A., and J. H. Weiner: "Theory of Thermal Stresses," John Wiley & Sons, Inc., New York, 1960.
5. Timoshenko, S. F.: "Strength of Materials, pt. II," 2d ed., D. Van Nostrand Company, Inc., Princeton, N.J., 1941.

6. Benscoter, S. U., and M. L. Gossard: Matrix Methods for Calculating Cantilever-beam Deflections, *NACA Tech. Note* 1827, March, 1949.
7. Bisplinghoff, R. L., H. Ashley, and R. L. Halfman: "Aeroelasticity," Addison-Wesley Publishing Company, Inc., Reading, Mass., 1955.

PROBLEMS

10-1. Obtain (*a*) the rotation at the right support and (*b*) the vertical displacement at the point of application of F_1 in Example 10-3.

10-2. Solve Prob. 7-4 by the virtual-work method.

10-3. Determine the tip rotation in Prob. 8-11 by the virtual-work method.

10-4. Show that the shear-deformation form factor $K_{yy} = 1.2$ for a beam with a thin deep rectangular cross section.

10-5. Show for small displacements that δW_e, for the unit-load system illustrated gives the displacement at point A relative a tangent to the beam at point B.

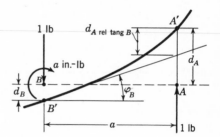

Fig. P10-5

10-6. Show that δW_e, for the unit-load system illustrated equals the slope at B relative to a line drawn through A and C.

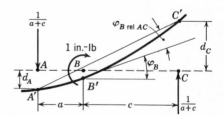

Fig. P10-6

10-7. Find (*a*) the vertical displacement and (*b*) the slope at the tip of the cantilevered curved beam illustrated when it is loaded by a tip force P which is normal to the plane of the beam.

> *Ans.*

(*a*)
$$d = \frac{PR^3}{4}\left(\frac{\pi}{EI} + \frac{3\pi - 8}{GJ}\right)$$

(*b*)
$$\varphi = \frac{PR^2}{2}\left(\frac{1}{EI} + \frac{1}{GJ}\right)$$

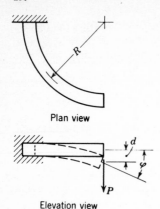

Plan view

Elevation view   **Fig. P10-7**

10-8. Obtain the vertical and horizontal deflections of point A of the pin-jointed truss illustrated when it is subjected to the forces and temperature changes shown.   The areas of all members are 1 in.^{2}, $E = 29 \times 10^6$ psi, and $\alpha = 6.3 \times 10^{-6}$ in./(in.)(°F).

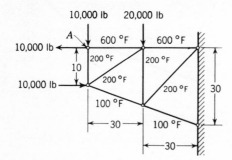

Fig. P10-8

10-9. Determine the tip deflection (including shear deformations) of the idealized beam shown. [*Ans.* $w = PL^3/3EI + PL/Ght.$]

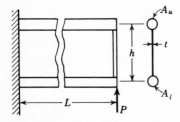

Fig. P10-9

10-10. Show that the torsional flexibility coefficients of a cantilevered member with a variable torsional rigidity are given by

$$c_{ij} = \begin{cases} \displaystyle\int_0^{x_i} \frac{dx}{G_1 J^*} & x_i \le x_j \\[2mm] \displaystyle\int_0^{x_i} \frac{dx}{G_1 J^*} & x_i \ge x_j \end{cases}$$

where x is measured from the root and c_{ij} is the rotation at x_i to a unit torque at x_j.

10-11. Determine (a) the flexibility matrix of the beam shown, (b) the stiffness matrix from Eq. (10-30), and (c) the flexibility matrix by inverting $\mathbf{k}$.

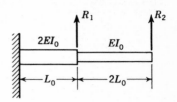

Fig. P10-11

10-12. Determine the general equation for c_{ij} of a uniform simply supported beam.

10-13. By noting that $p_z = p - \bar{k}w$ for a beam on an elastic foundation show that the displacements can be found from the flexibility matrix by

$$\mathbf{w} = (\mathbf{I} + \mathbf{c}\bar{\mathbf{W}}\bar{\mathbf{k}})^{-1}\mathbf{c}\bar{\mathbf{W}}\mathbf{p}$$

where $\mathbf{I}$ is the identity matrix and $\bar{\mathbf{k}}$ is a diagonal matrix of the stiffness of the elastic foundation at grid points along the beam.

10-14. Show that the aeroelastic deflections of a slender cantilevered wing (Example 8-9) can be determined from the matrix equation

$$\boldsymbol{\phi} = (\mathbf{I} - qc\bar{\mathbf{W}}\mathbf{C}_1\bar{\mathbf{c}}\mathbf{e})^{-1}\mathbf{c}\bar{\mathbf{W}}(q\mathbf{C}_1\bar{\mathbf{c}}\mathbf{e}\boldsymbol{\alpha}_r + q\mathbf{C}_2\mathbf{f} - n_z\,\mathbf{w}\mathbf{d})$$

where c is the matrix of torsional influence coefficients. The following matrices are diagonal: $\mathbf{C}_1$ with elements $(dC_n/d\alpha)_i$, $\bar{\mathbf{c}}$ with elements c_i, $\mathbf{e}$ with elements e_i, $\mathbf{C}_2$ with elements $(C_{m_{AC}})_i$, and $\mathbf{w}$ with elements w_i. The following are column matrices: $\boldsymbol{\alpha}_r$ with elements α_{ri}, $\mathbf{f}$ with elements $(c^2)_i$, and $\mathbf{d}$ with elements d_i. The notation is given in Example 8-9.

10-15. The beam shown is supported by rigid and elastic supports. Consider all possible displacements to determine the matrices in Eq. (10-30) if $\mathbf{k}$ relates the applied loads and lateral displacements at the load points.

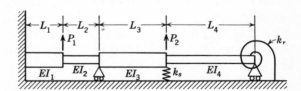

Fig. P10-15

11
Statically Indeterminate Structures

11-1 INTRODUCTION

The theories for beams (Chaps. 7 and 9) and torsion members (Chap. 8) apply only when the body is slender and does not have structural or loading discontinuities. The bending theory was predicated on the assumption that plane sections remain plane, and the torsion theory assumes that cross sections are free to warp. While these conditions are often satisfied in the components of flight structures, there are many instances when they are not. The theories are inadequate for low-aspect-ratio aerodynamic surfaces and are not sufficiently accurate in the root region of high-aspect-ratio aerodynamic surfaces, particularly if they have sweepback. They are unsatisfactory in the vicinity of cutouts or concentrated forces. In all these instances the fundamental stress or displacement assumptions that lead to the simplified bending and torsion theories have to be discarded, and the body must be analyzed by more general principles for statically indeterminate structures.

In Chaps. 4 and 6 we found that there are two alternate methods for analyzing elastic bodies. Displacements are taken as the unknowns in one formulation, while stresses are taken as the unknowns in the other. These same approaches can be used to analyze complex structures; however, it is usually more convenient to use forces (stress resultants) rather than stresses as unknowns.

The internal forces must be in equilibrium with the applied and reacting forces. In a *statically determinate* structure the equilibrium conditions are sufficient to determine the internal forces. In this case the stress-strain-temperature equations do not enter the analysis, and thermal stresses and prestressing cannot occur. In a *statically indeterminate* structure it is necessary to consider displacement compatibility in addition to equilibrium to obtain a solution. In this case we must introduce stress-strain-temperature (or force-displacement-temperature) relationships.

In the displacement formulation we consider only compatible deformation systems and find the true system by enforcing equilibrium, while in the force method we consider only equilibrium systems and determine the true system by enforcing compatibility. Equilibrium is imposed by means of the principle of the stationary value of the total potential (Sec. 6-9), and compatibility is ensured by using the principle of the stationary value of the total complementary potential (Sec. 6-10). In a given problem the number of unknowns in the displacement and force formulations may be different, and it is often simpler to use the method which involves the fewest unknowns.

In this chapter we shall apply the stationary principles of the total potential and the total complementary potential to indeterminate structures. The matrix formulations of these methods are given in Chap. 12. Further information on the use of energy methods may be found in Refs. 1 to 3. Though not employing the energy or matrix principles, Ref. 4 is suggested reading for its extensive theoretical and experimental investigation of the effects of shear lag, warping restraint, and cutouts in indeterminate stiffened-shell structures.

11-2 APPLICATION OF THE PRINCIPLE OF THE STATIONARY VALUE OF THE TOTAL POTENTIAL

It was shown in Sec. 6-9 that among all compatible displacement states of an elastic body, only the true one (which also satisfies equilibrium) has a stationary total potential for any virtual displacements. The potential $U + V$ is expressed in terms of the generalized displacements $q_1, q_2, \ldots, q_r$ that define the deformed configuration of the structure. To express U in terms of the displacements we use the force-displacement-temperature relationships for the members of the structures. Equation (6-17) applies for any virtual displacements when $q_1, q_2, \ldots, q_r$ satisfy equilibrium. The solution of the resulting r simultaneous equations gives the displacements q_i, which, when substituted into the force-displacement-temperature equations, give the internal forces in the structure.

The potential-energy, or displacement, method is especially useful when the number of unknown displacements in the structure is less than the number of indeterminate forces. The method is illustrated by the following example, which for generality includes the effects of initial strains resulting from intentional prestressing or manufactured lack of fit of the members.

Example 11-1 The structure in Fig. 11-1a consists of an arbitrary number of parallel cantilevered beams joined at their tips by a rib that is rigid within its plane but perfectly flexible with respect to forces normal to its plane. Each of the beams has an initial lack of fit with regard to vertical displacement and twist and is

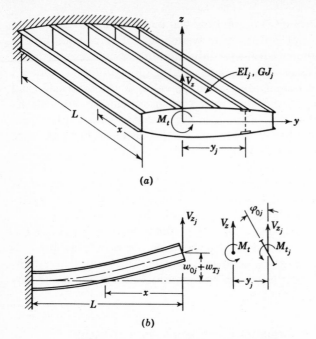

Fig. 11-1 Example 11-1. (*a*) Geometry and loading of structure; (*b*) loading and initial lack of fit of *j*th beam.

subjected to a temperature distribution that produces vertical deflections. The tip rib is subjected to a vertical force V_z and a torque M_t. Determine the tip deflection and twist and the internal forces in the structure.

We measure the displacements from the position where all beams fit perfectly with no initial lack of fit, applied loads, or temperature change. The vertical deflection w_j and twist φ_j of the tip of the *j*th beam from this position are

$$w_j = w + y_j\varphi \qquad \varphi_j = \varphi \qquad\qquad (a)$$

where w is the deflection of the tip rib at the point of application of V_z, φ is the rotation of the tip rib, and y_j is the distance from V_z to the *j*th beam. Deformations that satisfy Eqs. (*a*) are compatible with the restraints imposed by the tip rib.

To determine U we consider the *j*th beam which is loaded by a tip shear V_{z_j} and torque M_{t_j} that are applied by the tip rib (Fig. 11-1*b*). If we imagine this beam to be cut loose from the rib, the total tip displacement due to the internal forces, the lacks of fit, and the temperature distribution will be

$$w_j = \frac{V_{z_j}L^3}{3EI_j} + w_{0j} + w_{T_j} \qquad \varphi_j = \frac{M_{t_j}L}{GJ_j} + \varphi_{0j} \qquad (b)$$

where it is assumed that the beam is of constant cross section. In Eqs. (*b*), w_{0j} and φ_{0j} are the initial lacks of fit for vertical displacement and twist, and w_{T_j} is the tip displacement due to temperature. The unit-load method (Chap.

10) can be used to compute w_{T_j}. Taking a unit tip force, we find $\bar{M} = x$, from which

$$w_{T_j} = \frac{1}{EI_j} \int_0^L M_{T_j} x \, dx$$

The force-displacement-temperature equations from Eqs. (a) and (b) are

$$V_{z_j} = \beta_j(w + y_j\varphi - w_{0_j} - w_{T_j}) \qquad M_{t_j} = \gamma_j(\varphi - \varphi_{0_j}) \qquad (c)$$

where $\beta_j = 3EI_j/L^3$ and $\gamma_j = GJ_j/L$. The elastic strain energy is the work done by these forces in moving through the *elastic displacements* $w_j - w_{0_j} - w_{T_j}$ and $\varphi_j - \varphi_{0_j}$, or

$$U = \frac{1}{2} \sum_{j=1}^n [\beta_j(w + y_j\varphi - w_{0_j} - w_{T_j})^2 + \gamma_j(\varphi - \varphi_{0_j})^2] \qquad (d)$$

where n is the number of beams. The potential of the applied forces is

$$V = -V_z w - M_t \varphi \qquad (e)$$

The structure has two degrees of freedom, w and φ. Letting $q_1 = w$ and $q_2 = \varphi$, we find from Eqs. (6-17), (d), and (e)

$$\left(\sum_{j=1}^n \beta_j \right) w + \left(\sum_{j=1}^n \beta_j y_j \right) \varphi = V_z + \sum_{j=1}^n \beta_j(w_{0_j} + w_{T_j})$$

$$\left(\sum_{j=1}^n \beta_j y_j \right) w + \left[\sum_{j=1}^n (\beta_j y_j^2 + \gamma_j) \right] \varphi = M_t + \sum_{j=1}^n [\beta_j y_j(w_{0_j} + w_{T_j}) + \gamma_j\varphi_{0_j}]$$

$$(f)$$

The solution of these equations is simplified if V_z and M_t are determined with the line of action of V_z at the point where

$$\sum_{j=1}^n \beta_j y_j = 0 \qquad (g)$$

In this case the equations are uncoupled, and

$$w = \frac{V_z + \displaystyle\sum_{j=1}^n \beta_j(w_{0_j} + w_{T_j})}{\displaystyle\sum_{j=1}^n \beta_j}$$

$$(h)$$

$$\varphi = \frac{M_t + \displaystyle\sum_{j=1}^n [\beta_j y_j(w_{0_j} + w_{T_j}) + \gamma_j\varphi_{0_j}]}{\displaystyle\sum_{j=1}^n (\beta_j y_j^2 + \gamma_j)}$$

Substituting Eqs. (h) into Eqs. (a) and (c) gives the deflections and internal forces for each of the beams.

Equation (g) establishes the location of the shear center, for from the second of Eqs. (h) a shear force at this point produces no twist.

We note that the equilibrium equations (f) in the preceding example could have been written directly by using the stiffness coefficients

$$k_{11} = \sum_{j=1}^{n} \beta_j \qquad k_{12} = k_{21} = \sum_{j=1}^{n} \beta_j y_j \qquad k_{22} = \sum_{j=1}^{n} (\beta_j y_j{}^2 + \gamma_j)$$

where k_{11} and k_{21} are values of V_z and M_t for $w = 1$ and $\varphi = 0$, while k_{12} and k_{22} are these same loads for $w = 0$ and $\varphi = 1$. To determine the last terms on the right of Eqs. (f) we imagine the tip rib to be cut loose and determine the tip force and torque that must be applied to each beam to overcome the initial lacks of fit due to manufacturing inaccuracies and thermal gradients. The negative of the resultant of these forces is the last term on the right of the first equation, and the negative of the resultant moment of these forces and torques is the second term on the right of the second equation.

In general, the equilibrium equations will be of the matrix form

$$\mathbf{kq} = \mathbf{Q} - \mathbf{Q}_0 - \mathbf{Q}_T \qquad\qquad (11\text{-}1)$$

which can be solved for $\mathbf{q}$. In this equation:

$\mathbf{k}$ = stiffness matrix with k_{ij} equal to the force at, and in the direction of, q_i due to $q_j = 1$, while all other $q_i = 0$ $(i \neq j)$

$\mathbf{q}$ = column matrix of unknown displacements

$\mathbf{Q}$ = column matrix of applied forces in which Q_i is the work producing force at, and in the direction of, q_i

$\mathbf{Q}_0$ = column matrix of forces due to initial manufacturing inaccuracies in which Q_{0_i} is the resultant force at, and in the direction of, $-Q_i$ to overcome the manufacturing inaccuracies

$\mathbf{Q}_T$ = column matrix of forces due to temperature in which Q_{T_i} is the resultant force at, and in the direction of, $-Q_i$ to overcome the lacks of fit due to temperature

11-3 APPLICATION OF THE PRINCIPLE OF THE STATIONARY VALUE OF THE TOTAL COMPLEMENTARY POTENTIAL

In a statically indeterminate structure there are an infinite number of internal-force systems which may be in equilibrium with the applied forces. Only one of these systems, the true one, results in compatible deformations. It was shown in Sec. 6-10 that the compatible internal-force system has the unique property that $\delta U' + \delta V' = 0$ for any self-equilibrating variation of the internal and reacting forces.

To apply this principle we divide the internal and external forces into a (1) *basic system* consisting of the applied forces and *any* set of internal and reactive forces that equilibrate the applied forces and (2) independent *self-equilibrating redundant systems* of internal and reactive forces. The number of redundant-force systems is equal to the number of constraints (internal and external) that must be removed to make the structure statically determinate.

One method for determining a suitable basic system is to imagine a set of *releases* or *cuts* to be made in the structure until it is reduced to a determinate system. These releases may take the forms of hinges, sliding joints, etc., such as shown in Fig. 11-2, or they may be cuts that cause multiple releases. The effect of a release is to reduce one of the stress resultants in a member to zero. The number of releases to make the structure determinate is known as the *degree of redundancy* of the structure.

We designate the work-producing forces at the releases by X_1, $X_2, \ldots, X_r$, where r is the degree of redundancy. Each of these forces and *any* set of equilibrating internal and reactive forces may be used as a self-equilibrating redundant-force system. By considering the foundation to be a part of the structure we may include redundant-support reactions in the X_j. However, because it is rigid, the foundation will not contribute to $\delta U'$. An example of basic and redundant-force systems is given in Fig. 11-3; it shows a fuselage frame with load applied to the floor beam which is maintained in equilibrium by a reacting shear flow imposed

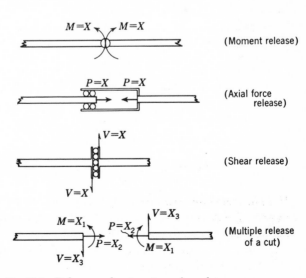

Fig. 11-2   Release and cut systems for a beam.

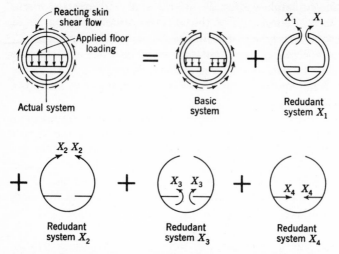

Fig. 11-3 Separation of a force system into basic and self-equilibrating redundant systems.

by the skin (Sec. 9-7). The internal shear forces at the cuts are zero because of symmetry in the structure and loading.

In the basic system we designate the stresses, reactions, and strains due to the applied loads and temperatures by σ_{xx_0}, σ_{yy_0}, . . . , σ_{zx_0}, R_{1_0}, R_{2_0}, . . . , R_{p_0} (where p is the number of reactions), and ϵ_{xx_0}, ϵ_{yy_0}, . . . , ϵ_{zx_0}. In the jth redundant-force system we designate the stresses, reactions, and strains due to $X_j = 1$ by σ_{xx_j}, σ_{yy_j}, . . . , σ_{zx_j}, R_{1_j}, R_{2_j}, . . . , R_{p_j}, and ϵ_{xx_j}, ϵ_{yy_j}, . . . , ϵ_{zx_j}. The true stresses, reactions, and strains can then be expressed as

$$\sigma_{mn} = \sigma_{mn_0} + \sum_{j=1}^{r} X_j \sigma_{mn_j} \qquad m \text{ and } n = x, y, z \qquad (11\text{-}2)$$

$$R_k = R_{k_0} + \sum_{j=1}^{r} X_j R_{k_j} \qquad k = 1, 2, \ldots, p \qquad (11\text{-}3)$$

$$\epsilon_{mn} = \epsilon_{mn_0} + \sum_{j=1}^{r} X_j \epsilon_{mn_j} \qquad m \text{ and } n = x, y, z \qquad (11\text{-}4)$$

We note the similarity of Eq. (11-2) to Eq. (6-78) in the Rayleigh-Ritz method for determining the stress function (Sec. 6-12). The basic stress system is analogous to φ_0, the jth redundant-stress system corresponds to φ_j, and the unknown redundant force X_j is equivalent to a_j.

We now imagine that the true stresses are given a variation $\delta\sigma_{xx} = \sigma_{xx_i}$, $\delta\sigma_{yy} = \sigma_{yy_i}$, . . . , $\delta\sigma_{zz} = \sigma_{zz_i}$, while the true reactions are varied by $\delta R_1 = R_{1_i}$, $\delta R_2 = R_{2_i}$, . . . , $\delta R_p = R_{p_i}$ (where the i subscripts refer to the ith redundant system). During this variation of the stresses and reactions we find from Eq. (6-65) that the variation in the complementary strain energy is

$$\delta U' = \int_V (\epsilon_{xx}\sigma_{xx_i} + \epsilon_{yy}\sigma_{yy_i} + \cdots + \epsilon_{zz}\sigma_{zz_i})\, dV \qquad (11\text{-}5)$$

Applying Eq. (11-4) to this result, we find

$$\delta U' = \sum_{j=1}^r \left[\int_V (\epsilon_{xx_j}\sigma_{xx_i} + \epsilon_{yy_j}\sigma_{xx_i} + \cdots + \epsilon_{zz_j}\sigma_{zz_i})\, dV\right] X_j$$
$$+ \int_V (\epsilon_{xx_0}\sigma_{xx_i} + \epsilon_{yy_0}\sigma_{yy_i} + \cdots + \epsilon_{zz_0}\sigma_{zz_i})\, dV \qquad (11\text{-}6)$$

The variation in the complementary potential of the reactions is

$$\delta V' = -\sum_{k=1}^p r_k R_{k_i} \qquad (11\text{-}7)$$

where r_k is the prescribed displacement at, and in the direction of, the reaction R_k. The stress and reaction variations that have been used are self-equilibrating, so that Eq. (6-66) applies to Eqs. (11-6) and (11-7). Noting that this result applies for $i = 1, 2, \ldots, r$, we find

$$\sum_{j=1}^r a_{ij}X_j + b_i - c_i = 0 \qquad i = 1, 2, \ldots, r \qquad (11\text{-}8)$$

where

$$a_{ij} = \int_V (\epsilon_{xx_j}\sigma_{xx_i} + \epsilon_{yy_j}\sigma_{yy_i} + \cdots + \epsilon_{zz_j}\sigma_{zz_i})\, dV \qquad (11\text{-}9)$$

$$b_i = \int_V (\epsilon_{xx_0}\sigma_{xx_i} + \epsilon_{yy_0}\sigma_{yy_i} + \cdots + \epsilon_{zz_0}\sigma_{zz_i})\, dV \qquad (11\text{-}10)$$

$$c_i = \sum_{k=1}^p r_k R_{k_i} \qquad (11\text{-}11)$$

Equation (11-8) may be written in the matrix form

$$\mathbf{a}\mathbf{X} = -\mathbf{b} + \mathbf{c} \qquad (11\text{-}12)$$

Solution of Eq. (11-8) or (11-12) gives the forces X_j, which when substituted into Eqs. (11-2) and (11-3) gives the true stresses and reactions. In most structures the deflections at the supports are assumed to be zero, and the c_i terms in Eqs. (11-8) and (11-12) vanish.

The coefficients a_{ij}, b_i, and c_i have physical significance. Recalling the discussion on relative displacements in Sec. 10-4 and noting that ϵ_{xx_j}, ϵ_{yy_j}, . . . ,ϵ_{zz_j} are the strains due to $X_j = 1$ while $\sigma_{xx_i}, \sigma_{yy_i}$, . . . , σ_{zz_i} are the stresses due to $X_i = 1$, we observe by comparing Eqs. (6-88) and (11-9) that a_{ij} is the *relative* displacement at the ith release due to a pair of unit forces $X_j = 1$ applied at the jth release.   The **a** matrix is therefore a flexibility matrix of the relative deflections at the releases due to the redundant forces at the releases.   Similarly b_i is the relative displacement at the ith release due to the applied loads and temperatures acting on the released structure.   Relative displacements that occur in the released structure because of initial lack of fit of members may also be included in b_i (Ref. 1).   Finally c_i is the relative displacement which occurs at the ith release as a result of the displaced supports in the released structure. Equation (11-8) is therefore a statement that the total relative motion due to (1) the redundant forces, (2) the applied loads, temperatures, and lack of member fit, and (3) the support displacements is zero at each of the releases, i.e., that continuity of deformation is preserved at each of the releases.

11-4 EQUATIONS FOR $\delta U'$ OF SIMPLE ELEMENTS

It is convenient to have equations for $\delta U'$ of simple elements such as beams, bars, shear webs, etc.   These can be obtained by expressing the stresses and strains in Eq. (11-5) in terms of stress resultants.   As an example, the true strain in a beam is given by Eq. (10-4) if principal axes are used.   In this equation P^*, M_y^*, and M_z^* are given by Eqs. (7-13), where P, M_y, and M_z are the *true* internal forces obtained by *superimposing* the stress resultants in the *basic* and *redundant* systems.   As a result

$$P = P_0 + \sum_{j=1}^{r} \bar{P}_j X_j \qquad M_y = M_{y_0} + \sum_{j=1}^{r} \bar{M}_{y_j} X_j \qquad M_z = M_{z_0} + \sum_{j=1}^{r} \bar{M}_{z_j} X_j$$

where P_0, M_{y_0}, and M_{z_0} are the stress resultants in the basic system and $\bar{P}_j$, $\bar{M}_{y_j}$, and $\bar{M}_{z_j}$ are the stress resultants due to $X_j = 1$.

From Eq. (7-23), the stress due to $X_i = 1$ is

$$\sigma_{xx_i} = \frac{E}{E_1} \left(\frac{\bar{P}_i}{A^*} - \frac{\bar{M}_{y_i} z}{I_{yy}^*} - \frac{\bar{M}_{z_i} y}{I_{zz}^*} \right) \qquad (11\text{-}13)$$

Substituting Eqs. (10-4) and (11-13) into Eq. (11-5), we find

$$\delta U' = \int_L \frac{P^* \bar{P}_i}{E_1 A^*} \, dx + \int_L \frac{M_y^* \bar{M}_{y_i}}{E_1 I_{yy}^*} \, dx + \int_L \frac{M_z^* \bar{M}_{z_i}}{E_1 I_{zz}^*} \, dx \qquad (11\text{-}14)$$

We note that this result is identical to δU from Eq. (10-6) if we add i subscripts to $\bar{P}$, $\bar{M}_y$, and $\bar{M}_z$ in the latter equation.   However, the physical interpretations of $\bar{P}$, $\bar{M}_y$, and $\bar{M}_z$ are different from those of $\bar{P}_i$, $\bar{M}_{y_i}$, and $\bar{M}_{z_i}$.   The first set of stress resultants is due to a unit force $Q = 1$ applied at the location of the desired displacement; whereas the second set of stress resultants is due to $X_i = 1$, a unit increment of the ith redundant-force system.

Equations for $\delta U'$ of other simple structural elements can be found in a similar manner. In each case the results are identical to the corresponding equations for δU that are given in Sec. 10-3 if an i subscript is appended to the stress-resultant symbols with bars.

11-5 NOTES ON BASIC AND REDUNDANT-FORCE SYSTEMS

Cuts or releases are easily visualized in trusses and frameworks but are often difficult to imagine in continuously joined structures such as stiffened shells. However, it is not necessary to use physical releases to obtain a basic system. The only condition that the basic system of internal forces has to satisfy is that it be in equilibrium with the applied loads. The concepts of cuts or releases were introduced merely to provide a simple method for obtaining such an equilibrium system.

We noted earlier that there are an infinite number of internal-force systems that may be in equilibrium with the applied loads in an indeterminate structure, so that the basic system is not unique. Another possibility for computing an admissible basic system is to compute the internal forces with the simplified theories for extension, bending, torsion, and shear of beams given in Chaps. 7 to 9.

Computing the basic system from these theories is often advantageous because it gives a result that is reasonably close to the true solution. The redundant systems then become small corrections that are superimposed on the basic system. Computational accuracy is seldom a problem in determining the internal forces in the basic system, but it can be troublesome in calculating the redundant forces, since they are found by solving a set of simultaneous equations. When the structure is highly redundant, the number of simultaneous equations is large, and sizable errors can occur in the $\mathbf{X}_j$ if the equations are ill conditioned. The overall errors are therefore reduced if the forces in the redundant systems are small compared to those in the basic system.

This same reasoning applies to the selection of releases. Thus it is desirable to make the releases where the redundant force at the release is small. An example of this occurs in the analysis of multicell structures subjected to shear. In Fig. 9-8 a release or cut is made in each cell to obtain a basic system (the $q^{(0)}$ shear flows). The $q^{(1)}$ flows are the redundant systems. When the applied load is a vertical shear, it is advisable to cut each cell near the center of the outer skin rather than to cut each of the vertical webs, since the former set of cuts results in a basic system that is closer to the true system.

Accuracy is also improved if the simultaneous equations are well conditioned. This is usually assured when the diagonal elements of $\mathbf{a}$ are large compared to the off-diagonal elements. The elements a_{ii} are the direct flexibilities for relative displacements, and the a_{ij} are the cross flexibilities. The releases should therefore be chosen so that they cause the largest deflections at the points where they are applied. As an example we refer to the continuous beam shown in Fig. 11-4a. The release system in which the supports are removed gives the redundancies shown in Fig. 11-4b. This is seen to be a poor choice because it results in large off-diagonal elements. For example, a_{41} (the displacement at X_4 due to $X_1 = 1$) is much larger than a_{11} (the displacement at X_1 due to $X_1 = 1$). The release system shown in Fig. 11-4c, which contains a moment release at each support, is a much better choice. This system, which is used in the well-known three-moment equation, reduces **a** to a tridiagonal matrix (all elements other than those on the main and two adjacent diagonals are zero). We also note that the redundant shear flows in the analysis of multicell structures in torsion and shear were selected to give a tridiagonal matrix (Sec. 9-5).

Ideally, the redundancies should be chosen so that **a** is diagonal; however, it is usually difficult to determine how such redundancies should be defined. An exception occurs in the case of a singly connected frame, which is treated in Sec. 11-6.

Redundancies do not have to be single forces but may be generalized forces consisting of groups of forces which are self-equilibrating. In these cases the releases are the generalized relative displacements corresponding to the generalized forces. Computational labor is reduced if the redundant-force systems are chosen so that they cause only internal forces in the vicinity of their application. Redundant-force systems that are useful in the analysis of stiffened shells are given in Ref. 1.

The application of the complementary-potential principle, which is

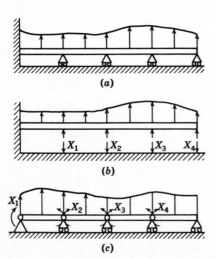

(a)

(b)

(c)

Fig. 11-4 Alternate release systems for a continuous beam. (a) Continuous beam; (b) poor choice of releases; (c) improved choice of releases.

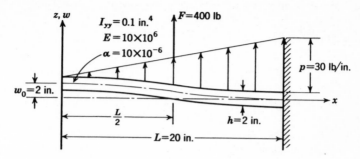

Fig. 11-5 Example 11-2.

illustrated in the following examples, forms the basis for the elastic-center and column-analogy methods given in Sec. 11-6.

Example 11-2 The beam in Fig. 11-5 is subjected to the loads that are shown and a temperature change $T = 400xz/hL$. Both ends of the beam are clamped, and the left end is vertically displaced a distance $w_0 = 2$ in. Determine the bending moments in the beam.

Substituting T into Eqs. (7-15), we find $M_{y_T} = -400\alpha EI_{yy}x/Lh$, $P_T = M_{z_T} = 0$. There are four reactions, and only two independent equi-

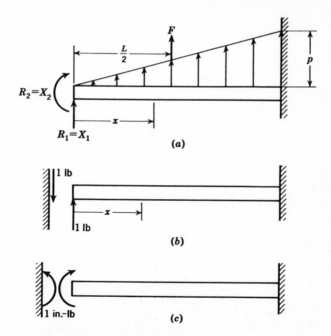

Fig. 11-6 Force systems for Example 11-2. (a) True force system; (b) force system $X_1 = 1$; (c) $X_2 = 1$.

librium equations can be written, so that the degree of redundancy is 2. To obtain a basic system we assume shear and moment releases at $x = 0$ and designate the released shear force and moment by X_1 and X_2, as shown in Fig. 11-6a. From this figure

$$M_y^* = \frac{px^3}{6L} + X_1 x + X_2 - \frac{400\alpha EI_{yy}x}{Lh} \qquad 0 \le x \le \frac{L}{2}$$

$$M_y^* = \frac{px^3}{6L} + F\left(x - \frac{L}{2}\right) + X_1 x + X_2 - \frac{400\alpha EI_{yy}x}{Lh} \qquad \frac{L}{2} \le x \le L \tag{a}$$

The self-equilibrating force system for $X_1 = 1$ is shown in Fig. 11-6b, from which $\bar{M}_{y_1} = x$. The displacement at the reaction $R_1 = X_1$ is w_0, so that for the force system of Fig. 11-6b, $R_{1_1} = 1$ and $r_1 = w_0$. From Eqs. (a), (11-14), (11-7), and (6-66) we find

$$\frac{p}{6EI_{yy}L}\int_0^L x^4\,dx + \frac{F}{EI_{yy}}\int_{L/2}^L \left(x - \frac{L}{2}\right)x\,dx + \frac{X_1}{EI_{yy}}\int_0^L x^2\,dx$$

$$+ \frac{X_2}{EI_{yy}}\int_0^L x\,dx - \frac{400\alpha}{Lh}\int_0^L x^2\,dx - w_0 = 0 \tag{b}$$

The $X_2 = 1$ force system is shown in Fig. 11-6c, which gives $\bar{M}_{y_2} = 1$. Since there is no rotation at the reaction R_2, $r_2 = 0$. Applying Eqs. (a), (11-14), (11-7), and (6-66) for the self-equilibrating force variation of Fig. 11-6c, we obtain

$$\frac{p}{6EI_{yy}L}\int_0^L x^3\,dx + \frac{F}{EI_{yy}}\int_{L/2}^L \left(x - \frac{L}{2}\right)dx + \frac{X_1}{EI_{yy}}\int_0^L x\,dx$$

$$+ \frac{X_2}{EI_{yy}}\int_0^L dx - \frac{400\alpha}{Lh}\int_0^L x\,dx = 0 \tag{c}$$

Carrying out the integrations and substituting the numerical constants into Eqs. (b) and (c), we find

$$2.667 \times 10^{-3}X_1 + 2 \times 10^{-4}X_2 = 1.773 \qquad 2 \times 10^{-4}X_1 + 2 \times 10^{-5}X_2 = -0.01$$

The simultaneous solution of these equations gives $X_1 = 2810$ lb and $X_2 = -28,600$ in.-lb. Substituting these results into Eq. (a) gives the equivalent bending moments in the beam.

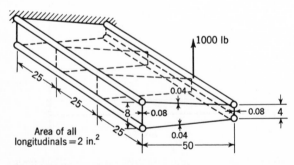

Fig. 11-7 Example 11-3 (see p. 301).

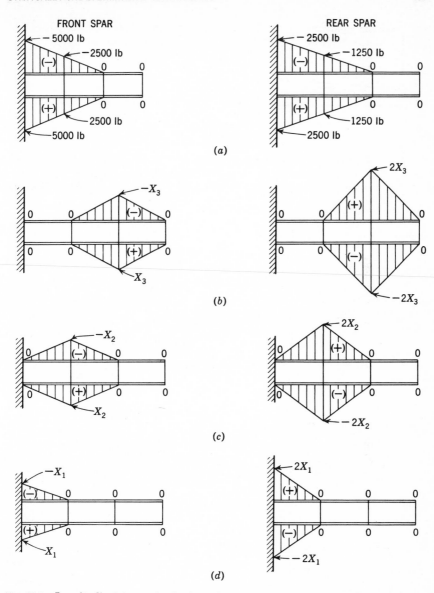

Fig. 11-8 Longitudinal forces for basic and redundant-force systems, Example 11-3. (a) Basic load system; (b) X_3 load system; (c) X_2 load system; (d) X_1 load system

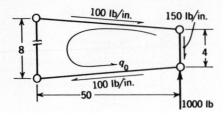

Fig. 11-9 Basic shear flows in two inboard bays, Example 11-3.

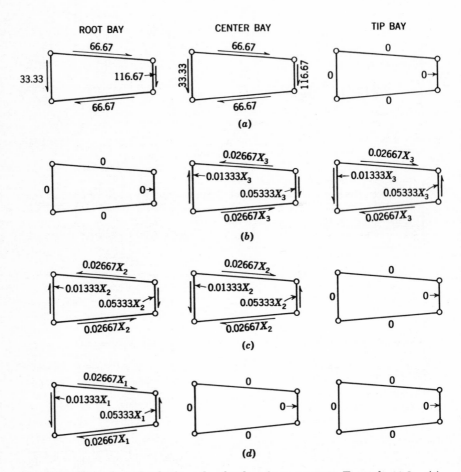

Fig. 11-10 Shear flows for basic and redundant-force systems, Example 11-3. (*a*) Basic system shear flows; (*b*) X_3 shear flows; (*c*) X_2 shear flows; (*d*) X_1 shear flows.

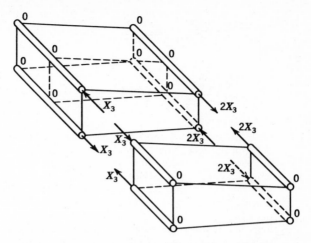

Fig. 11-11 Longitudinal forces for X_3 system, Example 11-3.

Example 11-3 Determine the internal forces in the idealized three-bay box beam in Fig. 11-7. Assume that the ribs are rigid within their planes.†

The beam has a small aspect ratio, there are torque and shear discontinuities 50 in. from the root, and the root cross section is restrained from warping. As a result, the theories of Chaps. 7 and 9 would be expected to give inaccurate results for the bending and shear stresses. However, we shall use these simplified theories to obtain the internal forces in the basic system.

The basic forces in the longitudinals, computed from $P = -(M_y z / I_{yy})A$, are shown in Fig. 11-8a. The basic shear flows in the tip bay are zero because $V_z = M_t = 0$ in this bay. To determine the basic shear flows in the two inboard bays we imagine the front spar web in these bays to be cut as shown in Fig. 11-9. The flows in the resulting open section, computed from Eq. (9-14) with $V_z = 1000$ lb, are shown on the outside of the cross section in this figure. The shear flow q_0 in the front web is found by torque equilibrium to be 33.33 lb/in. The resultant shear flows in each of the bays of the basic system are shown in Fig. 11-10a.

The structure has three redundancies, which is seen by cutting any one of the longitudinals at the root and at the two inboard ribs. Three unknown longitudinal forces would remain at each cross section, which can be found from equations for force equilibrium in the axial direction and moment equilibrium about horizontal and vertical axes in the cross section. The shear flow would be the same in the webs on either side of the cut longitudinal, giving only three unknown shear flows at each cross section. These can be found by force equilibrium in the vertical and horizontal directions and torque equilibrium about a spanwise axis. Since the structure can be made statically determinate by three releases, it has three redundancies.

While the basic system satisfies equilibrium, it does not assure compatibility. Let X_1, X_2, and X_3 be the redundant forces that must be added to the basic forces to obtain the true forces in the lower left longitudinal at distances

† The effect of this assumption will be studied in Example 12-2.

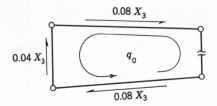

$0.08\,X_3$

$0.04\,X_3$

q_0

$0.08\,X_3$

Fig. 11-12 Shear flows in tip bay due to X_3 system, Example 11-3.

of 0, 25, and 50 in. from the root. At each of these spanwise locations we obtain a system of axial forces that are in equilibrium with the redundant force in the lower left longitudinal by determining the forces in the three remaining longitudinals from equations for force equilibrium in the axial direction and moment equilibrium about vertical and chordwise axes. The forces in the longitudinals due to the redundant force X_3 in the lower left longitudinal at a distance of 50 in. from the root are shown in Fig. 11-11. The equilibrating longitudinal forces for X_1 and X_2 are found in a similar manner. The longitudinal forces for each of the redundancies are summarized in Fig. 11-8b to d.

The shear flows that accompany the redundant longitudinal forces can be found by the fluid-flow analogy (Sec. 9-3). Consider, for example, the shear flows in the tip bay due to the X_3 load system. If we cut the rear spar web in Fig. 11-11 and apply Eqs. (9-18) and (9-20), we obtain the shear flows shown on the outside of the section in Fig. 11-12. A constant shear flow q_0 equal to the flow in the cut web acts around the cell. This is found by equating the sum of the torques to zero, which gives $q_0 = 0.05333X_3$. The resultant self-equilibrating shear flows are shown in Fig. 11-10b. The remaining redundant shear flows are found in the same manner and are summarized in Fig. 11-10c and d.

There are no displacements at the supports, and it follows from Eq. (11-7) that $\delta V' = 0$. From Eqs. (10-12) and (10-17) we find

$$\delta U' = \sum_k \left\{ \frac{L}{6EA} \left[\bar{P}_{Ai}(2P_A + P_B) + \bar{P}_{Bi}(P_A + 2P_B) \right] \right\}_k + \sum_k \left(\frac{q\bar{q}_i bL}{Gt} \right)_k = 0$$

$$i = 1, 2, 3 \quad (a)$$

where L is the bay length and the summations extend over all longitudinals and webs. In this equation P_A and P_B are the true forces in the longitudinals at the ends of each bay, while $\bar{P}_{Ai}$ and $\bar{P}_{Bi}$ are the corresponding forces for the $X_i = 1$ load system. The shear flows q are the true flows, and the $\bar{q}_i$ flows are due to $X_i = 1$.

The true forces are the sums of the forces in the basic and redundant systems. As examples, we find from Fig. 11-8 that the true end loads in the top left longitudinal of the root bay are

$$P_A = -5000 - X_1 \qquad P_B = -2500 - X_2 \qquad (b)$$

Referring to Fig. 11-10, we see that in the top skin of the root bay

$$q = 66.67 + 0.02667X_1 - 0.02667X_2 \qquad (c)$$

In a similar manner we find that for $X_1 = 1$ the forces in the same members are

$$\bar{P}_{A_1} = -1 \qquad \bar{P}_{B_1} = 0 \qquad \bar{q}_1 = 0.02667$$

The same method can be used to determine the forces in the remaining members for the true and unit redundant-force systems. Substituting these into Eqs. (a) gives the following set of equations for $G/E = 0.4$:

$$1.623X_1 - 1.003X_2 \qquad\qquad = -2611$$
$$-1.003X_1 + 3.256X_2 - 1.003X_3 = 0$$
$$-1.003X_2 + 3.256X_3 = 2611$$

which has the solution $X_1 = -1822.3$, $X_2 = -345.6$, $X_3 = 695.4$. Substituting these values into Eqs. (b) and (c) and similar equations for the remaining elements, we obtain the true internal forces shown in Fig. 11-13. The basic solution is also shown in this figure so that the true solution can be compared

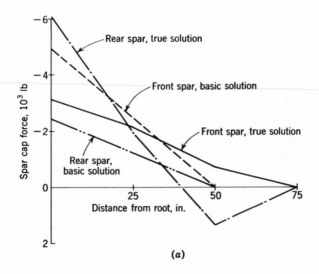

(a)

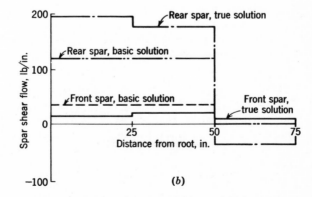

(b)

Fig. 11-13 Comparison of basic and true solutions, Example 11-3. (a) Spar cap forces; (b) spar shear flows.

with that obtained from elementary beam theory. We see that the elementary theory leads to gross errors in this case, as suspected.

11-6 ELASTIC-CENTER AND COLUMN-ANALOGY METHODS

It was pointed out in Sec. 7-6 that frames are used in body structures to distribute concentrated forces into the skin of shell structures and redistribute stresses around structural discontinuities. If the frame is very flexible, we must determine the internal forces in the skins, longitudinals, and frame simultaneously by an indeterminate analysis of the entire assembly. When the frame is relatively rigid, the analysis can be divided into two independent parts. The forces in the longitudinals and skin can be determined by assuming the frames are rigid and using the methods of analysis for indeterminate structures described in this chapter; or if the structure meets the restrictions imposed by beam theory, the skin shear flows can be determined by the method of Sec. 9-7. The resulting skin shear flows are used to determine the loads on the frames. The frames can then be analyzed independently by separate indeterminate analyses.

The complementary-potential principle described in Sec. 11-3 can be used for the frame analysis. Sufficient releases can be made in the frame to reduce it to a statically determinate basic system. The forces X_1, X_2, . . . , X_r at the releases can then be found from Eq. (6-66) by taking a unit variation of each of the X_i. The result is a set of simultaneous equations of the form of Eqs. (11-8) in which all the X_i will generally occur in each equation.

When the frame is singly connected, there are generally three redundancies. These may be taken as the axial force, bending moment, and shear force at an arbitrary cut in the frame (Fig. 11-14a). However, such a choice causes all unknowns to appear in each of the three simultaneous equations. By properly defining the redundancies the equations can be uncoupled, and an explicit solution for the redundancies can be determined. The procedure for accomplishing this was first proposed by Mohr and is known as the *redundant-* or *elastic-center* method.

We imagine the frame to be cut at an arbitrary point; however, instead of applying the redundant forces directly at the cut, we consider them to be applied at the ends of *rigid* members attached to the edges of the cut, as shown in Fig. 11-14b. Coordinate axes are taken so that the origin coincides with the free ends of the rigid bars.

The shape of the frame is defined by a reference line that passes through the elastic centroids of the cross sections of the frame. It is assumed that this line is a plane curve and that at each cross section one of the modulus-weighted principal axes of the cross section lies in the plane of this reference curve. It is further assumed that the applied

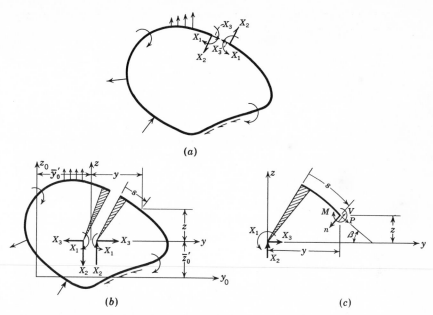

Fig. 11-14   Redundant forces applied at (a) cut and (b) at elastic center.   (c) Internal forces at s.

forces are in this plane as is the equivalent thermal bending moment. Under these circumstances the deflected position of the reference line remains in the original plane of the line.

The axial force and bending moment at any cross section of the frame (Fig. 11-14b) are given by

$$P = P_0 + X_2 \sin \beta - X_3 \cos \beta \qquad M = M_0 + X_1 - X_2 y + X_3 z$$
$$(11\text{-}15)$$

where y and z are the coordinates of the reference line at the cross section and β is the angle that the tangent to the reference line makes with the y axis.   The symbols P_0 and M_0 denote the axial force and bending moment in the basic (cut) frame due to the applied forces.   Positive moments are defined as those which cause compression in the inner fiber of the frame.

When there are no prescribed displacements or the prescribed displacements are zero, $\delta V' = 0$.   Neglecting shear deformations (these deformations are included in Ref. 5), we find from Eqs. (6-67), (10-8), and (10-9)

$$\delta U' = \oint \frac{P^* \bar{P}_i}{E_1 A^*} ds + \oint \frac{M^* \bar{M}_i}{E_1 I^*} ds = 0 \qquad i = 1, 2, 3 \quad (11\text{-}16)$$

where the integrals extend around the frame, and where

$$P^* = P + P_T = P + \int_A E\alpha T \, dA$$

$$M^* = M + M_T = M - \int_A E\alpha T n \, dA \tag{11-17}$$

The coordinate s is along the reference curve, and n is a coordinate measured in the direction of the inward normal to the curve. The fictitious rigid bars do not enter $\delta U'$ because $E_1 A^*$ and $E_1 I^*$ of these members are infinite.

The symbols $\bar{P}_i$ and $\bar{M}_i$ in Eq. (11-16) denote the force and moment due to $X_i = 1$. From Fig. 11-14b or Eqs. (11-15)

$$\bar{P}_1 = 0 \qquad \bar{M}_1 = 1$$

$$\bar{P}_2 = \sin \beta \qquad \bar{M}_2 = -y \tag{11-18}$$

$$\bar{P}_3 = -\cos \beta \qquad \bar{M}_3 = z$$

Substituting Eqs. (11-15), (11-17), and (11-18) into Eq. (11-16), we find the three equations

$$A'X_1 - \bar{y}'A'X_2 + \bar{z}'A'X_3 = -P'$$

$$-\bar{y}'A'X_1 + I'_{zz}X_2 - I'_{yz}X_3 = M'_z \tag{11-19}$$

$$\bar{z}'A'X_1 - I'_{yz}X_2 + I'_{yy}X_3 = -M'_y$$

where to simplify writing the equations the following symbols are defined

$$A' = \oint \frac{1}{E_1 I^*} \, ds \qquad \bar{y}' = \frac{1}{A'} \oint \frac{y}{E_1 I^*} \, ds \qquad \bar{z}' = \frac{1}{A'} \oint \frac{z}{E_1 I^*} \, ds$$

$$I'_{yy} = \oint \frac{z^2}{E_1 I^*} \, ds + \oint \frac{\cos^2 \beta}{E_1 A^*} \, ds \qquad I'_{zz} = \oint \frac{y^2}{E_1 I^*} \, ds + \oint \frac{\sin^2 \beta}{E_1 A^*} \, ds$$

$$I'_{yz} = \oint \frac{yz}{E_1 I^*} \, ds + \oint \frac{\sin \beta \cos \beta}{E_1 A^*} \, ds \tag{11-20}$$

$$P' = \oint \frac{M'_0}{E_1 I^*} \, ds$$

$$M'_y = \oint \frac{M'_0 z}{E_1 I^*} \, ds - \oint \frac{P'_0 \cos \beta}{E_1 A^*} \, ds \qquad M'_z = \oint \frac{M'_0 y}{E_1 I^*} \, ds - \oint \frac{P'_0 \sin \beta}{E_1 A^*} \, ds$$

In these definitions

$$P'_0 = P_0 + P_T \qquad M'_0 = M_0 + M_T \tag{11-21}$$

The solution of Eqs. (11-19) is greatly simplified if the origin for the coordinates (and therefore the ends of the rigid bars) is chosen so that $\bar{y}' = \bar{z}' = 0$. This point is found by selecting an arbitrary set of axes

y_0 and z_0, from which the distances $\bar{y}'_0$ and $\bar{z}'_0$ to the y and z axes are computed by

$$\bar{y}'_0 = \frac{1}{A'} \oint \frac{y_0}{E_1 I^*} \, ds \qquad \bar{z}'_0 = \frac{1}{A'} \oint \frac{z_0}{E_1 I^*} \, ds \qquad (11\text{-}22)$$

The origin of the y and z axes depends upon the $E_1 I^*$ distribution of the frame and is referred to as the *elastic* or *redundant center* of the frame.

For axes at the elastic center the first of Eqs. (11-19) gives $X_1 = -P'/A'$, and the simultaneous solution of the last two equations yields

$$X_2 = \frac{M'_z I'_{yy} - M'_y I'_{yz}}{I'_{yy} I'_{zz} - (I'_{yz})^2} \qquad X_3 = -\frac{M'_y I'_{zz} - M'_z I'_{yz}}{I'_{yy} I'_{zz} - (I'_{yz})^2}$$

Substituting these results into Eqs. (11-15) gives

$$P = P_0 + \frac{M'_z I'_{yy} - M'_y I'_{yz}}{I'_{yy} I'_{zz} - (I'_{yz})^2} \sin \beta + \frac{M'_y I'_{zz} - M'_z I'_{yz}}{I'_{yy} I'_{zz} - (I'_{yz})^2} \cos \beta \qquad (11\text{-}23)$$

$$M = M_0 + \left[-\frac{P'}{A'} - \frac{M'_z I'_{yy} - M'_y I'_{yz}}{I'_{yy} I'_{zz} - (I'_{yz})^2} y - \frac{M'_y I'_{zz} - M'_z I'_{yz}}{I'_{yy} I'_{zz} - (I'_{yz})^2} z \right] \qquad (11\text{-}24)$$

We note that these equations are further simplified if the y and z axes are chosen so that $I'_{yz} = 0$. For these *principal axes*

$$P = P_0 + \frac{M'_z}{I'_{zz}} \sin \beta + \frac{M'_y}{I'_{yy}} \cos \beta \qquad (11\text{-}25)$$

$$M = M_0 + \left[-\frac{P'}{A'} - \frac{M'_z y}{I'_{zz}} - \frac{M'_y z}{I'_{yy}} \right] \qquad (11\text{-}26)$$

Equations (11-25) and (11-26) appear easier to use than Eqs. (11-23) and (11-24). However, unless principal axes are obvious by inspection, Eqs. (11-23) and (11-24) involve less labor than determining the principal values of I'_{yy} and I'_{zz}. We note from the defining equation for I'_{yz} that an axis of geometric and elastic symmetry is a principal axis.

Once P and M are known at any cross section, the normal stresses for the cross section are given by

$$\sigma_{ss} = \frac{E}{E_1} \left(\frac{P^*}{A^*} - \frac{M^* n}{I^*} - E_1 \alpha T \right) \qquad (11\text{-}27)$$

where P^* and M^* are obtained from Eqs. (11-17).

The similarity of the bracketed terms in Eqs. (11-24) and (11-26) to the equations for stresses in a beam-column subjected to compression and bending [Eq. (7-22) and its simplification when $I_{yz} = 0$] is apparent. In the analogy we consider a column with a sandwich wall (Fig. 11-15) in which the faces of thickness $t = 1/2E_1 I^*$ are separated a distance $h = 2(I^*/A^*)^{1/2}$ by a core. The core is assumed to have negligible

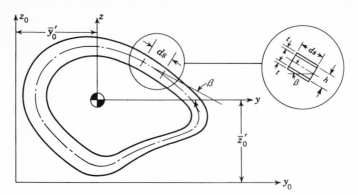

Fig. 11-15 Sandwich column analogy.

stiffness in the axial direction, so that the effective cross-sectional area of the analogous column is

$$A = \oint 2t \, ds = \oint \frac{1}{E_1 I^*} \, ds = A'$$

The distance from an arbitrary y_0 axis to the centroid of the analogous column is

$$\bar{y}_0 = \frac{2}{A} \oint y_0 t \, ds = \frac{1}{A'} \oint \frac{y_0}{E_1 I^*} \, ds = \bar{y}_0'$$

and in a similar manner $\bar{z}_0 = \bar{z}_0'$.

Examining a differential length ds of the wall (Fig. 11-15), we see that

$$dI_{yy} = \left[2t \left(\frac{h}{2}\right)^2 \cos^2 \beta + 2tz^2 \right] ds = \left(\frac{\cos^2 \beta}{E_1 A^*} + \frac{z^2}{E_1 I^*} \right) ds$$

Integrating around the wall of the column, we find $I_{yy} = I_{yy}'$. In a similar fashion $I_{zz} = I_{zz}'$ and $I_{yz} = I_{yz}'$. Thus the section properties of the analogous column are equal to the elastic-frame properties of Eqs. (11-20) when the midline of the column wall is identical to the reference line of the frame.

It is common practice to neglect the extensional deformations when there are no temperature changes. In doing so we assume $E_1 A^* = \infty$, so that $h = 0$, and the model reduces to the conventional column analogy, wherein the wall is homogeneous with a total thickness of $1/E_1 I^*$ (Refs. 6 and 7). In this case I_{yy}', I_{zz}', and I_{yz}' are computed by dropping the $E_1 A^*$ terms in Eqs. (11-20).

The loading on the analogous column is a compressive force per unit length $M_0'/E_1 I^*$ and a moment per unit length $P_0'/E_1 A^*$. The moment

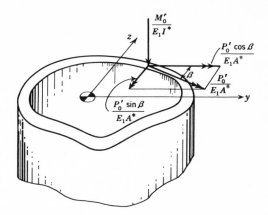

Fig. 11-16 Loading on analogous column.

per unit length is about an axis that is tangent to the midline of the column wall, as shown in Fig. 11-16. For these loadings, the resultant compressive force and bending moments on the analogous column are $P = P'$, $M_y = M'_y$, and $M_z = M'_z$, which completes the analogy. When extensional deformations are neglected, the E_1A^* terms drop out, and the only loading is the compressive force per unit length of the column wall.

The preceding method also applies to frames with built-in ends, as shown by the example in Fig. 11-17a. In this case the foundation is treated as a closing member with $E_1I^* = \infty$ (Fig. 11-17b), so that $t = 0$ for this section of the wall, and the analogous column appears as shown in Fig. 11-17c.

Doubly connected frames similar to that shown in Fig. 11-18a frequently occur in body structures where floor beams or the spars of aerodynamic surfaces are attached. When the redundancies in Fig. 11-18b are used, Eq. (11-8) will give six simultaneous equations in which all the unknowns will generally appear in each equation. However, if the redundancies in the ring are applied at the elastic center of the ring

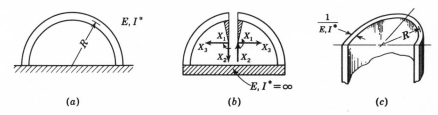

(a) (b) (c)

Fig. 11-17 Elastic center and column analogy for a frame with clamped ends. (a) Clamped frame; (b) redundancies at elastic center; (c) analogous column.

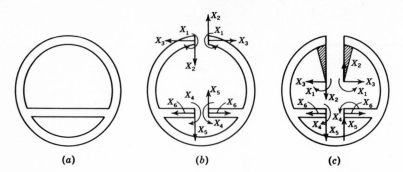

Fig. 11-18 Alternate redundant-force systems for a doubly connected frame. (*a*) Doubly connected frame; (*b*) redundant forces at cuts; (*c*) improved redundant system.

alone, as shown in Fig. 11-18*c*, the equations will be easier to solve because $a_{12} = a_{13} = a_{21} = a_{23} = a_{31} = a_{32} = 0$ in Eq. (11-8).

Example 11-4 The homogeneous frame in Fig. 11-19 is a transverse stiffening member in an external fuel tank. The temperature of the upper half of the frame increases by T_1 as a result of aerodynamic heating, while the lower half remains at the initial temperature because it is submerged in the fuel, which acts as a heat sink. Determine the stresses at the outer radius of the frame.

There are no applied loads, so that $P_0 = M_0 = 0$ in the cut frame. From Eqs. (11-17) and (11-21) we find $M'_0 = 0$, while $P'_0 = \alpha T_1 EA$ for $0 < \theta < \pi/2$ and $P'_0 = 0$ for $\pi/2 < \theta < \pi$. From symmetry, the axes shown are principal axes. From the last three of Eqs. (11-20), $P' = M'_z = 0$, and

$$M'_y = -2\alpha T_1 R \int_0^{\pi/2} \cos \theta \, d\theta = -2\alpha T_1 R$$

We also find from Eqs. (11-20) that

$$I'_{yy} = \frac{2R^3}{EI} \int_0^\pi \cos^2\theta \, d\theta + \frac{2R}{EA} \int_0^\pi \cos^2\theta \, d\theta = \frac{\pi R}{EI} \left(R^2 + \frac{I}{A} \right)$$

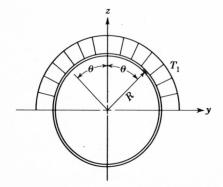

Fig. 11-19 Example 11-4.

Applying Eqs. (11-25) and (11-26), we obtain

$$P = -\frac{2\alpha T_1 EI \cos \theta}{\pi(R^2 + I/A)} \qquad M = \frac{2\alpha T_1 EIR \cos \theta}{\pi(R^2 + I/A)}$$

We note that P and M would have been zero if extensional deformations had been neglected. Substituting P and M into Eqs. (11-17) gives

$$P^* = \begin{cases} -\dfrac{2\alpha T_1 EI}{\pi(R^2 + I/A)} \cos \theta + \alpha T_1 EA & 0 \le \theta < \dfrac{\pi}{2} \\[3mm] -\dfrac{2\alpha T_1 EI}{\pi(R^2 + I/A)} \cos \theta & \dfrac{\pi}{2} \le \theta < \pi \end{cases}$$

$$M^* = \frac{2\alpha T_1 EIR}{\pi(R^2 + I/A)} \cos \theta \qquad 0 \le \theta \le \pi$$

At the outer radius $n = -h/2$, and from Eq. (11-27) we find

$$\sigma_{\theta\theta} = \frac{\alpha T_1 E(ARh - 2I)}{\pi(I + AR^2)} \cos \theta$$

which applies over the full range $0 \le \theta \le \pi$.

Example 11-5 Determine the bending moments in the fuselage frame in Fig. 11-20 that is used to diffuse the two concentrated forces into the skin. The concentrated forces on the frame are reacted by the shear flow that acts along the line of rivets joining the frame and skin.

The reacting shear flow is determined from Example 9-7 by letting $\Delta V_z = P$ and $\Delta V_y = \Delta M_t = 0$. This gives $q = (P/\pi R_s) \sin \beta$, where R_s is the skin radius. The elastic center is at the center of the frame, and the axes shown are the principal axes. If extensional deformations are neglected, the analogous section properties are $A' = 2\pi R/EI$ and $I'_{yy} = \pi R^3/EI$, where R is the radius to the centroid of the frame cross section.

A cut is imagined at $\beta = 0$ to determine M_0, the moment in the basic system. The method of sections is used in Fig. 11-21a to determine the moment at θ which is due to q. The skin force on a differential length $ds = R_s \, d\beta$ at β is $(P/\pi) \sin \beta \, d\beta$. The moment arm from a point on the reference curve at

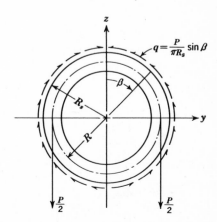

Fig. 11-20 Example 11-5.

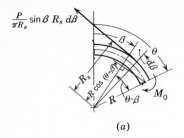

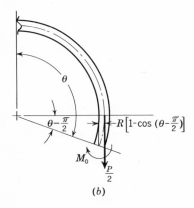

(a)

(b)

Fig. 11-21 Example 11-5. Geometry for moment of (a) shear flow and (b) concentrated force.

θ to the line of action of the skin force at β is $R_s - R\cos(\theta - \beta)$. Integrating the resulting moment from 0 to θ, we find

$$M_0 = \frac{P}{\pi} \int_0^\theta \sin\beta [R_s - R\cos(\theta - \beta)]\, d\beta$$

which reduces to

$$M_0 = \frac{PR_s}{\pi}\left(1 - \cos\theta - \frac{R\theta}{2R_s}\sin\theta\right) \qquad 0 \le \theta < \frac{\pi}{2}$$

For $\theta > \pi/2$ the moment due to $P/2$ must be added. This additional moment is found from Fig. 11-21b, where, by noting that $\cos(\theta - \pi/2) = \sin\theta$, we find

$$M_0 = \frac{PR_s}{\pi}\left(1 - \cos\theta - \frac{R\theta}{2R_s}\sin\theta\right) - \frac{PR}{2}(1 - \sin\theta) \qquad \frac{\pi}{2} < \theta \le \pi$$

From Eqs. (11-20)

$$P' = \frac{2R}{EI} \int_0^\pi M_0\, d\theta$$
$$= \frac{2R}{EI}\left[\frac{PR_s}{\pi} \int_0^\pi \left(1 - \cos\theta - \frac{R\theta}{2R_s}\sin\theta\right) d\theta - \frac{PR}{2} \int_{\pi/2}^\pi (1 - \sin\theta)\, d\theta\right]$$

which reduces to $P' = (2PR/EI)(R_s - \pi R/4)$. The value of M_y' for the analogous column is obtained from Eqs. (11-20) by noting that $z = R\cos\theta$.

Neglecting the EA term, we find

$$
\begin{aligned}
M'_y &= \frac{2R^2}{EI} \int_0^\pi M_0 \cos\theta \, d\theta \\
&= \frac{2R^2}{EI} \left[\frac{PR_s}{\pi} \int_0^\pi \left(1 - \cos\theta - \frac{R\theta}{2R_s} \sin\theta \right) \cos\theta \, d\theta \right. \\
&\qquad\qquad \left. - \frac{PR}{2} \int_{\pi/2}^\pi (1 - \sin\theta) \cos\theta \, d\theta \right]
\end{aligned}
$$

which simplifies to $M'_y = \dfrac{PR^2}{EI}\left(\dfrac{3R}{4} - R_s\right)$. We note that $M'_z = 0$ as a result of symmetry.

Substituting the previously determined results into Eq. (11-26), we obtain

$$
M = \begin{cases}
\dfrac{PR}{4\pi} (\pi - 3\cos\theta - 2\theta\sin\theta) & 0 \le \theta \le \dfrac{\pi}{2} \\[2mm]
\dfrac{PR}{4\pi} [2(\pi - \theta)\sin\theta - \pi - 3\cos\theta] & \dfrac{\pi}{2} \le \theta \le \pi
\end{cases}
$$

It is interesting to observe that M depends upon R but not R_s. As a result, the shear flow could have been assumed to act at the radius R to simplify the calculations.

REFERENCES

1. Argyris, J. H., and S. Kelsey: "Energy Theorems and Structural Analysis," Butterworth & Co. (Publishers) Ltd., London, 1960.
2. Hoff, N. J.: "The Analysis of Structures," John Wiley & Sons, Inc., New York, 1960.
3. Bruhn, E. F.: "Analysis and Design of Flight Vehicle Structures," Tri-state Offset Co., Cincinnati, Ohio, 1965.
4. Kuhn, P.: "Stresses in Aircraft and Shell Structures," McGraw-Hill Book Company, New York, 1956.
5. Argyris, J. H., and P. C. Dunn: Structural Analysis, in "Handbook of Aeronautics," vol. 1, Sir Isaac Pitman & Sons, Ltd., London, 1952.
6. Cross, H.: The Column Analogy, *Univ. Ill. Eng. Expt. Sta. Bull.* 215, 1930.
7. Du Plantier, D. A.: "Analysis of Fuselage Rings by the Column Analogy," *J. Aeron. Sci.*, **11**(2): 137–152 (April, 1944).

PROBLEMS

11-1. The plane truss shown is composed of bars with pinned ends. A typical bar of stiffness $(EA)_j$ and correct manufactured length L_j is d_j over length and has its temperature increased by T_j. In addition, the truss is subjected to the forces F_x and F_y at the common pin, as shown. Determine the vertical and horizontal components of displacement of the common pin from the perfectly manufactured and unheated position and the force in the jth member by the principle of the stationary value of the total potential.

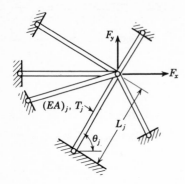

Fig. P11-1

11-2. A plane two-bay stiffened panel is subjected to the end loads shown. All longitudinals are uniform and of stiffness EA, while the shear webs are of constant thickness t. The transverse stiffeners are rigid in extension and perfectly flexible in bending. Determine the shear flows in the webs and the longitudinal forces at the ends of each bay.

11-3. The edge longitudinals in Prob. 11-2 are subjected to a temperature change T_0 instead of the forces. Determine the longitudinal forces and web flows.

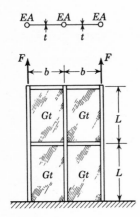

Fig. P11-2 and P11-3

11-4. The upper left longitudinal in Example 11-3 is subjected to a temperature change T_0. Determine the longitudinal forces and web flows.

11-5. All members of the pinned truss shown have the same value of EA. Determine the member forces.

11-6. The top members of the truss of Prob. 11-5 are subjected to a temperature change T_0. Determine the member forces that are due to the temperature change.

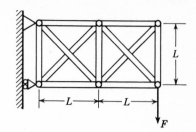

Fig. P11-5 and P11-6

11-7. A circular frame of rectangular cross section is subjected to a temperature change that varies linearly as shown. Determine the stress at the outside radius. [*Ans.* $\sigma_{\theta\theta} = \alpha E(T_1 - T_2)/2.$]

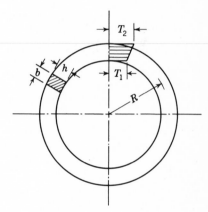

Fig. P11-7

11-8. A circular frame is loaded by a pair of radial forces, as shown. Determine the bending moment under the applied forces. [*Ans.* $M = -FR/\pi.$]

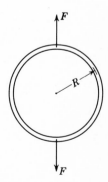

Fig. P11-8

11-9. A circular frame is loaded by moments, as shown. Determine the axial force and moment at A. [*Ans.* $P_A = -2M_1/\pi R$; $M_A = M_1(4 - \pi)/2\pi$.]

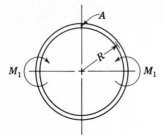

Fig. P11-9

11-10. A rectangular frame of uniform cross section is subjected to a pair of moments, as shown. Determine the bending moment at A. [*Ans.* $M_A = M_1ab/(a + b)$ $(3a + b)$.]

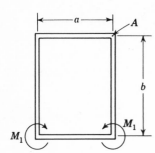

Fig. P11-10

12

Introduction to Matrix Methods of Structural Analysis

12-1 INTRODUCTION

The energy methods introduced in Chap. 6 were applied to the deflection and stress analysis of complex structures in Chaps. 10 and 11. While these are powerful analytical methods, the computations become tedious when the structure contains a large number of elements and is highly redundant. Some of the concepts of matrix methods have been introduced in Chaps. 10 and 11, but for the most part the matrix notation has been used only to write the final equations in a compact form. Digital computers are ideally suited for the manipulations of matrix algebra, and it is therefore desirable to formulate the entire analysis in matrix notation. By doing so, the analysis of complex problems can be reduced to a routine procedure involving relatively little effort on the part of the engineer.

The literature on matrix methods has grown rapidly in recent years as a result of the widespread use of computers. Some of the more important references on the subject are listed at the end of the chapter. The procedures used most frequently are the force method and the displacement method. In the *force method* the internal forces in the structure are chosen as the unknowns. The analysis utilizes the flexibilities of the component parts of the structure and is often referred to as the *flexibility method*. The deformations are taken as the unknowns in the *displacement, or stiffness method*, which uses the element stiffnesses. The complete duality of the two methods is discussed in detail in Ref. 1.

The methods were originally developed to analyze complex built-up structures; however, they have also proved useful in the analysis of two- and three-dimensional elasticity problems, plates, and shells. In contrast to the finite-difference matrix methods of Chap. 5, the force and displacement matrix methods are not based upon a numerical solution of the governing differential equations. Instead, they subdivide the continuous body into a finite number of discrete subelements which are joined at their boundaries. As a result, these methods are frequently referred to as *finite-element methods*.

A brief introduction to the force and displacement matrix methods is given in this chapter. The development generally follows that introduced by Argyris and Kelsey.[1] It is assumed for simplicity that structures are linearly elastic, deformations are small, and buckling does not occur. The resulting equations are therefore linear, and the solutions are unique. The methods have been extended to problems involving large deflections,[2,3] plasticity and creep,[2,4] and buckling.[2]

12-2 THE FORCE METHOD

As in the stress formulation of the theory of elasticity, the force method uses equations that enforce equilibrium, compatibility, and stress-strain-temperature relationships. These equations and the algebraic manipulations that lead to the solution are expressed in matrix notation. The basic theory is developed in this section, and the concepts are clarified by examples in the next section.

Consider an elastic structure under the action of applied generalized forces $R_1, R_2, \ldots, R_i, \ldots, R_m$. These may be conveniently represented by the column matrix

$$\mathbf{R} = \{R_1 \quad R_2 \quad \cdots \quad R_i \quad \cdots \quad R_m\} \qquad (12\text{-}1)$$

It is assumed that the structure is composed of an assemblage of s simple elements (bars, beams, shear webs, etc.) for which the displacement-force relationships are known. Internal generalized forces exist in the structure at the points where the elements join, and the juncture forces on the gth element are written as a column matrix $\mathbf{S}_g$. The set of internal forces for the entire structure is then given by the partitioned column matrix

$$\mathbf{S} = \{\mathbf{S}_a \quad \mathbf{S}_b \quad \cdots \quad \mathbf{S}_g \quad \cdots \quad \mathbf{S}_s\} \qquad (12\text{-}2)$$

Not all the juncture forces that act upon the gth element are included in $\mathbf{S}_g$. Some of the forces are considered to be applied forces on the element, and the remaining forces are treated as reactions; only the applied juncture forces are contained in $\mathbf{S}_g$. Since there is a choice of which juncture forces are to be taken as reactions, there are corresponding alternative possibilities for the element forces in $\mathbf{S}_g$. We include the foundation as part of the structure so that reaction forces for the structure become internal forces and may therefore be included in $\mathbf{S}_g$.

The juncture forces on the gth element are linearly related to the externally applied forces; this is expressed by

$$\mathbf{S}_g = \mathbf{b}_g \mathbf{R} \qquad (12\text{-}3)$$

where $\mathbf{b}_g$ is a rectangular matrix with as many rows as $\mathbf{S}_g$ and with m columns. The elements of the ith column of $\mathbf{b}_g$ are the juncture forces for $R_i = 1$. The internal and applied forces for the entire structure are related by

$$S = bR \qquad (12\text{-}4)$$

where b is the partitioned matrix

$$b = \begin{bmatrix} b_a \\ b_b \\ \cdot \\ \cdot \\ \cdot \\ b_g \\ \cdot \\ \cdot \\ \cdot \\ b_s \end{bmatrix} \qquad (12\text{-}5)$$

With a statically determinate structure the elements of b can be determined from equilibrium conditions alone. This is not the case for an indeterminate structure, since, in addition to equilibrium, the internal forces must also satisfy compatibility. In an indeterminate structure with r redundancies the internal forces are divided into a basic system of internal forces that is in equilibrium with the applied forces and r independent systems of internal forces, each in equilibrium with one of the self-equilibrating redundant forces in the structure. As a result we can write the equilibrium equation as

$$S = b_0 R + b_1 X \qquad (12\text{-}6)$$

where b_0 is a partitioned matrix given by

$$b_0 = \begin{bmatrix} b_{0_a} \\ b_{0_b} \\ \cdot \\ \cdot \\ \cdot \\ b_{0_g} \\ \cdot \\ \cdot \\ \cdot \\ b_{0_s} \end{bmatrix} \qquad (12\text{-}7)$$

The ith column of b_0 is a set of juncture forces that are in equilibrium with $R_i = 1$. As a result, b_0 gives m sets of internal forces that are in equilibrium with the m unit loads associated with

$$R = \{1 \quad 1 \quad \cdot \cdot \cdot \quad 1 \quad \cdot \cdot \cdot \quad 1\}$$

The submatrix b_{0_g} gives the juncture forces on the gth element.

The **X** matrix is a column matrix of the redundant generalized forces. For a system with r redundancies

$$\mathbf{X} = \{X_1 \quad X_2 \quad \cdots \quad X_j \quad \cdots \quad X_r\} \tag{12-8}$$

The foundation has been included as part of the structure; therefore if redundant-support reactions are present, they are included in **X**.

The $\mathbf{b}_1$ matrix is the partitioned matrix

$$\mathbf{b}_1 = \begin{bmatrix} \mathbf{b}_{1_a} \\ \mathbf{b}_{1_b} \\ \cdot \\ \cdot \\ \cdot \\ \mathbf{b}_{1_g} \\ \cdot \\ \cdot \\ \cdot \\ \mathbf{b}_{1_s} \end{bmatrix} \tag{12-9}$$

in which the jth column is a set of internal forces that are in equilibrium with $X_j = 1$. Therefore, the $\mathbf{b}_1$ matrix gives r sets of internal forces that are in equilibrium with unit increments $\delta\mathbf{X} = \{1 \quad 1 \quad \cdots \quad 1 \quad \cdots \quad 1\}$ of the r redundancies. The submatrix $\mathbf{b}_{1_g}$ gives the juncture forces on the gth element. The matrices $\mathbf{b}_0$ and $\mathbf{b}_1$ are analogous to the stress functions φ_0 and φ_j in Eq. (6-78) and the stress systems σ_{mn_0} and σ_{mn_j} in Eq. (11-2).

The redundant forces of Eq. (12-6) must be determined so that compatibility is satisfied. This is done by applying the complementary-potential principle; however, to use this principle we must develop force-displacement relationships. Letting r_i be the generalized work-producing displacement at R_i, we define a matrix **r** of external displacements at the applied forces by

$$\mathbf{r} = \{r_1 \quad r_2 \quad \cdots \quad r_i \quad \cdots \quad r_m\} \tag{12-10}$$

The elements of the structure deform as a result of the internal forces **S** and temperature changes. Letting $\mathbf{v}_g$ be a column matrix of the work-producing displacements at, and in the direction of, the juncture forces $\mathbf{S}_g$, we obtain the partitioned column matrix of internal deformations for the entire structure

$$\mathbf{v} = \{\mathbf{v}_a \quad \mathbf{v}_b \quad \cdots \quad \mathbf{v}_g \quad \cdots \quad \mathbf{v}_s\} \tag{12-11}$$

Instead of being absolute displacements, the elements of $\mathbf{v}_g$ are the displacements of the forces in $\mathbf{S}_g$ *relative* to supports at the remaining juncture forces that are considered to be reactions for the gth element. The elements of **v** are the resultant internal deformations due to elastic deformation associated with the forces $\mathbf{S}$ and any initial lack of fit in the junctures that results from thermal expansion and/or prescribed manufacturing tolerances.   Therefore

$$\mathbf{v} = \mathbf{v}_e + \mathbf{H} \tag{12-12}$$

where $\mathbf{v}_e$ is a column matrix of elastic deformations and $\mathbf{H}$ is a column matrix of initial lacks of fit of members of the structure.

In subdividing the structure it is assumed that the displacement-force relations are known or can be determined for each of the elements. For the gth element these relations can be expressed as

$$\mathbf{v}_{e_g} = \mathbf{c}_g \mathbf{S}_g \tag{12-13}$$

where $\mathbf{c}_g$ is a symmetric matrix of flexibility coefficients that relate the elastic deformations $\mathbf{v}_{e_g}$ to the forces $\mathbf{S}_g$.   These equations can be written for the entire structure in the form

$$\mathbf{v}_e = \mathbf{c}\mathbf{S} \tag{12-14}$$

where

$$\mathbf{c} = \begin{bmatrix} \mathbf{c}_a & & & & & & \\ & \mathbf{c}_b & & & & & \\ & & \cdot & & & & \\ & & & \cdot & & & \\ & & & & \mathbf{c}_g & & \\ & & & & & \cdot & \\ & & & & & & \cdot \\ & & & & & & & \mathbf{c}_s \end{bmatrix} \tag{12-15}$$

is a symmetric partitioned diagonal matrix which we shall call the *unassembled flexibility matrix*.   Substituting Eq. (12-14) into Eq. (12-12) gives

$$\mathbf{v} = \mathbf{c}\mathbf{S} + \mathbf{H} \tag{12-16}$$

which with Eq. (12-6) becomes

$$\mathbf{v} = \mathbf{c}\mathbf{b}_0 \mathbf{R} + \mathbf{c}\mathbf{b}_1 \mathbf{X} + \mathbf{H} \tag{12-17}$$

To this point we have utilized the equilibrium and displacement-force relations.   It remains to require the redundancies $\mathbf{X}$ to have magnitudes that make the deformations $\mathbf{v}$ compatible.   To enforce this condition we impose the condition that $\delta U' = 0$ while each of the redundancies is given a unit variation ($\delta V' = 0$ because the foundation is included in the structure, so that support reactions become internal forces).   The ith row of $\mathbf{b}_1'$ is a self-equilibrating set of internal forces for a unit increment of X_i, so that its product with the true internal displacements $\mathbf{v}$ gives $\delta U'$ for $\delta X_i = 1$. Thus $\mathbf{b}_1' \mathbf{v} = \mathbf{0}$ enforces compatibility for all the redundant forces. Substituting Eq. (12-17) into this equation gives

$$\mathbf{D}_{11}\mathbf{X} = -\mathbf{D}_{10}\mathbf{R} - \mathbf{b}_1'\mathbf{H} \tag{12-18}$$

where $\mathbf{D}_{ij}$ is defined by

$$\mathbf{D}_{ij} = \mathbf{b}_i'\mathbf{c}\mathbf{b}_j \tag{12-19}$$

The internal forces can be obtained by solving Eq. (12-18) for $\mathbf{X}$ and substituting the result into Eq. (12-6). It can be proved that the matrix product $\mathbf{A}'\mathbf{B}\mathbf{A}$ is symmetric if $\mathbf{B}$ is symmetric. Since $\mathbf{c}$ is symmetric, it follows from Eq. (12-19) that $\mathbf{D}_{11}$ is symmetric.

It is interesting to obtain a physical interpretation of Eq. (12-18). The matrix $\mathbf{b}_1$ gives sets of internal forces that are in equilibrium with unit values of each of the redundancies. The product $\mathbf{c}\mathbf{b}_1$ yields the internal displacements associated with unit values of the redundancies, and from the virtual-work method (Sec. 10-4) $\mathbf{b}_1'\mathbf{c}\mathbf{b}_1$ gives the relative displacements at the releases for unit values of the redundancies. The product $\mathbf{D}_{11}\mathbf{X}$ is therefore equal to the relative displacements at the releases that result from the application of the redundant forces to the releases. In a similar manner $\mathbf{D}_{10}\mathbf{R}$ and $\mathbf{b}_1'\mathbf{H}$ are the relative displacements at the releases that are respectively due to the applied forces and the released lacks of fit. Equation (12-18) is therefore equivalent to Eq. (11-8).

When the number of loads and temperature conditions to be analyzed is large, it may be convenient to solve Eq. (12-18) by inversion, to give

$$\mathbf{X} = -\mathbf{D}_{11}^{-1}(\mathbf{D}_{10}\mathbf{R} + \mathbf{b}_1'\mathbf{H}) \tag{12-20}$$

Substituting this into Eq. (12-6) produces the result

$$\mathbf{S} = \mathbf{b}\mathbf{R} + \mathbf{e}\mathbf{H} \tag{12-21}$$

where

$$\mathbf{b} = \mathbf{b}_0 - \mathbf{b}_1\mathbf{D}_{11}^{-1}\mathbf{D}_{10} \qquad \mathbf{e} = -\mathbf{b}_1\mathbf{D}_{11}^{-1}\mathbf{b}_1' \tag{12-22}$$

The displacement matrix $\mathbf{r}$ can be found by the virtual-work method. We may interpret $\mathbf{r}$ as a matrix in which the element r_i is δW_e for a unit external force $R_i = 1$ moving through the real displacements. The ith row of $\mathbf{b}_0'$ gives a set of internal forces in equilibrium with $R_i = 1$. The product of the ith row of $\mathbf{b}_0'$ with $\mathbf{v}$ is therefore $-\delta W_i$, the virtual work of the internal forces in equilibrium with $R_i = 1$ in moving through the real internal deformations. Applying this argument to each load point gives $\mathbf{r} = \mathbf{b}_0'\mathbf{v}$, which with Eqs. (12-17) and (12-19) becomes

$$\mathbf{r} = \mathbf{D}_{00}\mathbf{R} + \mathbf{D}_{10}'\mathbf{X} + \mathbf{b}_0'\mathbf{H} \tag{12-23}$$

In writing this equation we have used the rule for the transpose of matrix products that $(\mathbf{ABC})' = \mathbf{C}'\mathbf{B}'\mathbf{A}'$.   Applying this to $\mathbf{b}_0'\mathbf{cb}_1$ and noting that $\mathbf{c}$ is symmetric so that $\mathbf{c}' = \mathbf{c}$, we find $\mathbf{b}_0'\mathbf{cb}_1 = (\mathbf{b}_1'\mathbf{cb}_0)' = \mathbf{D}_{10}'$.   The displacements are found from Eq. (12-23) once $\mathbf{X}$ is found from Eq. (12-18).

If $\mathbf{X}$ is determined by inversion, we can substitute Eq. (12-20) into Eq. (12-23) and use Eq. (12-22) to obtain

$$\mathbf{r} = \mathbf{CR} + \mathbf{b}'\mathbf{H} \tag{12-24}$$

where

$$\mathbf{C} = \mathbf{D}_{00} - \mathbf{D}_{10}'\mathbf{D}_{11}{}^{-1}\mathbf{D}_{10} \tag{12-25}$$

is the *assembled flexibility matrix* of the structure.

In a statically determinate structure we find from Eqs. (12-6) and (12-16) that $\mathbf{S} = \mathbf{b}_0\mathbf{R}$ and $\mathbf{v} = \mathbf{cb}_0\mathbf{R} + \mathbf{H}$.   Substituting the latter equation into $\mathbf{r} = \mathbf{b}_0'\mathbf{v}$, we obtain $\mathbf{r} = \mathbf{CR} + \mathbf{b}_0'\mathbf{H}$, where $\mathbf{C} = \mathbf{D}_{00}$.

We see from the foregoing development that the internal forces and displacements in a structure can be obtained by simple matrix operations once $\mathbf{b}_0$, $\mathbf{b}_1$, and $\mathbf{c}$ have been determined and $\mathbf{R}$ and $\mathbf{H}$ are specified.   The matrix operations are easily handled by a digital computer.

12-3 DISCUSSION OF THE FORCE METHOD

To clarify the meaning of the matrices in the preceding section and show how they are found, we refer to the indeterminate beam in Fig. 12-1.   In this case the applied-force matrix is simply $\mathbf{R} = \{R_1 \quad R_2\}$, and the displacement matrix is $\mathbf{r} = \{r_1 \quad r_2\}$, where r_1 is the lateral displacement at R_1 and r_2 is the slope at R_2.

To determine $\mathbf{b}_0$ and $\mathbf{b}_1$ we must decide upon the internal forces $\mathbf{S}$ that define the loading upon each of the elements.   Two of the possible choices are shown in Fig. 12-2.   In Fig. 12-2a the end moments for each element are taken as the internal forces, while in Fig. 12-2b the shear and moment at one end of each element are chosen to define the internal forces. The remaining juncture forces, which can be determined from $\mathbf{S}$ by statics, do not enter the analysis and are simply shown as reacting supports in Fig. 12-2.   The juncture-force matrices for the elements are then

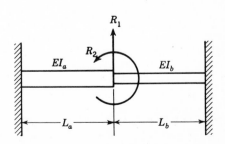

Fig. 12-1 Statically indeterminate beam.

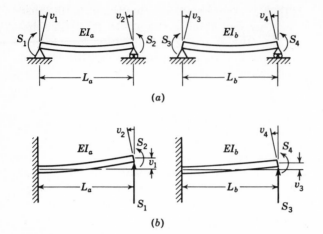

Fig. 12-2 Alternate methods for defining the juncture forces of the beam in Fig. 12-1. (a) Defined by juncture moments of both ends of each element; (b) defined by juncture shear and moment of one end of each element.

$\mathbf{S}_a = \{S_1 \quad S_2\}$ and $\mathbf{S}_b = \{S_3 \quad S_4\}$, and the internal-force matrix for the entire structure is

$$\mathbf{S} = \{\mathbf{S}_a \quad \mathbf{S}_b\} = \{S_1 \quad S_2 | S_3 \quad S_4\}$$

where the partitioning between element submatrices is shown by the dashed line.

Having defined a set of internal forces $\mathbf{S}$, the basic-force matrix $\mathbf{b}_0$ is simply *any* set of internal forces that are in equilibrium with $\mathbf{R}$. The $\mathbf{b}_0$ matrix can be found by making the structure statically determinate by any convenient system of releases; however, it is advisable to follow the suggestions for choosing the basic system that are given in Sec. 11-5. In Fig. 12-1, the built-in ends could be replaced by simple supports, or the beam could be cut where the loads or reactions are applied. For example, if a cut is made at the right end and the internal-force system of Fig. 12-2b is used, the $\mathbf{b}_0$ matrix is

$$
\mathbf{b}_0 =
\begin{array}{cc}
R_1 = 1 & R_2 = 1 \\
\end{array}
$$

$$
\mathbf{b}_0 =
\begin{bmatrix}
1 & 0 \\
0 & 1 \\
0 & 0 \\
0 & 0 \\
\end{bmatrix}
\begin{array}{l}
S_1 \\
S_2 \\
S_3 \\
S_4 \\
\end{array}
$$

In this case the first column gives the values of the internal forces for $R_1 = 1$, and the second column lists these forces for $R_2 = 1$.

The redundancies must be chosen before b_1 can be determined. Again, the suggestions of Sec. 11-5 are pertinent in selecting the redundancies. One choice of redundancies is the set of forces that is released to obtain the basic system. Thus, in Fig. 12-1 X_1 and X_2 could be taken as the moments at the built-in ends or the shear and moment at a release cut. If X_1 and X_2 are chosen as the shear and moment at the right wall and the internal-force system of Fig. 12-2b is used, we find

$$X_1 = 1 \quad X_2 = 1$$

$$b_1 = \begin{bmatrix} 1 & 0 \\ L_b & 1 \\ 1 & 0 \\ 0 & 1 \end{bmatrix} \begin{matrix} S_1 \\ S_2 \\ S_3 \\ S_4 \end{matrix}$$

The first column lists the internal forces for $X_1 = 1$, and the second column gives the forces for $X_2 = 1$.

The internal-deformation matrix v is the set of element deformations associated with the member forces S. The elements of v are shown in Fig. 12-2 for the two methods of defining S that have been described. The element-deformation matrices are $v_a = \{v_1 \quad v_2\}$ and $v_b = \{v_3 \quad v_4\}$, and the deformation matrix for the entire structure is

$$v = \{v_a \quad v_b\} = \{v_1 \quad v_2 | v_3 \quad v_4\}$$

The element-flexibility matrix c_g relates the elastic-deformation matrix v_{e_g} of the element to the element-force matrix S_g. The unit-load method provides a convenient procedure for determining the elements of c_g. Wehle and Lansing[5] have given the element-flexibility matrices for shear webs, and bars and beams with linearly varying section properties. The element-flexibility matrix for the internal-force system of Fig. 12-2b is found from Eqs. (a) and (b) of Example 10-2, which gives

$$c_g = \frac{L_g}{6EI_g} \begin{bmatrix} 2L_g^2 & 3L_g \\ 3L_g & 6 \end{bmatrix} \tag{12-26}$$

In a similar manner c_g for the force system of Fig. 12-2a is

$$c_g = \frac{L_g}{6EI_g} \begin{bmatrix} 2 & 1 \\ 1 & 2 \end{bmatrix} \tag{12-27}$$

If the beam elements have initial curvatures due to manufacturing tolerances or a thermal gradient through the depth of the beam, an H matrix must also be determined. The elements of H are the initial values of v when the structure is subdivided into its elements and $S = 0$, as shown in Fig. 12-2. Thus, if the system of Fig. 12-2a is used, H is a

column matrix of initial slopes at the ends of the beam elements.   On the
other hand, if the system of Fig. 12-2*b* is employed, **H** is a column matrix
listing the initial displacement and slope at the right end of each span.

Example 12-1   In the beam of Fig. 12-1, $L_a = L_b = L$, $EI_a = 2EI$, and $EI_b = EI$.
A temperature gradient through the depth of the left beam element produces
an equivalent thermal moment M_T that is constant over L_a.   Determine the
internal forces and the deflection and slope at the center of the beam when
the only applied force is R_1.

   The displacement at R_1 is r_1, and if the slope at this point is also desired,
it is necessary to apply an imaginary moment R_2 (which is subsequently set
equal to zero) at the point.   To find b_0 we assume hinges at the built-in ends
to obtain the basic system of Fig. 12-3.   Adopting the S forces of Fig. 12-2*a*
and using Fig. 12-3 to obtain the end moments for each of the elements, we
find

$$\begin{array}{cc} R_1 = 1 & R_2 = 1 \end{array}$$

$$\mathbf{b}_0 = \frac{1}{2} \begin{bmatrix} 0 & 0 \\ -L & 1 \\ -L & -1 \\ 0 & 0 \end{bmatrix} \begin{array}{c} S_1 \\ S_2 \\ S_3 \\ S_4 \end{array}$$

   If we choose the moments at the built-in ends as redundancies, we obtain
the X_1 and X_2 systems of Fig. 12-4.   From this figure, the element end moments
are

$$\begin{array}{cc} X_1 = 1 & X_2 = 1 \end{array}$$

$$\mathbf{b}_1 = \frac{1}{2} \begin{bmatrix} 2 & 0 \\ 1 & 1 \\ 1 & 1 \\ 0 & 2 \end{bmatrix} \begin{array}{c} S_1 \\ S_2 \\ S_3 \\ S_4 \end{array}$$

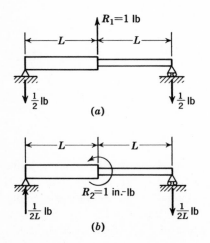

(a)

(b)

Fig. 12-3   Basic force systems for construction of $\mathbf{b}_0$, Example 12-1.   (*a*)
For $R_1 = 1$; (*b*) for $R_2 = 1$.

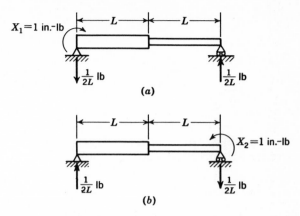

Fig. 12-4 Redundant-force systems for construction of $\mathbf{b}_1$, Example 12-1. (a) for $X_1 = 1$; (b) for $X_2 = 1$.

Applying Eqs. (12-15) and (12-27), we obtain the unassembled flexibility matrix

$$\mathbf{c} = \frac{L}{12EI} \begin{bmatrix} 2 & 1 & & \\ 1 & 2 & & \\ & & 4 & 2 \\ & & 2 & 4 \end{bmatrix}$$

Substituting $\mathbf{b}_0$, $\mathbf{b}_1$, and $\mathbf{c}$ into Eqs. (12-19), (12-22), and (12-25), we find

$$\mathbf{b} = \begin{bmatrix} 0.303L & -0.182 \\ -0.242L & 0.545 \\ -0.242L & -0.455 \\ 0.212L & 0.273 \end{bmatrix} \qquad \mathbf{e} = -\frac{144EI}{396L} \begin{bmatrix} 10 & 3 & 3 & -4 \\ 3 & 2 & 2 & 1 \\ 3 & 2 & 2 & 1 \\ -4 & 1 & 1 & 6 \end{bmatrix}$$

$$\mathbf{C} = \frac{L}{EI} \begin{bmatrix} 0.0303L^2 & 0.0152L \\ 0.0152L & 0.0909 \end{bmatrix}$$

The applied-load matrix is $\mathbf{R} = \{R_1 \quad 0\}$. To determine $\mathbf{H}$ we use the unit-load method and the load systems in Fig. 12-5, from which

$$H_1 = \int_0^{L_a} \frac{M^* \bar{M} \, dx}{EI_a} = \frac{M_T L}{4EI}$$

and $H_2 = H_1$ by symmetry. Member b is unheated, so that $H_3 = H_4 = 0$, and

$$\mathbf{H} = \frac{M_T L}{4EI} \{1 \quad 1 \quad 0 \quad 0\}$$

Equation (12-21) then gives the internal-force matrix

$$\mathbf{S} = \begin{Bmatrix} S_1 \\ S_2 \\ S_3 \\ S_4 \end{Bmatrix} = \begin{Bmatrix} 0.303R_1L - 1.182M_T \\ -0.242R_1L - 0.455M_T \\ -0.242R_1L - 0.455M_T \\ 0.212R_1L + 0.272M_T \end{Bmatrix}$$

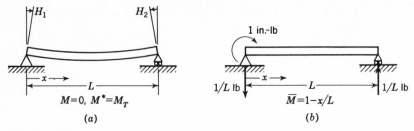

Fig. 12-5 Force systems for determining initial lack of fit due to M_T. (a) Actual displacements and force systems; (b) unit-load system.

and from Eq. (12-24) the displacements are

$$\mathbf{r} = \begin{Bmatrix} r_1 \\ r_2 \end{Bmatrix} = \frac{1}{EI} \begin{Bmatrix} 0.0303R_1L^3 + 0.0152M_TL^2 \\ 0.0152R_1L^2 + 0.0909M_TL \end{Bmatrix}$$

12-4 APPLICATION TO STIFFENED SHELLS

To apply the force method to idealized stiffened shells we need the flexibility matrices for longitudinals and shear webs. The longitudinal is assumed to be loaded by end forces and a constant shear flow along its length, as shown in Fig. 12-6a. Any one of these three forces can be considered to be the reaction and can be determined from the other two forces by equilibrium. As a result, we can use the force system of Fig. 12-6b to define the juncture forces. In this case, the associated element deformations are the end displacements v_A and v_B shown in Fig. 12-6b.

To obtain v_A we use the unit-load method with a 1-lb force at A, so that $\delta W_e = v_A$. To determine δU we use Eq. (10-12) with the real-load system of Fig. 12-6b. Noting that $\bar{P}_A = 1$ and $\bar{P}_B = 0$, we find from Eq. (10-1) that $v_A = (L/6EA)(2S_A + S_B)$, and in a similar manner $v_B = (L/6EA)(S_A + 2S_B)$. From these equations, the element-flexibility matrix is seen to be

$$\mathbf{c}_g = \frac{L_g}{6EA_g} \begin{bmatrix} 2 & 1 \\ 1 & 2 \end{bmatrix} \tag{12-28}$$

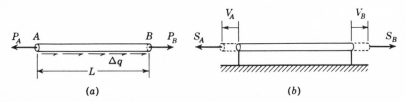

Fig. 12-6 (a) Forces on a longitudinal attached to shear webs; (b) juncture forces and deformations.

To determine c_g for a shear web we refer to Fig. 12-7a, which shows an initially rectangular web that is subjected to edge shear flows S_g.   The flow on one of the edges (the bottom edge in Fig. 12-7a) may be considered to be the reaction.   The unit-load system is shown in Fig. 12-7b.   Subjecting the unit-load system to the real displacements, we find $\delta W_e = Ld = v_g$.   From Eq. (10-17) $\delta U = S_g bL/Gt$, so that $\delta W_e = \delta U$ gives $v_g = (bL/Gt)S_g$.   Letting the surface area $bL = S$, we find the flexibility matrix consists of the single element

$$\mathbf{c}_g = \left(\frac{S}{Gt}\right)_g \tag{12-29}$$

Though only an approximation, this equation may be used for trapezoidal webs when the taper is small.

We next consider the $\mathbf{H}_g$ matrices for the longitudinal and web when there are temperature changes.   For the longitudinal of Fig. 12-6 we assume that the temperature varies linearly along the length from T_A to T_B, so that $P_A^* = EA\alpha T_A$ and $P_B^* = EA\alpha T_B$.   Applying a unit load at A to obtain the displacement v_A at that point, we find $\bar{P}_A = 1$ and $\bar{P}_B = 0$. From Eqs. (10-1) and (10-12) we obtain $v_A = (\alpha L/6)(2T_A + T_B)$, and in a similar manner $v_B = (\alpha L/6)(T_A + 2T_B)$.   From these equations we obtain

$$\mathbf{H}_g = \frac{\alpha L}{6}\begin{Bmatrix} 2T_A + T_B \\ T_A + 2T_B \end{Bmatrix} \tag{12-30}$$

Temperature changes do not produce shearing deformations in an isotropic web, so that $\mathbf{H}_g = 0$ for these elements.

Argyris and Kelsey[1,6] discuss the selection of redundant-force systems for stiffened shells at length.   These discussions are helpful in

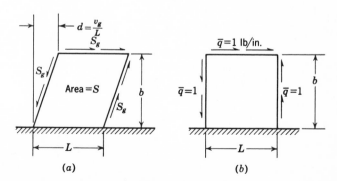

Fig. 12-7 Juncture forces and deformations for a shear web.
(a) Real forces and displacements; (b) unit-load system.

setting up the $\mathbf{b}_1$ matrices for complex wing and body structures, with and without cutouts.

The application of the force method to a stiffened shell is illustrated in the following example.

Example 12-2　Determine the **C** matrix for the structure in Fig. 11-7 (Example 11-3) if vertical forces are applied to the front and rear spars at each of the rib locations. Compute the internal forces for a 1000-lb force applied as shown in Fig. 11-7. Include the flexibility of the 0.04-in.-thick ribs in the analysis.

The numbering system for the internal forces is shown in Fig. 12-8. In each of the spars the forces in the lower longitudinals are equal and opposite in direction to those in the upper longitudinals, and the shear flows in the bottom skins are equal and opposite to those in the upper skins. As a result, it is only necessary to number the forces in the upper surface. The forces in the lower surface will be accounted for in setting up **c**.

The internal forces associated with bending and torsion theory could be used to determine the basic system for $\mathbf{b}_0$ (as was done in Example 11-3). Instead we use the simpler basic system shown in Fig. 12-9, which assumes that the entire load is carried by the spars. It is convenient to partition $\mathbf{b}_0$ and $\mathbf{b}_1$ in the following manner

$$\mathbf{b}_0 = \begin{bmatrix} \mathbf{b}_{0_{fl}} \\ \mathbf{b}_{0_{rl}} \\ \mathbf{b}_{0_{fw}} \\ \mathbf{b}_{0_{rw}} \\ \mathbf{b}_{0_s} \\ \mathbf{b}_{0_r} \end{bmatrix} \qquad \mathbf{b}_1 = \begin{bmatrix} \mathbf{b}_{1_{fl}} \\ \mathbf{b}_{1_{rl}} \\ \mathbf{b}_{1_{fw}} \\ \mathbf{b}_{1_{rw}} \\ \mathbf{b}_{1_s} \\ \mathbf{b}_{1_r} \end{bmatrix} \qquad (a)$$

where the subscripts fl refer to the front longitudinal, rl to the rear longitudinal, fw to the front web, rw to the rear web, s to the skin, and r to the ribs.

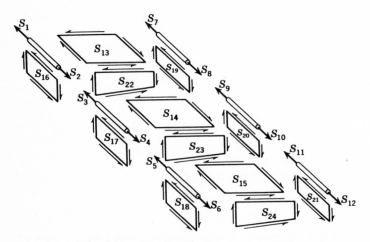

Fig. 12-8　Element forces for Example 12-2.

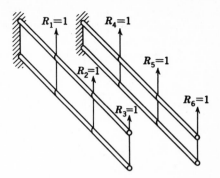

Fig. 12-9 Basic system for $\mathbf{b}_0$ matrix, Example 12-2.

The longitudinal forces in the basic system are easily found from $P = M/h$ and the web shear flows from $q = V/h$, where M is the bending moment, V is the shear, and h is the height of the spar. From Fig. 12-9 and the dimensions of Fig. 11-7

$$\mathbf{b}_{0fl} = -\frac{25}{8}\begin{array}{c} \begin{array}{cccccc} R_1 & R_2 & R_3 & R_4 & R_5 & R_6 \end{array} \\ \begin{bmatrix} 1 & 2 & 3 & 0 & 0 & 0 \\ 0 & 1 & 2 & 0 & 0 & 0 \\ 0 & 1 & 2 & 0 & 0 & 0 \\ 0 & 0 & 1 & 0 & 0 & 0 \\ 0 & 0 & 1 & 0 & 0 & 0 \\ 0 & 0 & 0 & 0 & 0 & 0 \end{bmatrix} \end{array} \begin{array}{l} S_1 \\ S_2 \\ S_3 \\ S_4 \\ S_5 \\ S_6 \end{array}$$

$$\mathbf{b}_{0rl} = -\frac{25}{4}\begin{array}{c} \begin{array}{cccccc} R_1 & R_2 & R_3 & R_4 & R_5 & R_6 \end{array} \\ \begin{bmatrix} 0 & 0 & 0 & 1 & 2 & 3 \\ 0 & 0 & 0 & 0 & 1 & 2 \\ 0 & 0 & 0 & 0 & 1 & 2 \\ 0 & 0 & 0 & 0 & 0 & 1 \\ 0 & 0 & 0 & 0 & 0 & 1 \\ 0 & 0 & 0 & 0 & 0 & 0 \end{bmatrix} \end{array} \begin{array}{l} S_7 \\ S_8 \\ S_9 \\ S_{10} \\ S_{11} \\ S_{12} \end{array}$$

$$\mathbf{b}_{0fw} = -\frac{1}{8}\begin{array}{c} \begin{array}{cccccc} R_1 & R_2 & R_3 & R_4 & R_5 & R_6 \end{array} \\ \begin{bmatrix} 1 & 1 & 1 & 0 & 0 & 0 \\ 0 & 1 & 1 & 0 & 0 & 0 \\ 0 & 0 & 1 & 0 & 0 & 0 \end{bmatrix} \end{array} \begin{array}{l} S_{16} \\ S_{17} \\ S_{18} \end{array}$$

$$\mathbf{b}_{0rw} = -\frac{1}{4}\begin{array}{c} \begin{array}{cccccc} R_1 & R_2 & R_3 & R_4 & R_5 & R_6 \end{array} \\ \begin{bmatrix} 0 & 0 & 0 & 1 & 1 & 1 \\ 0 & 0 & 0 & 0 & 1 & 1 \\ 0 & 0 & 0 & 0 & 0 & 1 \end{bmatrix} \end{array} \begin{array}{l} S_{19} \\ S_{20} \\ S_{21} \end{array}$$

$$\mathbf{b}_{0s} = \mathbf{b}_{0r} = 0$$

We choose the same redundant-force systems that are used in Example

11-3, and from Figs. 11-8 and 11-10 we find

$$
\mathbf{b}_{1_{fl}} = - \begin{array}{c} X_1 \quad X_2 \quad X_3 \\ \begin{bmatrix} 1 & 0 & 0 \\ 0 & 1 & 0 \\ 0 & 1 & 0 \\ 0 & 0 & 1 \\ 0 & 0 & 1 \\ 0 & 0 & 0 \end{bmatrix} \begin{array}{l} S_1 \\ S_2 \\ S_3 \\ S_4 \\ S_5 \\ S_6 \end{array} \end{array}
\qquad
\mathbf{b}_{1_{rl}} = 2 \begin{array}{c} X_1 \quad X_2 \quad X_3 \\ \begin{bmatrix} 1 & 0 & 0 \\ 0 & 1 & 0 \\ 0 & 1 & 0 \\ 0 & 0 & 1 \\ 0 & 0 & 1 \\ 0 & 0 & 0 \end{bmatrix} \begin{array}{l} S_7 \\ S_8 \\ S_9 \\ S_{10} \\ S_{11} \\ S_{12} \end{array} \end{array}
$$

$$
\mathbf{b}_{1_{fw}} = 0.01333 \begin{array}{c} X_1 \quad X_2 \quad X_3 \\ \begin{bmatrix} -1 & 1 & 0 \\ 0 & -1 & 1 \\ 0 & 0 & -1 \end{bmatrix} \begin{array}{l} S_{16} \\ S_{17} \\ S_{18} \end{array} \end{array}
\qquad
\mathbf{b}_{1_{rw}} = -0.05333 \begin{array}{c} X_1 \quad X_2 \quad X_3 \\ \begin{bmatrix} -1 & 1 & 0 \\ 0 & -1 & 1 \\ 0 & 0 & -1 \end{bmatrix} \begin{array}{l} S_{19} \\ S_{20} \\ S_{21} \end{array} \end{array}
$$

$$
\mathbf{b}_{1_s} = 0.02667 \begin{array}{c} X_1 \quad X_2 \quad X_3 \\ \begin{bmatrix} -1 & 1 & 0 \\ 0 & -1 & 1 \\ 0 & 0 & -1 \end{bmatrix} \begin{array}{l} S_{13} \\ S_{14} \\ S_{15} \end{array} \end{array}
$$

The rib shear flows (which were not determined in Example 11-3) are found by taking the differences in the shear flows in the adjacent skins in Fig. 11-10. This gives

$$
\mathbf{b}_{1_r} = -0.02667 \begin{array}{c} X_1 \quad X_2 \quad X_3 \\ \begin{bmatrix} 1 & -2 & 1 \\ 0 & 1 & -2 \\ 0 & 0 & 1 \end{bmatrix} \begin{array}{l} S_{22} \\ S_{23} \\ S_{24} \end{array} \end{array}
$$

The unassembled flexibility matrix can be partitioned as follows

$$
\mathbf{c} = \begin{bmatrix} \mathbf{c}_{fl} & & & & & \\ & \mathbf{c}_{rl} & & & & \\ & & \mathbf{c}_{fw} & & & \\ & & & \mathbf{c}_{rw} & & \\ & & & & \mathbf{c}_s & \\ & & & & & \mathbf{c}_r \end{bmatrix} \tag{b}
$$

By using Eqs. (12-28) and (12-29), the submatrices of $\mathbf{c}$ are found to be

$$
\mathbf{c}_{fl} = \frac{2 \times 25}{6 \times 2 \times 10^7} \begin{bmatrix} 2 & 1 & 0 & 0 & 0 & 0 \\ 1 & 2 & 0 & 0 & 0 & 0 \\ 0 & 0 & 2 & 1 & 0 & 0 \\ 0 & 0 & 1 & 2 & 0 & 0 \\ 0 & 0 & 0 & 0 & 2 & 1 \\ 0 & 0 & 0 & 0 & 1 & 2 \end{bmatrix} = \mathbf{c}_{rl}
$$

$$
\mathbf{c}_{fw} = \frac{8 \times 25}{4 \times 10^6 \times 0.08} \begin{bmatrix} 1 & 0 & 0 \\ 0 & 1 & 0 \\ 0 & 0 & 1 \end{bmatrix}
$$

$$
\mathbf{c}_{rw} = \frac{4 \times 25}{4 \times 10^6 \times 0.08} \begin{bmatrix} 1 & 0 & 0 \\ 0 & 1 & 0 \\ 0 & 0 & 1 \end{bmatrix}
$$

$$c_s = \frac{2 \times 50 \times 25}{4 \times 10^6 \times 0.04} \begin{bmatrix} 1 & 0 & 0 \\ 0 & 1 & 0 \\ 0 & 0 & 1 \end{bmatrix}$$

$$c_r = \frac{50(8+4)/2}{4 \times 10^6 \times 0.04} \begin{bmatrix} 1 & 0 & 0 \\ 0 & 1 & 0 \\ 0 & 0 & 1 \end{bmatrix}$$

where it is assumed that $E = 10^7$ psi and $G = 4 \times 10^6$ psi. The matrices c_{fl}, c_{rl}, and c_s have been multiplied by 2 to account for the fact that the forces and deformations associated with these flexibilities occur twice, once in the upper and once in the lower surfaces.

Assembling the b_0, b_1, and c matrices according to Eqs. (a) and (b) and substituting these into Eqs. (12-19) and (12-25), we find the assembled flexibility matrix

$$\mathbf{C} = 10^{-7} \begin{bmatrix} 168 & & & & & \\ 271 & 745 & & & \text{sym.} & \\ 371 & 1148 & 2127 & & & \\ 43 & 122 & 211 & 350 & & \\ 122 & 407 & 751 & 524 & 1369 & \\ 211 & 751 & 1461 & 654 & 1949 & 3538 \end{bmatrix}$$

To determine the internal forces for the assigned loading we use the loading matrix $\mathbf{R} = \{0 \ \ 0 \ \ 0 \ \ 0 \ \ 1000 \ \ 0\}$. Since there are no temperature changes or manufacturing lacks of fit, $\mathbf{H} = 0$. From Eq. (12-21) and the first of Eqs. (12-22) we find

$$\mathbf{S}_{fl} = -10^3 \{3.181 \quad 2.108 \quad 2.108 \quad 0.720 \quad 0.720 \quad 0\}$$

$$\mathbf{S}_{rl} = -10^3 \{6.138 \quad 2.034 \quad 2.034 \quad -1.439 \quad -1.439 \quad 0\}$$

$$\mathbf{S}_{fw} = -\{14.30 \quad 18.51 \quad 9.59\}$$

$$\mathbf{S}_{rw} = -\{192.8 \quad 175.9 \quad -38.4\}$$

$$\mathbf{S}_s = -\{28.6 \quad 37.0 \quad 19.1\}$$

$$\mathbf{S}_r = -\{-8.41 \quad 17.85 \quad 19.19\}$$

Comparing these results with those in Example 11-3 (Fig. 11-13), which neglected rib flexibility, we see that in this case the effects are small.

12-5 THE DISPLACEMENT METHOD

As in the displacement formulation of the theory of elasticity, the matrix displacement method considers only compatible deformation states. The true deformations are those which satisfy equilibrium, a condition we enforce with the total-potential principle. The basic equations of this method will be derived in this section, and the concepts involved will be clarified by the discussion and examples in the next section. The development and notation are essentially those given by Argyris and Kelsey.[1]

We again consider an elastic structure consisting of an assemblage of

s simple elements. The component members deform at the points where they join neighboring elements, and the juncture displacements of the gth element are given by the column matrix $\mathbf{v}_g$. In writing $\mathbf{v}_g$ we do not include the displacements of all the juncture points. Instead we consider some of the junctures to be supports for the gth element, and only the deformations at the remaining junctures are included in $\mathbf{v}_g$. As a result, the elements of $\mathbf{v}_g$ are not absolute displacements but are the deformations *relative* to the remaining junctures that are chosen as supports for the gth element. The choice of points used to define the deformations of the elements is arbitrary. The deformation state of the entire structure is given by the partitioned column matrix $\mathbf{v} = \{\mathbf{v}_a \quad \mathbf{v}_b \quad \cdots \quad \mathbf{v}_g \quad \cdots \quad \mathbf{v}_s\}$. The total number of elements in $\mathbf{v}$ is p.

We suppose that m external displacements, given by the column matrix $\mathbf{r} = \{r_1 \quad r_2 \cdots r_i \cdots r_m\}$, are specified at juncture points in the structure. For small displacements the element deformations are linearly related to the prescribed displacements, and so we may write

$$\mathbf{v} = \mathbf{ar} \tag{12-31}$$

where $\mathbf{a}$ is a rectangular *displacement transformation matrix* with p rows and m columns. If displacements are specified at *all* junctures, so that $m = p$, the matrix $\mathbf{a}$ can be determined by kinematic reasoning alone. In the usual case $m < p$, and $\mathbf{a}$ must be found from kinematic and equilibrium conditions.

For generality we assume that displacements are not specified at all junctures and define a column matrix

$$\mathbf{U} = \{U_1 \quad U_2 \quad \cdots \quad U_i \quad \cdots \quad U_r\} \tag{12-3 2}$$

with $r = p - m$ elements which consist of the unspecified joint displacements. The total of all joint displacements is then given by the partitioned column matrix $\{\mathbf{r} \quad \mathbf{U}\}$. We call the elements of $\mathbf{U}$ the *kinematic redundancies* since they play a role that is analogous to the elements of $\mathbf{X}$ in the force method.

The element deformations are linear functions of $\mathbf{r}$ and $\mathbf{U}$, and so we may write

$$\mathbf{v} = \mathbf{a}_0\mathbf{r} + \mathbf{a}_1\mathbf{U} \tag{12-33}$$

The matrix $\mathbf{a}_0$ has p rows and m columns. The ith column of $\mathbf{a}_0$ is a set of element deformations that is compatible with $r_i = 1$, while r_j $(j \neq i)$ and $\mathbf{U}$ are zero. It can be found by kinematic conditions alone. The matrix $\mathbf{a}_1$ has p rows and r columns. The ith column of $\mathbf{a}_1$ is a set of element deformations that are compatible with $U_i = 1$, while U_j $(j \neq i)$ and r are zero. It too can be determined by direct geometric reasoning. We can express $\mathbf{a}_0$ and $\mathbf{a}_1$ with the partitioned forms

$$\mathbf{a}_0 = \begin{bmatrix} \mathbf{a}_{0_a} \\ \mathbf{a}_{0_b} \\ \cdot \\ \cdot \\ \cdot \\ \mathbf{a}_{0_g} \\ \cdot \\ \cdot \\ \cdot \\ \mathbf{a}_{0_s} \end{bmatrix} \qquad \mathbf{a}_1 = \begin{bmatrix} \mathbf{a}_{1_a} \\ \mathbf{a}_{1_b} \\ \cdot \\ \cdot \\ \cdot \\ \mathbf{a}_{1_g} \\ \cdot \\ \cdot \\ \cdot \\ \mathbf{a}_{1_s} \end{bmatrix} \qquad (12\text{-}34)$$

where the subscript g refers to the gth element.

The elastic deformations of the gth element give rise to forces at the juncture points. These are related to the deformations by

$$\mathbf{S}_{g_e} = \mathbf{k}_g \mathbf{v}_g \qquad (12\text{-}35)$$

where $\mathbf{S}_{g_e}$ is a column of the elastic forces at, and in the direction of, the deformations $\mathbf{v}_g$ and $\mathbf{k}_g$ is a symmetric matrix of stiffness coefficients of the element. We have already noted that different choices may be made for the element deformations to be included in $\mathbf{v}_g$. Associated with a particular choice of deformations will be a related set of forces $\mathbf{S}_{g_e}$ and stiffnesses $\mathbf{k}_g$.

For generality, we introduce the possibility of internal lacks of fit in the structure due to manufacture or temperature change. We define these lacks of fits by the partitioned column matrix

$$\mathbf{J} = \{\mathbf{J}_a \quad \mathbf{J}_b \quad \cdots \quad \mathbf{J}_g \quad \cdots \quad \mathbf{J}_s\} \qquad (12\text{-}36)$$

where $\mathbf{J}_g$ is a column matrix of the forces that must be applied to the gth element to reduce the lacks of fit to zero. These forces act at, and in the direction of, the juncture forces in $\mathbf{S}_{g_e}$. The total forces on the gth element are therefore $\mathbf{S}_g = \mathbf{S}_{g_e} + \mathbf{J}_g$, or from Eq. (12-35)

$$\mathbf{S}_g = \mathbf{k}_g \mathbf{v}_g + \mathbf{J}_g \qquad (12\text{-}37)$$

By collecting these matrices for the entire structure we obtain

$$\mathbf{S} = \mathbf{k}\mathbf{v} + \mathbf{J} \qquad (12\text{-}38)$$

where

$$
\mathbf{k} = \begin{bmatrix}
\mathbf{k}_a & & & & & \\
& \mathbf{k}_b & & & & \\
& & \cdot & & & \\
& & & \cdot & & \\
& & & & \mathbf{k}_g & \\
& & & & & \cdot \\
& & & & & & \cdot \\
& & & & & & & \cdot \\
& & & & & & & & \mathbf{k}_s
\end{bmatrix}
\tag{12-39}
$$

is the *unassembled stiffness matrix* of the system. Substituting Eq. (12-33) into (12-38) gives

$$
\mathbf{S} = \mathbf{ka}_0\mathbf{r} + \mathbf{ka}_1\mathbf{U} + \mathbf{J}
\tag{12-40}
$$

It is necessary to apply a set of forces

$$
\mathbf{R} = \{R_1 \quad R_2 \quad \cdots \quad R_i \quad \cdots \quad R_m\}
$$

to the structure to maintain the specified displacements, where R_i acts at, and in the direction of, r_i.

The unspecified displacements $\mathbf{U}$ in Eq. (12-33) are determined from the condition that the internal-force system $\mathbf{S}$ must be in equilibrium with the applied-force system $\mathbf{R}$. Equilibrium exists if Eq. (6-63) is satisfied for r independent virtual displacements which do not violate the constraints of the system. The displacements $\mathbf{r}$ are prescribed; therefore the virtual displacements must vanish at these points. As a result $\delta V = 0$, and Eq. (6-63) reduces to $\delta U = 0$. As virtual displacements we take deformations that are compatible with a unit variation of each of the r independent U_i, while U_j ($j \neq i$) and $\mathbf{r}$ are held constant. The r columns of the $\mathbf{a}_1$ matrix give r sets of internal deformations, each compatible with a different one of the unit displacements. Since $\delta U = -\delta W_i = 0$, the equilibrium conditions may be written

$$
\mathbf{a}_1'\mathbf{S} = 0
\tag{12-41}
$$

where the product of the ith row of $\mathbf{a}_1'$ with $\mathbf{S}$ gives $-\delta W_i$ for the virtual-displacement system associated with $U_i = 1$. Substituting Eq. (12-40) into (12-41) gives

$$
\mathbf{C}_{11}\mathbf{U} = -\mathbf{C}_{10}\mathbf{r} - \mathbf{a}_1'\mathbf{J}
\tag{12-42}
$$

where the notation

$$
\mathbf{C}_{ij} = \mathbf{a}_i'\mathbf{ka}_j
\tag{12-43}
$$

is used. Equation (12-42) can be solved for $\mathbf{U}$ and the result substituted into Eq. (12-40) to obtain the internal forces.

If Eq. (12-42) is solved by inversion and the result is inserted into Eq. (12-33), we find

$$\mathbf{v} = \mathbf{ar} + \mathbf{dJ} \tag{12-44}$$

where

$$\mathbf{a} = \mathbf{a}_0 - \mathbf{a}_1\mathbf{C}_{11}{}^{-1}\mathbf{C}_{10} \qquad \mathbf{d} = -\mathbf{a}_1\mathbf{C}_{11}{}^{-1}\mathbf{a}_1' \tag{12-45}$$

To determine $\mathbf{R}$ we use Eq. (6-63) with the equilibrium set of external forces $\mathbf{R}$ and internal forces $\mathbf{S}$. By taking a virtual-displacement system compatible with $r_i = 1$, while r_j $(j \neq i)$ and $\mathbf{U}$ are zero, we find $\delta V = -R_i$. The ith column of $\mathbf{a}_0$ is a set of element deformations that is compatible with $r_i = 1$. The product of the transpose of this column with $\mathbf{S}$ is $-\delta W_i = \delta U$. If $-\delta V = \delta U$ is determined in this manner for a unit variation of each of the r_i, the resulting equations may be written in the form

$$\mathbf{R} = \mathbf{a}_0'\mathbf{S} \tag{12-46}$$

From Eqs. (12-40), (12-42), and (12-46) we find

$$\mathbf{R} = \mathbf{Kr} + \mathbf{a}'\mathbf{J} \tag{12-47}$$

where

$$\mathbf{K} = \mathbf{C}_{00} - \mathbf{C}_{10}'\mathbf{C}_{11}{}^{-1}\mathbf{C}_{10} \tag{12-48}$$

is the *assembled stiffness matrix* of the structure for the set of displacements $\mathbf{r}$. In deriving Eq. (12-47) we have used the identities

$$\mathbf{a}_0'\mathbf{ka}_1 = (\mathbf{a}_1'\mathbf{ka}_0)' \qquad \mathbf{C}_{10}'\mathbf{C}_{11}{}^{-1}\mathbf{a}_1' = \mathbf{a}_1\mathbf{C}_{11}{}^{-1}\mathbf{C}_{10}$$

which follow from the facts that $\mathbf{k}$ and $\mathbf{C}_{11}{}^{-1}$ are symmetric matrices and are therefore equal to their transposes. The internal forces are found from Eqs. (12-40) and (12-42), which give

$$\mathbf{S} = \mathbf{kar} + (\mathbf{I} + \mathbf{kd})\mathbf{J} \tag{12-49}$$

where $\mathbf{I}$ is the identity matrix.

In the preceding discussion we have assumed that $\mathbf{r}$ is given. In practice it is usually $\mathbf{R}$ that is specified and $\mathbf{r}$ that is unknown. In this case we find from Eq. (12-47) that

$$\mathbf{r} = \mathbf{K}^{-1}\mathbf{R} - \mathbf{K}^{-1}\mathbf{a}'\mathbf{J} \tag{12-50}$$

and from Eq. (12-49)

$$\mathbf{S} = \mathbf{kaK}^{-1}\mathbf{R} + (\mathbf{I} + \mathbf{kd} - \mathbf{kaK}^{-1}\mathbf{a}')\mathbf{J} \tag{12-51}$$

where we note that $\mathbf{K}^{-1} = \mathbf{C}$, the assembled flexibility matrix of the structure.

Equations (12-50) and (12-51) apply for the simultaneous application of load and temperature. If only loads are applied, the equations reduce to

$$\mathbf{r} = \mathbf{K}^{-1}\mathbf{R} \qquad \mathbf{S} = \mathbf{kaK}^{-1}\mathbf{R} \tag{12-52}$$

When only temperatures are applied, it is easier to determine $\mathbf{r}$ and $\mathbf{S}$ directly from Eqs. (12-42) and (12-49), which give

$$\mathbf{U} = -\mathbf{C}_{11}{}^{-1}\mathbf{a}_1'\mathbf{J} \qquad \mathbf{S} = (\mathbf{I} + \mathbf{kd})\mathbf{J} \qquad (12\text{-}53)$$

In using these equations $\mathbf{U}$, and therefore $\mathbf{a}_1'$, applies at all points where displacements are not prescribed.

We see that the displacements and internal forces in a structure can be determined by simple matrix operations once $\mathbf{a}_0$, $\mathbf{a}_1$, and $\mathbf{k}$ have been determined and $\mathbf{R}$ and $\mathbf{J}$ are specified. These matrices are discussed further in the next section.

12-6 DISCUSSION OF THE DISPLACEMENT METHOD

To clarify the concepts introduced in the last section we again refer to Figs. 12-1 and 12-2. We assume that only the single force R_1 is applied, as shown in Fig. 12-10, so that $\mathbf{r} = r_1$. To completely define the deformed state of the structure, the slope at the point of application of $\mathbf{R}_1$ must also be known. Since no moment is applied at this point, we denote this unknown slope by U_1, which gives $\mathbf{U} = U_1$.

Several choices are possible for the element deformations $\mathbf{v}$. As examples we may use the deformations in Fig. 12-2a or 12-2b to describe the deformed state. The matrices $\mathbf{a}_0$ and $\mathbf{a}_1$ depend upon the choice of $\mathbf{v}$. As an illustration we consider the deformations of Fig. 12-2b. The displacements for $r_1 = 1$, $U_1 = 0$ are shown in Fig. 12-11a, and the associated element deformations are shown in Fig. 12-11b. From these we find

$$r_1 = 1$$

$$\mathbf{a}_0 = \begin{bmatrix} 1 \\ 0 \\ -1 \\ 0 \end{bmatrix} \begin{matrix} v_1 \\ v_2 \\ v_3 \\ v_4 \end{matrix}$$

where the sign of the element a_{0_3} results from the fact that the right end of the element b undergoes a negative deflection *relative* to the left end, which has been selected as the support for the element.

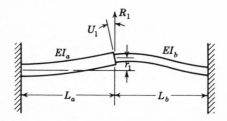

Fig. 12-10 Illustration of displacement method.

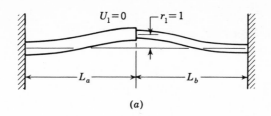

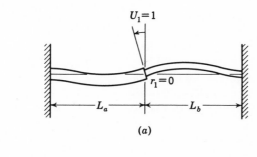

Fig. 12-11 (a) Displacements for $r_1 = 1$, $U_1 = 0$; (b) element deformations.

From Fig. 12-12, for $U_1 = 1$ and $r_1 = 0$ we find

$$U_1 = 1$$

$$\mathbf{a}_1 = \begin{bmatrix} 0 \\ 1 \\ -L_b \\ -1 \end{bmatrix} \begin{array}{l} v_1 \\ v_2 \\ v_3 \\ v_4 \end{array}$$

where we note that v_3 and v_4 are the deflection and slope of the right end of

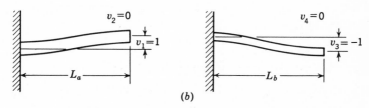

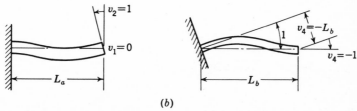

Fig. 12-12 (a) Displacements for $U_1 = 1$, $r_1 = 0$; (b) element deformations.

element b *relative* to the left end. The reader should verify that if the element deformations of Fig. 12-2a had been chosen to define the deformations,

$$r_1 = 1 \qquad\qquad U_1 = 1$$

$$\mathbf{a}_0 = \begin{bmatrix} \dfrac{1}{L_a} \\[2mm] -\dfrac{1}{L_a} \\[2mm] -\dfrac{1}{L_b} \\[2mm] \dfrac{1}{L_b} \end{bmatrix} \begin{matrix} v_1 \\[2mm] v_2 \\[2mm] v_3 \\[2mm] v_4 \end{matrix} \qquad\qquad \mathbf{a}_1 = \begin{bmatrix} 0 \\[2mm] 1 \\[2mm] -1 \\[2mm] 0 \end{bmatrix} \begin{matrix} v_1 \\[2mm] v_2 \\[2mm] v_3 \\[2mm] v_4 \end{matrix}$$

Different element forces $\mathbf{S}$ and stiffness matrices $\mathbf{k}$ are associated with the choices for $\mathbf{v}$. The element-stiffness matrices can be found from $\mathbf{k}_g = \mathbf{c}_g^{-1}$. Thus for the deformations and forces of Fig. 12-2a we find from Eq. (12-27)

$$\mathbf{k}_g = \frac{2EI_g}{L_g} \begin{bmatrix} 2 & -1 \\ -1 & 2 \end{bmatrix} \tag{12-54}$$

and from Eq. (12-26) we find for the forces and displacements of Fig. 12-2b that

$$\mathbf{k}_g = \frac{2EI_g}{L_g{}^3} \begin{bmatrix} 6 & -3L_g \\ -3L_g & 2L_g{}^2 \end{bmatrix} \tag{12-55}$$

Summaries of stiffness matrices for other simple structural elements can be found in Refs. 7 and 8.

The matrix $\mathbf{J}_g$ is the set of juncture forces that must be applied to the gth element to overcome the lacks of fit of the element. We again assume that the element a of Fig. 12-1 is subjected to a thermal gradient which results in a thermal equivalent moment M_T. The forces that must be applied to a beam element to suppress the initial displacements due to M_T are shown in Fig. 12-13 for the element-deformation systems of Fig.

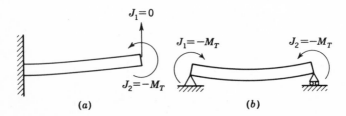

Fig. 12-13 Initial forces to suppress thermal deformations in beam elements. (a) For Fig. 12-2b; (b) for Fig. 12-2a.

12-2. Noting that $M_T = 0$ for element b, we find $\mathbf{J} = -M_T\{0 \quad 1 \quad 0 \quad 0\}$ if the deformations of Fig. 12-2b are used and $\mathbf{J} = -M_T\{1 \quad 1 \quad 0 \quad 0\}$ for the deformations of Fig. 12-2a.

Example 12-3 Solve Example 12-1 by the displacement method using the element-deformation and force system of Fig. 12-2b.

Assuming again that $L_a = L_b = L$, $EI_a = 2EI$, and $EI_b = EI$, we find from the foregoing discussion that

$$\mathbf{a}_0 = \begin{bmatrix} 1 \\ 0 \\ -1 \\ 0 \end{bmatrix} \qquad \mathbf{a}_1 = \begin{bmatrix} 0 \\ 1 \\ -L \\ -1 \end{bmatrix}$$

$$\mathbf{k} = \frac{12EI}{L^3} \begin{bmatrix} 2 & -L & 0 & 0 \\ -L & \dfrac{2L^2}{3} & 0 & 0 \\ 0 & 0 & 1 & -\dfrac{L}{2} \\ 0 & 0 & -\dfrac{L}{2} & \dfrac{L^2}{3} \end{bmatrix}$$

Substituting these into Eqs. (12-45) gives

$$\mathbf{a} = \begin{bmatrix} 1 \\ \dfrac{1}{2L} \\ -\dfrac{3}{2} \\ -\dfrac{1}{2L} \end{bmatrix} \qquad \mathbf{d} = -\frac{L}{12EI} \begin{bmatrix} 0 & 0 & 0 & 0 \\ 0 & 1 & -L & -1 \\ 0 & -L & L^2 & L \\ 0 & -1 & L & 1 \end{bmatrix}$$

From Eq. (12-48) we find that the $\mathbf{K}$ matrix has but the single element $K_{11} = 33EI/L^3$, so that $\mathbf{C} = \mathbf{K}^{-1} = L^3/33EI$.

Noting that $\mathbf{R} = R_1$ and $\mathbf{J} = -M_T\{0 \quad 1 \quad 0 \quad 0\}$, we find from Eqs. (12-50) and (12-51) that

$$\mathbf{r} = r_1 = 0.0303 \frac{R_1 L^3}{EI} + 0.0152 \frac{M_T L^2}{EI}$$

$$\mathbf{S} = \begin{bmatrix} 0.545R_1 - 0.728\dfrac{M_T}{L} \\ -0.242R_1 L - 0.455M_T \\ -0.455R_1 - 0.728\dfrac{M_T}{L} \\ 0.212R_1 L + 0.272M_T \end{bmatrix}$$

In comparing the $\mathbf{S}$ matrix with the results of Example 12-1, it must be recalled that S_1 and S_2 are defined differently in Example 12-1. However, the results can easily be shown to agree by applying equilibrium to each of the beam elements to determine the remaining juncture forces.

The slope at R_1 is U_1, which from Eq. (12-42) is found to be

$$\mathbf{U} = U_1 = 0.0152 \frac{R_1 L^2}{EI} + 0.0909 \frac{M_T L}{EI}$$

which agrees with r_2 in Example 12-1.

Example 12-4 Determine the internal forces and displacements in the pin-jointed truss illustrated in Fig. 12-14a when element 1 is subjected to a temperature change T and the applied forces are as shown.   All members have the same EA.

The elements of $\mathbf{r} = \{r_1 \quad r_2\}$, the displacements in the directions of the applied forces, and the free displacements $\mathbf{U} = \{U_1 \quad U_2\}$ are given in Fig. 12-14b.   As shown in Fig. 12-15, the element deformations $\mathbf{v}$ are taken as the elongations of the truss members, and the axial forces on the members are therefore the internal forces S.

The elements of $\mathbf{a_0}$ are obtained from geometry by determining the member deformations for $r_1 = 1$ and for $r_2 = 1$.   As an example, the geometry for $r_1 = 1$ is shown in Fig. 12-16a.   In a similar fashion, the elements of $\mathbf{a_1}$ are determined by obtaining the member deformations for $U_1 = 1$ and for $U_2 = 1$. The case of $U_2 = 1$ is shown in Fig. 12-16b.   In this manner we find

$$
\begin{array}{cc}
r_1 = 1 \quad r_2 = 1 & U_1 = 1 \quad U_2 = 1
\end{array}
$$

$$
\mathbf{a_0} =
\begin{bmatrix}
1 & 0 \\
0 & 0 \\
0 & -1 \\
0.707 & 0 \\
0 & -0.707
\end{bmatrix}
\begin{matrix}
v_1 \\ v_2 \\ v_3 \\ v_4 \\ v_5
\end{matrix}
\qquad
\mathbf{a_1} =
\begin{bmatrix}
0 & 0 \\
0 & 1 \\
1 & 0 \\
0.707 & 0 \\
0 & 0.707
\end{bmatrix}
\begin{matrix}
v_1 \\ v_2 \\ v_3 \\ v_4 \\ v_5
\end{matrix}
$$

The stiffness matrix of the gth member consists of the single element $\mathbf{k_g} = EA_g/L_g$.   As a result we find from Eq. (12-39) that

$$
\mathbf{k} = \frac{EA}{L}
\begin{bmatrix}
1 & & & & \\
 & 1 & & & \\
 & & 1 & & \\
 & & & 0.707 & \\
 & & & & 0.707
\end{bmatrix}
$$

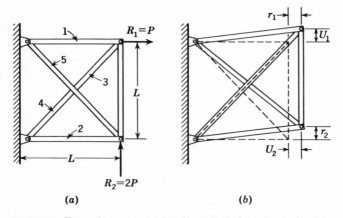

(a)   (b)

Fig. 12-14 Example 12-4.   (a) Loading; (b) designation of displacements.

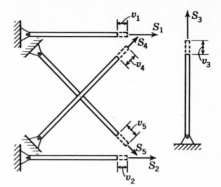

Fig. 12-15 Juncture forces and deformations, Example 12-4.

The matrix of applied loads is simply $\mathbf{R} = P\{1 \quad 2\}$. The matrices $\mathbf{C}_{00}$, $\mathbf{C}_{10}$, and $\mathbf{C}_{11}$ are determined from Eq. (12-43). With these and Eq. (12-45) we find

$$\mathbf{a} = \begin{bmatrix} 1 & 0 \\ 0 & 0.261 \\ -0.261 & -0.261 \\ 0.522 & 0.522 \\ 0 & -0.522 \end{bmatrix}$$

$$\mathbf{d} = -\frac{L}{1.354EA} \begin{bmatrix} 0 & 0 & 0 & 0 & 0 \\ 0 & 1 & 0 & 0 & 0.707 \\ 0 & 0 & 1 & 0.707 & 0 \\ 0 & 0 & 0.707 & 0.5 & 0 \\ 0 & 0.707 & 0 & 0 & 0.5 \end{bmatrix}$$

From Eq. (12-48) and $\mathbf{C} = \mathbf{K}^{-1}$ we obtain

$$\mathbf{K} = \frac{EA}{L}\begin{bmatrix} 1.261 & 0.261 \\ 0.261 & 0.522 \end{bmatrix} \qquad \mathbf{C} = \frac{L}{EA}\begin{bmatrix} 0.885 & -0.442 \\ -0.442 & 2.136 \end{bmatrix}$$

The $\mathbf{J}$ matrix is the set of axial forces that must be applied to suppress

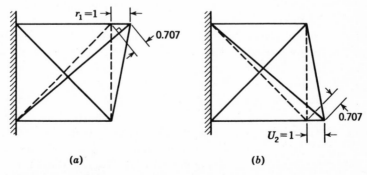

(a)   (b)

Fig. 12-16 Geometry for determining juncture deformations for
(a) $r_1 = 1$ and (b) $U_2 = 1$.

the thermal expansions of the members. For the gth element this is simply $J_g = -P_T = -(EA\alpha T)_g$, from which

$$J = -EA\alpha T\{1 \quad 0 \quad 0 \quad 0 \quad 0\}$$

The internal forces obtained from Eq. (12-51) are

$$S = P \begin{Bmatrix} 0 \\ 1.0 \\ -1.0 \\ 1.414 \\ -1.414 \end{Bmatrix} + EA\alpha T \begin{Bmatrix} -0.116 \\ -0.116 \\ -0.116 \\ 0.163 \\ 0.163 \end{Bmatrix}$$

and the displacements found from Eq. (12-50) are

$$r = \frac{PL}{EA} \begin{Bmatrix} 0 \\ 3.830 \end{Bmatrix} + \alpha TL \begin{Bmatrix} 0.885 \\ -0.442 \end{Bmatrix}$$

12-7 CONCLUDING REMARKS

By comparing the equations of Secs. 12-2 and 12-5 it is apparent that a complete duality exists between the force and displacement methods. This was first pointed out by Argyris and Kelsey,[1] who developed the equations and showed that with a change of symbols the equations for the two methods are identical. In view of this similarity it is natural to inquire about the considerations involved in choosing the method of analysis for a particular problem.

One factor to be considered is the number of unknowns involved. In general, the number of force and kinematic redundancies will be different, and some advantage is gained in using the method that involves the fewest unknowns. In the stiffened shells employed in flight vehicles, the number of kinematic redundancies is usually greater than the force redundancies. However, the number of unknowns is normally large in either case, and the question of the conditioning of the equations arises.

In the force method, special care must be exercised in choosing the redundancies to avoid ill-conditioned equations (Sec. 11-5). This requires experience and judgment or a sophisticated computer program (like that described in Ref. 9) which selects the redundancies automatically. On the other hand, the displacement method usually produces well-conditioned equations without special care.

The idealization of the structure should also be considered. In beams, frames, and trusses, both methods may be applied directly without introducing approximating assumptions about the internal stress or strain distributions within the elements. This is because the element stresses or strains are completely defined if the juncture forces or displacements are known. This is not the case for the elements of a stiffened shell, where in the force method the stress distribution was idealized as described in Sec. 7-6. In structures of this type, where the elements are continuously connected rather than being joined at discrete points, it is usually more natural to assume deformation distributions, as is done in the displacement method, than to assume stress distributions, as is done in the force method. It is for these reasons that the displacement method has found broader application to plates, shells, stiffened shells, and two- and three-dimensional elasticity problems.

Turner, Clough, Martin, and Topp have introduced an independently developed variation of the displacement method known as the *direct stiffness method*,[2,3,8,13] which has been widely used. The method is a systematic development of the procedure for deriving the stiffness matrix described in Sec. 10-5.

REFERENCES

1. Argyris, J. H., and S. Kelsey: "Energy Theorems and Structural Analysis," Butterworth & Co. (Publishers) Ltd., London, 1960.
2. de Veubeke, B. F.: "Matrix Methods of Structural Analysis," Pergamon Press, New York, 1964.
3. Turner, M. J., E. H. Dill, H. C. Martin, and R. J. Melosh: Large Deflection of Structures Subjected to Heating and External Loads, *J. Aerospace Sci.*, **27**(2): 97–106 (February, 1960).
4. Zienkiewicz, O. C., and G. S. Holister: "Stress Analysis," John Wiley & Sons, Inc., New York, 1965.
5. Wehle, L. B., Jr., and W. Lansing: A Method for Reducing the Analysis of Complex Redundant Structures to a Routine Procedure, *J. Aeron. Sci.*, **19**(10): 667–684 (October, 1952).
6. Argyris, J. H., and S. Kelsey: "Modern Fuselage Analysis and the Elastic Aircraft," Butterworth & Co. (Publishers) Ltd., London, 1963.
7. Pestel, E. C., and F. A. Leckie: "Matrix Methods in Elastomechanics," McGraw-Hill Book Company, New York, 1963.
8. Martin, H. C.: "Introduction to Matrix Methods of Structural Analysis," McGraw-Hill Book Company, New York, 1966.
9. Denke, P. H.: A General Digital Computer Analysis of Statically Indeterminate Structures, *NASA Tech. Note* D-1666, December, 1962.
10. Percy, J. H., T. H. Pian, S. Klein, and D. R. Navaratna: Application of Matrix Displacement Method to Linear Elastic Analysis of Shells of Revolution, *AIAA J.*, **3**(11): 2138–2145 (November, 1965).
11. Argyris, J. H.: Matrix Analysis of Three-dimensional Elastic Media, Small and Large Displacements, *AIAA J.*, **3**(1): 45–51 (January, 1965).
12. Wilson, E. L.: Stress Analysis of Axisymmetric Solids, *AIAA J.*, **3**(12): 2269–2274 (December, 1965).
13. Turner, M. J., R. W. Clough, H. C. Martin, and L. J. Topp: Stiffness and Deflection Analysis of Complex Structures, *J. Aeron. Sci.*, **23**(9): 805–824 (September, 1956).
14. Argyris, J. H.: On the Analysis of Complex Elastic Structures, *Appl. Mech. Rev.*, **11**(7): 331–338 (July, 1958).

15. Gallagher, R. H.: "A Correlation Study of Methods of Matrix Structural Analysis," Pergamon Press, New York, 1964.
16. Bruhn, E. F.: "Analysis and Design of Flight Vehicle Structures," Tri-state Offset Co., Cincinnati, Ohio, 1965.
17. Przemieniecki, J. S., et al.: Matrix Methods in Structural Mechanics, *Proc. Conf. Wright-Patterson Air Force Base, Ohio,* Oct. 26–28, 1965, AFFDL-TR-66-80, 1966.
18. Zienkiewicz, O. C., and Y. K. Chueng: "The Finite Element Method in Structural and Continuum Mechanics," McGraw-Hill Book Company, New York, 1967.
19. Przemieniecki, J. S.: "Theory of Matrix Structural Analysis," McGraw-Hill Book Company, New York, 1968.

PROBLEMS

12-1. Rework Example 12-1 by the force method using the member forces of Fig. 12-2*b*.

12-2. Use the force method to write the matrix equations for the deflections and element forces in Example 12-2 when the temperature of the upper longitudinal of the front spar is given an increase which varies linearly from 0°F at the tip to 300°F at the root. Assume $\alpha = 12.9 \times 10^{-6}$ in./(in.)(°F).

12-3. Rework Example 12-4 using the force method.

12-4. Rework Example 12-1 by the displacement method using the member displacements of Fig. 12-2*a*.

12-5. Use the force method to determine the internal forces in the pin-jointed truss shown. Find the displacements at the applied loads. All members have an extensional stiffness of $EA = 10^7$ lb.

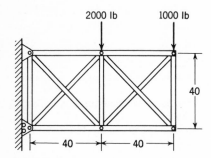

Fig. P12-5

12-6. Determine the internal forces in the frame shown and the deflection of P by the force method. Assume EI is constant for all members and neglect the axial flexibility of the members.

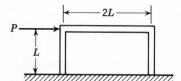

Fig. P12-6

12-7. Work Prob. 12-6 by the displacement method.

12-8. Use the force method to determine the flexibility matrix of the cantilevered structure shown.

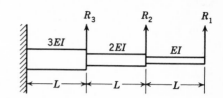

Fig. 12-8

12-9. Determine the member forces in the structure shown by the force method.   The longitudinal and transverse stiffeners have cross-sectional areas of 1 in.^{2}.   The elastic properties are $E = 10^7$ psi and $G = 4 \times 10^6$ psi.

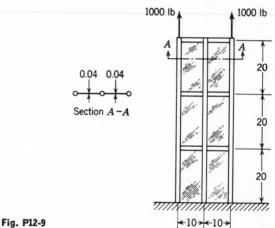

Fig. P12-9

12-10. Determine the member forces in Prob. 12-9 when the temperatures of the edge longitudinals are increased by 100°F if $\alpha = 12.9 \times 10^{-6}$ in./(in.)(°F).

13

The Bending and Extension
of Thin Plates

13-1 INTRODUCTION

An approximate theory for the bending and extension of slender beams
was derived in Chap. 7 to avoid the difficulties involved in solving the
equations of the theory of elasticity. In this chapter we generalize the
concepts of beam theory to two dimensions to obtain an approximate
theory for thin planar bodies, as illustrated in Fig. 13-1. We refer to
these bodies as *plates* when they are capable of resisting bending and as
membranes when they have negligible bending rigidity. The theory is
developed in a general form that includes the effects of nonhomogeneity,
thermal gradients, in-plane forces, and large lateral deflections.

As in beam theory, the deformation pattern is assumed. Where in
beam theory it is postulated that normals to the modulus-weighted centroidal axis remain normal to the axis and unchanged in length as the
beam deforms, in plate theory it is assumed that normals to an initially
plane modulus-weighted centroidal surface remain normal to the surface
and unchanged in length as the plate deforms. Stresses in the transverse
directions are neglected in beam theory, and in plate theory the stresses
in the direction of the thickness are ignored. This results in a uniaxial
stress-strain relationship for the beam and a biaxial relationship for the
plate.

In both theories equilibrium is satisfied only in the gross sense of
stipulating that the stress resultants must be in equilibrium with the

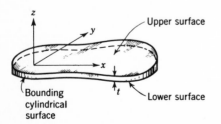

Fig. 13-1   Geometry of a plate.

applied forces, rather than requiring equilibrium of the stresses and body forces acting on an infinitesimal element, as in the theory of elasticity. Similarly, the boundary conditions are imposed only on a macroscopic scale rather than on the microscopic scale of elasticity theory.

Further information on the theory, which is based upon the contributions of Bernoulli, Navier, Kirchhoff, and von Kármán, may be found in Refs. 1 to 7. The theory neglects the effects of transverse shearing deformations and is therefore applicable only to thin plates. A more accurate theory which includes these effects has been developed by Reissner.[1,8]

13-2 GEOMETRY OF THE REFERENCE SURFACE

A plate occupies the domain enclosed by its upper and lower surfaces and a bounding cylindrical surface, as shown in Fig. 13-1. We shall find that in nonhomogeneous plates it is convenient to measure the z coordinate from a modulus-weighted reference plane. To locate this plane we assume an initial set of axes (x,y,z_0) such that the z_0 axis is in the direction of the thickness of the plate. Consider the points on a line AB that is parallel to z_0, where A and B have coordinates (x,y,z_{0_u}) and (x,y,z_{0_l}), as shown in Fig. 13-2. We weight each point on AB by E/E_1, the ratio of the modulus at the point to an arbitrary reference modulus E_1. The distance from the xy plane to the modulus-weighted centroid of the line AB is

$$\bar{z}_0(x,y) = \frac{1}{t^*} \int_{z_{0_l}(x,y)}^{z_{0_u}(x,y)} z_0 \, dz_0^* \qquad (13\text{-}1)$$

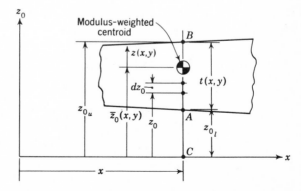

Fig. 13-2 Determination of modulus-weighted reference plane.

where $dz_0^* = (E/E_1)\,dz_0$, and t^* is the *modulus-weighted thickness*

$$t^* = \int_{z_{0l}(x,y)}^{z_{0u}(x,y)} dz_0^* \qquad (13\text{-}2)$$

The loci of points $(x,y,\bar{z}_0)$ establish a *modulus-weighted reference surface*. When this surface is curved, the body is called a *shell*, and when the surface is plane, the body is called a *plate*.† If we take a new set of axes (x,y,z) in which z is measured from the reference plane, the distance $\bar{z}$ from the new axes to the reference plane is zero, and it follows from Eq. (13-1) that

$$\int_{z_l(x,y)}^{z_u(x,y)} z\,dz^* = 0 \qquad (13\text{-}3)$$

In the remainder of the chapter we shall use the (x,y,z) coordinates to locate points in the plate. We note that when the plate is homogeneous, $t^* = t$, and the reference surface is the middle surface of the plate.

In general, the initially plane reference surface deforms into a curved surface when the plate is loaded and heated. We shall express the displacements of an arbitrary point (x,y,z) in terms of $u(x,y)$, $v(x,y)$, and $w(x,y)$, the displacements of a corresponding point $(x,y,0)$ in the reference plane (Fig. 13-3).

Let us consider some useful geometric properties of the deformed reference surface. The slopes of this surface in the x and y directions are $\partial w/\partial x$ and $\partial w/\partial y$. Applying the chain rule for derivatives, we find that the slope in an arbitrary x' direction is

$$\frac{\partial w}{\partial x'} = \frac{\partial w}{\partial x}\frac{dx}{dx'} + \frac{\partial w}{\partial y}\frac{dy}{dx'}$$

† A plate theory applicable for small initial deviations from a plane is given in Ref. 9.

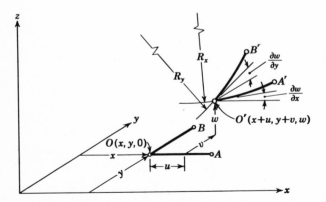

Fig. 13-3   Geometry of deformed reference surface.

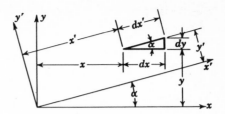

Fig. 13-4 Rotation of x and y coordinate axes.

From Fig. 13-4, this may be written

$$\frac{\partial w}{\partial x'} = \frac{\partial w}{\partial x} \cos \alpha + \frac{\partial w}{\partial y} \sin \alpha \qquad (13\text{-}4)$$

To find α_1, the angle for the maximum or minimum slope at a point, we differentiate Eq. (13-4) with respect to α and equate the result to zero. This gives

$$\tan \alpha_1 = \frac{\partial w/\partial y}{\partial w/\partial x} \qquad (13\text{-}5)$$

which has two solutions that differ by 180°. The triangle in Fig. 13-5 is constructed from Eq. (13-5). Substituting the trigonometric results from this figure into Eq. (13-4), we find

$$\left(\frac{\partial w}{\partial x'}\right)_{\text{max, min}} = \pm \sqrt{\left(\frac{\partial w}{\partial x}\right)^2 + \left(\frac{\partial w}{\partial y}\right)^2} \qquad (13\text{-}6)$$

Next let us consider the curvatures of the deformed surface. A point O with coordinates (x,y) in the reference plane undergoes displacements u, v, and w to O' in the deflected reference surface. A plane through O' that is parallel to the xz plane will intersect the reference surface in a curved line $O'A'$, and a plane through O' that is parallel to the yz plane will intersect in a curve $O'B'$ (Figs. 13-3 and 13-6). From calculus, the curvatures of $O'A'$ and $O'B'$ are

$$\frac{1}{R_x} = \frac{\partial^2 w/\partial x^2}{[1 + (\partial w/\partial x)^2]^{3/2}} \qquad \frac{1}{R_y} = \frac{\partial^2 w/\partial y^2}{[1 + (\partial w/\partial y)^2]^{3/2}} \qquad (13\text{-}7)$$

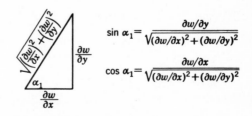

$$\sin \alpha_1 = \frac{\partial w/\partial y}{\sqrt{(\partial w/\partial x)^2 + (\partial w/\partial y)^2}}$$

$$\cos \alpha_1 = \frac{\partial w/\partial x}{\sqrt{(\partial w/\partial x)^2 + (\partial w/\partial y)^2}}$$

Fig. 13-5 Trigonometric relationships for maximum or minimum slope.

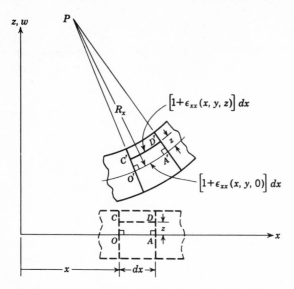

Fig. 13-6 Deformation geometry for thin-plate theory.

where R_x and R_y are the radii of curvature in the x and y directions. These are defined positive when the centers of curvature are in the positive w direction. A surface which has both centers of curvature on the same side of the surface is called *synclastic*, and one with the centers on opposite sides is called *anticlastic*.

We assume that the lateral deflections are small enough to neglect $(\partial w/\partial x)^2$ and $(\partial w/\partial y)^2$ compared to unity. Equations (13-7) then become

$$\frac{1}{R_x} = \frac{\partial^2 w}{\partial x^2} \qquad \frac{1}{R_y} = \frac{\partial^2 w}{\partial y^2} \tag{13-8}$$

The curvature in an arbitrary x' direction is

$$\frac{1}{R_{x'}} = \frac{\partial^2 w}{(\partial x')^2} = \frac{\partial}{\partial x'}\frac{\partial w}{\partial x'}$$

Utilizing the operator for $\partial/\partial x'$ from Eq. (13-4), we find

$$\frac{1}{R_{x'}} = \frac{\partial^2 w}{\partial x^2}\cos^2\alpha + \frac{\partial^2 w}{\partial y^2}\sin^2\alpha + 2\frac{\partial^2 w}{\partial x\,\partial y}\sin\alpha\cos\alpha \tag{13-9}$$

We see from Fig. 13-7 that $\partial^2 w/(\partial x\,\partial y)$ is the rate of change in the x direction of the slope in the y direction and note that

$$\frac{\partial^2 w}{\partial x\,\partial y} = \frac{\partial^2 w}{\partial y\,\partial x}$$

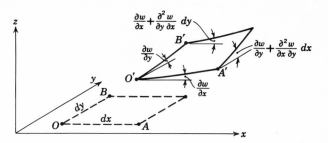

Fig. 13-7   Twist of the reference surface.

because the surface is continuous.   We define this term as the *twist* of the surface and denote it by the symbol $1/R_{xy}$, so that

$$\frac{1}{R_{xy}} = \frac{\partial^2 w}{\partial x \, \partial y} \tag{13-10}$$

Substituting Eqs. (13-8) and (13-10) into Eq. (13-9), we find

$$\frac{1}{R_{x'}} = \frac{1}{R_x} \cos^2 \alpha + \frac{1}{R_y} \sin^2 \alpha + \frac{2}{R_{xy}} \sin \alpha \cos \alpha \tag{13-11}$$

The slope of the surface in the y' direction is found by replacing α by $\alpha + \pi/2$ in Eq. (13-4), which gives

$$\frac{\partial w}{\partial y'} = - \frac{\partial w}{\partial x} \sin \alpha + \frac{\partial w}{\partial y} \cos \alpha \tag{13-12}$$

The twist with respect to the x' and y' axes is

$$\frac{1}{R_{x'y'}} = \frac{\partial}{\partial x'} \frac{\partial w}{\partial y'}$$

By using the operators for $\partial/\partial x'$ and $\partial/\partial y'$ from Eqs. (13-4) and (13-12)

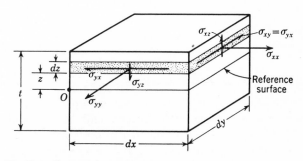

Fig. 13-8   Stresses in a plate.

and Eqs. (13-8) and (13-10), this becomes

$$\frac{1}{R_{x'y'}} = - \left(\frac{1}{R_x} - \frac{1}{R_y} \right) \sin \alpha \cos \alpha + \frac{1}{R_{xy}} (\cos^2 \alpha - \sin^2 \alpha) \quad (13\text{-}13)$$

Equations (13-11) and (13-13) are in a form that is identical to that of Eqs. (2-14a) and (2-14b). As a result, the equations of Sec. 2-5 can be applied to the curvatures and twist if we replace the normal stresses by the curvatures and the shearing stresses by the twist. From this we note that a set of orthogonal principal directions exists for the plate and that the curvatures have their maximum and minimum values in these directions while the twist is zero. The maximum twist occurs in directions that are at 45° to the principal axes. Since the transformation equations for rotation of axes are identical to those for plane stress, we can also obtain the principal curvatures and twist from a Mohr's circle construction.

13-3 STRESS RESULTANTS

The positively defined stresses that act upon an infinitesimal element of dimensions dx, dy, and dz are shown in Fig. 13-8. In plate theory it is convenient to use stress resultants *per unit length*, defined by the following equations:

$$N_x = \int_t \sigma_{xx} \, dz \qquad N_y = \int_t \sigma_{yy} \, dz \qquad N_{xy} = N_{yx} = \int_t \sigma_{xy} \, dz \tag{13-14}$$

$$M_x = - \int_t \sigma_{xx} z \, dz \qquad M_y = - \int_t \sigma_{yy} z \, dz \qquad M_{xy} = - M_{yx} = - \int_t \sigma_{xy} z \, dz \tag{13-15}$$

$$Q_x = - \int_t \sigma_{xz} \, dz \qquad Q_y = - \int_t \sigma_{yz} \, dz \tag{13-16}$$

where the integrals extend over the thickness t. The positive directions for the stress resultants that are consistent with Fig. 13-8 and Eqs. (13-14) to (13-16) are shown in Fig. 13-9. We note from this figure that M_{xy} and M_{yx} are twisting moments.

13-4 EQUILIBRIUM EQUATIONS

We assume that the plate has an applied loading *per unit area* of the reference surface. The components p_x, p_y, and p_z of this loading are parallel to, and positive in the directions of, the coordinate axes. In addition, the plate is subjected to a temperature change $T(x,y,z)$. As a result of the loads and temperatures, stresses are developed that lead to

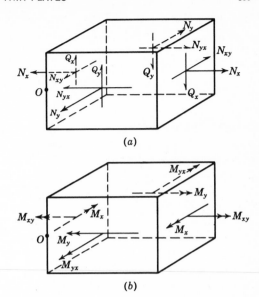

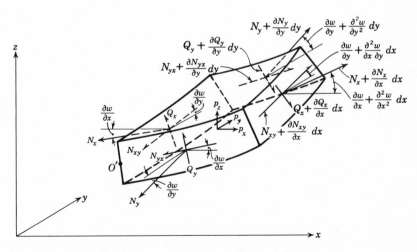

Fig. 13-9 Sign conventions for stress resultants. (*a*) Forces; (*b*) moments.

the stress resultants of Eqs. (13-14) to (13-16). Rather than requiring the stresses to satisfy the equations of equilibrium from the theory of elasticity, we stipulate only that the stress resultants be in equilibrium with the applied loads. Consider the element of Fig. 13-9, where the point O has the coordinates $(x,y,0)$. The applied forces and the reacting stress resultants that act upon this element in its deformed position are shown in Figs. 13-10 and 13-11. It is assumed that the components p_x,

Fig. 13-10 Free-body diagram of forces on a differential element.

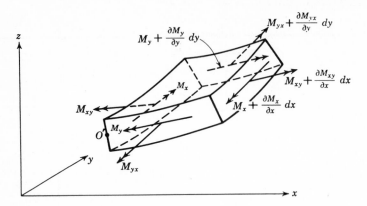

Fig. 13-11   Free-body diagram of moments on a differential element.

p_y, and p_z remain parallel to the x, y, and z axes as the plate deforms.

In writing the equilibrium equations we may assume that for small angles the cosine is unity and the sine equals the angle.   Summing forces in the x direction in Fig. 13-10, we find

$$-N_x\,dy + \left(N_x + \frac{\partial N_x}{\partial x}\,dx\right)dy - N_{yx}\,dx + \left(N_{yx} + \frac{\partial N_{yx}}{\partial y}\,dy\right)dx$$
$$+ p_x\,dx\,dy = 0$$

Simplifying this equation and following the same procedure for the y direction, we obtain

$$\frac{\partial N_x}{\partial x} + \frac{\partial N_{xy}}{\partial y} + p_x = 0 \qquad \frac{\partial N_{yx}}{\partial x} + \frac{\partial N_y}{\partial y} + p_y = 0 \qquad (13\text{-}17)$$

Equating moments about an axis that is parallel to the z axis and passes through the center of the element, we find $N_{xy} = N_{yx}$, a result that was already noted in Eqs. (13-14).

Summing moments about an axis that passes through the center of the element and is parallel to the y axis, we find

$$M_x\,dy - \left(M_x + \frac{\partial M_x}{\partial x}\,dx\right)dy + Q_x\,dy\,\frac{dx}{2} + \left(Q_x + \frac{\partial Q_x}{\partial x}\,dx\right)dy\,\frac{dx}{2}$$
$$- M_{yx}\,dx + \left(M_{yx} + \frac{\partial M_{yx}}{\partial y}\,dy\right)dx = 0$$

Following the same procedure for moment equilibrium about an axis parallel to the x axis, simplifying, and taking the limits as dx and dy approach zero, we obtain

$$\frac{\partial M_x}{\partial x} - \frac{\partial M_{yx}}{\partial y} - Q_x = 0 \qquad \frac{\partial M_{xy}}{\partial x} + \frac{\partial M_y}{\partial y} - Q_y = 0 \qquad (13\text{-}18)$$

Summing forces in the z direction leads to

$$\frac{\partial Q_x}{\partial x} + \frac{\partial Q_y}{\partial y} - N_x \frac{\partial^2 w}{\partial x^2} - N_y \frac{\partial^2 w}{\partial y^2} - 2N_{xy} \frac{\partial^2 w}{\partial x \, \partial y} - \frac{\partial N_x}{\partial x} \frac{\partial w}{\partial x} - \frac{\partial N_y}{\partial y} \frac{\partial w}{\partial y}$$
$$- \frac{\partial N_{yx}}{\partial y} \frac{\partial w}{\partial x} - \frac{\partial N_{xy}}{\partial x} \frac{\partial w}{\partial y} - p_z = 0$$

which, by using Eqs. (13-17) and (13-18), becomes

$$\frac{\partial^2 M_x}{\partial x^2} + 2 \frac{\partial^2 M_{xy}}{\partial x \, \partial y} + \frac{\partial^2 M_y}{\partial y^2} = p_z + N_x \frac{\partial^2 w}{\partial x^2} + N_y \frac{\partial^2 w}{\partial y^2} + 2N_{xy} \frac{\partial^2 w}{\partial x \, \partial y}$$
$$- p_x \frac{\partial w}{\partial x} - p_y \frac{\partial w}{\partial y} \quad (13\text{-}19)$$

There are eight stress resultants in Eqs. (13-14) to (13-16) but there are only five equilibrium equations that are not identically satisfied. Therefore, the stress resultants are statically indeterminate, and deformations must be considered.

13-5 STRAIN–DISPLACEMENT AND COMPATABILITY EQUATIONS

By assuming that normals to the reference surface remain normal to the surface and unchanged in length as the plate deforms we in effect assume that $\epsilon_{zz} = \epsilon_{yz} = \epsilon_{zz} = 0$. As a result, lateral shearing deformations are neglected. With these assumptions, the displacements and strains at all points in the plate can be determined in terms of the reference surface displacements u, v, and w.

The displacements of a point C with coordinates (x,y,z) are equal to the displacements at a corresponding point O in the reference plane with coordinates $(x,y,0)$ plus the displacements of C relative to O. Similarly, the strains at C are equal to the strains at O plus the strains at C relative to those at O. The strains in the reference can be found from the large-displacement equations (2-22a), (2-22b), and (2-22d). In using these equations we introduce the von Kármán assumption that the plate is more flexible in the transverse direction than in the in-plane directions, so that w is large compared to u and v. For this reason we neglect the terms that involve squares of the derivatives of u and v and find the following equations for strains at points in the reference surface:

$$\epsilon_{xx}(x,y,0) = \frac{\partial u}{\partial x} + \frac{1}{2}\left(\frac{\partial w}{\partial x}\right)^2 \qquad \epsilon_{yy}(x,y,0) = \frac{\partial v}{\partial y} + \frac{1}{2}\left(\frac{\partial w}{\partial y}\right)^2 \quad (13\text{-}20)$$

$$\epsilon_{xy}(x,y,0) = \frac{\partial v}{\partial x} + \frac{\partial u}{\partial y} + \frac{\partial w}{\partial x} \frac{\partial w}{\partial y} \quad (13\text{-}21)$$

These equations contain nonlinear terms involving squares and products of the derivatives of w. We shall find that these terms can be neglected when the deflected reference surface is a developable surface or when w is small relative to t.

To find $\epsilon_{xx}(x,y,z)$ we refer to Fig. 13-6, which shows normals to the reference surface, OC and AD, before and after deformation. The original length of OA is dx, and the points C and D are a distance z above the reference surface. In the deformed position the lengths of the line segments $O'A'$ and $C'D'$ are $[1 + \epsilon_{xx}(x,y,0)]\,dx$ and $[1 + \epsilon_{xx}(x,y,z)]\,dx$, respectively. From the similarity of the circular sectors $O'PA'$ and $C'PD'$ we find $O'P/C'P = O'A'/C'D'$, or

$$\frac{R_x}{R_x - z} = \frac{[1 + \epsilon_{xx}(x,y,0)]\,dx}{[1 + \epsilon_{xx}(x,y,z)]\,dx}$$

which upon rearrangement becomes

$$\epsilon_{xx}(x,y,z) = \epsilon_{xx}(x,y,0) - \frac{z}{R_x} - \frac{z}{R_x}\,\epsilon_{xx}(x,y,0) \tag{13-22}$$

We note that $\epsilon_{xx}(x,y,0)$ and z/R_x are both small, so that the last term on the right can be neglected relative to the first two terms. As a result of this simplification, we find from Eqs. (13-8), (13-20), and (13-22) that

$$\epsilon_{xx} = \frac{\partial u}{\partial x} - z\frac{\partial^2 w}{\partial x^2} + \frac{1}{2}\left(\frac{\partial w}{\partial x}\right)^2 \tag{13-23a}$$

and in a similar manner

$$\epsilon_{yy} = \frac{\partial v}{\partial y} - z\frac{\partial^2 w}{\partial y^2} + \frac{1}{2}\left(\frac{\partial w}{\partial y}\right)^2 \tag{13-23b}$$

To determine $(\epsilon_{xy})_{\text{rel}}$, the shearing strain at C relative to O, we refer to Fig. 13-12. From this figure $(\epsilon_{xy})_{\text{rel}} = -(\alpha + \beta)$, where the minus sign indicates an increase in the angle ECD. Determining the angles α and β from the figure, we find

$$(\epsilon_{xy})_{\text{rel}} = -\frac{z\left(\dfrac{\partial w}{\partial y} + \dfrac{\partial^2 w}{\partial x\,\partial y}\,dx\right) - z\dfrac{\partial w}{\partial y}}{[1 + \epsilon_{xx}(x,y,z)]\,dx} - \frac{z\left(\dfrac{\partial w}{\partial x} + \dfrac{\partial^2 w}{\partial y\,\partial x}\,dy\right) - z\dfrac{\partial w}{\partial x}}{[1 + \epsilon_{yy}(x,y,z)]\,dy}$$

which for $\epsilon_{xx}(x,y,z) \ll 1$ and $\epsilon_{yy}(x,y,z) \ll 1$ reduces to

$$(\epsilon_{xy})_{\text{rel}} = -2z\frac{\partial^2 w}{\partial x\,\partial y}$$

Adding this to Eq. (13-21), we find that the total shear strain at C is

$$\epsilon_{xy} = \frac{\partial v}{\partial x} + \frac{\partial u}{\partial y} - 2z\frac{\partial^2 w}{\partial x\,\partial y} + \frac{\partial w}{\partial x}\frac{\partial w}{\partial y} \tag{13-23c}$$

It is usually convenient to formulate the plate problem in terms of the lateral-displacement function $w(x,y)$ and a stress function $F(x,y)$

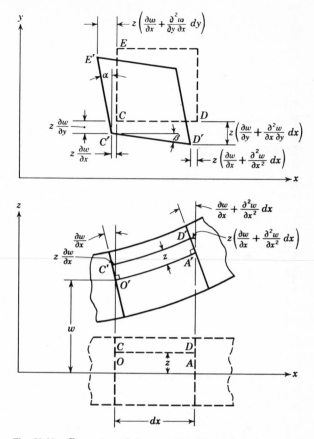

Fig. 13-12 Geometry of shear strains due to w.

rather than in terms of the three displacement functions u, v, and w. When this is done, we must introduce a compatibility condition to ensure that the stress function gives compatible strains. Equation (2-27a) may not be used because it was derived for infinitesimal strains, which do not contain the nonlinear terms in Eqs. (13-23). From Eqs. (13-23) we find

$$\frac{\partial^2 \epsilon_{yy}}{\partial x^2} + \frac{\partial^2 \epsilon_{xx}}{\partial y^2} = \frac{\partial^2 \epsilon_{xy}}{\partial x\, \partial y} + \left(\frac{\partial^2 w}{\partial x\, \partial y}\right)^2 - \frac{\partial^2 w}{\partial x^2} \frac{\partial^2 w}{\partial y^2} \qquad (13\text{-}24)$$

which reduces to Eq. (2-27a) in cases where the nonlinear terms may be neglected.

Equation (13-24) is insufficient to determine the three strain components, and it is necessary to use both equilibrium and compatibility equations to obtain a solution. This involves writing Eq. (13-24) in terms of the stress resultants, which requires the adoption of a stress-strain law.

13-6 STRESS–STRAIN EQUATIONS

Ideally we should use the three-dimensional stress-strain relations [Eqs. (3-18)]; however, these are inconsistent with our assumed deformation pattern for the plate. If we assume that the material is linearly elastic and isotropic in the xy plane (the extension of the theory to orthotropic plates is briefly described in Sec. 16-5), the generalized Hooke's law equations (3-17) become

$$\epsilon_{xx} = \frac{1}{E} \left(\sigma_{xx} - \nu\sigma_{yy} \right) - \frac{\nu_z}{E_z} \sigma_{zz} + \alpha T$$

$$\epsilon_{yy} = \frac{1}{E} \left(\sigma_{yy} - \nu\sigma_{xx} \right) - \frac{\nu_z}{E_z} \sigma_{zz} + \alpha T$$

$$\epsilon_{zz} = \frac{\sigma_{zz}}{E_z} - \frac{\nu_z}{E} \left(\sigma_{xx} + \sigma_{yy} \right) + \alpha_z T$$

$$\epsilon_{xy} = \frac{\sigma_{xy}}{G} \qquad \epsilon_{yz} = \frac{\sigma_{yz}}{G_z} \qquad \epsilon_{zx} = \frac{\sigma_{zx}}{G_z}$$

(13-25)

where E_z is the modulus in the z direction. The symbols ν_z and G_z denote the Poisson's ratio and shear modulus existing between the z direction and directions in the xy plane. In Sec. 13-5 we noted that assuming that normals to the reference surface remain normal and unchanged in length is equivalent to taking $\epsilon_{zz} = \epsilon_{yz} = \epsilon_{zz} = 0$. We see from Eqs. (13-25) that for this to be generally true the plate must have the properties of a fictitious material with $\nu_z = \alpha_z = 0$ and $E_z = G_z = \infty$. In this case, Eqs. (13-25) reduce to the first three plane-stress relations in Eqs. (3-20).

While it appears that assuming these unusual elastic properties would give erroneous results, in reality the errors are small for thin plates. This is because σ_{zz} (which is no greater than the lateral pressure on the plate) is small compared to σ_{xx} and σ_{yy}, and the shear deformations due to ϵ_{zz} and ϵ_{yz} are small compared to the bending deformations in a thin plate. The assumed stress-strain equations are exact when the loading is such that $\sigma_{xz} = \sigma_{yz} = \sigma_{zz} = 0$, since Eqs. (3-18) reduce to Eqs. (3-20) when this is the case. This situation occurs when the only loads are edge moments.

The stresses can be expressed in terms of displacements by substituting Eqs. (13-23) into the first three of Eqs. (3-20), to give

$$\sigma_{xx} = \frac{E}{1 - \nu^2} \left\{ \frac{\partial u}{\partial x} + \nu \frac{\partial v}{\partial y} - z \left(\frac{\partial^2 w}{\partial x^2} + \nu \frac{\partial^2 w}{\partial y^2} \right) \right.$$
$$\left. + \frac{1}{2} \left[\left(\frac{\partial w}{\partial x} \right)^2 + \nu \left(\frac{\partial w}{\partial y} \right)^2 \right] \right\} - \frac{E\alpha T}{1 - \nu} \quad (13\text{-}26a)$$

$$\sigma_{yy} = \frac{E}{1 - \nu^2} \left\{ \frac{\partial v}{\partial y} + \nu \frac{\partial u}{\partial x} - z \left(\frac{\partial^2 w}{\partial y^2} + \nu \frac{\partial^2 w}{\partial x^2} \right) \right.$$
$$\left. + \frac{1}{2} \left[\left(\frac{\partial w}{\partial y} \right)^2 + \nu \left(\frac{\partial w}{\partial x} \right)^2 \right] \right\} - \frac{E\alpha T}{1 - \nu} \quad (13\text{-}26b)$$

$$\sigma_{xy} = \frac{E}{2(1 + \nu)} \left(\frac{\partial v}{\partial x} + \frac{\partial u}{\partial y} - 2z \frac{\partial^2 w}{\partial x \, \partial y} + \frac{\partial w}{\partial x} \frac{\partial w}{\partial y} \right) \quad (13\text{-}26c)$$

13-7 FORMULATIONS OF THE PLATE EQUATIONS

The in-plane stress resultants can be expressed in terms of displacements by substituting Eqs. (13-26) into Eqs. (13-14) and using Eqs. (13-2) and (13-3). This gives

$$N_x = K^* \left\{ \frac{\partial u}{\partial x} + \nu \frac{\partial v}{\partial y} + \frac{1}{2} \left[\left(\frac{\partial w}{\partial x} \right)^2 + \nu \left(\frac{\partial w}{\partial y} \right)^2 \right] \right\} - N_T \quad (13\text{-}27a)$$

$$N_y = K^* \left\{ \frac{\partial v}{\partial y} + \nu \frac{\partial u}{\partial x} + \frac{1}{2} \left[\left(\frac{\partial w}{\partial y} \right)^2 + \nu \left(\frac{\partial w}{\partial x} \right)^2 \right] \right\} - N_T \quad (13\text{-}27b)$$

$$N_{xy} = \frac{K^*(1 - \nu)}{2} \left(\frac{\partial v}{\partial x} + \frac{\partial u}{\partial y} + \frac{\partial w}{\partial x} \frac{\partial w}{\partial y} \right) \quad (13\text{-}27c)$$

where

$$K^* = \frac{E_1 t^*}{1 - \nu^2} \quad (13\text{-}28)$$

is the extensional stiffness per unit length of the plate, which corresponds to $E_1 A^*$ for a beam, and

$$N_T = \frac{1}{1 - \nu} \int_t E\alpha T \, dz \quad (13\text{-}29)$$

is the thermal equivalent of a normal force per unit length of the plate. In deriving the preceding equations it is assumed that ν is constant, which is usually reasonable since ν differs only slightly for most structural materials.

Substituting Eqs. (13-26) into Eqs. (13-15) and using Eq. (13-3), we find

$$M_x = D^* \left(\frac{\partial^2 w}{\partial x^2} + \nu \frac{\partial^2 w}{\partial y^2} \right) - M_T \quad (13\text{-}30a)$$

$$M_y = D^* \left(\frac{\partial^2 w}{\partial y^2} + \nu \frac{\partial^2 w}{\partial x^2} \right) - M_T \tag{13-30b}$$

$$M_{xy} = D^*(1 - \nu) \frac{\partial^2 w}{\partial x \, \partial y} \tag{13-30c}$$

where the modulus-weighted bending rigidity per unit length of the plate

$$D^* = \frac{E_1}{1 - \nu^2} \int_t z^2 \, dz^* \tag{13-31}$$

corresponds to $E_1 I^*$ for a beam. The thermal equivalent bending moment M_T is defined by

$$M_T = -\frac{1}{1 - \nu} \int_t E\alpha T z \, dz \tag{13-32}$$

We note that when the plate is homogeneous, $t^* = t$, so that K^* and D^* reduce to K and D, defined by

$$K = \frac{Et}{1 - \nu^2} \qquad D = \frac{Et^3}{12(1 - \nu^2)} \tag{13-33}$$

The stresses can be expressed in terms of the stress resultants by combining Eqs. (13-26), (13-27), and (13-30). This gives

$$\sigma_{xx} = \frac{E}{E_1} \left(\frac{N_x^*}{t^*} - \frac{M_x^* z}{\bar{I}^*} - \frac{E_1 \alpha T}{1 - \nu} \right) \tag{13-34a}$$

$$\sigma_{yy} = \frac{E}{E_1} \left(\frac{N_y^*}{t^*} - \frac{M_y^* z}{\bar{I}^*} - \frac{E_1 \alpha T}{1 - \nu} \right) \tag{13-34b}$$

$$\sigma_{xy} = \frac{E}{E_1} \left(\frac{N_{xy}}{t^*} - \frac{M_{xy} z}{\bar{I}^*} \right) \tag{13-34c}$$

where

$$N_x^* = N_x + N_T \qquad N_y^* = N_y + N_T \tag{13-35a}$$

$$M_x^* = M_x + M_T \qquad M_y^* = M_y + M_T \tag{13-35b}$$

and $\bar{I}^*$ is the modulus-weighted moment of inertia per unit length of the plate, defined by

$$\bar{I}^* = \int_t z^2 \, dz^* \tag{13-36}$$

When the plate is homogeneous, $\bar{I}^* = \bar{I} = t^3/12$. The similarity of Eqs. (13-34a) and (13-34b) to Eq. (7-23) for stresses in a beam is apparent.

The strains can be written in terms of the stress resultants by substituting Eqs. (13-34) into the first, second, and fourth of Eqs. (3-19), which gives

$$\epsilon_{xx} = \frac{1}{E_1 t^*} (N_x^* - \nu N_y^*) - \frac{z}{E_1 \bar{I}^*} (M_x^* - \nu M_y^*) \qquad (13\text{-}37a)$$

$$\epsilon_{yy} = \frac{1}{E_1 t^*} (N_y^* - \nu N_x^*) - \frac{z}{E_1 \bar{I}^*} (M_y^* - \nu M_x^*) \qquad (13\text{-}37b)$$

$$\epsilon_{xy} = 2(1 + \nu) \left(\frac{N_{xy}}{E_1 t^*} - \frac{M_{xy} z}{E_1 \bar{I}^*} \right) \qquad (13\text{-}37c)$$

The compatibility equation can be expressed in terms of the stress resultants by inserting Eqs. (13-37) into Eq. (13-24) and using Eqs. (13-35) and (13-30). As a result

$$\frac{\partial^2}{\partial y^2} \left[\frac{1}{E_1 t^*} (N_x - \nu N_y) \right] + \frac{\partial^2}{\partial x^2} \left[\frac{1}{E_1 t^*} (N_y - \nu N_x) \right] - 2(1 + \nu) \frac{\partial^2}{\partial x\,\partial y} \frac{N_{xy}}{E_1 t^*}$$

$$= -(1 - \nu) \nabla^2 \frac{N_T}{E_1 t^*} + \left(\frac{\partial^2 w}{\partial x\,\partial y} \right)^2 - \frac{\partial^2 w}{\partial x^2} \frac{\partial^2 w}{\partial y^2} \qquad (13\text{-}38)$$

As in plane stress (Sec. 4-5), it is convenient to introduce a *stress function* $F(x,y)$ which identically satisfies the equilibrium equations (13-17). In doing so we assume that p_x and p_y are conservative forces that are derivable from a potential function $V(x,y)$ by the relationships

$$p_x = -\frac{\partial V}{\partial x} \qquad p_y = -\frac{\partial V}{\partial y} \qquad (13\text{-}39)$$

To identically satisfy Eqs. (13-17) we define F by the equations

$$N_x = \frac{\partial^2 F}{\partial y^2} + V \qquad N_y = \frac{\partial^2 F}{\partial x^2} + V \qquad N_{xy} = -\frac{\partial^2 F}{\partial x\,\partial y} \qquad (13\text{-}40)$$

Since F assures equilibrium in the xy plane, we need be concerned only with compatibility and equilibrium in the z direction. We can write the compatibility equation in terms of F by substituting Eqs. (13-40) into Eq. (13-38), which gives

$$\nabla^2 \left(\frac{1}{t^*} \nabla^2 F \right) - (1 + \nu) \left(\frac{\partial^2 F}{\partial y^2} \frac{\partial^2}{\partial x^2} \frac{1}{t^*} - 2 \frac{\partial^2 F}{\partial x\,\partial y} \frac{\partial^2}{\partial x\,\partial y} \frac{1}{t^*} + \frac{\partial^2 F}{\partial x^2} \frac{\partial^2}{\partial y^2} \frac{1}{t^*} \right)$$

$$= -(1 - \nu)\nabla^2 \left[\frac{1}{t^*} (V + N_T) \right] + E_1 \left[\left(\frac{\partial^2 w}{\partial x\,\partial y} \right)^2 - \frac{\partial^2 w}{\partial x^2} \frac{\partial^2 w}{\partial y^2} \right] \qquad (13\text{-}41)$$

When t^* is constant, this reduces to

$$\nabla^4 F = -(1 - \nu) \nabla^2 (V + N_T) + E_1 t^* \left[\left(\frac{\partial^2 w}{\partial x\,\partial y} \right)^2 - \frac{\partial^2 w}{\partial x^2} \frac{\partial^2 w}{\partial y^2} \right] \qquad (13\text{-}42)$$

The equilibrium equation for the z direction can be expressed in terms of w by substituting Eqs. (13-30) into (13-19), to obtain

$$\nabla^2(D^* \nabla^2 w) - (1 - \nu)\left(\frac{\partial^2 w}{\partial x^2}\frac{\partial^2 D^*}{\partial y^2} - 2\frac{\partial^2 w}{\partial x\,\partial y}\frac{\partial^2 D^*}{\partial x\,\partial y} + \frac{\partial^2 w}{\partial y^2}\frac{\partial^2 D^*}{\partial x^2}\right)$$

$$= p_z + \nabla^2 M_T + N_x\frac{\partial^2 w}{\partial x^2} + 2N_{xy}\frac{\partial^2 w}{\partial x\,\partial y} + N_y\frac{\partial^2 w}{\partial y^2} - p_x\frac{\partial w}{\partial x} - p_y\frac{\partial w}{\partial y}$$

$$(13\text{-}43)$$

If w and F are chosen as unknowns, we can substitute Eqs. (13-40) into Eq. (13-43) and find

$$\nabla^2(D^* \nabla^2 w) - (1 - \nu)\left(\frac{\partial^2 w}{\partial x^2}\frac{\partial^2 D^*}{\partial y^2} - 2\frac{\partial^2 w}{\partial x\,\partial y}\frac{\partial^2 D^*}{\partial x\,\partial y} + \frac{\partial^2 w}{\partial y^2}\frac{\partial^2 D^*}{\partial x^2}\right)$$

$$= p_z + \nabla^2 M_T + \frac{\partial^2 F}{\partial y^2}\frac{\partial^2 w}{\partial x^2} - 2\frac{\partial^2 F}{\partial x\,\partial y}\frac{\partial^2 w}{\partial x\,\partial y} + \frac{\partial^2 F}{\partial x^2}\frac{\partial^2 w}{\partial y^2} + \frac{\partial}{\partial x}\left(V\frac{\partial w}{\partial x}\right)$$

$$+ \frac{\partial}{\partial y}\left(V\frac{\partial w}{\partial y}\right)\quad(13\text{-}44)$$

When D^* is constant, Eqs. (13-43) and (13-44) reduce to

$$\nabla^4 w = \frac{1}{D^*}\left(p_z + \nabla^2 M_T + N_x\frac{\partial^2 w}{\partial x^2} + 2N_{xy}\frac{\partial^2 w}{\partial x\,\partial y} + N_y\frac{\partial^2 w}{\partial y^2} - p_x\frac{\partial w}{\partial x}\right.$$

$$\left. - p_y\frac{\partial w}{\partial y}\right)\quad(13\text{-}45)$$

$$\nabla^4 w = \frac{1}{D^*}\left[p_z + \nabla^2 M_T + \frac{\partial^2 F}{\partial y^2}\frac{\partial^2 w}{\partial x^2} - 2\frac{\partial^2 F}{\partial x\,\partial y}\frac{\partial^2 w}{\partial x\,\partial y} + \frac{\partial^2 F}{\partial x^2}\frac{\partial^2 w}{\partial y^2}\right.$$

$$\left. + \frac{\partial}{\partial x}\left(V\frac{\partial w}{\partial x}\right) + \frac{\partial}{\partial y}\left(V\frac{\partial w}{\partial y}\right)\right]\quad(13\text{-}46)$$

Equations (13-41) and (13-44) [or (13-42) and (13-46) when t^* and D^* are constant] are the governing differential equations for the *stress-function–lateral-displacement formulation* of plate theory.   They are often referred to as the *von Kármán plate equations*.

If u and v are prescribed on any portion of the boundary, it is more convenient to use a *displacement formulation* in which u, v, and w are the unknown functions.   As in the theory of elasticity, it is not necessary to use the compatibility equations in the displacement formulation.   However, the displacements must satisfy the equilibrium conditions.   For the x and y directions these are found by substituting Eqs. (13-27) into Eqs. (13-17).   This gives

$$\frac{\partial}{\partial x}\left\{K^*\left[\frac{\partial u}{\partial x} + \nu\frac{\partial v}{\partial y} + \frac{1}{2}\left(\frac{\partial w}{\partial x}\right)^2 + \frac{\nu}{2}\left(\frac{\partial w}{\partial y}\right)^2\right]\right\}$$

$$+ \frac{1 - \nu}{2}\frac{\partial}{\partial y}\left[K^*\left(\frac{\partial v}{\partial x} + \frac{\partial u}{\partial y} + \frac{\partial w}{\partial x}\frac{\partial w}{\partial y}\right)\right] = -p_x + \frac{\partial N_T}{\partial x}\quad(13\text{-}47a)$$

$$\frac{\partial}{\partial y}\left\{K^*\left[\frac{\partial v}{\partial y} + \nu\frac{\partial u}{\partial x} + \frac{1}{2}\left(\frac{\partial w}{\partial y}\right)^2 + \frac{\nu}{2}\left(\frac{\partial w}{\partial x}\right)^2\right]\right\}$$
$$+ \frac{1-\nu}{2}\frac{\partial}{\partial x}\left[K^*\left(\frac{\partial v}{\partial x} + \frac{\partial u}{\partial y} + \frac{\partial w}{\partial x}\frac{\partial w}{\partial y}\right)\right] = -p_y + \frac{\partial N_T}{\partial y} \quad (13\text{-}47b)$$

The third equilibrium equation is found by substituting Eqs. (13-27) into (13-43). The equilibrium equations may be simplified when K^* and D^* are constant.

We note that the differential equations in both formulations are nonlinear and coupled as a result of the nonlinear terms in the large-deflection–strain-displacement equations (13-23). These terms arise from the fact that the reference surface must undergo extensions when its deformed position is a nondevelopable surface or when the edges of the plate are restrained against in-plane motion. The extensions result in in-plane membrane forces N_x, N_y, and N_{xy}, which are functions of w.

Solutions to the nonlinear coupled equations are known for only a few special cases (see Refs. 1 and 7 for examples), and in most of these it is necessary to use approximate methods to obtain results. These solutions indicate the circumstances under which a linearized theory, obtained by discarding the nonlinear terms in all the preceding equations, is sufficiently accurate. The results show that the plate behaves in a manner that is similar to the beam with longitudinally restrained ends (Example 7-2). The restrained beam resists lateral loads by bending rigidity and by cable action which results from the stretching that accompanies lateral deflections. Similarly, the plate resists lateral loads by bending rigidity and by membrane action that results from the stretching of the reference surface that accompanies lateral deflections.

The nonlinear solutions show that if the deflected reference surface is nondevelopable or the edges of the plate are restrained against in-plane motion, the membrane forces due to w (and therefore the nonlinear terms) are negligible when w is small compared to t. In this case the lateral load is resisted by the bending rigidity. The bending and membrane resistances to lateral displacements are of comparable magnitudes when w is of the order of t, and all terms must be retained in the equations. On the other hand, when w is large relative to t, membrane action is predominant, and the terms containing D^* may be neglected.

The nonlinear terms may also be neglected when there are no in-plane edge restraints and the deflected reference surface is developable, even when w is large relative to t. However, in all cases the deflections must be small in the sense that $(\partial w/\partial x)^2 \ll 1$ and $(\partial w/\partial y)^2 \ll 1$, so that Eqs. (13-8) apply. For thin plates, the latter limitations are less restrictive than the requirement that w be small compared to t.

When the nonlinear terms are discarded, the differential equations become independent in the sense that the in-plane equations can be solved for F or for u and v and the result may be used to determine N_x, N_y, and N_{xy} in the equilibrium equation for the z direction. This equation can then be solved for w. The stress resultants N_x, N_y, and N_{xy} are negligible when w is small compared to t, there are no in-plane applied forces or restraints, and $\nabla^2 N_T = 0$. In this case we may discard Eq. (13-41) or Eqs. (13-47) and obtain w from Eq. (13-43), which reduces to

$$\nabla^2(D^* \nabla^2 w) - (1 - \nu)\left(\frac{\partial^2 w}{\partial x^2}\frac{\partial^2 D^*}{\partial y^2} - 2\frac{\partial^2 w}{\partial x\,\partial y}\frac{\partial^2 D^*}{\partial x\,\partial y} + \frac{\partial^2 w}{\partial y^2}\frac{\partial^2 D^*}{\partial x^2}\right)$$
$$= p_z + \nabla^2 M_T \quad (13\text{-}48)$$

13-8 BOUNDARY CONDITIONS

Before considering specific cases, let us discuss some of the problems involved in prescribing edge conditions. If boundary conditions are to be satisfied exactly, either the stresses must be in equilibrium with applied edge forces per unit area, or the displacements must satisfy prescribed conditions at all points on the edge surface. However, the first requirement is inconsistent with the equilibrium conditions in the interior of the plate, where we required equilibrium only between the stress resultants and the applied loads. Furthermore, we cannot arbitrarily specify displacements at all points on the edge of the plate without violating our assumption that normals to the reference surface remain normal and unchanged in length as the plate deforms. For these reasons we simply require equilibrium between the stress resultants and the applied edge loads per unit length and specify edge displacements only at points on the boundary curve of the reference surface.

Specifying the boundary conditions in this approximate manner compromises the accuracy of the stresses only in a narrow region at the edge of the plate, for, from St. Venant's principle, the stresses at several thicknesses from the edge depend only upon the resultants of the stresses and not upon how they are distributed through the thickness. As in beam theory, gross inaccuracies can result from using the derived results in the edge region. If stresses are required within these areas, it is necessary to resort to a more complicated theory for thick plates, such as that developed by Reissner,[8] or to the three-dimensional theory of elasticity.

Equations (13-41) and (13-44) are both fourth-order differential equations, so that two boundary conditions are required for F, and two are needed for w in the stress-resultant–lateral-displacement formulation. The displacement formulation uses the two second-order differential equations (13-47) and the fourth-order equation (13-43) [with N_x, N_y, and N_{xy} eliminated by using Eqs. (13-27)]. These require one boundary condition in u, one in v, and two in w.

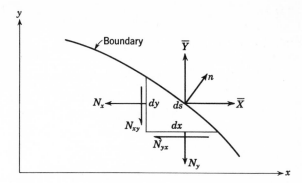

Fig. 13-13 Free-body diagram of a boundary element.

We first consider the in-plane forces and displacements. The stress-function formulation is convenient only when in-plane forces are prescribed on the entire boundary. We designate the in-plane boundary forces *per unit length* by $\bar{X}$ and $\bar{Y}$ (defined positive in the directions of the positive coordinate axes) and note that these must be in equilibrium with N_x, N_y, and N_{xy}. Referring to Fig. 13-13 and following the procedure used to derive the force boundary conditions in the theory of elasticity (Sec. 2-3), we find

$$N_x l + N_{xy} m = \bar{X} \qquad N_{yx} l + N_y m = \bar{Y} \qquad (13\text{-}49)$$

where l and m are the direction cosines of the outward normal to the boundary. These equations can be written in terms of F by using Eqs. (13-40), which give

$$\frac{\partial^2 F}{\partial y^2} l - \frac{\partial^2 F}{\partial x\, \partial y} m = \bar{X} - Vl \qquad -\frac{\partial^2 F}{\partial x\, \partial y} l + \frac{\partial^2 F}{\partial x^2} m = \bar{Y} - Vm \quad (13\text{-}50)$$

Noting the similarity of these relations to Eqs. (4-18), we can use the method used to derive Eqs. (4-22) and (4-23) to rewrite Eqs. (13-50) in the form

$$F = \int_0^s (Al + Bm)\, ds \qquad \frac{\partial F}{\partial n} = -Bl + Am \qquad (13\text{-}51)$$

where A and B are given by Eqs. (4-21).

In the displacement formulation we express the force boundary conditions in terms of u and v for portions of the boundary where in-plane edge forces are prescribed. Substituting Eqs. (13-27) into Eqs. (13-49), we find

$$\left[\frac{\partial u}{\partial x} + \nu\frac{\partial v}{\partial y} + \frac{1}{2}\left(\frac{\partial w}{\partial x}\right)^2 + \frac{\nu}{2}\left(\frac{\partial w}{\partial y}\right)^2\right] l$$

$$+ \frac{1 - \nu}{2}\left(\frac{\partial v}{\partial x} + \frac{\partial u}{\partial y} + \frac{\partial w}{\partial x}\frac{\partial w}{\partial y}\right) m = \frac{1}{K^*}(\bar{X} + N_T l) \quad (13\text{-}52a)$$

$$\frac{1 - \nu}{2}\left(\frac{\partial v}{\partial x} + \frac{\partial u}{\partial y} + \frac{\partial w}{\partial x}\frac{\partial w}{\partial y}\right) l$$

$$+ \left[\frac{\partial v}{\partial y} + \nu\frac{\partial u}{\partial x} + \frac{1}{2}\left(\frac{\partial w}{\partial y}\right)^2 + \frac{\nu}{2}\left(\frac{\partial w}{\partial x}\right)^2\right] m = \frac{1}{K^*}(\bar{Y} + N_T m) \quad (13\text{-}52b)$$

On portions of the boundary where displacements are prescribed, the boundary conditions are

$$u = u_1(x,y) \qquad v = v_1(x,y) \tag{13-53}$$

where u_1 and v_1 are prescribed functions of position on the boundary.

The two boundary conditions for w are obtained from prescribed values of either the lateral displacement or the shear, and either the slope or the bending moment normal to the edge of the plate. From Eq. (13-4), the slope in the normal direction can be written

$$\frac{\partial w}{\partial n} = \frac{\partial w}{\partial x} l + \frac{\partial w}{\partial y} m \tag{13-54}$$

and from Eq. (13-30a) the moment in the normal direction is

$$M_n = D^*\left(\frac{\partial^2 w}{\partial n^2} + \nu\frac{\partial^2 w}{\partial s^2}\right) - M_T \tag{13-55}$$

where s is in the direction of the boundary curve. By using Eq. (13-11) we can write the last equation as

$$M_n = D^*\left[(l^2 + \nu m^2)\frac{\partial^2 w}{\partial x^2} + (m^2 + \nu l^2)\frac{\partial^2 w}{\partial y^2} + 2lm(1 - \nu)\frac{\partial^2 w}{\partial x\,\partial y}\right] - M_T \tag{13-56}$$

The shear boundary condition is more involved and requires explanation. It would appear that the bending moment, twisting moment, and shear should be specified on an unrestrained edge (these conditions were actually proposed by Poisson). However, Eq. (13-43) is a fourth-order differential equation, which permits only two instead of three boundary conditions. This apparent paradox was resolved by Kirchhoff, who derived the Euler differential equation and natural boundary conditions for the plate by the variational method (Sec. 6-11). This derivation gives only two boundary conditions for an unrestrained edge.

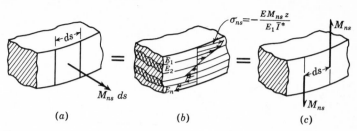

Fig. 13-14   Edge twisting moments.

One of these is Eq. (13-56), and the other combines the effects of the edge shear and twisting moment.

A physical explanation of the second condition which is based upon static considerations was given by Thompson and Tait.[1]   Figure 13-14a shows a differential length of the edge of the plate acted upon by a twisting moment per unit length M_{ns}.   This moment is the resultant of distributed forces applied to the edge surface.   By changing the coordinates in Eq. (13-34c) to n and s we see that the theory indicates that M_{ns} is related to shearing stresses σ_{ns} that are distributed through the thickness of the plate as shown in Fig. 13-14b.   Equilibrium is exactly satisfied if the edge loads that produce M_{ns} are distributed in this fashion.   The fact that this is seldom the case is of little concern, however, for we have seen that, except in the edge region, it is only the static resultant per unit length of the edge loading that is important.   Therefore, except for the boundary region, the vertical edge forces in Fig. 13-14c are equivalent to the twisting moments in Fig. 13-14a and b.

Next let us consider the distribution of M_{ns} along the edge of the plate.   A variable twisting moment is shown in Fig. 13-15a, and a statically equivalent system of couples due to lateral edge forces is illustrated in Fig. 13-15b.   It is seen that the M_{ns} portion of these lateral forces cancel, leaving an unbalanced edge shear per unit length equal to $\partial M_{ns}/\partial s$, as shown in Fig. 13-15$c$.   This can be combined with the prescribed edge shear Q_n to give a total shear V_n, where

$$V_n = Q_n + \frac{\partial M_{ns}}{\partial s} \tag{13-57}$$

By changing coordinates in the first of Eqs. (13-18) we find

$$Q_n = \frac{\partial M_n}{\partial n} - \frac{\partial M_{sn}}{\partial s}$$

Substituting this into Eq. (13-57) and noting from the last of Eqs. (13-15) that $M_{sn} = -M_{ns}$, we find $V_n = \partial M_n/\partial n + 2\partial M_{ns}/\partial s$.   From Eqs.

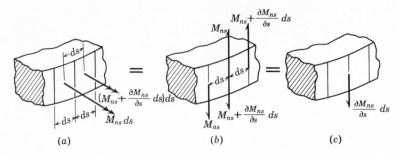

Fig. 13-15   Kirchhoff boundary condition for twisting moments.

(13-55) and (13-30c) (with coordinates changed to n and s) we obtain

$$V_n = \frac{\partial}{\partial n}\left[D^*\left(\frac{\partial^2 w}{\partial n^2} + \nu\,\frac{\partial^2 w}{\partial s^2}\right)\right] + 2(1 - \nu)\,\frac{\partial}{\partial s}\left(D^*\,\frac{\partial^2 w}{\partial n\,\partial s}\right) - \frac{\partial M_T}{\partial n}$$

(13-58)

When D^* is constant, this equation reduces to

$$V_n = D^*\left[\frac{\partial^3 w}{\partial n^3} + (2 - \nu)\,\frac{\partial^3 w}{\partial n\,\partial s^2}\right] - \frac{\partial M_T}{\partial n}$$

The equivalent edge shear $\partial M_{ns}/\partial s$ leads to a concentrated force when there is a corner on the boundary.   An example of this is given in Fig. 13-16a, which shows a rectangular plate with constant edge twisting moments M_{xy} and M_{yx}.   There are no running shears along the edge because $\partial M_{xy}/\partial y = \partial M_{yx}/\partial x = 0$ when the twisting moments are constant.   However, the equivalent edge shears add instead of canceling at the corner point, because $M_{xy} = -M_{yx}$.   This results in the corner force shown in Fig. 13-16b.

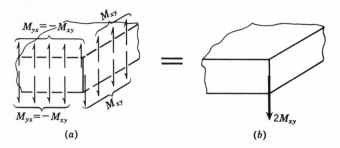

Fig. 13-16   Concentrated corner force due to edge twisting moments.

With the preceding equations we can write the following boundary conditions for the commonly encountered edge conditions:

1. Clamped edge:

$$w = 0 \qquad \frac{\partial w}{\partial n} = 0 \qquad (13\text{-}59)$$

2. Simply supported edge:

$$w = 0 \qquad \frac{\partial^2 w}{\partial n^2} = \frac{M_T}{D^*} \qquad (13\text{-}60)$$

3. Free edge:

$$\frac{\partial^2 w}{\partial n^2} + \nu \frac{\partial^2 w}{\partial s^2} = \frac{M_T}{D^*}$$

$$\frac{\partial}{\partial n}\left[D^*\left(\frac{\partial^2 w}{\partial n^2} + \nu \frac{\partial^2 w}{\partial s^2}\right)\right] + 2(1 - \nu)\frac{\partial}{\partial s}\left(D^* \frac{\partial^2 w}{\partial n\, \partial s}\right) = \frac{\partial M_T}{\partial n} \qquad (13\text{-}61)$$

Boundary conditions for elastically restrained edges are given in Ref. 1.

13-9 THE DIFFERENTIAL EQUATIONS FOR PLATES AND MEMBRANES

The equations of Secs. 13-7 and 13-8 can be used to write the governing differential equations and boundary conditions for special cases of plates and membranes. As in Secs. 7-9 and 8-9, the examples can be catagorized as equilibrium boundary-value problems or eigenvalue problems.

Examples of equilibrium problems

Example 13-1 A homogeneous rectangular plate rests upon an elastic foundation and is subjected to a lateral pressure p. The thickness of the plate tapers in the x direction, and the edges are simply supported at $x = 0$ and $x = a$, clamped at $y = 0$, and free at $y = b$. Write the governing differential equation and boundary conditions for small deflections.

There are no applied in-plane forces, and so for w small compared to t we may assume $N_x = N_y = N_{xy} = F = 0$. As a result, the only unknown function is w, and the differential equation can be found from Eq. (13-48).

We assume that the nature of the foundation is such that it provides an elastic restoring pressure at any point that is directly proportional to the displacement at the point. The elastic constant of the foundation is k psi per inch of displacement. The resultant force per unit area of the plate is therefore $p_z = p - kw$.

Noting that $D^* = D(x)$, Eq. (13-48) reduces to

$$D \nabla^4 w + 2 \frac{dD}{dx}\left(\frac{\partial^3 w}{\partial x^3} + \frac{\partial^3 w}{\partial x\, \partial y^2}\right) + \frac{d^2 D}{dx^2}\left(\frac{\partial^2 w}{\partial x^2} + \nu \frac{\partial^2 w}{\partial y^2}\right) + kw = p \qquad (a)$$

From Eqs. (13-59) to (13-61), the associated boundary conditions are $w = \partial^2 w/\partial x^2 = 0$ at $x = 0$ and $x = a$, $w = \partial w/\partial y = 0$ at $y = 0$, and

$$\frac{\partial^2 w}{\partial y^2} + \nu \frac{\partial^2 w}{\partial x^2} = 0 \qquad D\left[\frac{\partial^3 w}{\partial y^3} + (2 - \nu)\frac{\partial^3 w}{\partial x^2 \partial y}\right] + 2(1 - \nu)\frac{dD}{dx}\frac{\partial^2 w}{\partial x \partial y} = 0 \qquad (b)$$

at $y = b$. These equations will be used to illustrate the finite-difference method in Example 13-9.

Example 13-2 An arbitrarily shaped plate of constant thickness is subjected to a temperature change $T(z)$. The material properties are functions of z only. The edges are free to displace in the xy plane.   Write the governing differential equation and the boundary conditions for small displacements with free and clamped edges.   Determine the stresses in each case.

With E and α functions of z only and t constant, we find from Eqs. (13-2), (13-29), (13-31), and (13-32) that t^*, N_T, D^*, and M_T are constant.   There are no in-plane forces or restraints, so that $N_x = N_y = N_{xy} = 0$, and Eq. (13-45) reduces to $\nabla^4 w = 0$.

For free edges, w must satisfy Eqs. (13-61), the second of which reduces to

$$\frac{\partial^3 w}{\partial n^3} + (2 - \nu)\frac{\partial^3 w}{\partial n\,\partial s^2} = 0$$

It is easily verified that the differential equation and boundary conditions are satisfied by

$$w = \frac{M_T}{2(1 + \nu)D^*}(x^2 + y^2) \qquad (a)$$

where to prevent rigid-body motion it is assumed that $w = \partial w/\partial x = \partial w/\partial y = 0$ at the origin.   The solution describes a paraboloid of revolution which, because of its shallow shape, closely approximates a spherical surface.   The true deflection surface in this case is spherical, but the parabolic shape results from approximating the curvatures by Eqs. (13-8).   Substituting Eq. (a) into Eqs. (13-30), we find $M_x = M_y = M_{xy} = 0$, so that Eqs. (13-34) give $\sigma_{xy} = 0$ and

$$\sigma_{xx} = \sigma_{yy} = \frac{E}{E_1}\left(\frac{N_T}{t^*} - \frac{M_T z}{\bar{I}^*} - \frac{E_1 \alpha T}{1 - \nu}\right) \qquad (b)$$

It is easily verified that all stresses are zero when the plate is homogeneous and $T(z)$ is linear.

The boundary conditions are given by Eqs. (13-59) when the edges of the plate are clamped.   In this case the differential equation and boundary conditions are homogeneous, and the solution is $w = 0$.   From Eqs. (13-30) $M_x = M_y = -M_T$ and $M_{xy} = 0$, so that Eqs. (13-34) reduce to $\sigma_{xy} = 0$ and

$$\sigma_{xx} = \sigma_{yy} = \frac{E}{E_1}\left(\frac{N_T}{t^*} - \frac{E_1 \alpha T}{1 - \nu}\right) \qquad (c)$$

Example 13-3 A homogeneous rectangular plate of uniform thickness is simply supported on the edges $x = 0$ and L and free on the edges $y = 0$ and b.   The edges at $x = 0$ and L are restrained from in-plane motion while those at $y = 0$ and b are free of such restraints.   Write the differential equation and boundary conditions and determine the displacements for $p_z = p_0 \sin \pi x/L$.

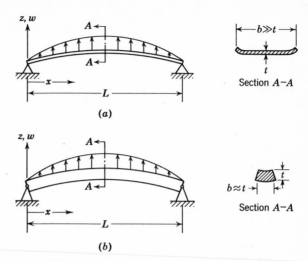

Fig. 13-17 Example 13-3. (a) Plate deformation; (b) beam deformation.

The deformation shape for $b \gg t$ is shown in Fig. 13-17a. Except for a narrow region along the free edges, the plate deforms into a cylindrical surface. The cylindrical surface is developable, and there are no edge restraints in the y direction, so that in this region u and w are functions of x only, and $N_y = N_{xy} = 0$. As a result, we find

$$\frac{Ebt^3}{12(1 - \nu^2)} \frac{d^4w}{dx^4} - N_x b \frac{d^2w}{dx^2} = p_0 b \sin \frac{\pi x}{L} \qquad (a)$$

by multiplying Eq. (13-45) by b and using the second of Eqs. (13-33). From Eqs. (13-60), the associated boundary conditions are $w = d^2w/dx^2 = 0$ at $x = 0$ and L.

From Eq. (13-27b), with $N_y = 0$, we find $\partial v/\partial y = \nu \left[du/dx + \frac{1}{2} (dw/dx)^2 \right]$. Applying this result to Eq. (13-47a), using Eq. (13-27c), and recalling that $N_{xy} = 0$, we find

$$\frac{d^2u}{dx^2} + \frac{dw}{dx} \frac{d^2w}{dx^2} = 0 \qquad (b)$$

which has the boundary conditions $u = 0$ at $x = 0$ and L. Except for the $1 - \nu^2$ in the denominator of the first term in Eq. (a), Eqs. (a), (b), and the boundary conditions are identical to Eqs. (a), (c), and the boundary conditions of Example 7-2 (where $I_{yy} = bt^3/12$, $P = N_x b$, and $p_0 = p_0 b$). As a result, all the equations of Example 7-2 and Fig. 7-12 are applicable if we replace E by $E/(1 - \nu^2)$, the effective modulus of the plate. Since ν is approximately 0.3 for most structural materials, we find that the plate is roughly 10 percent stiffer than would be computed by beam theory.

As already noted, except for a narrow edge region along the free edges, where anticlastic curvatures develop, the plate deflects into a cylindrical surface. In the cylindrical portion, $\partial^2w/\partial y^2 = 0$, and we find from Eqs. (13-30) and (13-34) that

$$\sigma_{xx} = \frac{N_x}{t} - \frac{Ez}{1-\nu^2}\frac{d^2w}{dx^2} \qquad \sigma_{yy} = -\frac{\nu Ez}{1-\nu^2}\frac{d^2w}{dx^2} \tag{c}$$

In a beam, where b is less than or approximately equal to t, the edge regions overlap, and anticlastic curvatures occur freely over the entire width of the beam, as shown in Fig. 13-17b.   In this case, strains due to Poisson's ratio are free to develop in the y direction, and $\sigma_{yy} = 0$.

When the supported edges of the plate are unrestrained against in-plane motion, or when w is small compared to t, $u = N_x = 0$, and the deflections can be computed from elementary linear beam theory if E is replaced by $E/(1 - \nu^2)$. The σ_{xx} stresses can be computed directly from beam theory, and the stresses in the y direction can be found from $\sigma_{yy} = -\nu\sigma_{xx}$.   These remarks are general and apply to other pressure distributions and edge conditions when $p_z = p_z(x)$ and the edges at $y = 0$ and b are free while the edges at $x = 0$ and L are supported so that $w = 0$.   These members are referred to as *wide beams*.   The remarks are also applicable to rectangular plates that are supported on all edges if $b \gg L$.   In these cases the supports on the edges $y = 0$ and b affect only the deflections and stresses in the vicinity of those edges, and in the central portion of the plate the deflections and stresses are the same as if these edges were free.

Example 13-4   Derive the differential equations and boundary conditions for a homogeneous membrane of uniform thickness under the action of edge forces and a lateral pressure p.

Noting that $D^* = 0$ for the membrane, we find from Eqs. (13-42) and (13-46) that

$$\nabla^4 F = Et\left[\left(\frac{\partial^2 w}{\partial x\,\partial y}\right)^2 - \frac{\partial^2 w}{\partial x^2}\frac{\partial^2 w}{\partial y^2}\right] \tag{a}$$

$$\frac{\partial^2 F}{\partial y^2}\frac{\partial^2 w}{\partial x^2} - 2\frac{\partial^2 F}{\partial x\,\partial y}\frac{\partial^2 w}{\partial x\,\partial y} + \frac{\partial^2 F}{\partial x^2}\frac{\partial^2 w}{\partial y^2} = -p \tag{b}$$

These coupled nonlinear equations, known as the *Föppl equations*,[1] are fourth- and second-order and require a total of three boundary conditions.   These are given by Eqs. (13-50) or (13-51) and $w = 0$.

When in-plane displacements are prescribed on the boundary, it is more convenient to use the displacement formulation.   In this case, two of the equations are found by setting $p_x = p_y = N_T = 0$ and treating $K^* = K$ as a constant in Eqs. (13-47).   The third equation, found by substituting Eqs. (13-27) into Eq. (13-45) and letting $D^* = M_T = 0$, is

$$\left[\frac{\partial u}{\partial x} + \nu\frac{\partial v}{\partial y} + \frac{1}{2}\left(\frac{\partial w}{\partial x}\right)^2 + \frac{\nu}{2}\left(\frac{\partial w}{\partial y}\right)^2\right]\frac{\partial^2 w}{\partial x^2} + (1-\nu)\left(\frac{\partial v}{\partial x} + \frac{\partial u}{\partial y} + \frac{\partial w}{\partial x}\frac{\partial w}{\partial y}\right)\frac{\partial^2 w}{\partial x\,\partial y}$$
$$+ \left[\frac{\partial v}{\partial y} + \nu\frac{\partial u}{\partial x} + \frac{1}{2}\left(\frac{\partial w}{\partial y}\right)^2 + \frac{\nu}{2}\left(\frac{\partial w}{\partial x}\right)^2\right]\frac{\partial^2 w}{\partial y^2} = -\frac{p}{K} \tag{c}$$

In this case the boundary conditions are given by Eqs. (13-53) and $w = 0$. As noted earlier, the membrane solution can be used for plates when the applied load is of such magnitude that $w \gg t$.

Solutions to the preceding equations are difficult to obtain because of the nonlinearity and coupling. The problem is greatly simplified when the membrane is preloaded by a uniform tensile force per unit length N along the boundary. It is readily verified that this produces an initially stressed condition with $N_x = N_y = N$ and $N_{xy} = 0$. If the initial tension is large enough, we may neglect the changes in N_x, N_y, and N_{xy} that result from the lateral deflections and assume that the membrane forces are constant. In this case Eq. (13-45) reduces to $\nabla^2 w = -p/N$, which, with a change of notation, agrees with Eq. (8-22) for the torsion-membrane analogy. If the membrane is restrained against lateral edge displacements, the boundary condition is $w = 0$. On the other hand, if N is applied at an angle θ to the xy plane, the boundary condition becomes $\partial w/\partial n = \tan \theta$.

Eigenvalue problems

Example 13-5 A uniform rectangular plate with boundaries at $x = 0$ and a and $y = 0$ and b is simply supported on all edges. The edges $x = 0$ and a are subjected to a compressive force per unit length N, and the edges $y = 0$ and b are unrestrained in the xy plane. If N is large enough, the plate will buckle laterally. The value of N at which neutral stability exists in which equilibrium is possible in either the flat or buckled position is designated N_{cr}. Write the differential equation and boundary conditions for equilibrium in the slightly buckled position.

At the neutral stability point, $N_x = N_{\mathrm{cr}}$ and $N_y = N_{xy} = 0$ in the flat plate. For bending deformations that are small relative to t we may assume that the in-plane stress resultants have the same values that exist immediately prior to buckling. Substituting these values into Eq. (13-45) and setting $p_x = p_y = p_z = M_T = 0$, we find

$$\nabla^4 w = -\frac{N_{\mathrm{cr}}}{D^*}\frac{\partial^2 w}{\partial x^2} \tag{a}$$

From Eqs. (13-60), the associated boundary conditions are $w = \partial^2 w/\partial x^2 = 0$ at $x = 0$ and a and $w = \partial^2 w/\partial y^2 = 0$ at $y = 0$ and b. These equations are solved in Sec. 15-3 for the eigenvalue $\lambda = N_{\mathrm{cr}}/D^*$ and the buckled mode shape $w(x,y)$.

Example 13-6 A homogeneous rectangular plate of uniform thickness is subjected to a temperature change $T = CT_0(x,y)$, where C is a constant that establishes the magnitude of the temperature change and $T_0(x,y)$ is a function that describes its distribution. Write the differential equations and boundary conditions that establish the deflections of the thermally buckled plate if the edges of the plate are free. Write the differential equation and boundary conditions that determine the value of C for buckling to occur.

It is easily verified that when T_0 is linear in x and y, Eqs. (13-42), (13-46), (13-50), and (13-61) are homogeneous and have the solution $F = w = 0$. In this case, $N_x = N_y = N_{xy} = 0$, and buckling cannot occur for any value C. On the other hand, when T_0 is not a linear function, in-plane stress resultants are developed. Since the plate is free, these forces must be self-equilibrating, and areas exist where the normal forces are compressive. As a result, buckling will occur if C is large enough.

In the postbuckling regime F and w must satisfy Eqs. (13-42) and (13-46), which reduce to

$$\nabla^4 F = -(1 - \nu) \nabla^2 N_T + Et \left[\left(\frac{\partial^2 w}{\partial x\, \partial y} \right)^2 - \frac{\partial^2 w}{\partial x^2} \frac{\partial^2 w}{\partial y^2} \right]$$

$$\nabla^4 w = \frac{1}{D} \left(\frac{\partial^2 F}{\partial y^2} \frac{\partial^2 w}{\partial x^2} - 2 \frac{\partial^2 F}{\partial x\, \partial y} \frac{\partial^2 w}{\partial x\, \partial y} + \frac{\partial^2 F}{\partial x^2} \frac{\partial^2 w}{\partial y^2} \right) \tag{a}$$

where N_T is obtained from Eq. (13-29). The related boundary conditions are found from Eqs. (13-51) and (13-61) by setting $A = B = M_T = 0$ and noting that $D^* = D$ is a constant. Equations (a) and the boundary conditions describe an equilibrium boundary-value problem. Gossard, Seide, and Roberts[10] have obtained approximate solutions to these equations by the Galerkin method.

To determine C_{cr}, the value of C at which the buckled configuration becomes a possible equilibrium state, we consider w small compared to t, neglect the nonlinear terms in Eqs. (a), and find

$$\nabla^4 \bar{F} = -(1 - \nu) \nabla^2 \bar{N}_T \tag{b}$$

$$\nabla^4 w = \frac{C_{cr}}{D} \left(\frac{\partial^2 \bar{F}}{\partial y^2} \frac{\partial^2 w}{\partial x^2} - 2 \frac{\partial^2 \bar{F}}{\partial x\, \partial y} \frac{\partial^2 w}{\partial x\, \partial y} + \frac{\partial^2 \bar{F}}{\partial x^2} \frac{\partial^2 w}{\partial y^2} \right) \tag{c}$$

where $\bar{F} = F/C_{cr}$ and $\bar{N}_T = N_T/C_{cr}$. From Eqs. (13-51), the boundary conditions for $\bar{F}$ are $\bar{F} = \partial \bar{F}/\partial n = 0$, and the boundary conditions for w remain the same as those given for the postbuckled case.

Since Eqs. (b) and (c) are not coupled, (b) can be solved for $\bar{F}$ and the result substituted into (c), which is then solved for the eigenvalue C_{cr}/D and the mode shape w. The Rayleigh-Ritz method has been used to obtain approximate solutions for the in-plane resultants[11] and the critical temperature level.[10] Examples 15-1 and 15-2 illustrate the application of the Rayleigh-Ritz and finite-difference methods to thermal buckling problems in plates.

Example 13-7 A simply supported rectangular plate of uniform thickness is disturbed from its plane and subsequently executes free vibrations. Write the differential equations that establish the mode shapes and natural frequencies for small free vibrations.

In this case $w = w(x,y,t)$, where t is the independent variable time. The only forces that are applied to the plate during free vibrations are the inertial forces $p_z = -\mu \ddot{w}$, where μ is the mass per unit area of the plate. Equation (13-45) therefore reduces to

$$\nabla^4 w = -\frac{\mu}{D^*} \frac{\partial^2 w}{\partial t^2} \tag{a}$$

and from Eqs. (13-60) the boundary conditions are $w = \partial^2 w/\partial x^2 = 0$ at $x = 0$ and a and $w = \partial^2 w/\partial y^2 = 0$ at $y = 0$ and b. The initial conditions are $w(x,y,0) = f(x,y)$ and $\dot{w}(x,y,0) = g(x,y)$, where f and g are prescribed functions.

Following Examples 7-8 and 8-8, we use the method of separation of variables and assume the solution to be of the form $w = W(x,y)T(t)$. Substituting this into Eq. (a) and using the procedure that was used in the cited examples, we find

$$\nabla^4 W = \frac{\omega^2 \mu}{D^*} W \tag{b}$$

where the separation constant ω^2 is the square of the frequency of the vibrations. The boundary conditions become $W = \partial^2 W/\partial x^2 = 0$ at $x = 0$ and a and $W = \partial^2 W/\partial y^2 = 0$ at $y = 0$ and b.

The reader may easily verify that Eq. (b) and its boundary conditions are satisfied by the mode shapes and frequencies

$$W_{mn} = a_{mn} \sin \frac{m\pi x}{a} \sin \frac{n\pi y}{b} \qquad (c)$$

$$\omega_{mn} = \pi^2 \sqrt{\frac{D^*}{\mu}} \left[\left(\frac{m}{a}\right)^2 + \left(\frac{n}{b}\right)^2 \right] \qquad (d)$$

where m and n are integers. A tabulation of natural frequencies for other boundary conditions is given in Ref. 7.

13-10 THE NAVIER SOLUTION

A useful method for the small-deflection analysis of simply supported uniform rectangular plates was developed by Navier. Consider a plate with edges at $x = 0$ and a and $y = 0$ and b and a pressure distribution $p(x,y)$ that is supported by a uniform elastic foundation of stiffness k, so that $p_z = p - kw$. For generality we also assume that the edges are subjected to uniform forces N_x and N_y along the edges $x = 0$ and a and $y = 0$ and b, respectively. The resulting in-plane stress resultants are equal to the edge forces in this case, and Eq. (13-45) becomes

$$D^* \nabla^4 w - N_x \frac{\partial^2 w}{\partial x^2} - N_y \frac{\partial^2 w}{\partial y^2} + kw = p \qquad (13\text{-}62)$$

Navier recommended expanding p into the double Fourier series

$$p(x,y) = \sum_{m=1}^{\infty} \sum_{n=1}^{\infty} p_{mn} \sin \frac{m\pi x}{a} \sin \frac{n\pi y}{b} \qquad (13\text{-}63)$$

in which case the form of the differential equation suggests the solution

$$w = \sum_{m=1}^{\infty} \sum_{n=1}^{\infty} w_{mn} \sin \frac{m\pi x}{a} \sin \frac{n\pi y}{b} \qquad (13\text{-}64)$$

which satisfies the boundary conditions for the simply supported edges.

Substituting Eqs. (13-63) and (13-64) into Eq. (13-62) and equating the coefficients of the corresponding sine terms on both sides of the equation, we find

$$w_{mn} = \frac{p_{mn}}{\pi^4 D^*[(m/a)^2 + (n/b)^2]^2 + k + \pi^2[N_x(m/a)^2 + N_y(n/b)^2]} \qquad (13\text{-}65)$$

To determine the p_{mn} coefficients we multiply both sides of Eq. (13-63) by $\sin (r\pi x/a) \sin (s\pi y/b)$ and integrate over the surface of the plate. Using the integration formula

$$\int_0^a \sin \frac{m\pi x}{a} \sin \frac{r\pi x}{a} \, dx = \begin{cases} 0 & m \neq r \\ \dfrac{a}{2} & m = r \end{cases} \tag{13-66}$$

and a corresponding expression for the y integration, we find

$$p_{mn} = \frac{4}{ab} \int_0^b \int_0^a p \sin \frac{m\pi x}{a} \sin \frac{n\pi y}{b} \, dx \, dy \tag{13-67}$$

In some cases it may be necessary to evaluate this integral numerically by using trapezoidal or Simpson's rule integration (Fig. 5-5).

Noting that N_x and N_y are positive when tensile and negative when compressive, we see from Eqs. (13-64) and (13-65) that tension on the edges stiffens the plate, and compression makes it more flexible. To evaluate the moments in the plate we substitute Eq. (13-64) into Eqs. (13-30) and find

$$M_x = -\pi^2 D^* \sum_{m=1}^{\infty} \sum_{n=1}^{\infty} \left[\left(\frac{m}{a} \right)^2 + \nu \left(\frac{n}{b} \right)^2 \right] w_{mn} \sin \frac{m\pi x}{a} \sin \frac{n\pi y}{b}$$

$$M_y = -\pi^2 D^* \sum_{m=1}^{\infty} \sum_{n=1}^{\infty} \left[\nu \left(\frac{m}{a} \right)^2 + \left(\frac{n}{b} \right)^2 \right] w_{mn} \sin \frac{m\pi x}{a} \sin \frac{n\pi y}{b} \tag{13-68}$$

$$M_{xy} = \pi^2 D^* (1 - \nu) \sum_{m=1}^{\infty} \sum_{n=1}^{\infty} \frac{mn}{ab} w_{mn} \cos \frac{m\pi x}{a} \cos \frac{n\pi y}{b}$$

The stresses are then determined by substituting Eqs. (13-68) into Eqs. (13-34), which reduce to

$$\sigma_{xx} = \frac{E}{E_1} \left(\frac{N_x}{t^*} - \frac{M_x z}{\bar{I}^*} \right) \qquad \sigma_{yy} = \frac{E}{E_1} \left(\frac{N_y}{t^*} - \frac{M_y z}{\bar{I}^*} \right) \qquad \sigma_{xy} = -\frac{E}{E_1} \frac{M_{xy} z}{\bar{I}^*} \tag{13-69}$$

The series for w usually converge rapidly, and satisfactory accuracy is frequently obtained with only a few terms. The series given by Eqs. (13-68) converge less rapidly, especially in the region of the edges. Convergence is also slow in the vicinity of a concentrated force, and divergence occurs at the point of application of the force. A method due to M. Levy which converges more rapidly in the edge regions is described in Ref. 1.

Example 13-8 Determine the deflections in a simply supported rectangular plate with an applied pressure $p = p_0 x / a$.

For this case Eq. (13-67) becomes

$$p_{mn} = \frac{4p_0}{a^2 b} \left(\int_0^a x \sin \frac{m\pi x}{a} \, dx \right) \left(\int_0^b \sin \frac{n\pi y}{b} \, dy \right)$$

which reduces to zero when n is even and to $p_{mn} = 8p_0(-1)^{m+1}/\pi^2 mn$ when n is odd. Substituting these into Eq. (13-65) and the results into Eq. (13-64), we find

$$w = \frac{8p_0}{\pi^6 D} \sum_{m=1,2,3}^{\infty} \sum_{n=1,3,5}^{\infty} \frac{(-1)^{m+1}}{mn[(m/a)^2 + (n/b)^2]^2} \sin \frac{m\pi x}{a} \sin \frac{n\pi y}{b}$$

The first term of the series gives $w = 0.002038 p_0 a^4/D$ for the center deflection of a square plate. Summing to $m = n = 3$ gives a center deflection of $w = 0.002024 p_0 a^4/D$, which differs little from the solution for a single term.

13-11 STRAIN ENERGY OF PLATES

We obtain the strain energy in a plate by observing that as a result of the assumed stress-strain equations, the stresses at a distance z from the reference surface are in a state of plane stress. The strain energy can therefore be found from

$$U = \iint_D \left(\int_t U_0 \, dz \right) dx \, dy$$

where U_0 is given by Eq. (6-52) and D is the domain of the plate in the xy plane. The strain energy can be expressed in terms of the displacements by substituting Eqs. (13-23) into U_0. Making this substitution, carrying out the integration with respect to z, and using Eqs. (13-3), (13-28), (13-29), (13-31), and (13-32), we find

$$U = U_B + U_M + U_T \tag{13-70}$$

where U_B, U_M, and U_T are the strain energies associated with the bending, membrane, and thermal strains respectively. These are given by

$$U_B = \frac{1}{2} \iint_D D^* \left\{ \left(\frac{\partial^2 w}{\partial x^2} + \frac{\partial^2 w}{\partial y^2} \right)^2 - 2(1-\nu) \left[\frac{\partial^2 w}{\partial x^2} \frac{\partial^2 w}{\partial y^2} - \left(\frac{\partial^2 w}{\partial x \, \partial y} \right)^2 \right] \right\} dx \, dy \tag{13-71}$$

$$U_M = \frac{1}{2} \iint_D K^* \left\{ \left(\frac{\partial u}{\partial x} + \frac{\partial v}{\partial y} \right)^2 - 2(1-\nu) \left[\frac{\partial u}{\partial x} \frac{\partial v}{\partial y} - \frac{1}{4} \left(\frac{\partial v}{\partial x} + \frac{\partial u}{\partial y} \right)^2 \right. \right.$$
$$+ \frac{\partial u}{\partial x} \left[\left(\frac{\partial w}{\partial x} \right)^2 + \nu \left(\frac{\partial w}{\partial y} \right)^2 \right] + \frac{\partial v}{\partial y} \left[\left(\frac{\partial w}{\partial y} \right)^2 + \nu \left(\frac{\partial w}{\partial x} \right)^2 \right]$$
$$+ (1-\nu) \frac{\partial w}{\partial x} \frac{\partial w}{\partial y} \left(\frac{\partial v}{\partial x} + \frac{\partial u}{\partial y} \right) + \frac{1}{4} \left[\left(\frac{\partial w}{\partial x} \right)^2 + \left(\frac{\partial w}{\partial y} \right)^2 \right]^2 \right\} dx \, dy \tag{13-72}$$

$$U_T = - \iint_D \left\{ N_T \left[\frac{\partial u}{\partial x} + \frac{\partial v}{\partial y} + \frac{1}{2} \left(\frac{\partial w}{\partial x} \right)^2 + \frac{1}{2} \left(\frac{\partial w}{\partial y} \right)^2 \right] \right.$$
$$\left. + M_T \left(\frac{\partial^2 w}{\partial x^2} + \frac{\partial^2 w}{\partial y^2} \right) \right\} dx \, dy + \iint_D \left[\int_t \frac{E(\alpha T)^2}{1-\nu} \, dz \right] dx \, dy \tag{13-73}$$

These equations can be used with the Rayleigh-Ritz method to obtain approximate solutions for the large deflections of plates with in-plane restraints.[1] Equation (13-72) can be used in this same manner to analyze the deflections of membranes without initial preloads.

The terms in Eq. (13-72) that involve the products of more than two derivatives are associated with the extension of the reference plane that accompanies large lateral displacements. These terms, which result in nonlinear equations, may be neglected when the deflected surface is developable or when w is small compared to t.

The term in Eq. (13-71) that is multiplied by $1 - \nu$ is zero for an arbitrarily shaped clamped plate when D^* is constant. This is proved by integrating the $\partial^2 w/(\partial x\,\partial y)$ term twice by parts. This gives

$$\iint\limits_{D} \frac{\partial^2 w}{\partial x\,\partial y} \frac{\partial^2 w}{\partial x\,\partial y}\, dx\, dy = \oint \frac{\partial^2 w}{\partial x\,\partial y} \frac{\partial w}{\partial x}\, dx - \oint \frac{\partial w}{\partial x} \frac{\partial^2 w}{\partial y^2}\, dy$$

$$+ \iint\limits_{D} \frac{\partial^2 w}{\partial x^2} \frac{\partial^2 w}{\partial y^2}\, dx\, dy \quad (13\text{-}74)$$

If the plate is clamped, $\partial w/\partial n$ and $\partial w/\partial s$ are zero on the boundary. Taking x' and y' in Eqs. (13-4) and (13-12) in the n and s directions, we see that $\partial w/\partial x = 0$ on the boundary, and the line integrals in Eq. (13-74) vanish. Substituting the remaining term of Eq. (13-74) into Eq. (13-71), we find

$$U_B = \frac{D^*}{2} \iint\limits_{D} \left(\frac{\partial^2 w}{\partial x^2} + \frac{\partial^2 w}{\partial y^2} \right)^2 dx\, dy \quad (13\text{-}75)$$

Equation (13-75) also applies to polygonal plates with $w = 0$ on the boundary. Consider a rectangular plate, for example. The slope $\partial w/\partial x = 0$ on the edges where y is constant, while $\partial w/\partial y = 0$ so that $\partial^2 w/(\partial x\,\partial y) = \partial^2 w/\partial y^2 = 0$ on the edges where x is constant. As a result the line integrals in Eq. (13-74) vanish, and Eq. (13-75) applies.

In applying the energy principles we also require the potential of the applied forces expressed in terms of displacements. For a concentrated force P which is positive in the z direction we find

$$V = -Pw(\xi,\eta) \quad (13\text{-}76)$$

where ξ and η are the coordinates of the point of application. The potential of the pressure p_z is

$$V = - \iint\limits_{D} p_z w\, dx\, dy \quad (13\text{-}77)$$

It is convenient to consider the potentials of the in-plane forces p_x, p_y, $\bar{X}$, and $\bar{Y}$ at the same time. To simplify the derivation we divide the problem into two parts, assuming first that lateral displacements are prevented during the application of the in-plane forces and then adding the increment of work that occurs during the w displacements.

When only u and v displacements are permitted, the potential of the in-plane forces is

$$V = - \oint (\bar{X}u + \bar{Y}v) \, ds - \iint_D (p_x u + p_y v) \, dx \, dy$$

which, with Eqs. (13-17) and (13-49), becomes

$$V = - \oint [(N_x l + N_{xy} m)u + (N_{yx} l + N_y m)v] \, ds$$
$$+ \iint_D \left[\left(\frac{\partial N_x}{\partial x} + \frac{\partial N_{xy}}{\partial y} \right) u + \left(\frac{\partial N_{yx}}{\partial x} + \frac{\partial N_y}{\partial y} \right) v \right] dx \, dy \quad (13\text{-}78)$$

The second integral can be rewritten through the use of Eq. (6-69) by noting that $\partial \gamma / \partial n = (\partial \gamma / \partial x)l + (\partial \gamma / \partial y)m$. Taking the first term of the second integral of Eq. (13-78) as an example and letting $a = 1$, $\beta = N_x$, $\partial \gamma / \partial x = u$, and $\partial \gamma / \partial y = 0$, we find

$$\iint_D \frac{\partial N_x}{\partial x} u \, dx \, dy = \oint N_x u l \, ds - \iint_D N_x \frac{\partial u}{\partial x} \, dx \, dy$$

Following the same procedure on the remaining terms, we find the second integral to be

$$\oint [(N_x l + N_{xy} m)u + (N_{yx} l + N_y m)v] \, ds$$
$$- \iint_D \left[N_x \frac{\partial u}{\partial x} + N_y \frac{\partial v}{\partial y} + 2N_{xy} \left(\frac{\partial v}{\partial x} + \frac{\partial u}{\partial y} \right) \right] dx \, dy$$

which, when substituted for the second integral of Eq. (13-78), reduces the potential to

$$V = - \iint_D \left[N_x \frac{\partial u}{\partial x} + N_y \frac{\partial v}{\partial y} + 2N_{xy} \left(\frac{\partial v}{\partial x} + \frac{\partial u}{\partial y} \right) \right] dx \, dy \quad (13\text{-}79)$$

We now permit the lateral displacements to occur and determine the additional work of the in-plane forces. The increment in the potential is equal to the work done by the internal forces, so that

$$V = \iint_D \left[\int_t (\sigma_{xx} \epsilon_{xx} + \sigma_{yy} \epsilon_{yy} + \sigma_{xy} \epsilon_{xy}) \, dz \right] dx \, dy \quad (13\text{-}80)$$

where the stresses are those associated with N_x, N_y, and N_{xy}. The $\frac{1}{2}$ factor is not present in this equation because the in-plane forces were applied during the first part of the derivation and are constant during

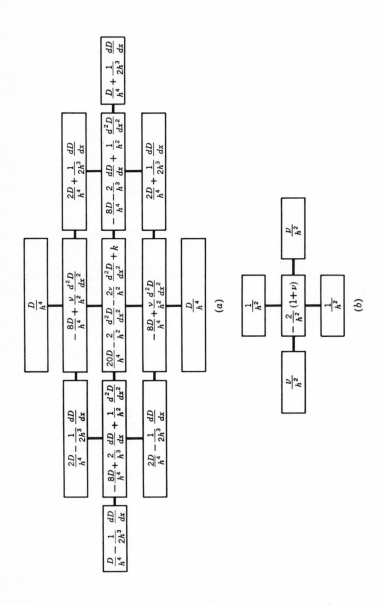

(a)

(b)

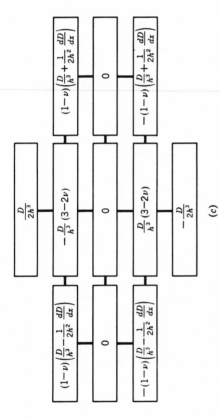

Fig. 13-18 Finite-difference modules for Example 13-9. (*a*) Differential equation module; (*b*) moment-boundary-condition module; (*c*) shear-boundary-condition module.

small lateral displacements.   From Eqs. (13-23), the strains due to w are

$$\epsilon_{xx} = -z \frac{\partial^2 w}{\partial x^2} + \frac{1}{2} \left(\frac{\partial w}{\partial x}\right)^2 \qquad \epsilon_{yy} = -z \frac{\partial^2 w}{\partial y^2} + \frac{1}{2} \left(\frac{\partial w}{\partial y}\right)^2$$

$$\epsilon_{xy} = -2z \frac{\partial^2 w}{\partial x \, \partial y} + \frac{\partial w}{\partial x} \frac{\partial w}{\partial y}$$

Using Eqs. (13-34), the stresses due to N_x, N_y, and N_{xy} are

$$\sigma_{xx} = \frac{E}{E_1} \frac{N_x}{t^*} \qquad \sigma_{yy} = \frac{E}{E_1} \frac{N_y}{t^*} \qquad \sigma_{xy} = \frac{E}{E_1} \frac{N_{xy}}{t^*}$$

Substituting the strains and stresses into Eq. (13-80), integrating with respect to z, and applying Eqs. (13-2) and (13-3), we find

$$V = \frac{1}{2} \iint_D \left[N_x \left(\frac{\partial w}{\partial x}\right)^2 + N_y \left(\frac{\partial w}{\partial y}\right)^2 + 2N_{xy} \frac{\partial w}{\partial x} \frac{\partial w}{\partial y} \right] dx \, dy \qquad (13\text{-}81)$$

The total potential of p_x, p_y, $\bar{X}$, and $\bar{Y}$ is then the sum of Eqs. (13-79) and (13-81).

The potential of the edge moments and shears is

$$V = - \oint \left(M_n \frac{\partial w}{\partial n} + M_{ns} \frac{\partial w}{\partial s} - Q_n w \right) ds \qquad (13\text{-}82)$$

13-12 APPROXIMATE METHODS

The differential equations for plates are simpler than those of the three-dimensional theory of elasticity.   Even so, it is often necessary to use approximate methods to obtain solutions.   This is especially true when the plate is of arbitrary planform or variable stiffness, rests on a non-uniform foundation, or has complicated boundary conditions.   In these cases the Rayleigh-Ritz and finite-difference methods are helpful.   The Rayleigh-Ritz method can also be used for stiffened plates,[12] and a modification of the method which reduces the problem to solving ordinary differential equations is frequently useful.[13]   The following examples illustrate the Rayleigh-Ritz and finite-difference methods; additional examples involving plate buckling are given in Sec. 15-5.

Example 13-9   Obtain the finite-difference modules for Example 13-1 and describe their application.

Equation (a) of Example 13-1 is of the form of Eq. (5-13), where

$$L_{2m} = D\nabla^4 + 2 \frac{dD}{dx} \left(\frac{\partial^3}{\partial x^3} + \frac{\partial^3}{\partial x \, \partial y^2}\right) + \frac{d^2 D}{dx^2} \left(\frac{\partial^2}{\partial x^2} + \nu \frac{\partial^2}{\partial y^2}\right) + k \qquad (a)$$

The analogous difference operator, obtained by superimposing the appropriate modules from Figs. 5-4 and 5-5, is shown in Fig. 13-18a.   The same method is used to obtain the difference operators that correspond to the boundary

operators B_i in Eq. (5-14).   The modules that are associated with the operators
of Eqs. (b) of Example 13-1 are shown in Fig. 13-18b and c.

We superimpose a difference mesh upon the plate, as shown in Fig. 13-19.
For simplicity it is assumed that the plate dimensions are such that boundary
points coincide with mesh points and that m rows and n columns of interior
mesh points are used.   We operate on w at the mesh points by successively
applying the module of Fig. 13-18a at each of the $m \times n$ interior points, equating
the result equal to p at the point.

In this way, $m \times n$ equations are obtained.   However, when the module
is applied to points next to the boundary, we involve the points on and outside
the boundary that are shown in Fig. 13-19.   Points on the boundaries $x = 0$
and a and $y = 0$ present no problems, since $w = 0$ for these edges.   The fictitious points outside of these same boundaries are also simple to handle and
add no additional unknowns to the problem.   On the edges $x = 0$ and a the
condition $\partial^2 w / \partial x^2 = 0$ is satisfied by noting that the corresponding difference

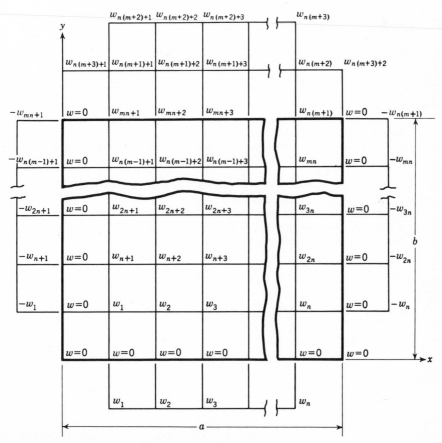

Fig. 13-19   Finite-different mesh for Example 13-9.

operator in Fig. 5-4 requires that w at each of the points outside of the boundary be equal to $-w$ at the corresponding point inside of the boundary.

In a similar fashion the $\partial w/\partial y = 0$ condition at $y = 0$ is satisfied by taking w at points below this edge equal to w at the corresponding point above this boundary. This is indicated in Fig. 13-19 by using the same symbol to designate w at matching exterior and interior points.

The module of Fig. 13-18c is applied to each of the boundary points at $y = b$, excluding the corner points. In doing this we involve w at fictitious points that are two mesh distances above the corresponding boundary points and the points that are one mesh space above the upper corner points. The unknowns now consist of $m \times n$ interior point displacements, n boundary point displacements on the boundary $y = b$, $2n$ fictitious displacements at points above the $y = b$ boundary, and the displacements at the two points above the upper corners. This gives a total of $n(m + 3) + 2$ unknowns.

The total number of equations written to this point is $n(m + 1)$. If we apply the module of Fig. 13-18b to each of the $n + 2$ points on the edge $y = b$ (including the corner points), we raise the number of equations to $n(m + 2) + 2$ and add no new unknowns. The n equations that we lack are obtained by applying the module of Fig. 13-18a to each of the points on the boundary $y = b$ except the corner points. We justify this by noting that the differential equation to which this module is equivalent must be satisfied at an infinitesimal distance inside of the boundary.

An alternate method is to use backward-difference formulas in constructing the modules so that points outside of the boundary are not involved.

Example 13-10 Use the Rayleigh-Ritz method to determine the deflections of a uniform simply supported rectangular plate with a concentrated force P applied at a point with coordinates (ξ, η).

No in-plane forces are applied, and so for small deflections we find from Eqs. (13-75) and (13-76) that

$$U + V = \frac{D^*}{2} \int_0^b \int_0^a \left(\frac{\partial^2 w}{\partial x^2} + \frac{\partial^2 w}{\partial y^2} \right)^2 dx\, dy - Pw(\xi, \eta) \qquad (a)$$

The assumed functions must satisfy the displacement boundary condition $w = 0$. A series that meets this requirement and the additional (though not essential) boundary condition $M_n = 0$ is

$$w = \sum_{m=1}^{\infty} \sum_{n=1}^{\infty} w_{mn} \sin \frac{m\pi x}{a} \sin \frac{n\pi y}{b} \qquad (b)$$

The w_{mn} coefficients define the equilibrium configuration and are therefore generalized coordinates. Letting $w_{rs} = q_i$ and applying Eq. (6-17), we find

$$D^* \int_0^b \int_0^a \left(\frac{\partial^2 w}{\partial x^2} + \frac{\partial^2 w}{\partial y^2} \right) \frac{\partial}{\partial w_{rs}} \left(\frac{\partial^2 w}{\partial x^2} + \frac{\partial^2 w}{\partial y^2} \right) dx\, dy - P \frac{\partial w(\xi, \eta)}{\partial w_{rs}} = 0$$

Substituting Eq. (b) into this result and rearranging, we find

$$\pi^4 D^* \sum_{m=1}^{\infty} \sum_{n=1}^{\infty} \left(\frac{m^2}{a^2} + \frac{n^2}{b^2} \right) \left(\frac{r^2}{a^2} + \frac{s^2}{b^2} \right) \left(\int_0^a \sin \frac{m\pi x}{a} \sin \frac{r\pi x}{b} dx \right)$$
$$\left(\int_0^b \sin \frac{n\pi y}{b} \sin \frac{s\pi y}{b} dy \right) w_{mn} = P \sin \frac{r\pi \xi}{a} \sin \frac{s\pi \eta}{b}$$

Applying Eq. (13-66) and a corresponding expression for the y integration, we obtain

$$w_{rs} = \frac{4P}{\pi^4 D^* ab[(r/a)^2 + (s/b)^2]^2} \sin \frac{r\pi\xi}{a} \sin \frac{s\pi\eta}{b}$$

which when substituted into Eq. (b) gives

$$w = \frac{4P}{\pi^4 D^* ab} \sum_{m=1}^{\infty} \sum_{n=1}^{\infty} \frac{\sin (m\pi x/a) \sin (n\pi y/b) \sin (m\pi\xi/a) \sin (n\pi\eta/b)}{[(m/a)^2 + (n/b)^2]^2} \quad (13\text{-}83)$$

This series converges rapidly, and satisfactory results can usually be obtained with only a few terms. The series solutions for the moments, obtained by substituting Eq. (13-83) into Eqs. (13-30), converge more slowly and are unsatisfactory in the vicinity of the applied force, where divergence occurs. Methods that are better suited for computing moments and shears in the vicinity of concentrated forces are described in Refs. 1 and 7. The concept of a concentrated force is fictitious, of course, since all forces are distributed over a finite though sometimes small area. When the dimensions of the area over which the force is applied are of the order of the plate thickness, shearing stresses become important, and stresses in this region should be computed by thick-plate or three-dimensional elasticity theories.[1]

We see from Example 13-9 that the number of simultaneous equations that must be solved in the finite-difference method is usually large, and the method is practical only when a digital computer is used. The method has the advantage of being able to handle variable stiffnesses in the plate and foundation, complicated boundary conditions, and arbitrary pressure distributions with little additional difficulty over the case in which these properties are constant.

When the plate is uniform, simply supported, and polygonal, it is often more convenient to reduce the fourth-order differential equation to two second-order equations. The manner in which this is accomplished and examples in which the finite-difference method is used to solve the resulting equations are given in Ref. 1.

When the influence function $c(x,y,\xi,\eta)$ [the displacement at (x,y) due to a unit force at (ξ,η)] is known, the displacements for a distributed force can be found by the principle of superposition, as described in Sec. 10-6. The deflection at (x,y) due to a pressure p_z acting over a differential area $d\xi\, d\eta$ at (ξ,η) is $cp_z\, d\xi\, d\eta$. Applying the principle of superposition, we find for an arbitrary pressure distribution

$$w(x,y) = \iint_D c(x,y,\xi,\eta) p_z(\xi,\eta)\, d\xi\, d\eta \quad (13\text{-}84)$$

For a uniform simply supported rectangular plate, $c(x,y,\xi,\eta)$ may be found by setting $P = 1$ in Eq. (13-83). It is sometimes necessary to use numerical integration to evaluate the integral in Eq. (13-84). An effi-

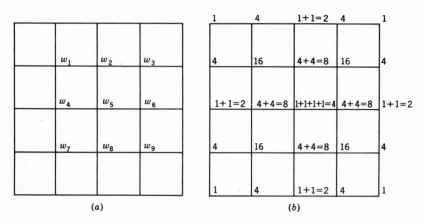

Fig. 13-20 (*a*) Finite-difference mesh for evaluation of Eq. (13-84); (*b*) Simpson's-rule module.

cient method for doing this is to superimpose a finite-difference grid on the plate and determine the influence coefficient c_{ij} at each of the mesh points. Applying the collocation method to the integral equation (13-84) reduces it to the matrix equation (10-35), where $\mathbf{w}$ and $\mathbf{p}$ are the deflections and pressures at the mesh points. When the plate is uniform, simply supported, and rectangular, the elements of $\mathbf{c}$ can be computed from Eq. (13-83). Experimentally determined influence coefficients can also be used for the $\mathbf{c}$ matrix.

The diagonal matrix $\overline{\mathbf{W}}$ contains the weighting numbers associated with the method that is used for the numerical integration. As an example, consider the plate in Fig. 13-20*a*, in which $w = 0$ at the boundary points. If we use Simpson's rule integration and superimpose the module of Fig. 5-5 on each of the sets of four difference blocks, we obtain the weighting numbers shown in Fig. 13-20*b*. As a result,

$$\overline{\mathbf{W}} = \frac{h^2}{9} \begin{bmatrix} 16 & 8 & 16 & 8 & 4 & 8 & 16 & 8 & 16 \end{bmatrix}$$

where the slashes in the matrix brackets denote a diagonal matrix written horizontally to conserve space.

REFERENCES

1. Timoshenko, S., and S. Woinowsky-Krieger: "Theory of Plates and Shells," 2d ed., McGraw-Hill Book Company, New York, 1959.
2. Love, A. E. H.: "A Treatise on the Mathematical Theory of Elasticity," 4th ed., Dover Publications, Inc., New York, 1927.

3. Mansfield, E. H.: "Bending and Stretching of Plates," Pergamon Press, New York, 1963.

4. Jaeger, J. G.: "Elementary Theory of Elastic Plates," Pergamon Press, New York, 1964.

5. Boley, B. A., and J. H. Weiner: "Theory of Thermal Stresses," John Wiley & Sons, Inc., New York, 1960.

6. Johns, D. J.: "Thermal Stress Analysis," Pergamon Press, New York, 1965.

7. Flugge, W.: "Handbook of Engineering Mechanics," McGraw-Hill Book Company, New York, 1962.

8. Reissner, E.: The Effect of Transverse Shear Deformation on the Bending of Elastic Plates, *J. Appl. Mech.*, **12**(3): A69–77 (June, 1945).

9. Marguerre, K.: Zur Theorie der gekrümmten Platte grosser Formänderung, *Proc. 5th Intern. Congr. Appl. Mech.*, Cambridge, Mass., pp. 93–101, 1938.

10. Gossard, M. L., P. Seide, and W. M. Roberts: Thermal Buckling of Plates, *NACA Tech. Note* 2771, August, 1952.

11. Heldenfels, R. R., and W. M. Roberts: Experimental and Theoretical Determination of Thermal Stresses in a Flat Plate, *NACA Tech. Note* 2769, August, 1952.

12. Rivello, R. M., and S. Durvasula: Deflection of Shaft Supported Aerodynamic Surfaces, *Aerospace Eng.*, **21**(3): 64–65, 92–100 (March, 1962).

13. Stein, M., J. Anderson, and J. M. Hedgepeth: Deflection and Stress Analysis of Thin Solid Wings of Arbitrary Planform with Particular Reference to Delta Wings, *NACA Rept.* 1131, 1954.

PROBLEMS

13-1. Write the equations for the principal curvatures and the maximum twist in terms of the curvature and twist in the x and y directions.

13-2. Show by the Navier method that the small deflections of a uniformly loaded rectangular plate with simply supported edges are given by

$$w = \frac{16p_0}{\pi^6 D^*} \sum_{m=1,3,5}^{\infty} \sum_{n=1,3,5}^{\infty} \frac{\sin{(m\pi x/a)} \sin{(n\pi y/b)}}{mn[(m/a)^2 + (n/b)^2]^2}$$

where p_0 is the magnitude of the lateral load. The edges of the plate are at $x = 0$ and a and $y = 0$ and b.

13-3. Determine the center deflection of a uniformly loaded square plate with simply supported edges by using the first four terms of the series of Prob. 13-2. [*Ans.* $w = 0.00416p_0a^4/D^*$.]

13-4. Solve Prob. 13-2 by the Rayleigh-Ritz method.

13-5. Obtain an approximate solution for the center deflection of the plate of Prob. 13-3 by the finite-difference method, taking $h = a/4$ and observing the symmetries of the problem.

13-6. The inverse method may be used with plate theory. Determine the lateral load, edge shears, and moments for the displacement functions $w = c_{20}x^2$ and $w = c_{11}xy$. Assume the plate is rectangular with boundaries at $x = \pm a/2$ and $y = \pm b/2$.

13-7. A uniform rectangular plate is heated so that the temperature on the upper surface is greater than that of the lower face by an amount ΔT, with a linear variation between the faces. The edges are simply supported and permit free in-plane expansion. Determine the center deflection by the Rayleigh-Ritz method.

13-8. Choose a suitable Rayleigh-Ritz series for the deflection of each of the following rectangular plates:

(a) The edge $x = 0$ clamped and the edges $x = a$ and $y = \pm b/2$ free.

(b) The edges $x = 0$ and a and $y = 0$ and b clamped.

13-9. A molybdenum plate coated with a layer of zirconium dioxide is used as a heat shield to protect the primary substructure of a lifting reentry vehicle. The panel supports provide unrestrained expansion, so that the plate may be considered free. The outer ZrO_2 layer is 0.10 in. thick, and the Mo layer is 0.050 in. During reentry the panel is heated until there is a linear gradient through the ZrO_2 ranging from 3500°F at its surface to 1500°F at its interface with the Mo. At this time the temperature of the Mo is a uniform 1500°F. The initial temperature of the panel is 70°F. Determine the stresses at each of the panel faces and in each of the materials at the interface assuming the following properties:

	E, 10^6 psi	α, 10^{-6} $in./(in.)(°F)$	ν
ZrO_2	13	3.1	0.3
Mo	34	2.7	0.3

13-10. Use Eqs. (4-25) and (13-48) to show that the differential equation for the small deflection of a homogeneous circular plate carrying an axisymmetric lateral pressure is given by

$$\frac{1}{r}\frac{d}{dr}\left\{ r\frac{d}{dr}\left[\frac{1}{r}\frac{d}{dr}\left(r\frac{dw}{dr}\right)\right]\right\} = \frac{p_z}{D}$$

Hint: Note that

$$\frac{d^2w}{dr^2} + \frac{1}{r}\frac{dw}{dr} = \frac{1}{r}\frac{d}{dr}\left(r\frac{dw}{dr}\right)$$

13-11. When p_z is known, the differential equation of Prob. 13-10 can be solved by successively separating variables and integrating. Show that the deflections for a plate with a uniform lateral pressure p_0 and clamped edges is

$$w = \frac{p_0}{64D}(a^2 - r^2)^2$$

where a is the radius to the edge of the plate.

14

Primary Bending Instability and Failure of Columns

14-1 INTRODUCTION

The thin webs and slender longitudinals of flight-vehicle structures are subject to failure by buckling at relatively low stress levels, frequently below the proportional limit and seldom appreciably above the yield stress. As a result, buckling rather than tensile rupture is the critical mode of failure for the major portion of the structure, and the prediction of buckling loads for columns, plates, and shells is a subject of vital concern to the aerospace engineer. In this chapter we consider the simplest of these elements, the column.

A column may buckle by *primary* or *secondary instability*. In the former there is no distortion of the cross section, and the wavelength of the buckle is on the order of the column length; whereas in secondary, or *local*, instability there is a change in the cross-sectional shape, and the wavelength of the buckle is on the order of the cross-sectional dimensions. Primary instability may occur by lateral bending or, if the section is torsionally flexible, by a combination of twisting and bending. In this chapter we consider only primary bending instability (for a discussion of torsional-flexure instability see Ref. 1). Plate buckling is treated in Chap. 15, and column and stiffened-plate failure by local instability is described in Chap. 16.

In spite of its relative simplicity, column theory has been the subject of controversy from the time of the first significant contributions by Euler in 1744 until the present day. Interesting accounts of the evolution of the theory are given in Refs. 1 to 3. Because of the fundamental role that the column has played in the development of structural-instability theory, the subject of column buckling will be presented in much the same manner that it developed.

We first consider the *small-deflection elastic theory* for *perfect columns* that buckle below the proportional limit. A perfect column is one in which the modulus-weighted centroidal axis is initially perfectly straight and in which the compressive force is precisely applied along this axis. The results of this theory will then be reexamined by considering the effects of initial imperfections and large deflections. Finally we discuss the theory for perfect and imperfect columns that buckle in the inelastic range.

14-2 SMALL DEFLECTIONS OF LINEARLY ELASTIC PERFECT COLUMNS

When a perfect column is subjected to a compressive axial force P, the only deformation that occurs is an axial shortening. To examine the stability of this compressed equilibrium position we imagine the column to be disturbed into a slightly bent position and released. For small values of P the column will return to the perfectly straight form when it is released, indicating that in this case the straight position is a stable equilibrium configuration. As P is increased, a point is reached where the column no longer returns to its initially straight position but remains in the slightly bent shape. At this critical value of P, which we designate by P_{cr}, a state of neutral stability exists. If a force $P > P_{cr}$ is applied to the perfect column and the column is disturbed and released, it will move away from the straight and disturbed positions. The perfect column is therefore unstable for $P > P_{cr}$, and the design load should not exceed P_{cr}.

A *bifurcation point* occurs at P_{cr}, where both the straight and slightly bent positions are equilibrium configurations. At this load the bent column is said to be buckled. The equilibrium equation that the lateral displacements and P must satisfy in the slightly bent position is found from Eq. (7-46). Assuming that the only loading is the axial force and noting that it is compressive and therefore negative, we find

$$(E_1 I^* w'')'' + P w'' = 0 \tag{14-1}$$

Letting $I^* = I_0^* g(x)$, where I_0^* is a constant that establishes the magnitude of I^* and $g(x)$ is a function that describes its distribution, we may write Eq. (14-1) as

$$g w^{iv} + 2g' w''' + g'' w'' = -\lambda w'' \tag{14-2}$$

where

$$\lambda = \frac{P}{E_1 I_0^*} \tag{14-3}$$

The two boundary conditions at each end depend upon the end restraints. At each end, conditions for either the deflection or shear and the slope or moment are known. In general, the boundary conditions for the ends $x = 0$ and L are of the form

$$
\begin{aligned}
B_1 \psi(0) &= \lambda C_1 \psi(0) & B_2 \psi(0) &= \lambda C_2 \psi(0) \\
B_1 \psi(L) &= \lambda C_1 \psi(L) & B_2 \psi(L) &= \lambda C_2 \psi(L)
\end{aligned}
\tag{14-4}
$$

where the B_i and C_i are differential operators.

The value of P for equilibrium in the bent position is not known, so that Eqs. (14-2) and (14-4) are of the form of Eqs. (5-18) and (5-19). The determination of P for a nontrivial solution for w is therefore an eigenvalue problem. It was shown in preceding chapters that this form is also typical of vibration and static-aeroelastic instability problems. These problems can then be solved by the same analytical techniques that are described in this and the next section for column buckling.

The solution of Eq. (14-2) is of the general form

$$w = C_1 f_1(\lambda,x) + C_2 f_2(\lambda,x) + C_3 f_3(\lambda,x) + C_4 f_4(\lambda,x) \tag{14-5}$$

where the coefficients C_i are constants of integration and the functions $f_i(\lambda,x)$ are independent solutions of Eq. (14-2). Substituting Eq. (14-5) into (14-4) results in the following set of equations that are homogeneous and linear in the C_i:

$$\begin{aligned}
a_{11}C_1 + a_{12}C_2 + a_{13}C_3 + a_{14}C_4 = 0 \\
a_{21}C_1 + a_{22}C_2 + a_{23}C_3 + a_{24}C_4 = 0 \\
a_{31}C_1 + a_{32}C_2 + a_{33}C_3 + a_{34}C_4 = 0 \\
a_{41}C_1 + a_{42}C_2 + a_{43}C_3 + a_{44}C_4 = 0
\end{aligned} \tag{14-6}$$

where the a_{ij} are functions of λ and x that are evaluated at $x = 0$ or L. These equations have the trivial solution $C_i = 0$ ($i = 1$ to 4). We find from Eq. (14-5) that $w = 0$ for this case, indicating that the perfectly straight position is an equilibrium configuration for all values of P. This solution is of no interest, however, for we are seeking the smallest value of P that can maintain equilibrium in the bent configuration. For this case, all the C_i cannot be zero.

Equations (14-6) are of the form $\mathbf{aC} = \mathbf{0}$. If we attempt to obtain C_i by Cramer's rule of determinants, we find

$$C_i = \frac{|\bar{a}|}{|a|}$$

where $|a|$ is the determinant of $\mathbf{a}$ and $|\bar{a}|$ is the determinant of $\mathbf{a}$ with the ith column replaced by zeros. Recalling that a determinant is zero if all the elements of a column are zero, we see that a nontrivial solution is possible only if

$$|a| = 0 \tag{14-7}$$

which gives the indeterminate form $C_i = 0/0$.

Expansion of $|a|$ in Eq. (14-7) leads to the *characteristic equation* of the problem, which will generally include transcendental functions with the argument $\lambda^{\frac{1}{2}}L$. Solutions to the characteristic equations are possible only for certain values of λ. The roots of the characteristic equation are the eigenvalues $\lambda_1, \lambda_2, \ldots$, and the smallest of these gives P_{cr} when it is substituted into Eq. (14-3).

For each λ_i there is an associated mode shape w_i found by substituting λ_i into Eq. (14-6). However, since $|\mathbf{a}| = 0$, the resulting equations are not linearly independent and cannot be solved for absolute values of the C_i. If we discard one of the four equations (taking care that no two of the remaining equations are linearly dependent), we shall have three equations and four unknown C_i. By transferring one of the unknowns (say C_j) to the right side, we can determine the remaining unknowns in terms of C_j. Substituting these into Eq. (14-5) then gives the mode shape multiplied by the undetermined constant C_j.

If both ends of the column are pinned or one end is free and the other is clamped, Eq. (14-1) can be reduced to a second-order differential equation. Two successive integrations of Eqs. (14-1) give

$$(E_1 I^* w'')' + Pw' + A = 0 \tag{14-8}$$

$$E_1 I^* w'' + Pw + Ax + B = 0 \tag{14-9}$$

where A and B are constants of integration. For the column with both ends pinned the boundary conditions are

$$w(0) = 0 \qquad w''(0) = 0 \qquad w(L) = 0 \qquad w''(L) = 0 \tag{14-10}$$

Evaluating Eq. (14-9) at $x = 0$ and L we find

$$E_1 I^*(0)w''(0) + Pw(0) + B = 0$$

$$E_1 I^*(L)w''(L) + Pw(L) + AL + B = 0$$

Substituting Eqs. (14-10) into these equations, we find $A = B = 0$, so that Eq. (14-9) reduces to

$$gw'' + \lambda w = 0 \tag{14-11}$$

The boundary conditions for this second-order differential equation are

$$w(0) = 0 \qquad w(L) = 0 \tag{14-12}$$

The boundary conditions associated with Eq. (14-1) for the column that is free at $x = 0$ and clamped at $x = L$ are

$$
\begin{aligned}
w(0) &= 0 & w''(0) &= 0 \\
w'(L) &= 0 & (E_1 I^* w'')'_{x=L} &= 0
\end{aligned}
\tag{14-13}
$$

if w is measured from the line of action of the applied force as shown in Fig. 14-1. Evaluating Eq. (14-8) at $x = L$ and Eq. (14-9) at $x = 0$ and using Eqs. (14-13), we again find that $A = B = 0$. Therefore Eq. (14-11) also applies to the free-clamped column, where the boundary conditions are

$$w(0) = 0 \qquad w'(L) = 0 \tag{14-14}$$

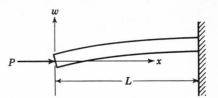

Fig. 14-1 Coordinate system for free-clamped column.

Equation (14-11) is easier to solve than Eq. (14-2) and is therefore preferable for pinned-pinned and free-clamped columns.

Example 14-1 Determine the buckling load of a uniform column with pinned ends. For a uniform column, $g(x) = 1$, and the solution to Eq. (14-11) is

$$w = C_1 \sin kx + C_2 \cos kx \qquad (a)$$

where $k^2 = \lambda$. Substituting Eq. (a) into Eqs. (14-12) results in the equations

$$(\sin 0)C_1 + (\cos 0)C_2 = 0$$
$$(\sin kL)C_1 + (\cos kL)C_2 = 0 \qquad (b)$$

For a nontrivial solution

$$|a| = \begin{vmatrix} 0 & 1 \\ \sin kL & \cos kL \end{vmatrix} = 0$$

which gives the characteristic equation $\sin kL = 0$, whose roots are $kL = n\pi$ ($n = 1, 2, \ldots$). The smallest value for P is obtained for $n = 1$, which gives $kL = L(P_{cr}/E_1 I^*)^{1/2} = \pi$ or

$$P_{cr} = P_E = \frac{\pi^2 E_1 I^*}{L^2}$$

where P_E is known as the *Euler load*.

To obtain the associated mode shape we substitute $k = \pi/L$ into Eqs. (b), which gives

$$(\sin 0)C_1 + (\cos 0)C_2 = 0$$
$$(\sin \pi)C_1 + (\cos \pi)C_2 = 0$$

We see that these equations are not linearly independent because the second equation is obtained by multiplying the first by -1. Discarding one of these equations and solving the other, we find $C_2/C_1 = 0$. Substituting this result and $k = \pi/L$ into Eq. (a) gives the mode shape $w = C_1 \sin (\pi x/L)$.

Example 14-2 Determine the buckling load of a uniform column that is clamped at one end and simply supported at the other.

When $E_1 I^*$ is constant, Eq. (14-2) reduces to

$$w^{iv} + k^2 w'' = 0 \qquad (a)$$

where, as before, $k^2 = P/E_1 I^*$. The solution of Eq. (a) is

$$w = C_1 \sin kx + C_2 \cos kx + C_3 x + C_4 \qquad (b)$$

The boundary conditions for the end restraints are

$$w(0) = 0 \qquad w'(0) = 0 \qquad w(L) = 0 \qquad w''(L) = 0 \tag{c}$$

Substituting Eq. (b) into Eqs. (c) gives

$$C_2 + C_4 = 0 \qquad kC_1 + C_3 = 0$$
$$(\sin kL)C_1 + (\cos kL)C_2 + LC_3 + C_4 = 0 \tag{d}$$
$$-(k^2 \sin kL)C_1 - (k^2 \cos kL)C_2 = 0$$

For a nontrivial solution of Eqs. (d)

$$|\mathbf{a}| = \begin{vmatrix} 0 & 1 & 0 & 1 \\ k & 0 & 1 & 0 \\ \sin kL & \cos kL & L & 1 \\ -k^2 \sin kL & -k^2 \cos kL & 0 & 0 \end{vmatrix} = 0$$

which gives the characteristic equation

$$k^2(kL \cos kL - \sin kL) = 0$$

The quantity $k^2 \neq 0$, since this would imply that $P = 0$; therefore the term in the parentheses must vanish. This gives the transcendental equation

$$\tan kL = kL \tag{e}$$

which can be solved for kL (in radians) by referring to a table of tangents. The solution to Eq. (e) can also be found by the graphical construction shown in Fig. 14-2. The curves $y = kL$ and $y = \tan kL$ are plotted on the same y and kL axes. The circled points of intersection give the roots of the characteristic equation, the smallest of which is at $kL = 4.49$, giving

$$P_{\text{cr}} = \frac{2.05\pi^2 E_1 I^*}{L^2} \tag{f}$$

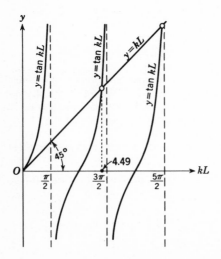

Fig. 14-2 Solution of transcendental equation, Example 14-2.

Discarding the third of Eqs. (d) and letting $k = 4.49/L$ in the remaining equations, we find

$$C_3 = -\frac{4.49}{L}C_1 \qquad C_2 = -4.49C_1 \qquad C_4 = 4.49C_1 \qquad (g)$$

Substituting these results into Eq. (b) gives the mode shape

$$w = 4.49C_1 \left(\frac{1}{4.49} \sin \frac{4.49x}{L} - \cos \frac{4.49x}{L} - \frac{x}{L} + 1 \right) \qquad (h)$$

We observe from the foregoing examples that the buckling loads of pinned-pinned and clamped-pinned columns are proportional to the parameter $\pi^2 E_1 I^*/L^2$. This is also true for other end conditions, so that in general we may write

$$P_{cr} = \frac{c\pi^2 E_1 I^*}{L^2} \qquad (14\text{-}15)$$

where c is a constant, known as the *coefficient of end fixity*, which depends upon the end restraints.

When the column is homogeneous, the compressive stress at P_{cr} is uniform over the cross section and is given by

$$\sigma_{cr} = \frac{c\pi^2 E}{(L/\rho)^2} \qquad (14\text{-}16)$$

where $\rho = (I/A)^{1/2}$ is the radius of gyration of the cross section. The dimensionless ratio L/ρ is known as the *slenderness ratio*. Equation (14-16) is often written in the form

$$\sigma_{cr} = \frac{\pi^2 E}{(L'/\rho)^2} \qquad (14\text{-}17)$$

where L' is the *effective length* defined by

$$L' = \frac{L}{\sqrt{c}} \qquad (14\text{-}18)$$

To obtain a physical interpretation of the effective length we return to the clamped-pinned column in Example 14-2. Differentiating Eq. (h) in this example twice, we obtain the curvature equation

$$w'' = -\frac{4.49^2 C_1}{L^2} \left(\sin \frac{4.49x}{L} - 4.49 \cos \frac{4.49x}{L} \right)$$

From this equation we find that $w'' = 0$ at $x = 0.30L$ and L. The moment is also zero at these points, and so the column between these points behaves like a pinned-ended column with a length of $0.7L = L/2.05^{1/2}$. The distance between inflection points of the deflected column is therefore equal to the effective length. Values of L' and c for frequently encountered end conditions are given in Fig. 14-3.

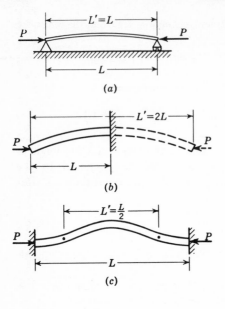

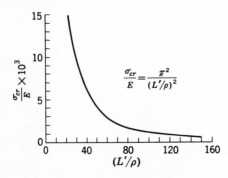

Fig. 14-3 Effective lengths and end-fixity coefficients of commonly encountered end conditions. (a) Pinned ends, $c = 1.0$; (b) free-clamped ends, $c = \frac{1}{4}$; (c) clamped-clamped ends, $c = 4.0$; (d) clamped-pinned ends, $c = 2.05$.

Equation (14-17) is plotted in Fig. 14-4 in terms of the nondimensional parameters σ_{cr}/E and L'/ρ. This curve is applicable to all materials and end conditions; however, caution should be observed in using Fig. 14-4 or Eq. (14-17) to be certain that the predicted stresses do not exceed the proportional limit. In such cases it is necessary to modify the

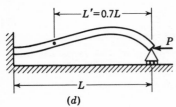

Fig. 14-4 Nondimensionalized Euler column curve.

results to account for the plastic behavior of the material.   The theory for inelastic buckling of columns is given in Sec. 14-6.

14-3 APPROXIMATE METHODS

It is frequently impossible to obtain an exact closed-form solution to Eq. (14-2) [or (14-11)] when the column has a variable cross section.   Even if the general solution indicated by Eq. (14-5) is known, it may be difficult to obtain the roots of the transcendental characteristic equation.   Determining the exact solution may also be difficult when the column is elastically restrained.   In these cases it is often more convenient (and frequently necessary) to resort to approximate methods.

In this section we shall apply the finite-difference and Rayleigh-Ritz methods to determine the column buckling load.   A more complete discussion of approximate methods for solving eigenvalue problems is given in Ref. 4.

Both the finite-difference and Rayleigh-Ritz methods reduce the continuum system with its infinite degrees of freedom to an approximating system with n degrees of freedom.   Both methods result in a set of homogeneous simultaneous equations of the form of Eqs. (5-21), which may be written in the matrix form of Eq. (5-22).   The n unknown ψ_i in these equations are the displacements at the mesh points in the finite-difference method and the undetermined coefficients of the series solution in the Rayleigh-Ritz method.   The eigenvalue λ is a function of the applied load.   Solutions to the matrix equation can be obtained by the methods described in Sec. 5-5.

As in equilibrium problems, it is usually possible to obtain acceptable accuracy with fewer unknowns by the Rayleigh-Ritz method than by the finite-difference method if physical intuition is used in choosing the functions.   Sometimes a single term suffices for the Rayleigh-Ritz solution.

The Rayleigh-Ritz method always produces matrices **a** and **b** that are positive definite;[4] this is not assured by the finite-difference method. This property is required in some eigenvalue computer programs to guarantee the convergence of the solution.   On the other hand, the difference method does not involve the evaluation of integrals, and when n is large, the only nonzero elements of the matrices are in a band about the main diagonal.

The assumption of functions in the Rayleigh-Ritz method imposes constraints on the deformations, so that the predicted buckling load is always an upper limit to the true buckling load.   In the difference method there are no assurances of whether the predicted buckling load is an upper or lower bound to the exact value.

Finite-difference method   Equation (14-11) applies when the column has both ends pinned or is free at one end and clamped at the other.   By using the computational module for d^2/dx^2 from Fig. 5-2, this equation becomes

$$\frac{I_i^*}{I_0^*} (w_{i-1} - 2w_i + w_{i+1}) = - \frac{PL^2}{n^2 E_1 I_0^*} w_i \qquad (14\text{-}19)$$

where n is the number of segments into which the column is divided by the mesh and $I_i^* = I^*(x_i)$.   For other end conditions Eq. (14-2) must be used.   Applying the appropriate computational modulus from Fig. 5-2 to this equation gives

$$\left(g - \frac{L}{n} g' \right) w_{i-2} - \left(4g - 2\frac{L}{n} g' - \frac{L^2}{n^2} g'' \right) w_{i-1}$$

$$+ 2 \left(3g - \frac{L^2}{n^2} g'' \right) w_i - \left(4g + 2\frac{L}{n} g' + \frac{L^2}{n^2} g'' \right) w_{i+1}$$

$$+ \left(g + \frac{L}{n} g' \right) w_{i+2} = - \frac{PL^2}{n^2 E_1 I_0^*} (w_{i-1} - 2w_i + w_{i+1}) \qquad (14\text{-}20)$$

where g and its derivatives are evaluated at $x = x_i$.   When g is not known analytically, its derivatives may be written in finite-difference form in terms of g_i at the mesh points.

If $E_1 I^*$ and the end-support conditions are symmetric about the center of the column, the buckle shape will also be symmetric about the center, and the same subscript should be used to designate points that are at equal distances from the center.   This decreases the number of unknowns for a given mesh size.

When Eq. (14-19) is applied to interior points next to the ends, w at boundary points becomes involved.   The application of Eq. (14-20) to points next to the ends requires the use of fictitious points outside of the boundary.   The additional equations required for a solution are obtained by applying the boundary conditions to the end points.   For the free-clamped column it is also necessary to apply Eq. (14-19) at the clamped end to obtain as many equations as there are unknowns.

Examples 5-2 and 5-3 illustrate the application of the difference method to column-buckling problems.

Rayleigh-Ritz method   The Rayleigh-Ritz method can be used with the principle of the stationary value of the total potential to obtain approximate solutions for the buckling load.   For generality we suppose that the column may rest on an elastic foundation and have elastic supports, as shown in Fig. 14-5.   The total strain energy in the buckled position,

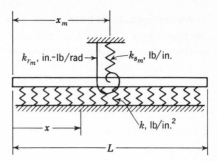

Fig. 14-5 Elastic supports.

obtained from Eqs. (7-55), (7-57), (7-59), and (7-61), is

$$U = \frac{1}{2} \int_0^L E_1 I^*(w'')^2 \, dx + \frac{1}{2} \int_0^L kw^2 \, dx$$

$$+ \frac{1}{2} \sum_{m=1}^M \{k_{s_m}[w(x_m)]^2 + k_{r_m}[w'(x_m)]^2\} \quad (14\text{-}21)$$

where the summation extends over all the elastic supports. The straight compressed position of the bar immediately prior to buckling has been taken as the zero-potential datum configuration in writing this equation. Because of this, terms involving u (the axial shortening due to direct compression) do not enter the problem.

Again for generality, we assume that the column may be loaded by both concentrated and distributed forces acting parallel to the x axis (Fig. 14-6). It is assumed that the column is restrained against longitudinal motion at $x = 0$. The potential of the applied loads that is associated with the bending displacements is found from Eqs. (7-64) and (7-66). These give

$$V = -\frac{1}{2} \sum_{k=1}^K P_k \int_0^{x_k} (w')^2 \, dx - \frac{1}{2} \int_0^L p_x \left\{ \int_0^x [w'(\eta)]^2 \, d\eta \right\} dx \quad (14\text{-}22)$$

where we have taken $F_x = -P_k$ and $q_x = -p_x$ to account for the fact that the forces are in the negative direction to produce compression. The

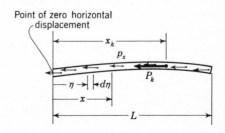

Fig. 14-6 Column with concentrated and distributed forces.

summation extends over all the concentrated forces. We do not include the u terms of Eqs. (7-63) and (7-65) in V because of our assumption that the straight compressed position is the datum configuration. The forces P_k and p_x do not change during the w displacements; therefore u does not vary during the bending from the datum configuration.

We assume that the applied forces increase simultaneously and that their relative values do not change during loading. We may then write Eq. (14-22) as

$$V = - \frac{P_1}{2} \left(\sum_{k=1}^{K} \frac{P_k}{P_1} \int_0^{x_k} (w')^2 \, dx + \frac{p_0}{P_1} \int_0^L h(x) \left\{ \int_0^x [w'(\eta)]^2 \, d\eta \right\} dx \right)$$

(14-23)

where $p_x = p_0 h(x)$ and the ratios P_k/P_1 and p_0/P_1 are known constants. The lateral displacements are approximated by the series

$$w = \sum_{j=1}^{n} c_j w_j(x)$$

(14-24)

where the c_j are undetermined coefficients and the $w_j(x)$ are assumed functions that must satisfy the displacement boundary conditions. The c_j coefficients define the configuration and therefore constitute a set of generalized coordinates. Applying the principle of the stationary value of the total potential in the form of Eq. (6-17), we find

$$\int_0^L E_1 I^* w'' \frac{\partial w''}{\partial c_i} \, dx + \int_0^L k w \frac{\partial w}{\partial c_i} \, dx + \sum_{m=1}^{M} \left[k_{s_m} w(x_m) \frac{\partial w(x_m)}{\partial c_i} \right.$$
$$\left. + k_{r_m} w'(x_m) \frac{\partial w'(x_m)}{\partial c_i} \right] - P_1 \left\{ \sum_{k=1}^{K} \frac{P_k}{P_1} \int_0^{x_k} w' \frac{\partial w'}{\partial c_i} \, dx \right.$$
$$\left. + \frac{p_0}{P_1} \int_0^L h(x) \left[\int_0^x w'(\eta) \frac{\partial w'(\eta)}{\partial c_i} \, d\eta \right] dx \right\} = 0 \qquad i = 1, 2, \ldots, n$$

(14-25)

Introducing Eq. (14-24) into Eq. (14-25) and rearranging, we find

$$\sum_{j=1}^{n} a_{ij} c_j = P_1 \sum_{j=1}^{n} b_{ij} c_j \qquad i = 1, 2, \ldots, n$$

(14-26)

or in matrix notation

$$\mathbf{ac} = P_1 \mathbf{bc}$$

(14-27)

where

$$a_{ij} = \int_0^L E_1 I^* w_i'' w_j'' \, dx + \int_0^L k w_i w_j \, dx$$
$$+ \sum_{m=1}^M [k_{s_m} w_i(x_m) w_j(x_m) + k_{r_m} w_i'(x_m) w_j'(x_m)] \quad (14\text{-}28)$$

$$b_{ij} = \sum_{k=1}^K \frac{P_k}{P_1} \int_0^{x_k} w_i' w_j' \, dx + \frac{p_0}{P_1} \int_0^L h(x) \left[\int_0^x w_i'(\eta) w_j'(\eta) \, d\eta \right] dx \quad (14\text{-}29)$$

Equation (14-27) can be solved by the methods of Sec. 5-5 to determine the magnitude of P_1 for buckling and the c_j/c_i ratios, which when substituted into Eq. (14-24) give the mode shape.

Equations (14-28) and (14-29) can be simplified when there are no elastic supports or foundation and when the only loading is a single concentrated force or a distributed loading. It is only necessary to set the appropriate terms equal to zero. When there are no elastic supports and the applied loading is a single force applied at the ends, Eqs. (14-28) and (14-29) reduce to

$$a_{ij} = \int_0^L E_1 I^* w_i'' w_j'' \, dx \qquad b_{ij} = \int_0^L w_i' w_j' \, dx \quad (14\text{-}30)$$

If the column has both ends pinned or one end free and the other clamped, a_{ij} may be expressed in terms of w_i and w_j instead of w_i'' and w_j''. In these cases we find from Eqs. (14-3) and (14-11) that

$$E_1 I^* w'' = -Pw$$

where we recall that for the free-clamped column w and x must be measured as shown in Fig. 14-1. Solving this equation for w'' and substituting the result into the first of Eqs. (14-30), we find

$$a_{ij} = P^2 \int_0^L \frac{w_i w_j}{E_1 I^*} \, dx$$

Substituting this and the second of Eqs. (14-30) into Eq. (14-27) gives

$$\mathbf{bc} = P \bar{\mathbf{a}} \mathbf{c} \quad (14\text{-}31)$$

where

$$\bar{a}_{ij} = \int_0^L \frac{w_i w_j}{E_1 I^*} \, dx \quad (14\text{-}32)$$

Equation (14-31) is preferable to Eq. (14-27) for pinned-pinned or free-clamped columns because $\bar{a}_{ij}$ does not involve derivatives and is therefore easier to evaluate than a_{ij}. More important is the fact that it gives a more accurate approximation than Eq. (14-27) if the same functions are used in both equations. This is because the difference between the exact and approximating solution is increased by differentiation.

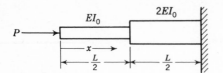

Fig. 14-7 Example 14-3.

Example 14-3 Determine the buckling load of the nonuniform column shown in Fig. 14-7.

The column has free-clamped ends, so that the buckling load can be computed from Eq. (14-31). To satisfy the displacement boundary conditions $w(0) = w'(L) = 0$ we assume

$$w = \sum_{j=1}^{n} c_j \sin \frac{(2j-1)\pi x}{2L}$$

Taking only one term in the series, we find $b_{11}c_1 = P\bar{a}_{11}c_1$ or $P = b_{11}/\bar{a}_{11}$ where from the second of Eqs. (14-30) and Eq. (14-32)

$$b_{11} = \frac{\pi^2}{4L^2} \int_0^L \cos^2 \frac{\pi x}{2L}\, dx = \frac{\pi^2}{8L}$$

$$\bar{a}_{11} = \frac{1}{EI_0} \int_0^{L/2} \sin^2 \frac{\pi x}{2L}\, dx + \frac{1}{2EI_0} \int_{L/2}^{L} \sin^2 \frac{\pi x}{2L}\, dx = \frac{L}{EI_0}\left(\frac{3}{8} - \frac{1}{4\pi}\right)$$

With these we obtain the approximation $P_{cr} = 4.18EI_0/L^2$ for the buckling load.

The exact solution can be obtained with considerably more difficulty by solving the differential equations

$$w'' + \frac{P_{cr}}{EI_0} w = 0 \qquad 0 < x < \frac{L}{2}$$

$$w'' + \frac{P_{cr}}{2EI_0} w = 0 \qquad \frac{L}{2} < x < L$$

applying the boundary conditions $w(0) = w'(L) = 0$, and equating the deflections and slopes in the two regions at $x = L/2$. The solution obtained by this procedure is $P_{cr} = 4.135EI_0/L^2$,[5] only 0.97 percent less than the approximate solution obtained by using a single term in the series.

If two terms are used, we find that for a nontrivial solution

$$|b - P\bar{a}| = \begin{vmatrix} \dfrac{\pi^2}{8L} - \dfrac{PL}{EI_0}\left(\dfrac{3}{8} - \dfrac{1}{4\pi}\right) & -\dfrac{PL}{4\pi EI_0} \\[3mm] -\dfrac{PL}{4\pi EI_0} & \dfrac{9\pi^2}{8L} - \dfrac{PL}{EI_0}\left(\dfrac{3}{8} + \dfrac{1}{12\pi}\right) \end{vmatrix} = 0$$

Expansion of the determinant gives the characteristic equation

$$13.70\beta^2 - 3.775\beta + 0.1123 = 0$$

where $\beta = EI_0/PL^2$. The solutions for this equation are

$$\beta = \frac{3.775 \pm \sqrt{3.775^2 - 4 \times 13.70 \times 0.1123}}{2 \times 13.70}$$

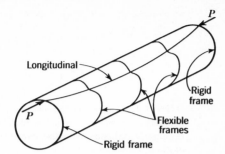

Fig. 14-8   Example 14-4.

The smallest value of P is associated with the largest value of β, which is $\beta = 0.2415$. This gives $P_{cr} = 4.14EI_0/L^2$, which agrees with the exact solution to three significant figures.

Example 14-4   A compressively loaded longitudinal is continuous over five equally spaced transverse ring frames (Fig. 14-8). The stiff end rings provide rigid support against lateral displacements, while the three relatively flexible intermediate rings give elastic lateral support. An idealization of the structure is shown in Fig. 14-9, where the intermediate rings are shown as elastic supports of stiffness k_s. Determine the axial force at which the longitudinal will buckle.

The deflection shape of the buckled longitudinal can be represented by the series

$$w = \sum_{j=1}^{n} c_j \sin \frac{j\pi x}{L} \tag{a}$$

which satisfies the displacement boundary conditions $w(0) = w(L) = 0$. To simplify the calculations we truncate the series at $n = 4$, the smallest value of n that permits the longitudinal to buckle between intermediate frames without deforming them. The four assumed functions are shown in Fig. 14-10.

From Eqs. (14-28) and (14-29) we find

$$a_{ij} = \frac{i^2 j^2 \pi^4 EI}{L^4} \int_0^L \sin \frac{i\pi x}{L} \sin \frac{j\pi x}{L}\, dx$$
$$+ k_s \left(\sin \frac{i\pi}{4} \sin \frac{j\pi}{4} + \sin \frac{i\pi}{2} \sin \frac{j\pi}{2} + \sin \frac{3i\pi}{4} \sin \frac{3j\pi}{4} \right)$$

$$b_{ij} = \frac{ij\pi^2}{L^2} \int_0^L \cos \frac{i\pi x}{L} \cos \frac{j\pi x}{L}\, dx$$

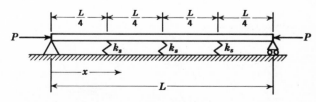

Fig. 14-9   Idealization of Example 14-4.

which with the aid of Eq. (13-66) give

$$a = \frac{\pi^4 EI}{2L^3} \begin{bmatrix} 1 & 16 & 81 & 256 \end{bmatrix} + 2k_s \begin{bmatrix} 1 & 1 & 1 & 0 \end{bmatrix}$$

$$b = \frac{\pi^2}{2L} \begin{bmatrix} 1 & 4 & 9 & 16 \end{bmatrix}$$

where the slashes on the brackets indicate diagonal matrices written horizontally to conserve space. It should be noted that the fact that a is diagonal is coincidental, since off-diagonal terms involving k_s appear in a for larger values of n.

For a nontrivial solution of Eq. (14-27) we find $|a - Pb| = 0$, which gives the characteristic equation

$$\frac{\pi^4 EI}{2L^3} \left(1 + \beta - \frac{P}{P_E} \right) \left(16 + \beta - 4\frac{P}{P_E} \right) \left(81 + \beta - 9\frac{P}{P_E} \right) \left(256 - 16\frac{P}{P_E} \right) = 0$$

where the dimensionless parameter $\beta = 4k_s L^3 / \pi^4 EI$ and $P_E = \pi^2 EI / L^2$ is the Euler load of the longitudinal without intermediate supports. The roots of

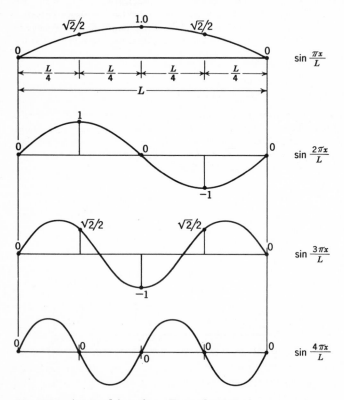

Fig. 14-10   Assumed functions, Example 14-4.

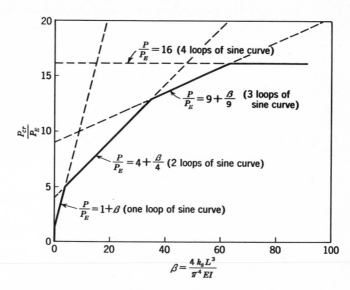

Fig. 14-11 Results for Example 14-4.

the characteristic equation are

$$\frac{P}{P_E} = 1 + \beta \qquad \frac{P}{P_E} = 4 + \frac{\beta}{4} \qquad \frac{P}{P_E} = 9 + \frac{\beta}{9} \qquad \frac{P}{P_E} = 16$$

These equations are plotted in Fig. 14-11, where it is seen that the equation that gives the smallest value of P depends upon β. The ratio P_{cr}/P_E is the lower envelope of the plots of the equations, which is shown by the solid lines. Each of the lines is associated with a different value of j and therefore a different number of loops of a sine curve, as shown in the figure. It is seen that the intermediate frames prevent lateral displacements at the supports if $\beta > 63$, that is, if $k_s > 1535EI/L^3$.

14-4 SMALL DEFLECTIONS OF IMPERFECT ELASTIC COLUMNS

In Secs. 14-2 and 14-3 it was assumed that the column was perfectly straight before the application of the perfectly aligned axial load. Under these circumstances the only possible equilibrium position was found to be the straight configuration when $P < P_{cr}$. At P_{cr} a bifurcation occurs, and both the straight and slightly bent configurations satisfy equilibrium. However, the magnitudes of the lateral displacements are indeterminate because of the neutral state of equilibrium at P_{cr}.

The perfect-column assumption is unrealistic, however, and we shall now investigate the effects that imperfections have upon the column deformations.

Consider a column with pinned ends that is subjected to end compression. Let us assume that the initial shape of the modulus-weighted centroidal axis is given by $w_0(x)$, the deviation from the perfectly straight position. The application of the compressive force results in a bending moment $-Pw_0$ due to the initial lack of straightness. This moment gives rise to bending displacements w, measured from the initially imperfect position, so that the total bending moment applied to the column is $-P(w_0 + w)$. The elastic restoring moment depends only upon w and is therefore $E_1 I^* w''$. Equating the applied and resisting moments, we find $E_1 I^* w'' = -P(w_0 + w)$, or

$$w'' + \frac{P}{E_1 I^*} w = -\frac{Pw_0}{E_1 I^*} \tag{14-33}$$

The boundary conditions which w must satisfy are $w(0) = 0$ and $w(L) = 0$. Comparing this differential equation and its boundary conditions with Eqs. (14-11) and (14-12) for the perfect column, we note that the problems are of surprisingly different forms. The equations for the perfect column describe an eigenvalue problem with no solution except $w = 0$ unless P is an eigenvalue. In contrast, the equations for the imperfect column describe an equilibrium problem which has a nontrivial solution for all values of P.

The complete solution to Eq. (14-33) is the sum of the complementary and particular solutions. The complementary solution for constant $E_1 I^*$ is

$$w_c = C_1 \sin kx + C_2 \cos kx \tag{14-34}$$

where $k = (P/E_1 I^*)^{\frac{1}{2}}$. To obtain the particular solution we express w_0 by the Fourier series

$$w_0(x) = \sum_{n=1}^{\infty} w_n \sin \frac{n\pi x}{L} \tag{14-35}$$

If $w_0(x)$ is known, we can obtain the Fourier coefficients by multiplying both sides of Eq. (14-35) by $\sin (m\pi x/L)$, integrating from 0 to L, and using Eq. (13-66). This gives

$$w_n = \frac{2}{L} \int_0^L w_0(x) \sin \frac{n\pi x}{L}\, dx \tag{14-36}$$

Equation (14-33) then becomes

$$w'' + \frac{P}{E_1 I^*} w = -\frac{P}{E_1 I^*} \sum_{n=1}^{\infty} w_n \sin \frac{n\pi x}{L} \tag{14-37}$$

Using the method of undetermined coefficients, the particular solution of this equation is

$$w_p = \sum_{n=1}^{\infty} c_n \sin \frac{n\pi x}{L} \qquad (14\text{-}38)$$

Substituting w_p into Eq. (14-37) and equating the coefficients of sine terms with the same argument on both sides of the equation, we find

$$c_n = \frac{w_n}{n^2 P_E/P - 1} \qquad (14\text{-}39)$$

where $P_E = \pi^2 E_1 I^*/L^2$ is the Euler load for the perfect column.

The complete solution, obtained by substituting Eq. (14-39) into (14-38) and adding the result to Eq. (14-34), is

$$w = C_1 \sin kx + C_2 \cos kx + \sum_{n=1}^{\infty} \frac{w_n}{n^2 P_E/P - 1} \sin \frac{n\pi x}{L}$$

Applying the boundary conditions $w(0) = 0$ and $w(L) = 0$, we find $C_1 = C_2 = 0$, so that

$$w = \sum_{n=1}^{\infty} A_n w_n \sin \frac{n\pi x}{L} \qquad (14\text{-}40)$$

where

$$A_n = \frac{1}{n^2 P_E/P - 1} \qquad (14\text{-}41)$$

is the amplification factor that applies to the nth component of w_0 as a result of the application of P. The total lack of straightness is then

$$w_0 + w = \sum_{n=1}^{\infty} (1 + A_n) w_n \sin \frac{n\pi x}{L} \qquad (14\text{-}42)$$

In practice w_0, and therefore w_n, is not likely to be known. However, it would be expected that w_n would become smaller as n becomes larger, and a reasonable assumption might be that the amplitude of the Fourier component is proportional to the wavelength of the component. This gives $w_n = \bar{w}(L/n)$, where $\bar{w}$ is assumed to be the same for all components. Equation (14-42) then becomes

$$\frac{w_0 + w}{L} = \bar{w} \sum_{n=1}^{\infty} \frac{1 + A_n}{n} \sin \frac{n\pi x}{L} \qquad (14\text{-}43)$$

Values of A_n as a function of P/P_E, computed from Eq. (14-41) for $n = 1$ to 3, are given in Table 14-1. We observe that unless P/P_E is small, the first term of Eq. (14-43) dominates the series and provides a good approximation for it.

Table 14-1

$\dfrac{P}{P_E}$	A_1	A_2	A_3
0.0	0.0	0.0	0.0
0.4	0.667	0.111	0.047
0.8	4.00	0.25	0.08
0.9	9.50	0.29	0.11
0.95	20.0	0.33	0.12
1.0	∞	0.33	0.13

Substituting Eq. (14-41) into the first term of Eq. (14-43) and evaluating the result at $x = L/2$, we find the nondimensionalized lack of straightness at the center of the column is

$$\left(\frac{w_0 + w}{L}\right)_{x=L/2} = \frac{\bar{w}}{1 - P/P_E} \tag{14-44}$$

This equation is plotted in Fig. 14-12 for values of the initial lack-of-straightness parameter $\bar{w} = w_1/L = 0.01$, 0.001, and 0.0001. These

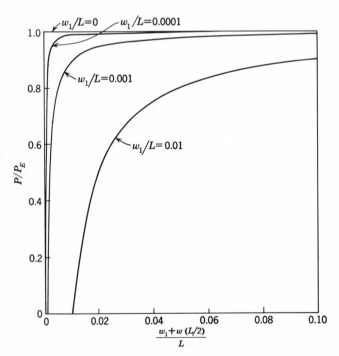

Fig. 14-12 Linear-theory results for an imperfect elastic column.

values are representative of columns of poor, average, and above average workmanship in manufacture.   We note that when P/P_E is small, the bending deformations are small, but as P/P_E approaches unity, the bending deflections increase rapidly and become very large.   As the initial imperfection approaches zero, the behavior of the imperfect column *approaches* that of the perfect column, which can have no deflection until $P/P_E = 1$.   However, we see that instead of remaining straight, the column with vanishingly small imperfections bends as the bifurcation point of the perfect column is approached, and thereafter it closely follows the solution for the buckled perfect column.

Equation (14-44) indicates that the center deflection approaches infinity as P/P_E approaches unity.   However, this result requires further examination because when the deflections become large, the approximations of the curvature by d^2w/dx^2 become inaccurate, and the linearized theory of bending breaks down.   The theory also fails at large center deflections because the stresses due to compression and bending exceed the proportional limit of the material.

While we have investigated only the effects of lack of straightness, it can be shown that the conclusions also apply to perfectly straight columns with small eccentricities in the column load or with small lateral loads.

14-5 LARGE DEFLECTIONS OF COLUMNS

We have seen that the bending displacements of imperfect columns increase rapidly as the axial force approaches the Euler load and soon exceed the range in which the curvature may be approximated by d^2w/dx^2. In this section we shall examine the behavior of perfect and imperfect columns with large deflections.   The exact solution for the postbuckling behavior of a perfect column is known.[5]   However, the solution involves elliptic integrals, and the results are in an inconvenient form for studying the relationship between axial load and bending displacements.   For this reason we shall obtain an approximate solution by the Rayleigh-Ritz method, which is sufficiently accurate over the range of bending deformations that are of practical interest.

When the bending deformations are large, we must reexamine our equations for U and V.   Consider a column with an initial lack of straightness $w_0(s)$, where s is a coordinate measured along the curved line that joins the modulus-weighted centroids of the cross section (Fig. 14-13).   When P is applied, the column undergoes a bending displacement $w(s)$ so that the total deviation from straightness is $w_t = w_0 + w$. During bending the distance between the ends of the column shortens by

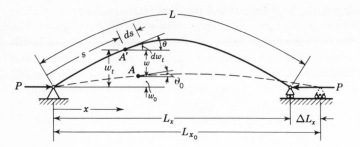

Fig. 14-13 Deformation geometry of an imperfect column.

an amount ΔL_x, and the compressive force does work. As a result,

$$V = -P\,\Delta L_x \tag{14-45}$$

The distance ΔL_x is the difference in the horizontal projections of the column in its initial and deformed positions. The projected length of the loaded imperfect column is

$$L_x = \int_0^L \cos\theta \, ds = \int_0^L (1 - \sin^2\theta)^{1\!/\!2} \, ds$$

where L is the developed length of the column and θ is the angle that the tangent to the column makes with the x axis. We can express θ in terms of w_t by noting from Fig. 14-13 that

$$\sin\theta = \frac{dw_t}{ds} \tag{14-46}$$

therefore

$$L_x = \int_0^L \left[1 - \left(\frac{dw_t}{ds}\right)^2\right]^{1\!/\!2} ds$$

This equation may be written in a series form by using the binomial expansion (Sec. 2-6), which gives

$$L_x = \int_0^L \left[1 - \frac{1}{2}\left(\frac{dw_t}{ds}\right)^2 - \frac{1}{8}\left(\frac{dw_t}{ds}\right)^4 + \cdots \right] ds$$

Noting that dw_t/ds is a small quantity, we can obtain an approximation of L_x by neglecting terms higher than the fourth power, to give

$$L_x = \int_0^L \left[1 - \frac{1}{2}\left(\frac{dw_t}{ds}\right)^2 - \frac{1}{8}\left(\frac{dw_t}{ds}\right)^4\right] ds \tag{14-47}$$

The initial horizontal projection of the imperfect column may be found by replacing w_t by w_0 in Eq. (14-47). However, because $w_0 \ll w_t$

when P is near P_E, we neglect the fourth-power term and find

$$L_{x_0} = \int_0^L \left[1 - \frac{1}{2}\left(\frac{dw_0}{ds}\right)^2 \right] ds \qquad (14\text{-}48)$$

Noting that $\Delta L_x = L_{x_0} - L_x$, we find from Eqs. (14-45), (14-47), and (14-48) that

$$V = -\frac{P}{2}\int_0^L \left[\left(\frac{dw_t}{ds}\right)^2 - \left(\frac{dw_0}{ds}\right)^2 + \frac{1}{4}\left(\frac{dw_t}{ds}\right)^4 \right] ds \qquad (14\text{-}49)$$

For small displacements of a perfect column we note that $(dw_t/ds)^4$ is much less than $(dw_t/ds)^2$, $w_t = w$, and $ds \approx dx$, in which case Eq. (14-49) reduces to Eq. (7-64).

In deriving the equation for U we must account for the initial curvature. Figure 14-14 shows a differential length of the column in the initial and deformed positions. The bending strain energy in the differential element is equal to the work done by the bending moments M in moving through the rotations at the ends of the element. Noting that the average moment during the rotation is $M/2$, we find

$$dU = -\frac{M}{2}(\theta - \theta_0) + \frac{M}{2}\left(\theta + \frac{d\theta}{ds}ds - \theta_0 - \frac{d\theta_0}{ds}ds\right)$$

which reduces to

$$dU = \frac{M}{2}\left(\frac{d\theta}{ds} - \frac{d\theta_0}{ds}\right)ds \qquad (14\text{-}50)$$

where θ_0 is the angle between the tangent to the initially curved column and the x axis. We see from Fig. 14-14 that

$$\frac{d\theta}{ds} = \frac{1}{R} \qquad \text{and} \qquad \frac{d\theta_0}{ds} = \frac{1}{R_0}$$

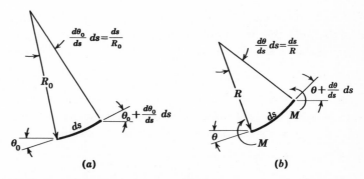

Fig. 14-14 Geometry for determination of change in curvature. (a) Initial geometry; (b) deformed geometry.

the final and initial curvatures, respectively. It was shown in Chap. 7 that the internal moment is E_1I^* times the change in curvature, so that for the imperfect column $M = E_1I^*(d\theta/ds - d\theta_0/ds)$. Substituting this result into Eq. (14-50) and integrating over the length of the column, we find

$$U = \frac{1}{2} \int_0^L E_1I^* \left(\frac{d\theta}{ds} - \frac{d\theta_0}{ds}\right)^2 ds \qquad (14\text{-}51)$$

For small displacements of a perfect column $\theta_0 = 0$, $d\theta/ds = 1/R \approx w''$, $ds \approx dx$, in which case Eq. (14-51) reduces to the first term of Eq. (14-21).

Using Eq. (14-46), we can write

$$\frac{d\theta}{ds} = \frac{d}{ds}\left(\sin^{-1}\frac{dw_t}{ds}\right) = \left[1 - \left(\frac{dw_t}{ds}\right)^2\right]^{-\frac{1}{2}} \frac{d^2w_t}{ds^2}$$

which, by the binomial expansion, becomes

$$\frac{d\theta}{ds} = \left[1 + \frac{1}{2}\left(\frac{dw_t}{ds}\right)^2 + \cdots\right] \frac{d^2w_t}{ds^2}$$

Neglecting higher-order terms, we obtain the approximation

$$\frac{d\theta}{ds} = \left[1 + \frac{1}{2}\left(\frac{dw_t}{ds}\right)^2\right] \frac{d^2w_t}{ds^2} \qquad (14\text{-}52)$$

for the curvature of the deformed column. The initial curvature may be found from Eq. (14-52) by replacing w_t by w_0. However, because $w_0 \ll w_t$, we may neglect the second term and write $d\theta_0/ds = d^2w_0/ds^2$. Substituting this result and Eq. (14-52) into Eq. (14-51) gives

$$U = \frac{1}{2} \int_0^L E_1I^* \left[\frac{d^2w_t}{ds^2} - \frac{d^2w_0}{ds^2} + \frac{1}{2}\left(\frac{dw_t}{ds}\right)^2 \frac{d^2w_t}{ds^2}\right]^2 ds \qquad (14\text{-}53)$$

It was shown in Sec. 14-4 that when w_0 is expanded into a Fourier series, only the first term of the series is of practical importance as P approaches P_E. We therefore assume that the initial imperfection is

$$w_0 = w_1 \sin\frac{\pi s}{L} \qquad (14\text{-}54)$$

To simplify the analysis we approximate the deflections by a single term. Guided by the small-deflection analysis, we assume

$$w_t = c \sin\frac{\pi s}{L} \qquad (14\text{-}55)$$

where c is an undetermined constant.

Applying the Rayleigh-Ritz method, $\partial(U + V)/\partial c = 0$, which with Eqs. (14-49) and (14-53) to (14-55) gives

$$\left(4 - 3\frac{P}{P_E}\right)\left(\frac{c}{L}\right)^3 - 3\frac{w_1}{L}\left(\frac{c}{L}\right)^2 + \frac{8}{\pi^2}\left(1 - \frac{P}{P_E}\right)\frac{c}{L} - \frac{8}{\pi^2}\frac{w_1}{L} = 0 \quad (14\text{-}56)$$

This equation, which relates $c/L = [w_1 + w(L/2)]/L$, w_1/L, and P/P_E, is plotted as the solid curves in Fig. 14-15 for $w_1/L = 0.01, 0.001$, and 0.0001. For a perfect column $w_1/L = 0$, and Eq. (14-56) becomes

$$\frac{c}{L}\left[\left(4 - 3\frac{P}{P_E}\right)\left(\frac{c}{L}\right)^2 + \frac{8}{\pi^2}\left(1 - \frac{P}{P_E}\right)\right] = 0$$

which has two solutions, the straight configuration given by $c/L = 0$, and the bent configuration given by

$$\frac{c}{L} = \frac{2\sqrt{2}}{\pi}\sqrt{\frac{P/P_E - 1}{4 - 3P/P_E}} \quad (14\text{-}57)$$

This equation is plotted as the $w_1/L = 0$ curve in Fig. 14-15.

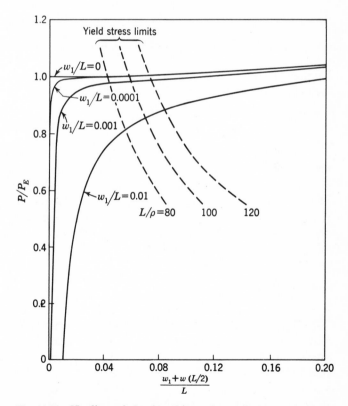

Fig. 14-15 Nonlinear behavior of imperfect columns.

Equation (14-57) has a horizontal tangent at $P/P_E = 1$, which accounts for the neutral stability and indeterminacy of the deflections in the linear small-deflection theory. We see from Fig. 14-15 that while the straight configuration is theoretically possible for a perfect column at $P/P_E > 1$, columns with vanishingly small imperfections approach the bent configuration given by Eq. (14-57). Since all columns have some small imperfections, Eq. (14-57) is the only solution of practical importance.

The nonlinear theory which has been given fails when the stresses due to direct compression and bending exceed the proportional limit. To obtain an indication of when this effect becomes important in a homogeneous column we shall idealize the material behavior by assuming it to be linearly elastic and perfectly plastic (Fig. 3-5f). Up to the yield stress the material is linearly elastic, so that the stresses can be computed from $\sigma_{xx} = P/A - Mz/I$. The maximum stress occurs at $x = L/2$ and $z = h/2$, where h is the height of the cross section. Noting that

$$M(L/2) = -P[w_1 + w(L/2)]$$

we find that

$$\sigma_{\max}A = P\left\{1 + \frac{Ah}{2I}[w_1 + w(L/2)]\right\}$$

To nondimensionalize this equation we divide by $P_E = \pi^2 EI/L^2$, and by letting $\sigma_{\max} = \sigma_{cy}$ (the compressive yield stress) we obtain

$$\frac{1}{\pi^2}\frac{\sigma_{cy}}{E}\left(\frac{L}{\rho}\right)^2 = \frac{P}{P_E}\left\{1 + \frac{h}{2\rho^2}[w_1 + w(L/2)]\right\} \tag{14-58}$$

We see from Eq. (14-58) that the value of P/P_E at which yielding occurs depends upon ρ. To obtain an indication of the combinations of P/P_E and w_1/L that cause yielding we assume that the column has the idealized H section shown in Fig. 14-16. For this section we assume that the web has negligible resistance in bending and extension but is rigid in shear. With this idealization $\rho = h/2$, so that Eq. (14-58) can be written

$$\frac{P}{P_E} = \frac{(1/\pi^2)(\sigma_{cy}/E)(L/\rho)^2}{1 + (L/\rho)\{[w_1 + w(L/2)]/L\}} \tag{14-59}$$

Equation (14-59) is plotted as the dotted curves in Fig. 14-15 for several values of L/ρ with $\sigma_{cy}/E = 6.66 \times 10^{-3}$. This corresponds to 7075-T6 aluminum alloy, which has a relatively large value of σ_{cy}/E. When the fibers on the concave side of the bent beam exceed the yield stress, the bending rigidity decreases, the deflections become greater than indicated by Fig. 14-15, and collapse soon follows. As a result, the ultimate value of P/P_E which the column can sustain is approximately at the point of intersection of the w_1/L and L/ρ curves that apply to the column.

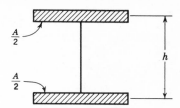

Fig. 14-16 Idealized section.

The 7075-T6 alloy is actually a linearly elastic-strain-hardening plastic material (Fig. 3-5g), so that the ultimate load obtained from Fig. 14-15 is an approximation; however, because of the flatness of the w_1/L curves the errors are small.

For columns that fail at *average* stresses within the linearly elastic range, we may draw the following conclusions from Figs. 14-12 and 14-15: (1) The straight position is the only equilibrium configuration for a column with vanishingly small imperfections until $P = P_E$.   (2) At $P = P_E$ the deflections of a column with vanishingly small imperfections grow rapidly and are approximately given by Eq. (14-57) until the concave fiber exceeds the proportional limit.   (3) Columns with practical imperfections do not bend appreciably until P is very nearly equal to P_E.   The deflections grow rapidly as P approaches P_E, and at large deflections they tend to follow the curve for the column with vanishingly small imperfections [Eq. (14-57)].   (4) The rapidly increasing bending deformations soon exceed the yield stress, and the practical column collapses at $P \approx P_E$. (5) The deflections at failure are small enough to permit analysis by the linear theory, in which the curvature is approximated by d^2w/dx^2.

The physical coincidence that the ultimate load-carrying capability of the imperfect column can be predicted by the linear buckling theory for a perfect column is fortunate.   It means that columns that fail at average stresses in the elastic range may be designed by the simple equation (14-17) rather than by tedious nonlinear calculations which account for large deflections and plasticity.   However, one should not conclude that this is true in all buckling problems.   In plates and shells the failing load may be considerably different from that predicted by the neutral-stability analysis for small deformations.   Indeed, we shall see in the next section that neutral-stability condition for a perfect column does not adequately predict the failing load of an imperfect column if the average stress at failure exceeds the proportional limit.

14-6 INELASTIC COLUMNS

When L'/ρ is large, the critical stress occurs in the linearly elastic range, and the failing stress can be predicted by Eq. (14-17) or the methods of

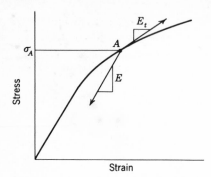

Fig. 14-17 Stress-strain behavior in the inelastic range.

Sec. 14-3. However, as L'/ρ decreases, the average stress at buckling becomes greater than the proportional limit, and the preceding methods are not applicable. In this section we shall consider the buckling and failure of homogeneous columns in the inelastic range.

In the derivation of Eq. (14-17) it was assumed that the bending rigidity of the column is EI and that this rigidity does not change with stress level. This assumption was valid because $d\sigma/d\epsilon = E$ is constant and single-valued in the elastic range. In the inelastic range $d\sigma/d\epsilon$ depends upon the stress level and whether the strain is increasing or decreasing. Referring to Fig. 14-17, we see that at point A, $d\sigma/d\epsilon = E_t$ when the strain is increasing and $d\sigma/d\epsilon = E$ when it is decreasing. As a result, the bending rigidity depends upon the magnitude of P and whether it increases, decreases, or remains constant as the column bends.

Consider a beam with a plane of symmetry that coincides with the plane of bending. We assume that the beam is subjected to a compressive force P and that P/A exceeds the linearly elastic limit. We suppose that the beam is subjected to a small bending moment M and that P is simultaneously given an increment ΔP. Depending upon the sign and magnitude of $\Delta P/M$, the compressive stresses may (1) increase throughout the cross section, (2) decrease throughout the cross section, or (3) increase on the concave side and decrease on the convex side of the bent beam. The latter case is shown in Fig. 14-18, where the compressive stresses decrease in A_1 (the area above the line AB) and increase in A_2. Referring to Fig. 14-17, we note that $d\sigma/d\epsilon = E$ in A_1, while

$$\frac{d\sigma}{d\epsilon} = E_t(\sigma_A)$$

in A_2. Because of this, the homogeneous beam behaves under incremental loads as if it were nonhomogeneous with moduli E and E_t in A_1 and A_2, respectively.

To analyze the beam we take a set of y_0 and z_0 axes through the geometric centroid and a second set of y and z axes through the modulus-weighted centroid (Fig. 14-18). Taking $E_1 = E$, we find from Eq. (7-16b) that the distance between the y_0 and y axes is

$$\bar{z}_0^* = \frac{\int_{A_1} z_0 \, dA + \dfrac{E_t}{E} \int_{A_2} z_0 \, dA}{A_1 + E_t A_2/E}$$

Substituting $z_0 = \bar{z}_0^* + z$ into this equation and simplifying, we obtain

$$EQ_1 + E_t Q_2 = 0 \tag{14-60}$$

where Q_1 and Q_2 are the static moments of A_1 and A_2 about the y axis.

From Eqs. (7-25) and (7-12) the weighted area and moment of inertia of the cross section are

$$A^* = A_1 + \frac{E_t}{E} A_2 \tag{14-61a}$$

$$I^* = I_1 + \frac{E_t}{E} I_2 \tag{14-61b}$$

where I_1 and I_2 are the moments of inertia of A_1 and A_2 about the y axis. The incremental stress that occurs in A_1 during the load increments ΔP and M can be found from Eq. (7-23), which gives $\Delta\sigma = \Delta P/A^* + Mz/I^*$, where we have noted that M is negative if the stress is to decrease in A_1. To determine e, the distance from the y_0 axis to the line AB where there

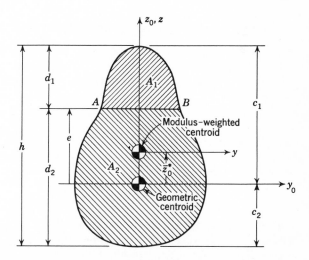

Fig. 14-18   Geometry for determining the bending rigidity of a compressed beam in the inelastic range.

is no change in stress during the load increments ΔP and M, we set $\Delta\sigma = 0$ and $z = e - \bar{z}_0^*$, to find

$$\frac{\Delta P}{M} = -\frac{A^*}{I^*}(e - \bar{z}_0^*) \tag{14-62}$$

Three special cases of Eq. (14-62) are of particular importance, namely, those in which (1) the compressive stresses increase over the entire section, (2) the compressive stresses decrease over the entire section, and (3) the axial force remains constant during bending. If the stresses increase over the whole section, the beam behaves like a homogeneous beam with modulus E_t during the incremental loading. We note from Fig. (14-18) that $e \geq c_1$ and $\bar{z}_0^* = 0$ for this case, so that

$$\frac{\Delta P}{M} \geq -\frac{Ac_1}{I} \tag{14-63}$$

For this case the bending rigidity $E_1 I^* = E_t I$, and the incremental bending displacements w are related to M by

$$E_t I w'' = M \tag{14-64}$$

In a similar manner, if the compressive stresses decrease over the entire section, the modulus E applies throughout the beam, so that $e \leq -c_2$ and $\bar{z}_0^* = 0$. Equation (14-62) then gives

$$\frac{\Delta P}{M} \leq \frac{Ac_2}{I} \tag{14-65}$$

For this case the increments in moment and displacement are related by

$$EI w'' = M \tag{14-66}$$

When $\Delta P = 0$, we find from Eq. (14-62) that $e = \bar{z}_0^*$, and Eq. (14-60) gives

$$EQ_1 = -E_t Q_2 \tag{14-67}$$

This result can be used with the condition that $d_1 + d_2 = h$ (Fig. 14-18) to determine e for particular cross sections and materials. In this case we find from Eq. (14-61b) and $E_1 = E$ that

$$E_1 I^* = EI_1 + E_t I_2 \tag{14-68}$$

The bending rigidity is customarily written in terms of I by defining a modulus E_r, known as the *reduced* or *double modulus*, by the relationship $E_r I = E_1 I^*$. From Eq. (14-68) we find

$$E_r = \frac{EI_1 + E_t I_2}{I} \tag{14-69}$$

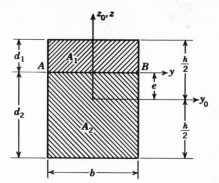

Fig. 14-19 Geometry for determining the reduced modulus of a rectangular cross section.

which gives the incremental moment–displacement equation

$$E_r I w'' = M \tag{14-70}$$

We note from Eq. (14-69) that E_r is a function of the material, the stress level, and the cross-sectional shape. However, by determining E_r for different materials and cross sections it can be shown that it is only slightly dependent upon the cross-sectional geometry.

To fix ideas let us determine E_r for a rectangular cross section (Fig. 14-19). From Eq. (14-67) we find $E d_1^2 = E_t d_2^2$. From this equation and $d_1 + d_2 = h$ we find

$$d_1 = \frac{h E_t^{\frac{1}{2}}}{E^{\frac{1}{2}} + E_t^{\frac{1}{2}}} \qquad d_2 = \frac{h E^{\frac{1}{2}}}{E^{\frac{1}{2}} + E_t^{\frac{1}{2}}}$$

Equation (14-69) then gives

$$E_r = \frac{E b d_1^3 / 3 + E_t b d_2^3 / 3}{b h^3 / 12} = \frac{4 E E_t}{(E^{\frac{1}{2}} + E_t^{\frac{1}{2}})^2}$$

We see from the preceding discussion that the bending rigidity of a compressively loaded beam in the inelastic regime depends upon $\Delta P / M$. If we consider the bending of a perfect column, the moment increment results from the lateral displacement and is given by $M = -Pw$. Substituting this relationship into Eqs. (14-64), (14-66), and (14-70), we find that each of the equations can be written in the form

$$w'' + k^2 w = 0 \tag{14-71}$$

where

$$k^2 = \frac{P}{E_t I} \qquad \text{for } \Delta P \geq \frac{P w A c_1}{I} \tag{14-72a}$$

$$k^2 = \frac{P}{E_r I} \qquad \text{for } \Delta P = 0 \tag{14-72b}$$

$$k^2 = \frac{P}{E I} \qquad \text{for } \Delta P \leq -\frac{P w A c_2}{I} \tag{14-72c}$$

The limits of applicability of Eqs. (14-72a) and (14-72c) are determined by substituting $M = -Pw$ into Eqs. (14-63) and (14-65).

For pinned ends, the boundary conditions for Eq. (14-71) are $w(0) = w(L) = 0$. The parameter k in Eq. (14-71) is independent of x when the cross section is constant. By following the method of Example 14-1 we find that the lowest compressive force at which an initially perfect column can be in equilibrium in the slightly bent position is

$$P_{cr} = \frac{\pi^2 E_t I}{L^2} \quad \text{for } \Delta P \geq \frac{PwAc_1}{I} \tag{14-73a}$$

$$P_{cr} = \frac{\pi^2 E_r I}{L^2} \quad \text{for } \Delta P = 0 \tag{14-73b}$$

$$P_{cr} = \frac{\pi^2 EI}{L^2} \quad \text{for } \Delta P \leq -\frac{PwAc_2}{I} \tag{14-73c}$$

where ΔP is the change in P during the displacement w. Below the elastic limit $E_t = E_r = E$, and all the equations reduce to the Euler equation.

We see from the foregoing that in the inelastic regime there is not a unique value of P at which the bent position becomes an equilibrium position. Instead, there is a range of possible buckling loads which depend upon whether P increases, remains constant, or decreases during the bending. The smallest value of P at which bending is possible is the tangent-modulus load given by Eq. (14-73a), which requires P to increase by at least $PwAc_1/I$ during the displacement w. The largest possible buckling load is given by the Euler load, which requires P to decrease by at least $PwAc_2/I$ during the displacement w.

The question naturally arises, "Which of these loads should be used in column design?" A brief account of the history of inelastic-column theory is interesting in this connection.[1] Euler derived his equation for the buckling of a perfect column in 1759, but doubts existed about its validity because it did not accurately predict the buckling stress of short (inelastic) columns. In 1845, Lamarle pointed out that Euler's equation was valid only in the elastic range, and Considère suggested in 1889 that E_t should be used in place of E in the inelastic range. Engesser introduced the double-modulus concept in 1889, when he made the observation that when buckling occurs at a constant load (which is the case for small deflections of a perfect column), a strain reversal must occur on the concave side of the bent column.

In 1908 von Kármán reported the results of carefully made tests on steel columns which indicated that the failing load was very nearly equal to the reduced-modulus buckling load. There was extensive testing of aluminum alloy columns by the aircraft industry in the 1940s. Unlike von Kármán's experiments, these tests indicated that the failing load is approximately equal to the tangent-modulus load.

These lower test results were usually blamed upon initial imperfections and poorly controlled test conditions, but the test conditions were more typical of operating conditions than von Kármán's tests were.   As a result, the tangent-modulus theory was usually used in design.   Shanley[6] resolved the problem in 1947, when he pointed out that buckling of a perfect column can occur at the tangent-modulus load if the axial force is permitted to increase during buckling.

The engineer is interested in finding the ultimate load of columns with small imperfections rather than the bifurcation load for perfect columns.   We have seen in Secs. 14-4 and 14-5 that compression and bending of imperfect columns proceed simultaneously.   Because the axial force increases as bending occurs, we would expect the imperfect column to develop large deflections and fail before reaching the reduced-modulus load.   This has been analytically confirmed for the idealized H section (Fig. 14-16) in Ref. 7, where it is concluded that the column with vanishingly small imperfections bends at the tangent-modulus load and fails below the double-modulus load.   Furthermore, it was found that when the imperfections are typical of real columns, the maximum load is approximately equal to the tangent-modulus load.

The tangent-modulus theory can also be used to predict the failing load for other end conditions.   In this case we write $P_{cr} = c\pi^2 E_t I/L^2$, or

$$\sigma_{cr} = \frac{\pi^2 E_t}{(L'/\rho)^2} \tag{14-74}$$

Below the proportional limit $E_t = E$, and this relationship reduces to Eq. (14-17), so that Eq. (14-74) may be used at all stress levels.

We note that the solution of Eq. (14-74) involves successive approximations because σ_{cr} depends upon E_t, which in turn is a function of σ_{cr}. This difficulty can be avoided by constructing column curves.   Figure 14-20 shows one type of column curve in which E_t is plotted against σ.

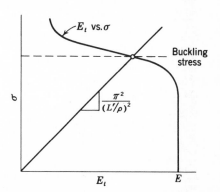

Fig. 14-20   Determination of the buckling stress from tangent-modulus curves.

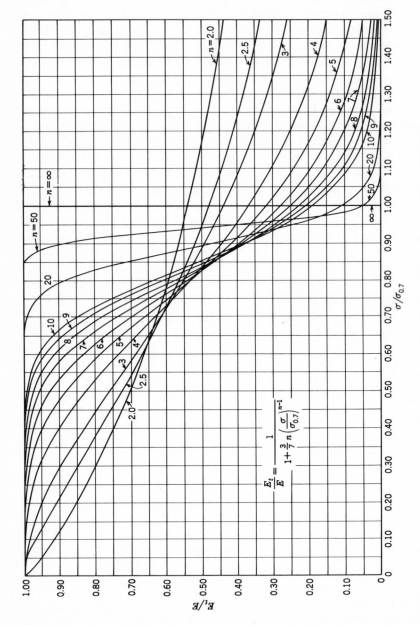

Fig. 14-21 Tangent-modulus curves as a function of the Ramberg-Osgood parameters (Ref. 9).

$$\frac{E_t}{E} = \cfrac{1}{1 + \cfrac{3}{7} n \left(\cfrac{\sigma}{\sigma_{0.7}} \right)^{n-1}}$$

Equation (14-74) becomes a straight line through the origin with a slope of $\pi^2/(L'/\rho)^2$ when it is plotted on the σ and E_t axes. The buckling stress occurs at the point of intersection of the E_t-versus-σ curve and the straight line. Tangent-modulus–stress curves for commonly used aerospace materials are given in Ref. 8. However, these curves are typical results and must be reduced to minimum guaranteed properties before they can be used in design. The Ramberg-Osgood method for describing the stress-strain curve provides a convenient method for constructing tangent-modulus curves. These curves, which are shown in Fig. 14-21, are obtained by plotting Eq. (3-10).

The Ramberg-Osgood parameters can also be used to construct the nondimensional column curves shown in Fig. 14-22. These curves are obtained by dividing both sides of Eq. (14-74) by $\sigma_{0.7}$ and using Eq. (3-10), which gives

$$\frac{\sigma_{cr}}{\sigma_{0.7}} = \frac{\pi^2}{(\sigma_{0.7}/E)(L'/\rho)^2[1 + (3n/7)(\sigma_{cr}/\sigma_{0.7})^{n-1}]}$$

To use the curves we (1) determine E, $\sigma_{0.7}$, and n for the material, (2) compute $\frac{1}{\pi}\left(\frac{\sigma_{0.7}}{E}\right)^{\frac{1}{2}}\frac{L'}{\rho}$ and enter the figure on the horizontal axis, (3) read $\sigma_{cr}/\sigma_{0.7}$ from the proper n curve, and (4) compute $\sigma_{cr} = \sigma_{0.7}(\sigma_{cr}/\sigma_{0.7})$.

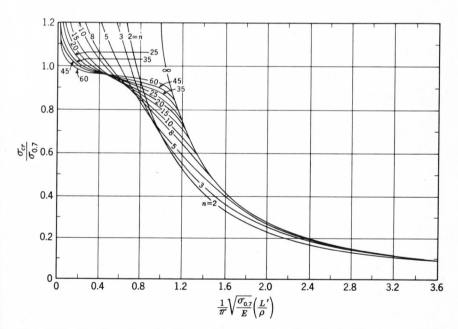

Fig. 14-22 Nondimensional column curves as a function of the Ramberg-Osgood parameters (Ref. 10).

Few data are available on the proportional limits of materials (Ref. 8 does not list this property). However, Figs. 14-21 and 14-22 are applicable with very little error in both the elastic and inelastic regimes. It is therefore suggested that the column failing stress be determined from either of these figures for all values of L'/ρ, which avoids the danger of using the Euler equation in the inelastic range.

The inelastic buckling of nonuniform columns is more complicated. In this case $\sigma = P/A$ is variable along the length, so that E_t in Eqs. (14-71) and (14-72a) is a function of x. A method of successive approximations can be used in which an initial estimate of P_{cr} is made. Using this estimate, $\sigma = P_{cr}/A$ is computed, and $E_t(\sigma)$ is determined at mesh points along the column. With these values of E_t an improved estimate of the buckling load can be computed from Eq. (14-71) by the finite-difference method if both ends of the column are pinned or if one end is free and the other is clamped. The process is repeated until P_{cr} at the end of a cycle is equal to P_{cr} at the beginning of the cycle. For other boundary conditions, Eq. (14-1) (with E_1I^* replaced by E_tI) must be solved with the appropriate boundary conditions.

The Rayleigh-Ritz method can also be used to predict the inelastic buckling stress for nonuniform columns. Since there is no strain reversal in the tangent-modulus theory, we can obtain the strain energy from

$$U = \frac{1}{2} \int_0^L E_t(x)I(x)(w'')^2 \, dx$$

The buckling load must again be found by successive approximations because E_t depends on $\sigma = P_{cr}/A$. This method was found by Newton[11] to be in acceptable agreement with experimental results.

14-7 EMPIRICAL COLUMN EQUATIONS

Empirical column equations have long been used to predict the failing loads of short columns. In view of the good correlation between the tangent-modulus theory and experimental data, it appears that the use of empirical equations is unnecessary for columns that fail by primary bending instability. However, a few of these equations have been used in the aerospace industry and are given in Ref. 8 as an alternative to the tangent-modulus equation. In Chap. 15 we shall see that the tangent-modulus theory does not apply to columns that fail by secondary instability, and it is here that the empirical equations have their greatest usefulness.

The most commonly used empirical relationships employ a simple power law of the form

$$\sigma_c = \sigma_{co} - \beta \left(\frac{L'}{\rho}\right)^n \tag{14-75}$$

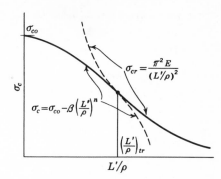

Fig. 14-23 Empirical short-column curves.

where σ_{co} is the stress intercept at $L'/\rho = 0$ (Fig. 14-23) and n is a parameter that establishes the shape of the empirical curve. The parameter σ_{co}, called the *column yield stress*, is given as a function of the compressive yield stress in Ref. 8 for the more commonly used materials. The coefficient β is found by requiring the empirical and Euler curves to be tangent at the *transitional slenderness ratio* $(L'/\rho)_{tr}$.

By equating the stresses predicted by Eqs. (14-75) and (14-17) at $(L'/\rho)_{tr}$ we find

$$\sigma_{co} - \beta \left(\frac{L'}{\rho}\right)_{tr}^{n} = \frac{\pi^2 E}{(L'/\rho)_{tr}^{2}}$$

and for equal slopes $d\sigma/d(L'/\rho)$ at this point we obtain

$$-\beta n \left(\frac{L'}{\rho}\right)_{tr}^{n-1} = -\frac{2\pi^2 E}{(L'/\rho)_{tr}^{3}}$$

The simultaneous solution of these equations gives

$$\beta = \frac{2E}{n\pi^n[(E/\sigma_{co})(1 + 2/n)]^{(n+2)/2}} \tag{14-76}$$

$$\left(\frac{L'}{\rho}\right)_{tr} = \pi \left[\frac{E}{\sigma_{co}}\left(1 + \frac{2}{n}\right)\right]^{\frac{1}{2}} \tag{14-77}$$

If $L'/\rho \leq (L'/\rho)_{tr}$, the failing stress can be computed from Eq. (14-75); otherwise it is calculated from the Euler equation (14-17).

The parameters σ_{co} and n are chosen so that the empirical equation fits the test data. To simplify the equation, n is usually taken as an integer. The most commonly used values are $n = 1$, which gives a straight line, and $n = 2$ which gives a parabola. With $n = 1$ we find from Eqs. (14-75) to (14-77) that

$$\sigma_c = \sigma_{co}\left[1 - \frac{0.385(L'/\rho)}{\pi(E/\sigma_{co})^{\frac{1}{2}}}\right] \tag{14-78}$$

for $L'/\rho \leq \pi \sqrt{3} \, (E/\sigma_{co})^{1/2}$. In a similar manner, for $n = 2$ we obtain

$$\sigma_c = \sigma_{co} \left[1 - \frac{\sigma_{co}(L'/\rho)^2}{4\pi^2 E} \right] \tag{14-79}$$

which is applicable when $L'/\rho \leq \pi \sqrt{2} \, (E/\sigma_{eo})^{1/2}$. Equations (14-78) and (14-79) are known respectively as the *straight-line* and *Johnson's parabola* short-column equations.

REFERENCES

1. Bleich, F.: "Buckling Strength of Metal Structures," McGraw-Hill Book Company, New York, 1952.
2. Hoff, N. J.: Buckling and Stability, *J. Roy. Aeron. Soc.*, **58**(517): 3–52 (January, 1954).
3. Shanley, F. R.: "Weight-Strength Analysis of Aircraft Structures," McGraw-Hill Book Company, New York, 1952.
4. Crandall, S. H.: "Engineering Analysis," McGraw-Hill Book Company, New York, 1956.
5. Timoshenko, S. P., and J. M. Gere: "Theory of Elastic Stability," 2d ed., McGraw-Hill Book Company, New York, 1961.
6. Shanley, F. R.: Inelastic Column Theory, *J. Aeron. Sci.*, **14**: 261–267 (1947).
7. Wilder, T. W., W. A. Brooks, and E. E. Mathauser: The Effect of Initial Curvature on the Strength of Inelastic Columns, *NACA Tech. Note* 3872, 1953.
8. Metallic Materials and Elements for Flight Vehicle Structures, *Military Handbook* MIL-HDBK-5A, Feb. 8, 1966.
9. Ramberg, W., and W. R. Osgood: Description of Stress-Strain Curves by Three Parameters, *NACA Tech. Note* 902, July, 1943.
10. Cozzone, F. P., and M. A. Melcon: Non-dimensional Buckling Curves: Their Development and Application, *J. Aeron. Sci.*, **13**(10): 511–517 (October, 1946).
11. Newton, R. E.: Experimental Study of Inelastic Buckling of Columns of Varying Section, *Proc. 1st U.S. Natl. Congr. Appl. Mech.*, pp. 625–629, 1952.

PROBLEMS

14-1. Derive the equation for the buckling load of a free-clamped uniform column. [*Ans.* $P_{cr} = \pi^2 EI/4L^2$.]

14-2. Derive the equation for the buckling load of a clamped-clamped column. [*Ans.* $P_{cr} = 4\pi^2 EI/L^2$.]

14-3. Determine the temperature rise that will cause a uniformly heated column of constant cross section and end fixity c to buckle if both ends are rigidly restrained against axial motion. [*Ans.* $T_{cr} = c\pi^2/\alpha(L/\rho)^2$]

14-4. Determine the dynamic pressure and mode shape for torsional divergence of a uniform wing by applying the method used to determine the eigenvalue and mode shape of Example 14-1 to the differential equation and boundary conditions of

Example 8-9. $\left[Ans. \quad q_D = \frac{\pi^2 G_1 J^*}{4L^2} \middle/ \frac{dC_n}{d\alpha} \, ce \text{ and } \varphi = C_1 \sin (\pi x/2L). \right]$

14-5. Show that the natural frequencies and mode shapes of torsional vibration of a uniform wing are $\omega_n = \dfrac{n\pi}{2L} \left(\dfrac{G_1 J^*}{I_x} \right)^{1/2}$ and $\Phi_n = C_1 \sin (n\pi x/2L)$, where $n = 1, 3, 5,$

. . . , by applying the method used to determine the eigenvalue and eigenfunction in Example 14-1 to the differential equation and boundary conditions of Example 8-8 (with $I_T = 0$).

14-6. The general solution of Eq. (d) in Example 7-8 is

$$W = C_1 \sin \lambda x + C_2 \cos \lambda x + C_3 \sinh \lambda x + C_4 \cosh \lambda x$$

if EI_{yy} is constant and I_0 is negligible, where $\lambda^4 = \omega^2 m / EI_{yy}$. Use the method of Example 14-2 to show that the natural frequencies and mode shapes of a beam with simply supported ends are $\omega_n = \left(\dfrac{n\pi}{L}\right)^2 \left(\dfrac{m}{EI_{yy}}\right)^{1/2}$ and $W_n = C_1 \sin (n\pi x/L)$, where $n = 1, 2, \ldots$.

14-7. Use the Rayleigh-Ritz method to determine the buckling load of a constant-cross-section column with pinned ends that rests upon a uniform elastic foundation.

14-8. Use the Rayleigh-Ritz method to obtain an approximate solution for the buckling load of a uniform column with ends which have elastic rotational restraints of stiffness $k_r = 2EI/L$. Assume that

$$w = C_1 \sin \frac{\pi x}{L} + C_2 \left(1 - \cos \frac{2\pi x}{L}\right)$$

[*Ans.* $P_{cr} = 1.675\pi^2 EI/L^2$.]

14-9. Obtain an approximate solution to Example 5-2 by the Rayleigh-Ritz method using the single term $w = c_1 \sin (\pi x/L)$ for the deflection curve. Compare the result with that of Example 5-3.

14-10. Use the Rayleigh-Ritz method to obtain the buckling load of a column with pinned ends which is loaded by an axial compressive force at the center of its span.

14-11. Use the finite-difference method to determine the buckling load of a free-clamped column with the following EI distribution

x/L	0	0.333	0.667	1
I/I_0	1.0	0.9	0.7	0.5

14-12. Determine the buckling load of a clamped-free 17-7 PH stainless steel (Fig. 3-9) column with a 1-in. diameter and a 10-in. length at a temperature of 400°F.

15

Instability and Failure of Plates

15-1 INTRODUCTION

A major portion of the structure of a flight vehicle is composed of thin-plate elements. These are found in the webs and flanges of spars, ribs, frames, floor beams, longitudinals, etc. In addition, the curvatures of the skin panels are usually small enough to permit them to be analyzed as plates. The loads and temperature gradients that act upon the structure can cause buckling of the plate elements. The prediction of critical loads and temperatures is important because the aerodynamic smoothness, stress distribution, and stiffness of the structure are affected by buckling.

In studying the stability and failure of plates we shall follow the approach used for columns in Chap. 14 and begin with the small-deflection elastic buckling theory for perfectly flat plates. The Rayleigh-Ritz and finite-difference methods will be developed to handle cases where exact solutions are not possible. The effects of initial imperfections, large deflections, and inelastic-material behavior are then considered. Finally, a semiempirical method for predicting the failing loads of plates will be described. The material covered in this chapter forms the basis for the local instability and failure analysis of thin-walled columns and stiffened plates in Chap. 16.

Additional information on the buckling and failure of plates may be found in Refs. 1 to 5. The latter reference is especially valuable because it summarizes the important work in the field up to 1957. References 5 to 7 contain useful summaries of results for the buckling of plates with various loading and boundary conditions.

15-2 FORMULATION OF THE BUCKLING PROBLEM

Plate buckling is more complicated than column buckling, and for generality we consider the problem in its broadest sense. Consider a non-homogeneous plate of arbitrary planform and variable thickness subjected

430

to forces that lie in the modulus-weighted reference plane. In the notation of Chap. 13 we assume that there are edge forces per unit length $\bar{X}$ and $\bar{Y}$ and body forces per unit area p_x and p_y. In addition, the plate is subjected to a temperature change in which $M_T = 0$. We assume that the forces and temperatures increase simultaneously and proportionally, so that we can write

$$\bar{X} = \lambda \bar{X}_0 \qquad \bar{Y} = \lambda \bar{Y}_0 \qquad p_x = \lambda p_{x_0} \qquad p_y = \lambda p_{y_0} \qquad T = \lambda T_0$$
$$(15\text{-}1)$$

where λ is a constant that establishes the magnitudes of the forces and temperatures and $\bar{X}_0$, $\bar{Y}_0$, etc., are functions that describe their distributions.

We seek the smallest value of λ for which the loads and temperatures produce a state of neutral stability in which the plate can be in equilibrium in the slightly bent position as well as in the perfectly flat position. Before determining λ we must find the stress function F or the in-plane stress resultants N_x, N_y, and N_{xy} that are associated with the applied loads and temperatures. The function F must satisfy the linearized form of Eq. (13-41) and Eqs. (13-50) or (13-51). Noting that F and the body-force potential V are linear functions of λ, we may write $F = \lambda F_0$ and $V = \lambda V_0$. Substituting these into Eq. (13-44), we obtain the homogeneous differential equation that must be satisfied by λ and w

$$\nabla^2(D^* \nabla^2 w) - (1 - \nu)\left(\frac{\partial^2 D^*}{\partial y^2} \frac{\partial^2 w}{\partial x^2} - 2 \frac{\partial^2 D^*}{\partial x \, \partial y} \frac{\partial^2 w}{\partial x \, \partial y} + \frac{\partial^2 D^*}{\partial x^2} \frac{\partial^2 w}{\partial y^2} \right)$$
$$= \lambda \left[\frac{\partial^2 F_0}{\partial y^2} \frac{\partial^2 w}{\partial x^2} - 2 \frac{\partial^2 F_0}{\partial x \, \partial y} \frac{\partial^2 w}{\partial x \, \partial y} + \frac{\partial^2 F_0}{\partial x^2} \frac{\partial^2 w}{\partial y^2} \right.$$
$$\left. + \frac{\partial}{\partial x}\left(V_0 \frac{\partial w}{\partial x} \right) + \frac{\partial}{\partial y}\left(V_0 \frac{\partial w}{\partial y} \right) \right] \quad (15\text{-}2)$$

Depending upon the type of edge support, λ and w must also satisfy one of the sets of boundary conditions from Eqs. (13-59) to (13-61). These are also homogeneous equations, since $M_T = 0$ and there are no applied lateral edge forces. The homogeneous differential equation and boundary conditions have the trivial solution $w = 0$ for all values of λ, indicating that the perfectly flat position is an equilibrium configuration for all values of the loads and temperatures. A nontrivial solution, i.e., the bent configuration, is possible only when λ is an eigenvalue. Substituting the smallest eigenvalue into Eq. (15-1) gives the conditions for buckling.

When in-plane displacements are specified on any of the edges, it is usually more convenient to take u and v as the unknowns rather than F. These displacement functions must satisfy the linearized forms of Eqs.

(13-47) in the domain of the plate and Eqs. (13-52) or (13-53) on the edges. The displacements u and v are directly proportional to λ, and it follows that N_x, N_y, and N_{xy} found from the linearized form of Eqs. (13-27) may be written

$$N_x = \lambda N_{x_0} \qquad N_y = \lambda N_{y_0} \qquad N_{xy} = \lambda N_{xy_0} \qquad (15\text{-}3)$$

Substituting these into Eq. (13-43), we obtain

$$\nabla^2(D^* \nabla^2 w) - (1 - \nu)\left(\frac{\partial^2 D^*}{\partial y^2}\frac{\partial^2 w}{\partial x^2} - 2\frac{\partial^2 D^*}{\partial x\, \partial y}\frac{\partial^2 w}{\partial x\, \partial y} + \frac{\partial^2 D^*}{\partial x^2}\frac{\partial^2 w}{\partial y^2}\right)$$

$$= \lambda\left(N_{x_0}\frac{\partial^2 w}{\partial x^2} + 2N_{xy_0}\frac{\partial^2 w}{\partial x\, \partial y} + N_{y_0}\frac{\partial^2 w}{\partial y^2} - p_{x_0}\frac{\partial w}{\partial x} - p_{y_0}\frac{\partial w}{\partial y}\right) \qquad (15\text{-}4)$$

which is an alternative form to Eq. (15-2) that can be solved with the appropriate boundary conditions described for that equation.

Except for simple special cases, it is usually impossible to obtain exact closed-form solutions to Eqs. (15-2) or (15-4). However, some of the simple cases are of practical interest. One of these is examined in the next section.

15-3 ELASTIC BUCKLING OF A SIMPLY SUPPORTED PLATE IN UNIAXIAL COMPRESSION

Consider a simply supported rectangular plate of width b and length a that is subjected to a uniform compressive force per unit length N on the edges $x = 0$ and a while the boundaries $y = 0$ and b are unrestrained against in-plane motion (Fig. 15-1). The plate may be nonhomogeneous, but it is assumed that D^* is constant and that the modulus-weighted reference surface is plane prior to buckling.

It is easily verified that $N_x = -N$ and $N_y = N_{xy} = 0$ in this simple case, so that Eqs. (15-3) and (15-4) reduce to the differential equation $\nabla^4 w = -(N/D^*)(\partial^2 w/\partial x^2)$ (see also Example 13-5). From Eqs. (13-60) the boundary conditions are $w = \partial^2 w/\partial x^2 = 0$ at $x = 0$ and a and $w = \partial^2 w/\partial y^2 = 0$ at $y = 0$ and b. The boundary conditions are satisfied

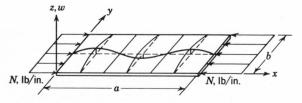

Fig. 15-1 Compressive buckling of a simply supported plate.

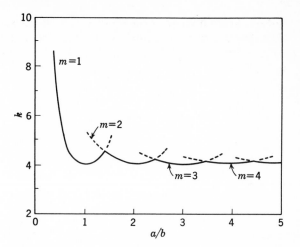

Fig. 15-2 Compressive-buckling coefficients for simply supported plates.

by the deflection mode shape

$$w = w_{mn} \sin \frac{m\pi x}{a} \sin \frac{n\pi y}{b} \qquad \begin{array}{l} m = 1, 2, \ldots \\ n = 1, 2, \ldots \end{array} \qquad (15\text{-}5)$$

Substituting this into the differential equation, we find that N must satisfy the characteristic equation

$$\left(\frac{m^2}{a^2} + \frac{n^2}{b^2} \right)^2 - \frac{m^2 N}{\pi^2 a^2 D^*} = 0$$

for Eq. (15-5) to be a nontrivial solution for all values of x and y.

The smallest value of N that satisfies the last equation is the buckling load N_{cr}, so that we may write

$$N_{cr} = \frac{k\pi^2 D^*}{b^2} \qquad (15\text{-}6)$$

where the *plate buckling coefficient* k is the minimum value of

$$k = \left(\frac{mb}{a} + \frac{an^2}{mb} \right)^2 \qquad (15\text{-}7)$$

obtained for a given a/b by the proper selection of m and n. We note from Eq. (15-5) that the integers m and n establish the wavelength of the buckle. From Eq. (15-7), the minimum value of k occurs when $n = 1$; that is, when the plate buckles into one loop of a sine wave in the direction normal to the applied load. To determine m, we plot k as a function of a/b for different values of m, as shown by the dotted curves in Fig. 15-2. The minimum value of k, which is used in Eq. (15-6), is obtained from the lower envelope of the curves; that is given by the solid line in the figure.

We see from the figure that m depends upon a/b and that k and the buckling load have their smallest values at $a/b = 1, 2, \ldots$, where $k = 4$. We also note that k approaches 4 as a/b becomes large and that it differs little from this value when $a/b \geq 3$.

When the plate is homogeneous, $D^* = D$, so that from the second of Eqs. (13-33), Eq. (15-6), and $\sigma_{cr} = N_{cr}/t$ we find

$$\sigma_{cr} = \frac{k\pi^2 E}{12(1 - \nu^2)} \left(\frac{t}{b}\right)^2 \tag{15-8}$$

This is known as the *Bryan equation*, after the English naval engineer who derived it in 1891 to predict the buckling stress of the plates in the hulls of steel ships. For a *long plate* $(a/b \geq 3)$ $k \approx 4$, and this equation reduces to

$$\sigma_{cr} = 3.62E \left(\frac{t}{b}\right)^2 \tag{15-9}$$

if it is assumed that $\nu = 0.3$.

The case in which a/b is small is also interesting. We observe from Fig. 15-2 that $m = 1$ when $a/b \leq 1$, so that Eq. (15-7) reduces to $k = (b/a)^2[1 + (a/b)^2]^2$. When $(a/b)^2 \ll 1$, the buckling coefficient further reduces to $k = (b/a)^2$, and Eq. (15-8) becomes

$$\sigma_{cr} = \frac{\pi^2 E}{12(1 - \nu^2)} \left(\frac{t}{a}\right)^2 \tag{15-10}$$

It is worth noting that if we consider the plate to behave as a simply supported column with $L' = a$ and $\rho = t/12^{1/2}$, Eq. (14-17) gives $\sigma_{cr} = \pi^2 E t^2/12a^2$. Comparing this result with Eq. (15-10), we find the only difference is that E in the column solution is replaced by $E/(1 - \nu^2)$. As noted in Example 13-3, this is due to the fact that $\sigma_y = 0$ in the column, whereas the restraint against anticlastic bending in the plate causes biaxial stresses and results in the $1 - \nu^2$ term. Because of the similarity to the column solution, compressively loaded plates with small values of a/b are called *wide columns*.

15-4 BUCKLING OF UNIFORM RECTANGULAR PLATES WITH SIMPLE EDGE LOADINGS

The buckling loads of uniform rectangular plates with constant or linearly varying normal edge forces or with constant shear forces on the edges have been determined for various boundary conditions by solving the appropriate differential equations or by using the Rayleigh-Ritz method (Sec. 15-5). In these simple cases the in-plane stress-resultant forces equal the applied edge forces.   The results are summarized in Ref. 5.   In general, the buckling load can be computed from Eq. (15-6), and for the homogeneous plate the buckling stress can be determined from Eq. (15-8). However, the value of k depends upon the type of loading and the edge restraints.   The buckling coefficient for a plate with a free edge, often referred to as a *flange*, also depends upon v, because this parameter appears in the boundary condition [Eq. (13-61)].

Figure 15-3 gives the value of k as a function of a/b for uniaxial compression with various combinations of simply supported, clamped, and free edges.   In the free-edge cases it is assumed that $v = 0.3$.   This figure illustrates several points that are worth noting.   The scalloped appearance of the curves is due to the fact that the number of longitudinal

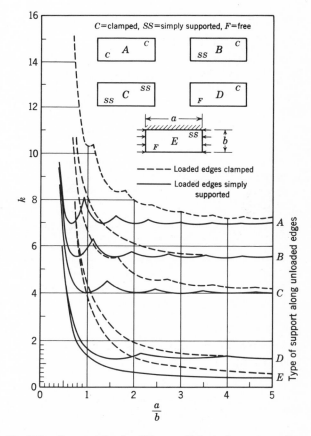

Fig. 15-3   Compressive-buckling coefficients for rectangular plates with various edge conditions (Ref. 5).

buckles depends upon a/b. In some cases, however, such as the flange with three simply supported edges, the buckle pattern is only a single loop of a sine function in the longitudinal direction for all values of a/b.

We observe that k is essentially independent of the restraint on the loaded edges when $a/b > 3$. However, in these cases k is strongly dependent upon the restraint on the unloaded edges. The buckling coefficient is nearly constant for long plates ($a/b \geq 3$). In this case N_{cr} and σ_{cr} do not depend upon a and are inversely proportional to b^2. This is in contrast to the column or the wide column $[(a/b)^2 \ll 1]$, where the length rather than the width is the critical dimension and the important edge support conditions are at the loaded instead of the unloaded edges.

Curves for k to be used for predicting the bending buckling of plates with the loaded edges simply supported and the unloaded edges simply supported or clamped are given in Fig. 15-4. The buckling force per unit length and the buckling stress that are predicted by using these values of k in Eqs. (15-6) and (15-8) are those which occur at $y = 0$ and b as shown by the insert in Fig. 15-4.

Equations (15-6) and (15-8) can also be used to determine the edge shear flow $N_{cr} = N_{xy} = q$ and the shear stress $\sigma_{cr} = \sigma_{xy}$ at which buckling

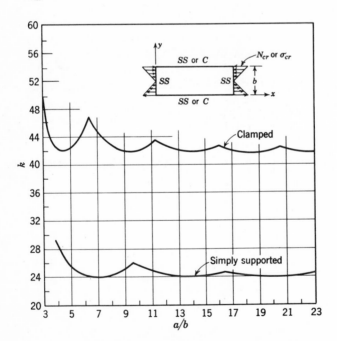

Fig. 15-4 Bending-buckling coefficients for rectangular plates (Ref. 5).

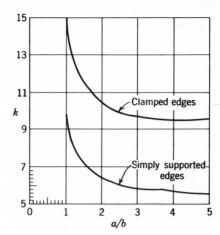

Fig. 15-5 Shear-buckling coefficients for rectangular plates (Ref. 5).

occurs.   In using these equations with the curves for k in Fig. 15-5, b is always the smaller dimension of the plate.

We see from Figs. 15-2 to 15-5 that k is essentially constant for $a/b > 3$.   Large values of a/b are common in the structures of flight vehicles, for an effective method of increasing the buckling stress of the skin is to use longitudinal stiffeners to divide the skin into panels with small values of b.   Values of k for long plates with simply supported, clamped, and free edges are summarized in Table 15-1.

The values of k for simply supported and clamped-edge conditions are idealizations of the boundary conditions in practical applications where the edges are supported by structural members that provide elastic edge restraint.   Curves for k as a function of the elastic rotational edge

Table 15-1   Values of k for long $(a/b \geq 3)$ rectangular plates[5]

Loading	Edge support	k
1. Compression	a. All edges simply supported	4.00
	b. All edges clamped	6.98
	c. Three edges simply supported, one unloaded edge free	0.43
	d. Three edges clamped, one unloaded edge free	1.28
2. Shear	a. All edges simply supported	5.35
	b. All edges clamped	8.98
3. Bending	a. All edges simply supported	23.9
	b. All edges clamped	41.8

restraint are given in Ref. 5.   These values fall between those for simply supported and clamped edges.

In the early stages of structural design, k is usually estimated from the limiting cases of simply supported and clamped edges.   Gerard[5] developed the chart shown in Fig. 15-6 to assist in estimating k for compressive buckling.   These curves, which are based upon tests of long plates stiffened by edge longitudinals of practical proportions, show the effect that the torsional rigidity of the stiffeners has upon k.

Kuhn[8] has obtained similar data for shear buckling.   He finds that k for elastically restrained edges is given by the semiempirical equation

$$k = k_{ss}\left[R_a + \tfrac{1}{2}(R_b - R_a)\left(\frac{b}{a}\right)^3 \right] \tag{15-11}$$

where k_{ss} is the theoretical value of the shear-buckling coefficient for simply supported edges found from Fig. 15-5.   The empirical coefficients R_a and R_b depend upon the elastic restraint along the edges of lengths a and b, respectively.   These are found from Fig. 15-7 and depend upon t_a/t and t_b/t, where t_a and t_b are the thicknesses of the restraining stiffeners along the a and b dimensions of the plate.   The upper curve is applicable to angle stiffeners attached to both sides of the plate or a T-section stiffener attached by a double row of fasteners.   The lower curve applies to an angle stiffener attached by a single line of fasteners.   Caution should be

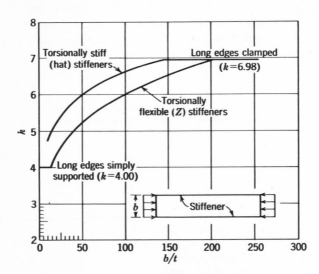

Fig. 15-6   Effect of stiffener torsional rigidity upon the compressive-buckling coefficient (Ref. 5).

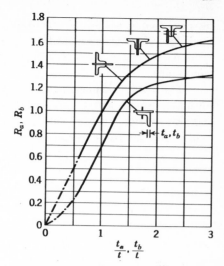

Fig. 15-7 Effect of stiffener rigidity upon shear-buckling coefficient (Ref. 8).

used in applying the method when the stiffeners are thinner than the plate, since this region of Fig. 15-7 has not been fully covered by tests. In this area the curves may predict buckling stresses below that of the unstiffened plate, in which case the unstiffened result should be used.

The buckling of parallelogram and triangular plates, which frequently occur in the skins of swept or delta-shaped aerodynamic surfaces, is treated in Ref. 5. The instability of circular plates, with or without center holes, is covered in Ref. 1. These elements appear in the bulkheads and frames of body structures.

15-5 APPROXIMATE METHODS

It is usually necessary to resort to approximate methods when the plate is of unusual planform or variable thickness or is subjected to nonuniform temperatures or edge forces. The Rayleigh-Ritz and finite-difference methods can be used in these cases. The procedures are similar to those described in Sec. 14-3, the only difference being that the energy expressions or differential equation and boundary conditions for the column are replaced by those of the plate.

Rayleigh-Ritz method   The equations for U and V of the plate are derived in Sec. 13-11. In determining these energies we use the strained position of the plate immediately prior to buckling as the datum configuration and measure u, v, and w from this state. The in-plane stress resultants N_x, N_y, and N_{xy} in this datum state are due to the applied in-plane forces and the prescribed edge displacements and temperature changes.   During the small lateral displacements that occur at the onset of buckling we may assume that N_x, N_y, and N_{xy} remain constant and equal to their prebuckling values.

The u and v displacements in the slightly bent position are negligible, since they are measured from the strained position that precedes buckling and do not change appreciably for small lateral displacements.   As a result, we find from Eqs. (13-70) to (13-73) and (15-1) that the strain energy of the slightly bent plate relative to the datum configuration is

$$U = \frac{1}{2} \iint_D D^* \left\{ \left(\frac{\partial^2 w}{\partial x^2} + \frac{\partial^2 w}{\partial y^2} \right)^2 - 2(1 - \nu) \left[\frac{\partial^2 w}{\partial x^2} \frac{\partial^2 w}{\partial y^2} \right. \right.$$
$$\left. \left. - \left(\frac{\partial^2 w}{\partial x \, \partial y} \right)^2 \right] \right\} dx \, dy - \frac{\lambda}{2} \iint_D N_{T_0} \left[\left(\frac{\partial w}{\partial x} \right)^2 \right.$$
$$\left. + \left(\frac{\partial w}{\partial y} \right)^2 \right] dx \, dy + \lambda \iint_D \left[\int_t \frac{E(\alpha T_0)^2}{1 - \nu} \, dz \right] dx \, dy \quad (15\text{-}12)$$

To obtain this result we have neglected the terms in Eq. (13-72) that involve the product of four derivatives of w.   We have seen in Chap. 13 that it is permissible to neglect these nonlinear terms when w is small compared to t.

Making similar assumptions, we find from Eqs. (13-79), (13-81), and (13-82) that the only nonzero potential-energy term is given by Eq. (13-81).   From Eqs. (15-3), this equation becomes

$$V = \frac{\lambda}{2} \iint_D \left[N_{x_0} \left(\frac{\partial w}{\partial x} \right)^2 + N_{y_0} \left(\frac{\partial w}{\partial y} \right)^2 + 2N_{xy_0} \frac{\partial w}{\partial x} \frac{\partial w}{\partial y} \right] dx \, dy \quad (15\text{-}13)$$

We approximate the buckling deflections by

$$w(x,y) = \sum_{j=1}^{n} c_j w_j(x,y) \quad (15\text{-}14)$$

where the c_j coefficients are undetermined constants and the w_j are assumed functions that must satisfy the prescribed displacement boundary conditions.   The coefficients c_j define the displacements in the bent configuration and therefore constitute a set of generalized coordinates. Substituting Eq. (15-14) into Eqs. (15-12) and (15-13) and applying Eq. (6-17) (with $q_i = c_i$), we obtain a set of homogeneous linear algebraic equations of the form of Eqs. (5-21), where $\psi_j = c_j$.   These may be written in the matrix form of Eq. (5-22), where

$$a_{ij} = \iint\limits_D D^* \left[\nabla^2 w_i \, \nabla^2 w_j - (1 - \nu) \left(\frac{\partial^2 w_i}{\partial x^2} \frac{\partial^2 w_j}{\partial y^2} + \frac{\partial^2 w_i}{\partial y^2} \frac{\partial^2 w_j}{\partial x^2} \right. \right.$$
$$\left. \left. - 2 \frac{\partial^2 w_i}{\partial x \, \partial y} \frac{\partial^2 w_j}{\partial x \, \partial y} \right) \right] dx \, dy \quad (15\text{-}15)$$

$$b_{ij} = - \iint\limits_D \left[(N_{x_0} - N_{T_0}) \frac{\partial w_i}{\partial x} \frac{\partial w_j}{\partial x} + (N_{y_0} - N_{T_0}) \frac{\partial w_i}{\partial y} \frac{\partial w_j}{\partial y} \right.$$
$$\left. + N_{xy_0} \left(\frac{\partial w_i}{\partial x} \frac{\partial w_j}{\partial y} + \frac{\partial w_i}{\partial y} \frac{\partial w_j}{\partial x} \right) \right] dx \, dy \quad (15\text{-}16)$$

When D^* is constant and the plate is polygonal and $w = 0$ on all edges, Eq. (13-71) reduces to Eq. (13-75), as shown in Sec. 13-11. In this case Eq. (15-15) is replaced by

$$a_{ij} = D^* \iint\limits_D \nabla^2 w_i \, \nabla^2 w_j \, dx \, dy \quad (15\text{-}17)$$

Equation (5-22) can be solved for λ and $\psi = c$ by the methods described in Sec. 5-5. The in-plane stress resultants at buckling can then be found from Eqs. (15-3).

Example 15-1   The applied loads and thermal gradients in the stiffened-shell structure of a flight vehicle induce the in-plane stress resultants

$$N_x = - \left[N_1 + (N_2 - N_1) \sin \frac{\pi y}{b} \right] \qquad N_y = N_{xy} = 0 \qquad (a)$$

in a typical skin panel, where N_1 and N_2 are defined in Fig. 15-8. The stiffening members provide simple support to the edges and the boundaries $y = 0$ and b are unrestrained with respect to in-plane motion. Determine the magnitudes of N_1 and N_2 that will produce buckling if the plate is long.

Guided by the results for the buckling of a rectangular plate under a uniform uniaxial loading, we assume that the buckle shape is given by a single function of the form of Eq. (15-5). Using the results for the uniformly loaded long plate, we take $m = a/b$ and $n = 1$, which give

$$w = c_1 \sin \frac{\pi x}{b} \sin \frac{\pi y}{b} \qquad (b)$$

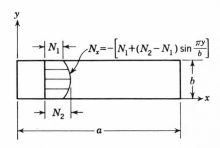

Fig. 15-8   Example 15-1.

Substituting this into Eq. (15-17), we find

$$a_{11} = \frac{4\pi^4 D}{b^4} \int_0^b \int_0^a \sin^2 \frac{\pi x}{b} \sin^2 \frac{\pi y}{b} \, dx \, dy = \frac{\pi^4 D a}{b^3} \tag{c}$$

With $\lambda = N_2$ we find from Eqs. (15-3) and (a) that

$$N_{x_0} = -\beta - (1 - \beta) \sin \frac{\pi y}{b} \qquad N_{y_0} = N_{xy_0} = 0 \tag{d}$$

where $\beta = N_1/N_2$. Substituting Eqs. (b) and (d) into Eq. (15-16), we obtain

$$b_{11} = \frac{\pi^2}{b^2} \int_0^b \int_0^a \left[1 + (1 - \beta) \sin \frac{\pi y}{b} \right] \cos^2 \frac{\pi x}{b} \sin^2 \frac{\pi y}{b} \, dy \, dy$$

$$= \frac{\pi^2 a}{4b} \left[\beta + \frac{8}{3\pi} (1 - \beta) \right]$$

Placing this result and Eq. (c) into Eq. (5-22), we find that for a nontrivial solution

$$N_{2_{cr}} = \frac{k\pi^2 D}{b^2} \tag{e}$$

where

$$k = \frac{4}{\beta + (8/3\pi)(1 - \beta)} \tag{f}$$

Values of k as a function of β are shown in Fig. 15-9. More accurate results can be obtained by adding additional terms to Eq. (b). Van der Neut[9] has determined k by the Galerkin method with the two-term approximation

$$w = c_1 \sin \frac{\pi x}{b} \sin \frac{\pi y}{b} + c_2 \sin \frac{\pi x}{b} \sin \frac{3\pi y}{b} \tag{g}$$

His results are given by the dotted line in Fig. 15-9. When the assumed functions satisfy both force and displacement boundary conditions, as in Eq. (g), the Rayleigh-Ritz and Galerkin methods give identical results.[13] It is therefore seen from Fig. 15-9 that little improvement is obtained by using the second

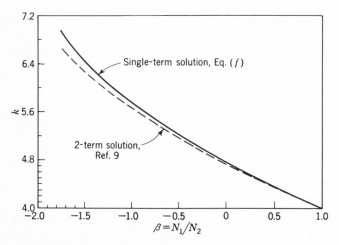

Fig. 15-9 Buckling coefficient, Example 15-1.

term of Eq. (g) in the Rayleigh-Ritz method. Equation (f) gives $k = 4$ when $\beta = 1$, which agrees with the previously obtained result for the long plate under uniform compression.

Buckling can occur even though the resultant compressive load on the plate is zero. To obtain β for this case we integrate N_x in Eq. (a) from $y = 0$ to b and equate the resulting force to zero. This leads to $\beta = -2/(\pi - 2)$, which gives $k = 6.9$ when it is substituted into Eq. (f).

Finite-difference method   The in-plane stress-resultant distributions and their magnitude for buckling can be determined by applying the finite-difference method to the differential equations and boundary conditions derived in Sec. 15-2. Equation (15-4) can be reduced to a second-order differential equation when the plate is uniform, polygonal, simply supported, and loaded by a uniform compressive force per unit length†️ $N_n = -N$. In this case $N_x = N_y = -N$ and $N_{xy} = 0$, so that Eq. (15-4) becomes

$$\nabla^2(\nabla^2 w) = -\frac{N}{D*} \nabla^2 w \tag{15-18}$$

At the edges $\partial^2 w/\partial n^2 = 0$ and $\partial^2 w/\partial s^2 = 0$, so that $\nabla^2 w = 0$. Introducing a new function $M(x,y)$, defined by $\nabla^2 w = M$, Eq. (15-18) becomes

$$\nabla^2 M = -\frac{N}{D*} M \tag{15-19}$$

which is subject to the boundary condition $M = 0$. When applicable, Eq. (15-19) is preferable to Eq. (15-4) because the ∇^2 finite-difference operator is more compact than the ∇^4 operator.

Example 15-2   A homogeneous uniform square plate has simply supported edges that are free to displace in the plane of the plate. The plate is subjected to a temperature change $T(x,y)$ that is doubly symmetric with respect to the x and y axes (Fig. 15-10) and vanishes on the boundary. Use the difference mesh of Fig. 15-10 (which makes use of the symmetries) to determine the buckling temperature when

$$\mathbf{T} = \lambda \mathbf{T_0} = \lambda\{1.0 \quad 0.7 \quad 0.5\} \tag{a}$$

where λ is a constant that establishes the magnitude of the temperature change and $\mathbf{T_0}$ is a matrix that gives the temperature at each of the mesh points relative to the temperature at point 1.

Noting from Eq. (13-29) that $N_T = Et\alpha T/(1 - \nu)$ and letting $T = \lambda T_0(x,y)$ and $F = \lambda F_0(x,y)$, we find that the linearized form of Eq. (13-41) becomes

$$\nabla^4 F_0 = -\alpha Et \, \nabla^2 T_0 \tag{b}$$

For unrestrained edges, $\bar{X} = \bar{Y} = 0$, and from Eqs. (13-51) the boundary conditions become $F_0 = 0$ and $\partial F_0/\partial n = 0$. The first of these conditions gives $F_{0_4} = 0$, and the second gives $F_{0_5} = F_{0_2}$ and $F_{0_6} = F_{0_3}$ (Fig. 15-10).

---

† This is the case in a uniformly heated plate with edges restrained against expansion, for which $N_n = -N_T$.

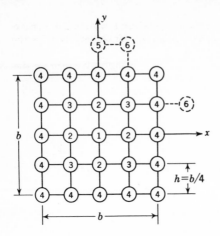

Fig. 15-10 Example 15-2.

Substituting the ∇^4 and ∇^2 modules from Fig. 5-5 into Eq. (b) and applying the resulting difference equation to points 1, 2, and 3, we find

$$\begin{bmatrix} 5 & -8 & 2 \\ -8 & 26 & -16 \\ 2 & -16 & 24 \end{bmatrix} \begin{Bmatrix} F_{0_1} \\ F_{0_2} \\ F_{0_3} \end{Bmatrix} = \frac{\alpha E b^2 t}{16} \begin{bmatrix} 1 & -1 & 0 \\ -1 & 4 & -2 \\ 0 & -2 & 4 \end{bmatrix} \begin{Bmatrix} T_{0_1} \\ T_{0_2} \\ T_{0_3} \end{Bmatrix} \qquad (c)$$

where the equation for point 1 has been divided by 4 to make the square matrices symmetric. In writing the difference equations we have used the given condition $T_{0_4} = 0$. Substituting Eq. (a) into Eq. (c) and solving for $\mathbf{F}_0$, we find

$$\mathbf{F}_0 = \frac{\alpha E b^2 t}{16} \{0.3989 \quad 0.2517 \quad 0.1596\} \qquad (d)$$

For our problem, Eq. (15-2) reduces to

$$\nabla^4 w = \frac{\lambda}{D} \left(\frac{\partial^2 F_0}{\partial y^2} \frac{\partial^2 w}{\partial x^2} - 2 \frac{\partial^2 F_0}{\partial x \, \partial y} \frac{\partial^2 w}{\partial x \, \partial y} + \frac{\partial^2 F_0}{\partial x^2} \frac{\partial^2 w}{\partial y^2} \right) \qquad (e)$$

and the associated boundary conditions for simply supported edges are $w = 0$ and $\partial^2 w / \partial n^2 = 0$. The first of these gives $w_4 = 0$, and the other requires that $w_5 = -w_2$ and $w_6 = -w_3$. Using these, we obtain the following equation by applying the finite-difference approximation of Eq. (e) at points 1, 2, and 3.

$$\begin{bmatrix} 5 & -8 & 2 \\ -8 & 24 & -16 \\ 2 & -16 & 20 \end{bmatrix} \begin{Bmatrix} w_1 \\ w_2 \\ w_3 \end{Bmatrix}$$

$$= \frac{2\lambda}{D} \begin{bmatrix} F_{0_1} - F_{0_2} & -F_{0_1} + F_{0_2} & 0 \\ -F_{0_2} + F_{0_3} & -F_{0_1} + 4F_{0_2} - 2F_{0_3} & F_{0_1} - 2F_{0_2} \\ -\dfrac{F_{0_1}}{16} & F_{0_2} - 2F_{0_3} & -2F_{0_2} + 4F_{0_3} \end{bmatrix} \begin{Bmatrix} w_1 \\ w_2 \\ w_3 \end{Bmatrix}$$

The equation for point 1 has again been divided by 4. Substituting the values of F_{0_i} from Eq. (d) into this equation, we find

$$\begin{bmatrix} 5 & -8 & 2 \\ -8 & 24 & -16 \\ 2 & -16 & 20 \end{bmatrix} \begin{Bmatrix} w_1 \\ w_2 \\ w_3 \end{Bmatrix} = \frac{\lambda \alpha E b^2 t}{8D} \begin{bmatrix} 0.1472 & -0.1472 & 0 \\ -0.0921 & 0.2887 & -0.1045 \\ -0.0249 & -0.0675 & 0.1350 \end{bmatrix} \begin{Bmatrix} w_1 \\ w_2 \\ w_3 \end{Bmatrix}$$

To solve this equation for the smallest value of λ we use the matrix-iteration method (Sec. 5-5). Multiplying both sides of the equation by the inverse of the square matrix on the left gives

$$\begin{bmatrix} 2.399 & -0.1774 & -0.322 \\ 1.051 & 0.5692 & -0.174 \\ 0.561 & 0.3651 & 0.109 \end{bmatrix} \begin{Bmatrix} w_1 \\ w_2 \\ w_3 \end{Bmatrix} = \bar{\lambda} \begin{Bmatrix} w_1 \\ w_2 \\ w_3 \end{Bmatrix} \qquad (f)$$

where

$$\bar{\lambda} = \frac{256D}{\lambda \alpha E b^2 t} \qquad (g)$$

To start the iteration we assume the deflection shape $w = \cos(\pi x/2b) \cos(\pi y/2b)$, which gives

$$w_0 = \{1.000 \quad 0.707 \quad 0.500\}$$

Substituting this into the left side of Eq. (f) gives

$$\bar{\lambda}_1 \mathbf{w}_1 = 2.11\{1.000 \quad 0.647 \quad 0.414\}$$

Successively repeating this process, we find after four iterations that the results have converged to three significant figures, giving

$$\bar{\lambda}\mathbf{w} = 2.17\{1.000 \quad 0.617 \quad 0.382\}$$

Substituting $\bar{\lambda} = 2.17$ into Eq. (g) and noting from Eq. (a) that $T_1 = \lambda T_{0_1} = \lambda$, we find that buckling occurs when

$$\alpha E T_1 = \frac{11.93\pi^2 E}{12(1 - \nu^2)}$$

15-6 COMBINED LOADS AND INTERACTION CURVES

Plates are frequently subjected to the combined actions of compression, shear, and bending. When this happens, the loads at which buckling takes place are less than those that occur when the loads are applied separately. As an example, consider the case of a uniform simply supported rectangular plate that is subjected to constant biaxial compressive forces N_x and N_y. In this case Eq. (15-4) becomes

$$D^* \nabla^4 w = -N_x \frac{\partial^2 w}{\partial x^2} - N_y \frac{\partial^2 w}{\partial y^2}$$

The mode shape of Eq. (15-5), which satisfies simply supported boundary conditions, gives the characteristic equation

$$D^* \left[\left(\frac{m\pi}{a}\right)^2 + \left(\frac{n\pi}{b}\right)^2 \right]^2 = \left(\frac{m\pi}{a}\right)^2 N_x + \left(\frac{n\pi}{b}\right)^2 N_y$$

when it is substituted into the differential equation. This may be written in the form

$$\frac{N_x}{(mb/a + n^2a/mb)^2 \pi^2 D^*/b^2} + \frac{N_y}{(m^2b/na + na/b)^2 \pi^2 D^*/a^2} = 1$$

where it is seen that the values of N_x and N_y that cause buckling depend upon a/b.

The square plate has been treated by Timoshenko.[1] If we let $a = b$, the last equation becomes

$$\frac{N_x}{(m + n^2/m)^2\pi^2D^*/a^2} + \frac{N_y}{(m^2/n + n)^2\pi^2D^*/a^2} = 1 \qquad (15\text{-}20)$$

To determine when buckling occurs, we must find the values of m and n that minimize N_x and N_y. To obtain these we successively assume pairs of integers for m and n and plot Eq. (15-20) as shown in Fig. 15-11. We see from this figure that N_x and N_y have their smallest values when $m = n = 1$, so that Eq. (15-20) becomes

$$\frac{N_x}{4\pi^2D^*/a^2} + \frac{N_y}{4\pi^2D^*/a^2} = 1$$

The denominators of the terms on the left of this equation are equal to the buckling load of the square plate in uniaxial compression (Sec. 15-3). We may therefore write the equation as

$$R_x + R_y = 1 \qquad\qquad (15\text{-}21)$$

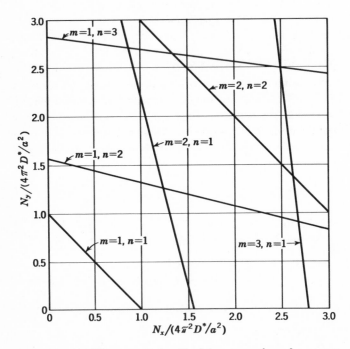

Fig. 15-11 Buckling curves for biaxial compression of a square plate.

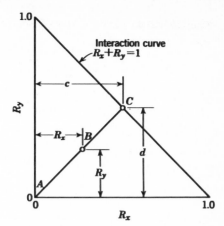

Fig. 15-12 Interaction curve for square and long plates in biaxial compression.

where $R_x = N_x/N_{x_{cr}}$, $R_y = N_y/N_{y_{cr}}$, and $N_{x_{cr}}$ and $N_{y_{cr}}$ are the uniaxial critical compressive loads when each of the loadings acts separately. This same method can be applied to determine the critical combination of loads for other values of a/b. When a/b is large, the plate buckles into square panels, and Eq. (15-21) again applies.

Equation (15-21) is known as an *interaction equation* because it describes the interacting effects of the two loadings. The equation can be plotted as an *interaction curve*, as shown in Fig. 15-12. Buckling occurs when N_x and N_y plot as a point that is on or above the interaction curve.   To determine the buckling margin of safety for design points that fall above or below the interaction curve, we assume that loads increase proportionally, so that N_y/N_x remains constant during loading. As a result, R_y/R_x is constant, and the design point proceeds along the line ABC of Fig. 15-12 as the loads are increased.   The margin of safety at B is therefore

$$MS = \frac{c}{R_x} - 1 \qquad\qquad (15\text{-}22)$$

From similar triangles $R_x/c = R_y/d$, and from Eq. (15-21) $c + d = 1$ for points on the interaction curve. As a result, $c = R_x/(R_x + R_y)$, and Eq. (15-22) becomes

$$MS = \frac{1}{R_x + R_y} - 1 \qquad\qquad (15\text{-}23)$$

Interaction equations can be determined for other combinations of loads. In general these may be written

$$R_a{}^\alpha + R_b{}^\beta + R_c{}^\gamma + \cdots = 1 \qquad\qquad (15\text{-}24)$$

where the *load* or *stress ratio* R_i is defined by

$$R_i = \frac{i\text{th applied load or stress acting with combined loads}}{i\text{th buckling load or stress when acting alone}}$$

The exponents can be determined theoretically or experimentally. In the general case the exponents depend upon the loadings, the edge restraint, and a/b.

Gerard[5,14] has summarized the interaction equations and curves for plate buckling under combinations of compression, tension, bending, and shear. Some of the frequently encountered cases that result in simple equations are given in Table 15-2. An interaction surface is described by Eq. (15-24) when three loads act simultaneously. A family of curves formed by the intersection of parallel cutting planes and the interaction surface is required for the graphical representation of these cases. An example which gives the interaction between transverse compression, shear, and bending is shown in Fig. 15-13.

Example 15-3   A $4 \times 10 \times 0.040$ in. aluminum alloy ($E = 10.5 \times 10^6$ psi, $\nu = 0.3$) skin panel of an airplane wing is subjected to a longitudinal compressive stress of 3000 psi and a shear flow of 100 lb/in. at the limit load. Determine the margin of safety if, to preserve aerodynamic smoothness, no buckling is permitted at the limit load.

Conservatively assuming that the edges are simply supported, we find from Figs. 15-3 and 15-5 with $a/b = 1\frac{9}{4} = 2.5$ that $k_c = 4.1$ and $k_s = 6.0$, where the subscripts refer to compression and shear. From Eq. (15-8), the

Table 15-2   Interaction equations for buckling of rectangular plates under combined loads[5]

Type of loading	Interaction equation	Buckling margin of safety
1. Biaxial compression†	$R_x + R_y = 1$	$\dfrac{1}{R_x + R_y} - 1$
2. Longitudinal compression and shear‡	$R_c + R_s{}^2 = 1$	$\dfrac{2}{R_c + (R_c{}^2 + 4R_s{}^2)^{\frac{1}{2}}} - 1$
3. Bending and shear§	$R_b{}^2 + R_s{}^2 = 1$	$\dfrac{1}{(R_b{}^2 + R_s{}^2)^{\frac{1}{2}}} - 1$

† Square or long simply supported plates.

‡ Simply supported plates with $a/b \geq 1$ and long plates with elastically restrained edges.

§ Applicable for all a/b.

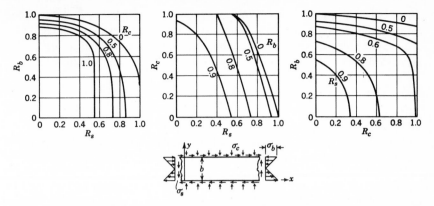

Fig. 15-13 Interaction curves for long simply supported plates under combinations of transverse compression, shear, and bending (Ref. 15).

critical stresses in compression and shear are

$$\sigma_{cr_c} = \frac{4.1 \times \pi^2 \times 10.5 \times 10^6}{12(1 - 0.3^2)} \left(\frac{0.040}{4}\right)^2 = 3890 \text{ psi}$$

$$\sigma_{cr_s} = \frac{6.0 \times \pi^2 \times 10.5 \times 10^6}{12(1 - 0.3^2)} \left(\frac{0.040}{4}\right)^2 = 5690 \text{ psi}$$

The stress ratios are $R_c = 3000/3890 = 0.771$ and $R_s = 100/0.040 \times 5690 = 0.440$. From Table 15-2, the interaction equation is $R_c + R_s^2 = 1$ and

$$MS = \frac{2}{0.771 + (0.771^2 + 4 \times 0.44^2)^{1/2}} - 1 = 0.03$$

15-7 EFFECTS OF LARGE DEFLECTIONS AND INITIAL IMPERFECTIONS

For an initially perfect plate, the only equilibrium configuration when $\sigma < \sigma_{cr}$ is the stable flat position. Neutral stability exists at $\sigma = \sigma_{cr}$, and the plate is in equilibrium in either the flat or the slightly bent position. The flat position is unstable when $\sigma > \sigma_{cr}$, and only the stable bent position is of interest. Large lateral displacements on the order of t occur at stresses that are only slightly greater than σ_{cr}. As a result, nonlinear large-deformation theory must be used in any postbuckling analysis. This introduces mathematical difficulties, and exact solutions for the deflection and stress distributions are not known for $\sigma > \sigma_{cr}$.

Approximate postbuckling solutions have been obtained by several investigators, and bibliographies and summaries of results are given in Refs. 1 and 16 to 19. The derivations are lengthy, and no attempt will be made to reproduce them here. Instead, the major methods of analysis will be outlined, and the significant results for uniform homogeneous rectangular plates in uniaxial compression will be summarized.

The Rayleigh-Ritz method has been used to compute the undetermined coefficients in an assumed series solution for the displacements.[1] However, a set of nonlinear algebraic equations results from using the large-deflection equations (13-72) and (13-75) to determine U. The difficulties involved in solving the nonlinear simultaneous equations severely limits the number of terms that can be taken in the approximating series.

An alternate method of solution[20] is to assume a series solution for w which contains undetermined coefficients. Substituting this into Eq. (13-42) gives a linear differential equation for F. A series solution for F in terms of the undetermined coefficients is obtained from the solution to this differential equation. Substituting w and F into Eq. (13-46) gives a nonlinear equation that is solved for the undetermined coefficients by the method of successive approximations. The membrane forces associated with w are then found by substituting F into Eqs. (13-40). Other approximate methods that combine the solution of Eq. (13-42) with an energy solution for w have also been used. The results obtained from the theoretical studies and test results for plates that are subjected to uniform compressive edge shortening are summarized in the following paragraphs.

The plate can sustain loads greater than the buckling load when lateral bending is prevented on the unloaded edges but the axial compressive stresses are not uniformly distributed across the width of the plate, as they are when $\sigma < \sigma_{cr}$. The stress distribution depends upon the in-plane restraint on the unloaded edges; distributions for edges that are held straight (but are free to move uniformly so that there are no resultant forces on the unloaded edges) and edges that are stress-free are shown in Fig. 15-14. The former case is typical of the skin panels of stiffened-shell structures, and the latter is representative of the web elements of thin-walled columns.

We note from Fig. 15-14a that tensile stresses are developed in the y direction on the center portion of the unloaded edges when they are constrained to remain straight. These membrane stresses, together with the fact that the unloaded edges are restrained against lateral bending, explain why the plate, unlike the column, can carry loads that are much greater than the buckling load. A contraction occurs at the center of the unloaded edges of a plate when they are stress-free (Fig. 15-14b). The lack of membrane edge forces in this case accounts for the fact that such a plate is unable to carry postbuckling loads that are as large as those of a plate with edges that are constrained to remain straight.

The bending deformations of perfect and imperfect plates are shown in Fig. 15-15. It is interesting to compare these curves with the corresponding column curves in Fig. 14-15. In both cases, we see that the

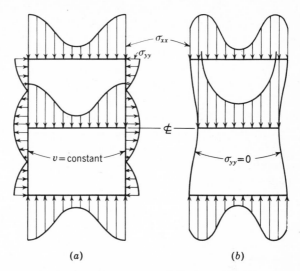

Fig. 15-14 Postbuckling stress and displacement distributions for plates with uniform end shortening (Ref. 21). (a) Straight unloaded edges; (b) stress-free unloaded edges free to warp in plane of plate.

member with small imperfections closely follows the theory for the perfect member in the prebuckling and postbuckling regions.   It is only in the immediate vicinity of σ_{cr} that the deflections of the member with practical imperfections differ appreciably from those of the perfect member.   However, as with the perfect member, the bending deflections of the practical member grow rapidly in the region of σ_{cr}.   We conclude from this that practical initial imperfections have small influence on the buckling and failing loads of plates and columns.   One should not infer that this is always the case in elastic-instability problems, for small initial imperfections have very large effects upon the failing loads of compressively loaded thin-walled cylinders.

We see from Fig. 15-15 that, unlike the column, sizable postbuckling stresses are possible.   Even though the stiffness of the plate decreases after buckling, failure does not occur until the axial stress at the unloaded edges reaches the yield stress.

The behavior of a plate at large postbuckling loads is complicated by the fact that the buckle shape may alter as the load is increased, and in a long plate the number of buckles may change.   The center portion of the plate approaches a developable surface with curvature in the direction of the loading only.   Double curvature is essentially restricted to the region of the unloaded edges.   In this case the strain energy in the central por-

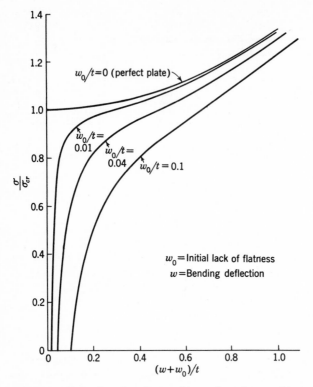

Fig. 15-15   Behavior of perfect and imperfect plates subjected
to uniform edge shortening (Ref. 22).

tion is mainly due to bending, while in the edge region it is mainly due to
membrane action.

It is impractical to use the actual nonuniform postbuckling stress
distributions of Fig. 15-14 in routine stress analysis.   It is more convenient to assume that the stress σ_e at the unloaded edges is uniformly
distributed over a fictitious *effective width* b_e adjacent to the edges (Fig.
15-16).   The effective width is determined from the condition that the
resultant force associated with the actual and assumed stress distributions
must be the same.   From this condition

$$P = 2\sigma_e b_e t = t \int_0^b \sigma_{xx} \, dy$$

which gives

$$b_e = \frac{1}{2\sigma_e} \int_0^b \sigma_{xx} \, dy \tag{15-25}$$

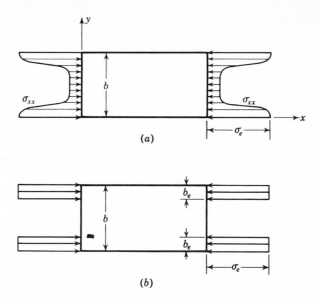

Fig. 15-16 (a) Actual and (b) assumed stress distributions in a buckled plate.

The distribution of σ_{xx} depends upon the rotational and in-plane restraints at the unloaded edges and the a/b ratio of the plate.   However, the stress distribution and b_e are independent of a/b when this ratio is greater than 3.   Several investigators have given equations for b_e that are based upon theoretical or experimental results.   Argyris and Dunne[18] give the family of solid curves shown in Fig. 15-17 for long plates with simply supported edges that are constrained to remain straight (that is, $v = $ const) but with elastic restraint against in-plane expansions due to Poisson's ratio.   In an actual structure this restraint is provided by the transverse stiffening members, i.e., the ribs or frames.   This effect is contained in the elastic-restraint parameter A_r/at, where A_r is the area of the transverse stiffening member.   The limiting cases of $A_r/at = 0$ and ∞ correspond to unrestrained and fully restrained straight unloaded edges.

In using Fig. 15-17, the effect of the elastic in-plane edge restraint must be taken into account in determining σ_{cr}.   Resisting the Poisson's ratio expansion in the y direction causes compressive stresses to develop in that direction.   It was seen in Sec. 15-6 that these stresses reduce the σ_{xx} stress at which buckling occurs.   This effect can be accounted for by determining k in Eq. (15-8) from Fig. 15-18.

The curves of Argyris and Dunne are restricted to small postbuckling loads ($\sigma_e/\sigma_{\mathrm{cr}} \leq 3$) because they do not contain the effect of buckle

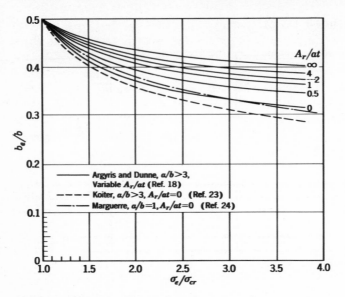

Fig. 15-17 Effective width of plates in compression.

change with load.   For long plates at large postbuckling loads the Koiter equation[23]

$$b_e = \frac{b}{2}\left[1.2\left(\frac{\sigma_{cr}}{\sigma_e}\right)^{0.4} - 0.65\left(\frac{\sigma_{cr}}{\sigma_e}\right)^{0.8} + 0.45\left(\frac{\sigma_{cr}}{\sigma_e}\right)^{1.2}\right] \qquad (15\text{-}26)$$

may be used.   This equation, which is shown dotted in Fig. 15-17, assumes that $A_r/at = 0$.   It has been shown to be satisfactory for simply supported, clamped, or elastically restrained edges.   The equation

$$b_e = \frac{b}{2}\left[0.19 + 0.81\left(\frac{\sigma_{cr}}{\sigma_e}\right)^{\frac{1}{2}}\right] \qquad (15\text{-}27)$$

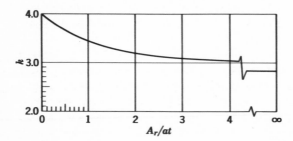

Fig. 15-18 Compressive-buckling coefficients for long plates with elastic restraint against in-plane motion at unloaded edges (Ref. 18).

which was developed by Marguerre[24] for large postbuckling loads of square plates with $A_r/at = 0$, is also shown in Fig. 15-17. The three methods for determining b_e are seen to be in reasonably good agreement when $A_r/at = 0$ even though they apply to different ranges of a/b and σ_{cr}/σ_e and different conditions of rotational edge restraint.

Initial imperfections have been found to have little influence on b_e at large values of σ_e/σ_{cr} but may sensibly reduce b_e in the region of $\sigma_{cr}/\sigma_e = 1$. The postbuckling behavior of rectangular plates in shear has been investigated by Kuhn, Peterson, and Levin[8] and by van der Neut.[16]

Example 15-4 The 0.064-in. aluminum alloy ($E = 10.5 \times 10^6$) skin of the interstage structure of a launch rocket is divided into 5×15 in. panels by longitudinal stiffeners and transverse-ring frames. Boost loads cause axial compressive stresses in the skin and stiffeners. The ring has an area 0.8 in.[2], and the structure is proportioned so that the skin buckles before the stiffeners or frames. Determine the effective width of the skin and the axial force carried by the skin when the stress in the stiffeners is 15,000 psi.

Conservatively assuming that the stiffeners and frames provide simple support to the skin, we find from Fig. 15-18 that $k = 3.48$ for $A_r/at = 0.8/15 \times 0.064 = 0.835$. From Eq. (15-8)

$$\sigma_{cr} = \frac{3.48 \times \pi^2 \times 10.5 \times 10^6}{12(1 - 0.3^2)} \left(\frac{0.064}{5}\right)^2 = 5460 \text{ psi}$$

so that $\sigma_e/\sigma_{cr} = 15,000/5460 = 2.75$. The plate is long and simply supported, and $\sigma_e/\sigma_{cr} < 3$; therefore b_e can be found from the results of Argyris and Dunne. From Fig. 15-17 we find $b_e/b = 0.375$, so that $b_e = 0.375 \times 5 = 1.875$ in. The axial force that is carried by the skin is therefore

$$P = 2b_e t\sigma_e = 2 \times 1.875 \times 0.064 \times 15,000 = 3600 \text{ lb}$$

15-8 INELASTIC BUCKLING OF PLATES

When b/t is small, σ_{cr} may exceed the proportional limit. Equation (15-8) is not directly applicable in these cases because the stress is no longer related to the strain by E, because ν depends upon the stress, and because in a flange the free-edge boundary condition Eq. (13-61) contains ν, so that k also depends upon the stress.

It is common practice to include all' these effects in a *plasticity correction factor* $\eta = \sigma_{cr}/\sigma_{cr}(\text{elastic})$, where σ_{cr} is the buckling stress corrected for plastic effects and $\sigma_{cr}(\text{elastic})$ is the elastic buckling stress computed from Eq. (15-8). With this notation

$$\sigma_{cr} = \frac{\eta k\pi^2 E}{12(1 - \nu_e^2)} \left(\frac{t}{b}\right)^2 \tag{15-28}$$

which is applicable at all stress levels, since $\eta = 1$ in the linearly elastic region. The notation ν_e is used in Eq. (15-28) to make it explicit that the elastic value of Poisson's ratio is used in this equation because the inelastic effect of ν is contained in η.

While there is general agreement upon the elastic stress-strain equations, there are no commonly accepted inelastic stress-strain equations that are applicable to all histories and states of stress. Two approaches have been widely used. In the *deformation theory* the stresses and strains are related by

$$\sigma_i = E_s \epsilon_i \tag{15-29}$$

for increasing stresses, while in the *incremental theory* they are related by

$$d\sigma_i = E_t \, d\epsilon_i \tag{15-30}$$

Unloading is assumed to occur elastically in both cases. The symbols σ_i and ϵ_i denote the *effective stress* and *strain intensities*. Various equations for σ_i and ϵ_i have been suggested by different investigators. One of these is the *distortional-energy* or *octahedral-shear theory* of Huber, Mises, and Henky, which for plane stress gives

$$\sigma_i = \sqrt{\sigma_{xx}^2 + \sigma_{yy}^2 - \sigma_{xx}\sigma_{yy} + 3\sigma_{xy}^2} \tag{15-31}$$

$$\epsilon_i = \frac{2}{\sqrt{3}} \sqrt{\epsilon_{xx}^2 + \epsilon_{yy}^2 + \epsilon_{xx}\epsilon_{yy} - \left(\frac{\epsilon_{xy}}{2}\right)^2} \tag{15-32}$$

We note that for a uniaxial loading $\sigma_{yy} = \sigma_{xy} = 0$, $\epsilon_{yy} = -\nu\epsilon_{xx}$, and $\epsilon_{xy} = 0$ and Eqs. (15-29) and (15-30) reduce to Eqs. (3-3) and (3-2) if ν is taken equal to $\nu_p = \frac{1}{2}$, the value of Poisson's ratio for an isotropic perfectly plastic material.

Plate-buckling theories based upon the deformation theory of Eq. (15-29) have been found to be in better agreement with experimental results than those obtained from incremental theories.[5] The question of whether an elastic or plastic modulus should be used on the convex side of the slightly buckled plate also arises, as it did in inelastic-column theory. If the plate were to remain perfectly flat until σ_{cr}, elastic unloading would occur on the convex side, and a reduced-modulus theory would be applicable. On the other hand, if there are small initial imperfections in flatness, bending and compression proceed simultaneously, and at small bending deformations there is no elastic unloading. As in columns, the theory with no strain reversal has been found to be in agreement with experimental evidence. Inelastic-plate buckling theory may therefore be thought of as a two-dimensional generalization of the tangent-modulus column theory.

Consider a simply supported rectangular plate in uniaxial compression (Fig. 15-1). Our derivation will follow that given by Gerard,[3] which is a simplified presentation of Stowell's theory.[25] The prebuckling state of strain is

$$\epsilon_{xx} = \frac{\sigma_{xx}}{E_s} \qquad \epsilon_{yy} = -\nu \frac{\sigma_{xx}}{E_s} \qquad \epsilon_{xy} = 0 \qquad (15\text{-}33)$$

A more complex set of strains is superimposed upon this simple state when the plate buckles. These additional strains require biaxial plastic stress-strain equations. To obtain these we set $E = E_s$, $\nu = \nu_p = \frac{1}{2}$, and $T = 0$ in Eqs. (3-20) to obtain

$$\sigma_{xx} = \tfrac{4}{3}E_s\left(\epsilon_{xx} + \frac{\epsilon_{yy}}{2}\right) \qquad \sigma_{yy} = \tfrac{4}{3}E_s\left(\epsilon_{yy} + \frac{\epsilon_{xx}}{2}\right) \qquad \sigma_{xy} = \frac{E_s\epsilon_{xy}}{3} \quad (15\text{-}34)$$

These equations are identical to those obtained from the Huber-Mises-Henky hypothesis and deformation theory.

For small buckling displacements the changes in the membrane strains are negligible, and the increments of strain superimposed upon the prebuckling strains of Eq. (15-33) are due only to bending and twisting. From Eqs. (13-23) for small displacements the strain increments due to w are

$$d\epsilon_{xx} = -z\frac{\partial^2 w}{\partial x^2} \qquad d\epsilon_{yy} = -z\frac{\partial^2 w}{\partial y^2} \qquad d\epsilon_{xy} = -2z\frac{\partial^2 w}{\partial x\,\partial y} \quad (15\text{-}35)$$

These are accompanied by stress increments $d\sigma_{xx}$, $d\sigma_{yy}$, and $d\sigma_{xy}$ found from Eqs. (15-34). Taking $d\sigma_{xx}$ as an example, we find from the first of Eqs. (15-34) that

$$d\sigma_{xx} = \tfrac{4}{3}E_s\left(d\epsilon_{xx} + \frac{d\epsilon_{yy}}{2}\right) + \tfrac{4}{3}\left(\epsilon_{xx} + \frac{\epsilon_{yy}}{2}\right)dE_s \qquad (15\text{-}36)$$

where dE_s is the change in E_s due to the strain increments.

Using the first of Eqs. (15-33), we have

$$dE_s = \frac{\partial E_s}{\partial \sigma_{xx}}\frac{d\sigma_{xx}}{d\epsilon_{xx}}d\epsilon_{xx} + \frac{\partial E_s}{\partial \epsilon_{xx}}d\epsilon_{xx} = \left(\frac{1}{\epsilon_{xx}}\frac{d\sigma_{xx}}{d\epsilon_{xx}} - \frac{\sigma_{xx}}{\epsilon_{xx}^2}\right)d\epsilon_{xx}$$

Noting that in the prebuckled state $d\sigma_{xx}/d\epsilon_{xx} = E_t$ and $\sigma_{xx}/\epsilon_{xx} = E_s$, the tangent and secant moduli at the buckling stress, we find

$$dE_s = (E_t - E_s)\frac{d\epsilon_{xx}}{\epsilon_{xx}}$$

Substituting this into Eq. (15-36) and using the prebuckling strain condition $\epsilon_{yy} = -\nu_p\epsilon_{xx} = -\epsilon_{xx}/2$, we find

$$d\sigma_{xx} = \tfrac{4}{3}E_s\left[\left(\frac{1}{4} + \frac{3}{4}\frac{E_t}{E_s}\right)d\epsilon_{xx} + \tfrac{1}{2}d\epsilon_{yy}\right]$$

which with Eqs. (15-35) gives

$$d\sigma_{xx} = -\tfrac{2}{3}E_s\left[\left(\frac{1}{4} + \frac{3}{4}\frac{E_t}{E_s}\right)\frac{\partial^2 w}{\partial x^2} + \frac{1}{2}\frac{\partial^2 w}{\partial y^2}\right]z$$

The bending-moment stress resultant M_x that results from the w displacements is found by replacing σ_{xx} in the first of Eqs. (13-15) by $d\sigma_{xx}$. This gives

$$M_x = \frac{E_s t^3}{9}\left[\left(\frac{1}{4} + \frac{3}{4}\frac{E_t}{E_s}\right)\frac{\partial^2 w}{\partial x^2} + \frac{1}{2}\frac{\partial^2 w}{\partial y^2}\right] \tag{15-37a}$$

In a similar fashion we can show that

$$M_y = \frac{E_s t^3}{9}\left(\frac{\partial^2 w}{\partial y^2} + \frac{1}{2}\frac{\partial^2 w}{\partial x^2}\right) \tag{15-37b}$$

$$M_{xy} = \frac{E_s t^3}{18}\frac{\partial^2 w}{\partial x\, \partial y} \tag{15-37c}$$

Equations (15-37) for the plastic plate replace Eqs. (13-30) for the elastic plate.

Substituting Eqs. (15-37) into Eq. (13-19) and letting

$$N_x = -N \qquad N_y = N_{xy} = p_x = p_y = p_z = 0$$

gives the lateral-equilibrium equation

$$\left(\frac{1}{4} + \frac{3}{4}\frac{E_t}{E_s}\right)\frac{\partial^4 w}{\partial x^4} + 2\frac{\partial^4 w}{\partial x^2\, \partial y^2} + \frac{\partial^4 w}{\partial y^4} = -\frac{9N}{E_s t^3}\frac{\partial^2 w}{\partial x^2} \tag{15-38}$$

which corresponds to $\nabla^4 w = -\dfrac{N}{D}\dfrac{\partial^2 w}{\partial x^2}$ for the elastic plate. Applying the method of Sec. 15-3 to Eq. (15-38), we find

$$N_{cr} = \frac{k_p \pi^2 E_s t^3}{9b^2} \tag{15-39}$$

where

$$k_p = \left(\frac{1}{4} + \frac{3}{4}\frac{E_t}{E_s}\right)\left(\frac{mb}{a}\right)^2 + 2 + \left(\frac{a}{mb}\right)^2 \tag{15-40}$$

The buckling coefficient k_p can be plotted against a/b for different values of m to give a family of curves similar to Fig. 15-2 for the elastic plate. As in the elastic case, k_p becomes nearly constant for large values of a/b. This asymptotic value can be obtained by minimizing k_p with respect to m, which gives

$$\left(\frac{a}{mb}\right)^2 = \frac{1}{2}\left(1 + 3\frac{E_t}{E_s}\right)^{1/2} \tag{15-41}$$

We note that, unlike the long elastic plate, where $a/mb = 1$, the inelastic plate does not buckle into square panels.

Substituting Eq. (15-40) into (15-39) and using Eq. (15-41), we find after division by t that

$$\sigma_{cr} = \frac{2\pi^2 E_s}{9} \left[1 + \frac{1}{2} \left(1 + 3 \frac{E_t}{E_s} \right)^{1/2} \right] \left(\frac{t}{b} \right)^2 \qquad (15\text{-}42)$$

Inserting this into Eq. (15-28) along with $\nu_p = \frac{1}{2}$ and $k = 4$, we obtain

$$\eta = \frac{1 - \nu_e^2}{1 - \nu_p^2} \frac{E_s}{E} \left[\frac{1}{2} + \frac{1}{2} \left(\frac{1}{4} + \frac{3E_t}{4E_s} \right)^{1/2} \right] \qquad (15\text{-}43)$$

The accuracy of Eq. (15-43) can be improved by replacing ν_p by ν, the actual value of Poisson's ratio at σ_{cr}. Gerard and Wildhorn[26] have suggested using the approximation

$$\nu = \frac{1}{2} - \frac{E_s}{E} \left(\frac{1}{2} - \nu_e \right) \qquad (15\text{-}44)$$

which reduces to ν_e below the proportional limit and approaches $\frac{1}{2}$ at large plastic strains. We note that when this is done, Eq. (15-43) gives $\eta = 1$ below the proportional limit.

Stowell[25] used the preceding method to derive the differential equations and boundary conditions for the plastic buckling of plates with simply supported, clamped, free, and elastically restrained edges. He also determined the energy integrals to be used with the Rayleigh-Ritz method in the plastic range. With these he determined the equations for η of plates, flanges, and wide columns for various edge conditions. Experimental evidence is in good agreement with the theoretical results in all cases except buckling under edge shear. Gerard found that in shear buckling a maximum-shear plasticity law gives more accurate results than the distortional-energy theory.[5]

The Ramberg-Osgood parameters provide a convenient means for reducing the number of design curves required to predict plate buckling. The method used to construct these curves (similar to Fig. 15-19) will be explained by using the simply supported plate in compression as an example. To determine a point on the curve for a particular n, we assume a value $\sigma_{cr}/\sigma_{0.7}$ and use Eqs. (3-10) and (3-11) to determine E_t and E_s (with $\sigma = \sigma_{cr}$). These are used with $\nu_e = 0.3$ to compute ν from Eq. (15-44). Next Eq. (15-43) is applied to compute η with $\nu_p = \nu$. The $\epsilon_{cr}E/\sigma_{0.7}$ coordinate of the point on the curve is then determined from Eq. (15-28) written in the form $\sigma_{cr}/\sigma_{0.7} = \eta(\epsilon_{cr}E/\sigma_{0.7})$, where

$$\epsilon_{cr} = \frac{k\pi^2}{12(1 - \nu_e^2)} \left(\frac{t}{b} \right)^2 \qquad (15\text{-}45)$$

The parameters $\sigma_{cr}/\sigma_{0.7}$ and $\epsilon_{cr}E/\sigma_{0.7}$ are plotted, and the process is repeated with new values of $\sigma_{cr}/\sigma_{0.7}$ until the curve for the particular n is established.

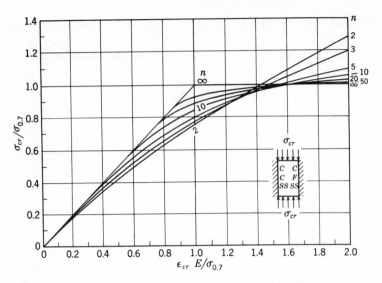

Fig. 15-19 Nondimensional compressive-buckling stress for long clamped flanges, and plates with simply supported and rotationally restrained edges (Ref. 5).

Gerard found that there is little difference in the Ramberg-Osgood curves for long compressively loaded plates with the unloaded edges simply supported or clamped and flanges with the supported edge clamped. Figure 15-19 is actually an average of these three cases and may be used for any one of them. The compressive-buckling stresses for long flanges with a simply supported edge can be determined from Fig. 15-20, and those for long plates in shear can be found from Fig. 15-21.

The errors that result from using Figs. 15-19 to 15-21 below the proportional limit are small, and it is suggested that they be used instead of Eq. (15-8) at all stress levels unless it is certain that the buckling stresses are in the linearly elastic range. The use of these curves is illustrated in the following example.

Example 15-5 The 0.080-in. HK31A-H24 magnesium alloy ($E = 6.5 \times 10^6$ psi, $\sigma_{0.7} = 17.3$ ksi, $n = 6.2$) skin of a fuselage is divided by Z-section stiffeners into long panels that are 4 in. wide. Determine the compressive buckling stress of the panels.

With $b/t = 4.0/0.08 = 50$, we find from the lower curve of Fig. 15-6 that $k \doteq 5.2$. Applying Eq. (15-45),

$$\epsilon_{cr} = \frac{5.2\pi^2}{12(1 - 0.3^2)} \left(\frac{0.08}{4.0}\right)^2 = 1.88 \times 10^{-3}$$

so that

$$\frac{\epsilon_{cr}E}{\sigma_{0.7}} = \frac{1.88 \times 10^{-3} \times 6.5 \times 10^6}{17.3 \times 10^3} = 0.705$$

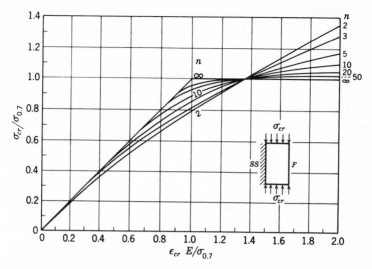

Fig. 15-20 Nondimensional compressive-buckling stress for long simply supported flanges (Ref. 5).

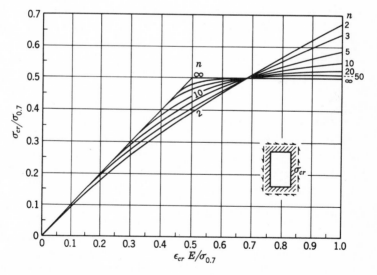

Fig. 15-21 Nondimensional shear-buckling stress for plates with simply supported and rotationally restrained edges (Ref. 5).

With this value and $n = 6.2$ we find $\sigma_{cr}/\sigma_{0.7} = 0.63$ from Fig. 15-19, which gives $\sigma_{cr} = 0.63 \times 17.3 = 10.9$ ksi.

The preceding method must be modified when the plate is made of a clad aluminum alloy, where the proportional limit σ_{cl} of the nearly pure aluminum cladding is much lower than the proportional limit σ_{co} of the alloy core.   As a result, the bending rigidity is significantly reduced when $\sigma_{cr} > \sigma_{cl}$.   Gerard[5] has derived an additional plasticity correction factor $\bar{\eta}$, to be used in computing the buckling stress $\bar{\sigma}_{cr}$ for clad materials from

$$\bar{\sigma}_{cr} = \bar{\eta}\sigma_{cr} \tag{15-46}$$

where σ_{cr} is obtained from Eq. (15-28).   The equations for $\bar{\eta}$ are

$$\bar{\eta} = \begin{cases} \dfrac{1 + 3f(\sigma_{cl}/\sigma_{cr})}{1 + 3f} & \sigma_{cl} < \sigma_{cr} < \sigma_{co} \\[4mm] \dfrac{1}{1 + 3f} & \sigma_{cr} > \sigma_{co} \end{cases} \tag{15-47}$$

where f is the ratio of the total cladding thickness to the total thickness. In using Ramberg-Osgood curves, the nominal buckling stress σ_{cr} is determined in the usual manner for a homogeneous plate of the same thickness as the clad plate but with the mechanical properties of the core.   The buckling stress of the clad plate is then found from Eq. (15-46).

Plasticity also changes the effective width of a buckled plate.   Van der Neut[16] gives experimental evidence that Eq. (15-26) can be used in the inelastic range if σ_{cr}/σ_e is replaced by the strain ratio ϵ_{cr}/ϵ_e.

15-9 THE FAILURE OF PLATES

We have seen that, unlike a column, a plate may carry loads that are considerably in excess of the buckling load.   In fact, the ultimate load is not reached until a sizable portion of the plate is plastically deformed. The theoretical prediction of the failing load is difficult, for in addition to the nonlinearity that results from the large deflections, the inelastic stress-strain relationships also lead to nonlinear behavior.

A theoretical solution exists for flanges that is in good agreement with experimental results,[27] but it has been necessary to resort to semi-empirical methods to predict the failing loads of plates with all edges supported.   The following method, which was proposed by Gerard,[28] is supported by an impressive number of room- and elevated-temperature tests on plates of different materials and geometric proportions.   It provides the basis for the method of predicting the failing stresses of thin-walled columns and stiffened plates described in Chap. 16.

We see from Eqs. (15-26) and (15-27) that $2b_e/b$ is a function of σ_{cr}/σ_e, and so we can write the approximate relation $2b_e/b = \alpha(\sigma_e/\sigma_{cr})^r$, where α and r are empirical constants to be obtained from test data. Designating the average stress in the plate by $\bar{\sigma}$, we find $\bar{\sigma} = 2\sigma_e b_e/b = \alpha\sigma_e(\sigma_e/\sigma_{cr})^r$, so that $\bar{\sigma}/\sigma_{cr} = \alpha(\sigma_e/\sigma_{cr})^{r+1}$. Theoretical and experimental results indicate that the ultimate load is reached when $\sigma_e \approx \sigma_{cy}$, the compressive yield stress of the material. Defining the *crippling stress* $\bar{\sigma}_f$ as the average stress in the plate at failure, we find $\bar{\sigma}_f/\sigma_{cr} = \alpha(\sigma_{cy}/\sigma_{cr})^n$, where $n = r + 1$. This equation applies to plates that buckle elastically. Experimental results indicate that $\bar{\sigma}_f \approx \sigma_{cr}$ for plates that buckle in the inelastic range. On the basis of this, Gerard suggested using the equations

$$\frac{\bar{\sigma}_f}{\sigma_{cr}} = \alpha \left(\frac{\sigma_{cy}}{\sigma_{cr}}\right)^n \qquad \sigma_{cr} \leq \alpha^{1/n}\sigma_{cy} \qquad (15\text{-}48a)$$

$$\frac{\bar{\sigma}_f}{\sigma_{cr}} = 1 \qquad \sigma_{cr} > \alpha^{1/n}\sigma_{cy} \qquad (15\text{-}48b)$$

The nondimensional parameters $\bar{\sigma}_f/\sigma_{cr}$ and σ_{cy}/σ_{cr} can be used to correlate experimental results. Gerard found that Eqs. (15-48) can be used to predict the failing stress of plates and flanges within ± 10 percent when the values of α and n in Table 15-3 are used. It is seen from this table that the in-plane restraints on the unloaded supported edges have a strong influence upon the failing stresses of plates and flanges. When the edges of the plate are constrained to remain straight (Fig. 15-14a), the resulting membrane forces assist in resisting the w displacements and thereby increase $\bar{\sigma}_f$. When the edges of the plate are free to warp (Fig. 15-14b), these do not exist, and failure occurs at a lower average stress.

It would be expected that when a plate is divided into panels by longitudinal stiffeners, as in the stiffened skins of aerodynamic surfaces

Table 15-3   Values of α and n for plate failure[17,28]

Condition	α	n
1. Theory for simply supported plate with straight unloaded edges (Fig. 15-14a)	0.78	0.80
2. Tests for simply supported and clamped plates, edges free to warp (Fig. 15-14b)	0.80	0.58
3. Test data for three-bay plates	0.80	0.65
4. Test data for simply supported flange with straight supported edge	0.81	0.80
5. Test data for simply supported flange with supported edge free to warp	0.68	0.58

and body structures, adjacent panel edges would constrain each other to remain straight. However, in the three-bay plate (Table 15-3), the free edges of the outer panels prevent the full development of the straight-edge condition. We note though that n in this case lies between the values for straight edges and edges free to warp.

It is frequently more convenient to write Eq. (15-48a) in a form that contains the dimensionless parameter t/b. Substituting Eq. (15-8) into Eq. (15-48a), we find

$$\frac{\bar{\sigma}_f}{\sigma_{cy}} = \beta \left[\frac{t}{b} \left(\frac{E}{\sigma_{cy}} \right)^{\frac{1}{2}} \right]^m \qquad (15\text{-}49)$$

where

$$\beta = \alpha \left[\frac{k\pi^2}{12(1 - \nu_e^2)} \right]^{1-n} \qquad m = 2(1 - n) \qquad (15\text{-}50)$$

This form of the equation is used extensively in Chap. 16.

When a rectangular plate is subjected to edge shear, failure occurs by tearing in the direction of the tensile principal stress in the buckled plate. The problem is complicated by the fact that the tensile and compressive stresses are not equal, as they are in the unbuckled plate. As a result, the members that support the edges of the plate apply tensile membrane forces to the boundary. A semiempirical method for predicting the failing load in shear is given in Ref. 8.

Plate failures in compression and in shear occur at large lateral displacements; for this reason small initial imperfections have a negligible effect upon the failing stress.

REFERENCES

1. Timoshenko, S. P., and J. M. Gere: "Theory of Elastic Stability," 2d ed., McGraw-Hill Book Company, New York, 1961.
2. Bleich, F.: "Buckling Strength of Metal Structures," McGraw-Hill Book Company, 1952.
3. Gerard, G.: "Introduction to Structural Stability Theory," McGraw-Hill Book Company, New York, 1962.
4. Cox, H. L.: "The Buckling of Plates and Shells," The Macmillan Company, New York, 1963.
5. Gerard, G., and H. Becker: Handbook of Structural Stability, pt. I: Buckling of Flat Plates, NACA Tech. Note 3781, 1957.
6. Roark, R. J.: "Formulas for Stress and Strain," 4th ed., McGraw-Hill Book Company, New York, 1965.
7. Flugge, W.: "Handbook of Engineering Mechanics," McGraw-Hill Book Company, New York, 1962.
8. Kuhn, P., J. P. Peterson, and L. R. Levin: A Summary of Diagonal Tension, pt. I vol. I: Methods of Analysis, NACA Tech. Note 2661, 1952.

9. van der Neut, A.: Buckling Caused by Thermal Stresses, in "High Temperature Effects in Aircraft Structures," (edited by N. J. Hoff), Pergamon Press, New York, 1958.

10. Boley, B. A., and J. H. Weiner: "Theory of Thermal Stresses," John Wiley & Sons, Inc., New York, 1960.

11. Hoff, N. J.: Thermal Buckling of Supersonic Wing Panels, *J. Aeron. Sci.*, **23**(11): 1019–1028, 1050 (November, 1956).

12. Gossard, M. L., P. Seide, and W. M. Roberts: Thermal Buckling of Plates, *NACA Tech. Note* 2771, 1952.

13. Hoff, N. J.: "The Analysis of Structures," John Wiley & Sons, Inc., New York, 1956.

14. Steinbacher, F. R., and G. Gerard: "Aircraft Structural Mechanics," Pitman Publishing Corporation, New York, 1952.

15. Johnson, A. E., Jr., and K. P. Buchert: Critical Combinations of Bending, Shear, and Transverse Compressive Stresses for Buckling of Infinitely Long Flat Plates, *NACA Tech. Note.* 2536, 1951.

16. van der Neut, A.: Post-buckling Behavior of Structures, *AGARD Rept.* 60, 1956.

17. Gerard, G.: Handbook of Structural Stability, pt. IV: Failure of Plates and Composite Elements, *NACA Tech. Note* 3784, 1957.

18. Argyris, J. H., and P. C. Dunne: "Handbook of Aeronautics, no. 1, Structural Principles and Data, pt. 2: Structural Analysis," 4th ed., Pitman Publishing Corporation, New York, 1952.

19. Mayers, J., and B. Budiansky: Analysis of Behavior of Simply-supported Flat Plates Compressed beyond the Buckling Load into the Plastic Range, *NACA Tech. Note* 3368, 1955.

20. Levy, S.: Bending of Rectangular Plates with Large Deflections, *NACA Tech. Rept.* 737, 1942.

21. Coan, J. M.: Large Deflection Theory for Plates with Small Initial Curvature Loaded in Edge Compression, *J. Appl. Mech.*, **18**(2): 143–151 (June, 1951).

22. Hu, P. C., E. E. Lundquist, and S. B. Batdorf: Effect of Small Deviations from Flatness on Effective Width and Buckling of Plates in Compression, *NACA Tech. Note* 1124, 1946.

23. Koiter, W. T.: The Effective Width of Flat Plates for Various Longitudinal Edge Conditions at Loads Far Beyond the Buckling Load, *Natl. Luchtvaart Lab. (NLL) Rept.* S287, 1943.

24. Marguerre, K.: The Apparent Width of the Plate in Compression, *NACA Tech. Memorandum* 833, 1937.

25. Stowell, E. Z.: A Unified Theory of Plastic Buckling of Columns and Plates, *NACA Tech. Rept.* 898, 1948.

26. Gerard, G., and S. Wildhorn: A Study of Poisson's Ratio in the Yield Region, *NACA Tech. Note* 2561, 1952.

27. Stowell, E. Z.: Compressive Strength of Flanges, *NACA Tech. Note* 2020, 1950.

28. Gerard, G.: The Crippling Stress of Compression Elements, *J. Aeron. Sci.*, **25**(1): 37–52 (January, 1958).

PROBLEMS

15-1. A simply supported plate is subjected to a compressive edge force of N lb/in. at $x = 0$ and a and a tensile edge force of $N/2$ lb/in. at $y = 0$ and b. The thickness varies linearly from t_0 at $x = 0$ to t_a at $x = a$. Write the differential equation and boundary conditions which determine the buckling load N_{cr}.

15-2. A $0.051 \times 10 \times 15$ .in. molybdenum ($E = 35 \times 10^6$ psi) simply supported rectangular plate is protected from oxidation by a 0.10-in. coating of zirconium dioxide ($E = 13 \times 10^6$ psi). The composite is subjected to a uniformly distributed compressive force per unit length on the 10-in. edges. Determine the buckling force per unit length N_{cr}.

15-3. The 0.04-in. web of a 7075-T6 aluminum alloy ($E = 10.5 \times 10^6$ psi, $\sigma_{0.7} = 70$ ksi, $n = 9.2$) spar is divided into 30×10 in. panels by vertical stiffeners. Bending stresses are applied along the 10-in. sides. Compute the buckling stress assuming that the spar caps and stiffeners provide simple support to the edges of the panel.

15-4. The panel of Prob. 15-3 is subjected to edge shear. Compute the critical shear stress.

15-5. The spar web of Probs. 15-3 and 15-4 is subjected to a bending stress that is one-half of that required to buckle the plate. Determine the shear stress that will cause buckling when applied simultaneously with the bending stress.

15-6. The bending stiffness of a square simply supported plate varies linearly from D_0 at $x = 0$ to D_a at $x = a$. A uniform compressive load is applied parallel to the x axis. Determine N_{cr} by the Rayleigh-Ritz method by approximating the displacements by the series $w = c_1 \sin (\pi x/a) \sin (\pi y/a) + c_2 \sin (2\pi x/a) \sin (\pi y/a)$.

15-7. The edges of the plate that is shown are simply supported and completely restrained against in-plane motion. The plate is uniformly heated. Use the finite-difference mesh shown to obtain an approximation of the buckling temperature.

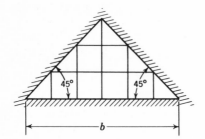

Fig. P15-7

15-8. A $0.04 \times 4 \times 16$ in. simply supported aluminum plate ($E = 10.5 \times 10^6$ psi) is subjected to uniform edge shortening in the 16-in. direction. The unloaded edges are constrained to remain straight but are free to displace uniformly. Determine the force that is carried by the plate when the edge stress is (a) 1 ksi, (b) 3 ksi, and (c) 12 ksi.

15-9. Determine the ultimate load of the plate of Prob. 15-8 if $\sigma_{cy} = 40$ ksi.

16

Instability and Failure of Thin-walled Columns and Stiffened Plates

16-1 INTRODUCTION

The column theory in Chap. 14 assumes that the cross section does not distort during buckling and failure and that the wavelength of the buckle is on the order of the column length. This primary-instability theory, which accurately describes the behavior of columns with solid or thick-walled cross sections, must be reexamined for thin-walled columns.

The failing load of thin-walled columns can also be predicted by the Euler equation when L'/ρ is greater than approximately 80 (Fig. 16-1). When $L'/\rho < 20$, the thin-walled column buckles by local instability, in which the cross section is distorted and the buckle length is on the order of the cross-sectional dimensions. Local failure occurs by crushing at a stress below that predicted by the tangent-modulus theory. This type of instability and failure is more closely related to plate behavior than to

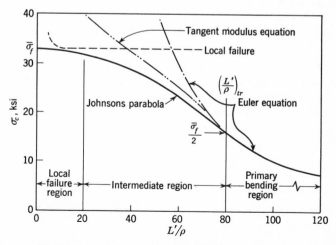

Fig. 16-1 Typical column curve for a thin-walled section.

primary column action.   In the intermediate range of L'/ρ from 20 to approximately 80, the buckling and failure occur by a combination of the primary and secondary modes.

Plates with large values of b/t buckle and fail at low stress levels and are therefore inefficient.   As a result, stiffeners are usually used to reduce the panel size and provide rotational edge restraints.   The plate theory of Chap. 15 assumes that the plate elements buckle and fail before the stiffening members.   However, in an optimum design all elements buckle at the same stress, and the plate and stiffeners must be considered simultaneously in predicting failure.

In this chapter we extend the methods of Chaps. 14 and 15 to the buckling and failure of thin-walled columns and stiffened plates.   The literature on these subjects is extensive, and a complete treatment of the topics is beyond the scope of this book.   For further information, the reader is referred to the comprehensive survey of Gerard and Becker,[1-7] from which most of the material in this chapter is taken.   The important contributions of Argyris and Dunne[8] are also recommended reading. Heller[9] has compiled an extensive bibliography on the behavior of stiffened plates.

16-2 SECONDARY INSTABILITY OF COLUMNS

Longitudinal stiffeners are usually fabricated by the extrusion process or by forming from flat sheet material with a bend brake or forming rolls. The cross section of these column members is usually composed of flat plate elements arranged to form an angle, channel, Z, H, or hat-shaped section, as shown in Fig. 16-2.   The plate elements may be divided into two categories: *flanges*, which have a free unloaded edge, and *webs*, which are supported by adjoining plate elements on both unloaded edges.

The secondary-instability mode is entirely different from that for primary bending instability.   The flange and web elements of the column buckle like plates, and, as a result, the cross-sectional shape is deformed. The wavelength of the buckles is on the order of the widths of the plate elements of the cross section.   The stress at which secondary, or local, instability occurs is essentially independent of the length of the column when the length is three or more times greater than the width of the largest plate element of the column cross section.

Adjoining plate elements of the cross section are usually perpendicular to each other.   As a result, their in-plane rigidities restrain the edges of the adjoining elements against lateral displacements.   In addition, the bending rigidities of adjoining elements may provide elastic rotational edge restraints.   The angle between adjoining elements is not a critical

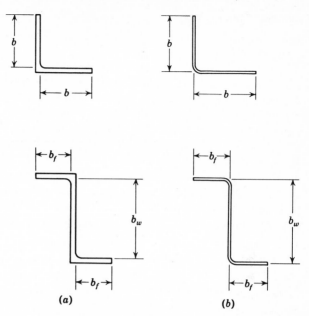

Fig. 16-2   Typical (*a*) extruded and (*b*) formed stiffeners (Ref. 2).

factor, however, for it has been found[10] that the local buckling stress is unchanged for angles between 30 and 120°.

Local instability occurs when the weakest plate element reaches its buckling stress. The simplest cases to analyze are those in which the properties of the elements are such that they reach their buckling stresses simultaneously. When this occurs, none of the elements can provide rotational edge restraint to its adjoining elements, and all elements behave as if they were simply supported along the lines where they join other elements. This situation exists for equal-legged angle, T, and cruciform sections. It is also the case for a square tube of constant thickness.

In these cases, the local buckling stress σ_{cr} can be determined from the direct application of the plate-buckling theory in Chap. 15. The critical strain is determined from Eq. (15-45), where for $a/b > 3$ we find from Table 15-1 that $k = 0.43$ for equal-legged flanged sections and $k = 4$ for the square tube. Taking $\nu_e = 0.3$, we obtain

$$\epsilon_{cr} = 0.388 \left(\frac{t}{b}\right)^2 \tag{16-1}$$

for the flanged section with equal legs and

$$\epsilon_{cr} = 3.62 \left(\frac{t}{b}\right)^2 \tag{16-2}$$

for the square tube. Figure 15-20 can be used to determine σ_{cr} for the flanged section and Fig. 15-19 can be used for the square tube.

Extruded sections contain fillets at reentrant corners which provide a small amount of rotational edge restraint that is not present in formed sections. This is approximately accounted for by measuring b as an inside dimension for extrusions and a midline dimension for formed sections, as shown in Fig. 16-2.

The preceding method can also be used on composite web-flange sections if the proportions of the elements are such that they buckle simultaneously. For sections that buckle elastically (so that $\eta = 1$), this

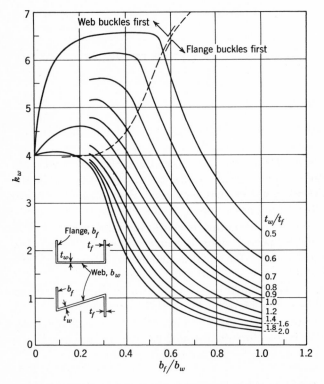

Fig. 16-3 Buckling coefficients for channel and Z-section columns (Ref. 12).

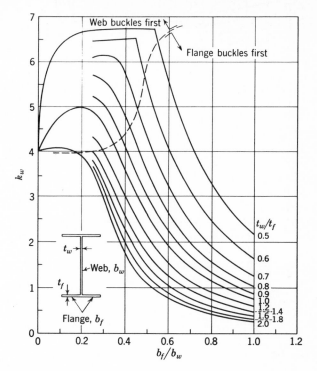

Fig. 16-4 Buckling coefficient for H-section columns (Ref. 12).

requirement is satisfied if $0.388(t_f/b_f)^2 = 3.62(t_w/b_w)^2$, where the subscripts f and w refer to flange and web elements, respectively.

The buckling stresses of other sections are difficult to compute, for the stiffer elements provide elastic rotational restraint to the more flexible elements, and the simply supported edge conditions do not apply. Theoretical methods of computing σ_{cr} for general web-flange sections have been given by Lundquist, Stowell, and Schuette[11] and by Kroll, Fisher, and Heimerl.[12] These methods are tedious and inconvenient for routine stress analysis of arbitrary sections; however, they have been used to develop the simple design charts in Figs. 16-3 to 16-6 for channel, Z, H, rectangular-tube, hat, and lipped Z-section stiffeners.

To use these charts, ϵ_{cr} is computed from Eq. (15-45) with k and t/b as listed in Table 16-1. The buckling stress is determined from ϵ_{cr} by using Fig. 15-19 if the web buckles first or Fig. 15-20 if the flange buckles first. The range in which buckling initially occurs in the web or the flange is indicated in Figs. 16-3 to 16-6.

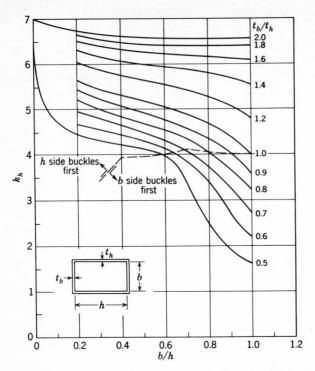

Fig. 16-5 Buckling coefficient for rectangular tubes (Ref. 12).

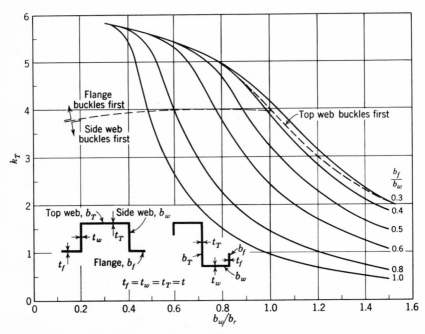

Fig. 16-6 Buckling coefficients for hat and lipped Z-section columns (Ref. 13).

Table 16-1   Values of k and t/b to be used with Eq. (15-45)

Shape	k	*Figure*	$\dfrac{t}{b}$
Channel	k_w	16-3	$\dfrac{t_w}{b_w}$
Z section	k_w	16-3	$\dfrac{t_w}{b_w}$
H section	k_w	16-4	$\dfrac{t_w}{b_w}$
Rectangular tube	k_h	16-5	$\dfrac{t_h}{h}$
Hat section	k_T	16-6	$\dfrac{t}{b_T}$
Lipped Z section	k_T	16-6	$\dfrac{t}{b_T}$

Example 16-1   Determine the local buckling stress of the 7075-T6 aluminum alloy ($E = 10.5 \times 10^6$ psi, $\sigma_{0.7} = 72$ ksi, $n = 16.6$) H-section column shown in Fig. 16-7.

With $b_f/b_w = 0.8125/1.25 = 0.65$ and $t_w/t_f = 1$ we find from Fig. 16-4 that $k_w = 1.75$.   From Eqs. (15-45) and Table 16-1

$$\frac{\epsilon_{cr}E}{\sigma_{0.7}} = \frac{1.75 \times \pi^2 \times 10.5 \times 10^6}{72 \times 10^3 \times 12(1 - 0.3^2)} \left(\frac{0.125}{1.25}\right)^2 = 2.31$$

With this value and $n = 16.6$ we find from a slight extrapolation of Fig. 15-20 (since from Fig. 16-4 the flange buckles first) that $\sigma_{cr}/\sigma_{0.7} = 1.06$ or that $\sigma_{cr} = 1.06 \times 72 \times 10^3 = 76.3$ ksi.

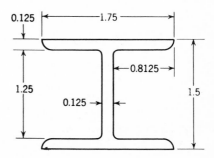

Fig. 16-7   Example 16-1.

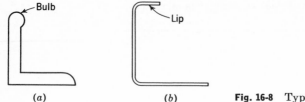

Fig. 16-8 Typical (a) bulb and (b) lip.

A bulb is often used on the outstanding leg of an extrusion to stabilize the free edge and increase its buckling stress from that for a flange to that for a web. A lip is frequently bent up on the outstanding leg of a formed section for the same purpose. Typical examples are shown in Fig. 16-8. If the bulb or lip is too small, it will buckle below the web-buckling stress of the leg which it is supposed to stabilize. Gerard has derived the curves in Fig. 16-9 to determine the minimum bulb or lip that is required to produce web behavior in the outstanding leg. With b_f/t of the outstanding leg known, the value of d/t for a bulb or b_L/b_f for a lip can be determined for web action. The curves can also be used to determine ϵ_{cr} of the outstanding leg when the proportions of the bulb or lip are less than required for web action.

16-3 CRIPPLING OF COLUMNS

It was pointed out in Sec. 15-9 that when a plate buckles elastically, the failing stress is greater than σ_{cr}. As would be expected, the same is true of a thin-walled short column when the cross section consists of flat-plate elements. The failing stress of such a column is independent of the length when it is greater than three times the width of the widest plate element and $L'/\rho < 20$ (Fig. 16-1). The average stress at failure in this range is known as the *crippling, crushing,* or *local-failing stress* and will be designated by $\bar{\sigma}_f$.

Stowell[14] has developed a theory for determining $\bar{\sigma}_f$ for cruciform sections that is in good agreement with experimental evidence. Bending deflections become large after the flanges buckle, and failure occurs when the stress at the supported edge of the flanges reaches the compressive yield stress σ_{cy}. The nonlinear behavior associated with large displacements and plasticity has prevented the development of a satisfactory theoretical solution for $\bar{\sigma}_f$ of arbitrary sections. There has been a great deal of testing to determine $\bar{\sigma}_f$ for specific cross sections and materials, and the data have been used to develop semiempirical methods. Some of these methods divide the section into web and flange elements,[15] while others use imaginary cuts to reduce the cross section to a series of angle

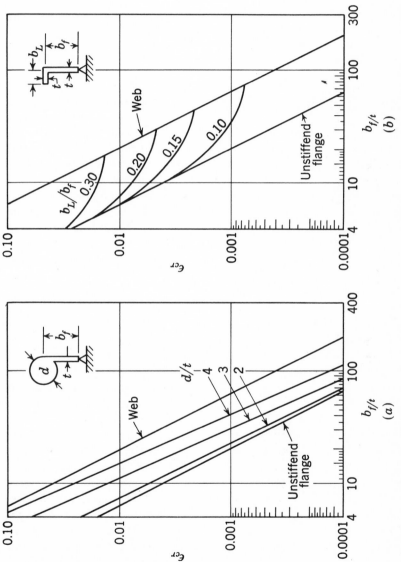

Fig. 16-9 Minimum bulb diameter and lip length to produce web action of an outstanding leg (Ref. 2). (a) Bulb flanges; (b) lip flanges.

sections.[16]   The crippling stress is taken as a weighted average of the failing stresses of the plate or angle elements in these cases.

Probably the most comprehensive semiempirical investigation has been made by Gerard,[4,5,7,17] who extended his method for predicting plate failure (Sec. 15-9) to thin-walled columns and stiffened plates. He studied test data for many different materials and cross sections at room and elevated temperatures and found that the results could be correlated by a generalization of Eq. (15-49) of the form

$$\frac{\bar{\sigma}_f}{\sigma_{cy}} = \beta_g \left[\left(\frac{gt^2}{A} \right) \left(\frac{E}{\sigma_{cy}} \right)^{1/2} \right]^m \tag{16-3}$$

where A is the cross-sectional area.   The parameter g in this equation is the number of cuts that would be required to reduce the cross section to a series of flanged sections plus the number of flanges that would exist after the cuts are made.   Figure 16-10 shows how g is determined for some typical sections.   The parameters β_g and m are empirical constants determined from test data.

For plates with supported edges and flanges, $A = bt$, and we find from Eqs. (15-49) and (16-3) that $\beta_g = \beta/g^m$.   Noting that $g = 3$ for a

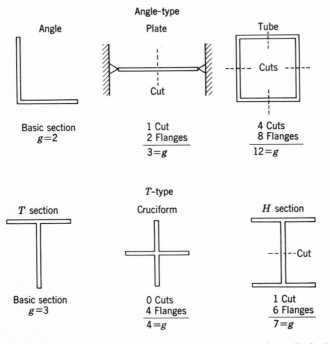

Fig. 16-10   Method of determining g for typical sections (Ref. 5).

simply supported plate (Fig. 16-10), we find from Table 15-3 and Eqs. (15-50) that Eq. (16-3) can be used to predict $\bar{\sigma}_f$ for plates if we take $m = 0.40$ and $\beta_g = 0.65$ for plates with straight unloaded edges and $m = 0.85$ and $\beta_g = 0.56$ for plates with edges that are free to warp. In a similar manner, $g = 1$ for simply supported flanges, and we find $m = 0.40$ and $\beta_g = 0.67$ for a straight supported edge, while $m = 0.85$ and $\beta_g = 0.56$ for a supported edge that is free to warp.

The opposing flanges of a cruciform, T, or H section constrain the supported edges to remain straight. We would therefore expect the $\bar{\sigma}_f$ of these sections to be the same as that of a flange with a straight supported edge. On the other hand, the flanges of an angle do not oppose each other, and the supported edge is free to warp. It would be anticipated that $\bar{\sigma}_f$ of an angle section would be the same as that of a flange with a supported edge that is free to warp. The adjacent plate elements of a rectangular tube or a multicorner section (a section with more than two corners) are perpendicular and cannot restrain edge warping in each other. We would also expect that $\bar{\sigma}_f$ for these sections would be the same as that of a flange with a supported edge that is free to warp. Gerard used this reasoning and found it to be supported by test data on extruded columns having the sections described. However, he found that the test data on two-corner sections, i.e., channel and Z sections, do not correlate with Eq. (16-3). He suggested that the equation

$$\frac{\bar{\sigma}_f}{\sigma_{cy}} = \beta \left[\frac{t^2}{A} \left(\frac{E}{\sigma_{cy}} \right)^{\frac{1}{2}} \right]^{\frac{3}{4}} \tag{16-4}$$

be used in these cases.

In the preceding discussion, all the material effects are included in the parameter E/σ_{cy}. Experimental results on extrusions with small corner radii are in good agreement with the empirical equations, but test data from formed sections indicate that the method must be modified if it is to be applied to such sections. Severe strain hardening occurs in the corners of formed sections when they are fabricated. This increases the yield stress in corners over that of the rest of the section if there is no subsequent heat treatment. This affects $\bar{\sigma}_f$, since failure occurs when $\bar{\sigma}_{cy}$, the compressive yield stress of the corners, is reached. On the basis of test results, Gerard has recommended that β in Eq. (16-4) be multiplied by the strain-hardening correction factor given in Fig. 16-11. The correction factor is not the same for all materials because of their varying work-hardening characteristics. The increased corner properties do not appear to affect $\bar{\sigma}_f$ for formed multicorner sections.

Formed sections are often made from clad aluminum-alloy sheet material. Gerard suggests that E be replaced by $\bar{\eta}E$ in these cases, where $\bar{\eta}$ is the cladding correction factor given in Eqs. (15-47). When the wall

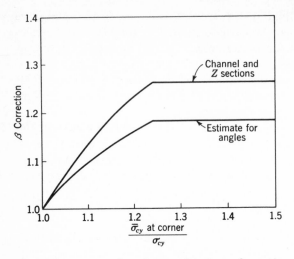

Fig. 16-11 β-correction factor for formed sections (Ref. 17).

thickness of an extruded section is not constant, a mean wall thickness

$$\bar{t} = \frac{\Sigma b_i t_i}{\Sigma b_i} \tag{16-5}$$

should be used in place of t in Eqs. (16-3) and (16-4). The equations for Gerard's method, which are summarized in Table 16-2 for various sections,

Table 16-2 Summary of equations and constants for Gerard's method

Section	Equation	g	β or β_g	m	Cutoff stress
1. Extruded angle	(16-3)	2	0.56	0.85	$0.8\sigma_{cy}$
2. Plate with edges free to warp (Fig. 15-14b)	(16-3)	3	0.56	0.85	$0.9\sigma_{cy}$
3. Extruded rectangular tube	(16-3)	12	0.56	0.85	$0.75\sigma_{cy}$
4. Formed multicorner section	(16-3)	†	0.55	0.85	$0.75\sigma_{cy}$
5. Straight-edge plate (Fig. 15-14a)	(16-3)	3	0.65	0.40	$0.8\sigma_{cy}$
6. Extruded T	(16-3)	3	0.67	0.40	$0.8\sigma_{cy}$
7. Extruded cruciform	(16-3)	4	0.67	0.40	$0.8\sigma_{cy}$
8. Extruded H	(16-3)	7	0.67	0.40	$0.8\sigma_{cy}$
9. Two-corner sections	(16-4)	—	3.2	0.75	‡

† g = number of cuts plus number of flanges.

‡ Cutoff stress = $2(t/b_w)^{1/3}\sigma_{cy}$.

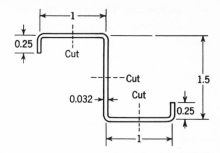

Fig. 16-12 Example 16-2.

agree with experimental results within ± 10 percent. As in the case of plates, Eqs. (16-3) and (16-4) apply only below a cutoff stress. Above this cutoff

$$\bar{\sigma}_f = \sigma_{cr} \tag{16-6}$$

The cutoff stresses for the various sections are listed in Table 16-2.

Example 16-2 Determine $\bar{\sigma}_f$ for the formed 2024-T3 aluminum alloy ($\sigma_{cy} = 40$ ksi, $E = 10.3 \times 10^6$ psi) lipped Z section shown in Fig. 16-12.

From Fig. 16-12, the number of cuts is 3 and the number of flanges is 8, so that $g = 11$. The developed length of the section is approximately 4 in.; therefore $A = 0.032 \times 4 = 0.128$ in.^{2} From Table 16-2 we find that $\beta_g = 0.55$ and $m = 0.85$ for a multicorner section, and from Eq. (16-3)

$$\frac{\bar{\sigma}_f}{\sigma_{cy}} = .055 \left[\frac{11 \times 0.032^2}{0.128} \left(\frac{10.3 \times 10^6}{40 \times 10^3} \right)^{\frac{1}{2}} \right]^{0.85} = 0.738$$

This is less than the cutoff value of 0.75, so that $\bar{\sigma}_f = 0.738 \times 40 \times 10^3 = 29.5$ ksi.

16-4 FAILURE OF THIN-WALLED COLUMNS

The failure mode of a thin-walled column depends upon its length, as shown in Fig. 16-1. When $L'/\rho < 20$, local failure occurs; this stress is essentially independent of L'/ρ and is equal to $\bar{\sigma}_f$ until the length becomes so short that $a/b < 1$ for the plate elements. At this point the increase in the plate buckling coefficient causes the failing stress to rise as shown. The small values of L'/ρ in this region are of little practical interest, however, and $\bar{\sigma}_f$ provides a conservative estimate of the failing stress for $0 \le L'/\rho \le 20$.

Primary bending instability occurs at large values of L'/ρ, where the failing stress is given by the Euler equation (14-17). Test results are in good agreement with the crippling stress at small values of L'/ρ and the Euler stress at large L'/ρ. However, experimental data fall below the stresses predicted by both of these theories at intermediate slenderness ratios.

In the intermediate range, the failure is a combination of the primary and secondary modes of deformation. At the lower end of the range, failure may be initiated by local buckling, which reduces the effective radius of gyration and causes collapse of the reduced section by primary bending. At the upper end of the range, failure may begin by primary flexural buckling, and local buckling may develop on the compression side of the bent column and precipitate failure.

A primary instability mode in which the column bends and twists simultaneously can also occur in the intermediate range when the section has a low torsional rigidity.[18,19] The resulting buckle length is on the order of the column length; and while the column cross section may warp out of its plane as it twists, the cross-sectional shape does not change. If the section is doubly symmetric, so that the shear center and the centroid coincide, the buckling may occur by twisting alone. As in primary bending instability, the buckling and failing stresses are essentially the same in the torsion-bending instability mode. Thin-walled open sections are particularly susceptible to this type of instability, and because these sections are also subject to local instability, failure may occur by a combination of bending, twisting, and local buckling.

The complex buckling and failure modes that occur in the intermediate range have resisted the development of a satisfactory general theory of failure for this region. As a result, empirical equations are generally used in the intermediate range. The Johnson's parabola (Sec. 14-7) is frequently used from $L'/\rho = 0$ to the transitional slenderness ratio $(L'/\rho)_{tr}$, where the parabola joins the Euler curve (Fig. 16-1). In applying the Johnson's parabola to sections that are subject to local failure, $\bar{\sigma}_{co}$ is replaced by $\bar{\sigma}_f$. As a result, Eqs. (14-79) and (14-77) become

$$\sigma_c = \bar{\sigma}_f \left[1 - \frac{\bar{\sigma}_f (L'/\rho)^2}{4\pi^2 E} \right] \tag{16-7}$$

$$\left(\frac{L'}{\rho}\right)_{tr} = \pi \sqrt{2} \left(\frac{E}{\bar{\sigma}_f}\right)^{\frac{1}{2}} \tag{16-8}$$

We note from Fig. 16-1 that below $L'/\rho = 20$, σ_c is nearly constant and is approximately equal to $\bar{\sigma}_f$.

Example 16-3 The column in Example 16-1 (Fig. 16-7) has an area of 0.594 in.^{2} and a minimum moment of inertia of 0.1023 in.^{4}. The compressive yield stress of the material is 70 ksi. Determine the failing load for column lengths of 20 and 40 in. if the end fixity coefficient is 1.5.

Using Gerard's method, we find from Table 16-2 and Eq. (16-3) that

$$\frac{\bar{\sigma}_f}{\sigma_{cy}} = 0.67 \left[\frac{7 \times 0.125^2}{0.594} \left(\frac{10.5 \times 10^6}{70 \times 10^3} \right)^{\frac{1}{2}} \right]^{0.4} = 0.928$$

This exceeds the cutoff value of 0.8; so that from Eq. (16-6) and Example 16-1, $\bar{\sigma}_f = 76.3$ ksi.

The radius of gyration is $\rho = (I/A)^{1/2} = (0.1023/0.594)^{1/2} = 0.415$. From Eq. (14-18), $L'/\rho = 20/1.5^{1/2} \times 0.415 = 39.3$ for the 20-in. length and 78.6 for the 40-in. length. Using Eq. (16-8), we find

$$\left(\frac{L'}{\rho}\right)_{tr} = 1.414\pi \left(\frac{10.5 \times 10^6}{76.3 \times 10^3}\right)^{1/2} = 52.2$$

For the 20-in. length, $L'/\rho < (L'/\rho)_{tr}$, and from Eq. (16-7) the failing stress is

$$\sigma_c = 76,300 \left(1 - \frac{76,300 \times 39.3^2}{4 \times \pi^2 \times 10.5 \times 10^6}\right) = 54,600 \text{ psi}$$

which results in a failing load of $P = 54,600 \times 0.594 = 32,600$ lb. For the 40-in. length, $L'/\rho > (L'/\rho)_{tr}$, and from Eq. (14-17) the failing stress is

$$\sigma_c = \frac{\pi^2 \times 10.5 \times 10^6}{78.6^2} = 16,800 \text{ psi}$$

which gives an ultimate load of $P = 16,800 \times 0.594 = 9800$ lb.

16-5 COMPRESSIVE BUCKLING OF STIFFENED PANELS

A stiffened panel is a composite structure consisting of a plate and stiffening members. The stiffeners may be used to increase the buckling stress of the plate, carry part of the compressive load, or perform both of these functions. The buckling stress of a plate can be increased by reducing its width so that it functions as a long plate with a small value of b/t. Alternatively, the critical stress can be increased by reducing the length of the plate so that it behaves like a wide column with a small value of a/t. Longitudinal stiffeners which are parallel to the applied load subdivide the panel into plates with small values of b/t and carry a portion of the applied load. Transverse stiffeners may be used to partition the plate into panels with small values of a/t, but these members are ineffective in resisting the applied load. Longitudinal and transverse stiffeners may be used together to form a grid-stiffened structure.

In early applications of stiffened-skin construction, the proportions of the skin and stiffeners were such that the stiffening members were sturdy relative to the thin skin. As a result, the skin buckled at a much lower stress level than the stiffening members, and other than enforcing nodal lines in the buckled skin there was little interaction between the skin and stiffeners. In such uses, the buckling stress of the skin can be conservatively predicted by assuming that the stiffeners provided a simply supported edge condition. Improved estimates of the edge restraints can be found from Fig. 15-6.

Current designs are more likely to have the skin and stiffeners proportioned to buckle at about the same stress level, since structures of this type are more efficient and maintain aerodynamic smoothness to higher load levels. They are characterized by closely spaced stiffeners of comparable thickness to the skin. Because they buckle at nearly the same stress level, there is significant interaction, and any theoretical analysis must deal with the skin and stiffeners as a unit. Buckling may occur by primary instability with a wavelength which is on the order of the panel length or by local instability with a wavelength which is on the order of the width of the plate elements of the skin and stiffeners.

Seide and Stein[21] used the Rayleigh-Ritz method to derive curves for k in Eq. (15-8) to predict the primary buckling stress of panels with one, two, three, and an infinite number of longitudinal stiffeners. The theory considers only the bending rigidity of the stiffeners and neglects their torsional rigidity. Similar curves, which include the torsional rigidity, have been obtained by Budiansky and Seide for panels with transverse stiffeners.[22] In addition to their usefulness in computing buckling stresses, the results of these investigations are helpful in determining the minimum stiffness required in longitudinal or transverse stiffeners to provide simply supported edge conditions to the skin. The curves of Ref. 21 are for stiffened plates in which the centroid of the stiffener is in the midplane of the plate. This is seldom the case in practice, where the stiffeners are usually attached to one side of the plate. Seide[23] has given a method for correcting the results of Ref. 21 when the stiffeners are on one side of the plate.

Gerard and Becker[2,7] have summarized this and related work on the buckling of stiffened plates. The results show that the buckling stress of a plate with three or more stiffeners differs little from that of a plate with an infinite number of stiffeners if EI/dD does not approach zero (where EI is the bending rigidity of the stiffener, D is the bending rigidity of the plate, and d is the stiffener spacing). As a result, the primary buckling stress of a plate with three or more stiffeners can be predicted with orthotropic-plate theory.

The bending and twisting moments in an orthotropic plate are related to the curvatures and twist by the relationships[26]

$$M_x = D_x \frac{\partial^2 w}{\partial x^2} + D_1 \frac{\partial^2 w}{\partial y^2} \tag{16-9a}$$

$$M_y = D_y \frac{\partial^2 w}{\partial y^2} + D_1 \frac{\partial^2 w}{\partial x^2} \tag{16-9b}$$

$$M_{xy} = 2D_{xy} \frac{\partial^2 w}{\partial x \, \partial y} \tag{16-9c}$$

which take the place of Eqs. (13-30) in isotropic-plate theory. The lateral-equilibrium equation (13-19) is the same for orthotropic and isotropic plates. Substituting Eqs. (16-9) into Eq. (13-19), we find for uniaxial compression that the differential equation for the orthotropic

Table 16-3   Values of C for Eq. (16-11)[24]

Loaded edges	Unloaded edges	C
Simply supported	Simply supported	2
	Clamped	2.40
Clamped	Simply supported	2
	Clamped	2.46

plate is

$$D_x \frac{\partial^4 w}{\partial x^4} + 2(D_1 + 2D_{xy}) \frac{\partial^4 w}{\partial x^2 \partial y^2} + D_y \frac{\partial^4 w}{\partial y^4} = -N \frac{\partial^2 w}{\partial x^2} \quad (16\text{-}10)$$

Wittrick[24] has shown that the smallest eigenvalue of Eq. (16-10) is

$$N_{cr} = \frac{\pi^2 (D_x D_y)^{\frac{1}{2}}}{b^2} \left\{ k - C \left[1 - \frac{D_1 + 2D_{xy}}{(D_x D_y)^{\frac{1}{2}}} \right] \right\} \quad (16\text{-}11)$$

where C is given in Table 16-3.   The value of k can be found from the curves in Fig. 15-3 that correspond to the boundary conditions in Table 16-3.   However, a/b in Fig. 15-3 must be replaced by an effective length-to-width ratio $(a/b)_e$ given by

$$\left(\frac{a}{b}\right)_e = \frac{a}{b} \left(\frac{D_y}{D_x}\right)^{\frac{1}{4}} \quad (16\text{-}12)$$

Equation (16-11) reduces to Eq. (15-6) when the plate is isotropic.

The theoretical determination of D_x, D_y, D_1, and D_{xy} is not a simple problem, especially when the stiffeners are on only one side of the plate. In some cases it may be necessary to determine these rigidities experimentally by applying known moments and measuring the curvatures and twist.   Theoretical rigidities for wafflelike stiffened panels are given in Ref. 25.   For panels with stiffeners that are symmetric about the mid-plane of the plate, the rigidities can be determined from the approximate relations[26]

$$D_x = D + \frac{(EI)_x}{d_y} \qquad D_y = D + \frac{(EI)_y}{d_x} \qquad D_1 + 2D_{xy} = D \quad (16\text{-}13)$$

if the torsional rigidities of the stiffeners are neglected.   In these equations $D = Et^3/12(1 - \nu^2)$, while d_x and d_y are the stiffener spacings in the x and y directions.

Example 16-4   A 7075-T6 aluminum-alloy ($E = 10.5 \times 10^6$ psi) simply supported panel is 25 in. wide and 50 in. long.   The 0.125-in. plate is stiffened on both sides by angle-section longitudinal stiffeners on 5-in. centers.   The properties

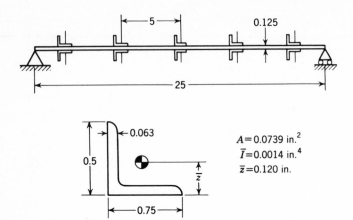

Fig. 16-13 Example 16-4.

of the angle sections are given in Fig. 16-13. Determine the compressive stress and force to produce primary buckling of the panel.

The bending rigidity of a pair of stiffeners about the midplane of the plate is

$$(EI)_x = 2 \times 10.5 \times 10^6 \left[0.0014 + 0.0739 \left(0.120 + \frac{0.125}{2} \right)^2 \right]$$
$$= 8.10 \times 10^4 \text{ lb-in.}^2$$

and the plate rigidity is

$$D = \frac{10.5 \times 10^6 \times 0.125^3}{12(1 - 0.3^2)} = 1.88 \times 10^3 \text{ lb-in.}$$

From Eqs. (16-13) the orthotropic-plate rigidities are

$$D_x = 1.88 \times 10^3 + \frac{8.10 \times 10^4}{5} = 18.08 \times 10^3 \text{ lb-in.}$$

$$D_y = D_1 + 2D_{xy} = D = 1.88 \times 10^3 \text{ lb-in.}$$

Equation (16-12) then gives

$$\left(\frac{a}{b} \right)_e = \frac{50}{25} \left(\frac{1.88 \times 10^3}{18.08 \times 10^3} \right)^{\frac{1}{4}} = 1.136$$

and from Fig. 15-3, $k = 4.1$. With $C = 2$ from Table 16-3, we find from Eq. (16-11)

$$N_{cr} = \frac{\pi^2 (18.08 \times 1.88 \times 10^6)^{\frac{1}{2}}}{5^2} \left\{ 4.1 - 2 \left[1 - \frac{1.88 \times 10^3}{(18.08 \times 1.88 \times 10^6)^{\frac{1}{2}}} \right] \right\}$$
$$= 6310 \text{ lb/in.}$$

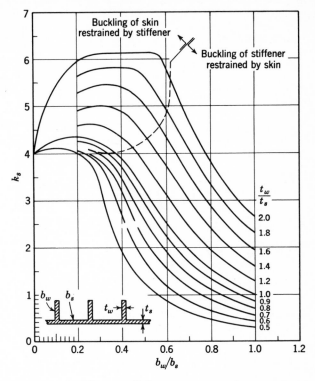

Fig. 16-14 Buckling coefficients for web-stiffener panels (Ref. 27).

This gives a total compressive force of $P_{cr} = 6310 \times 25 = 158,000$ lb.   The average thickness of the panel is

$$\bar{t} = \frac{5 \times 0.125 + 2 \times 0.0739}{5} = 0.1545 \text{ in.}$$

which gives a buckling stress of $\sigma_{cr} = 6310/0.1545 = 40,900$ psi.

Boughan, Baab, and Gallaher[27,28] have used the methods of Refs. 11 and 12 to determine the local buckling stresses of idealized web-, Z-, and T-stiffened panels.   The assumed geometries and results are shown in Figs. 16-14 to 16-16.   The theory assumes monolithic construction and is therefore applicable to integrally stiffened extruded panels.   If inter-fastener buckling (Sec. 16-7) does not occur, the results can be used to give approximate values for riveted or spot-welded panels with angle-, Z-, or H-section stiffeners.   To determine the local buckling stress, ϵ_{cr} is computed from Eq. (15-45) with $t/b = t_s/b_s$ and k from the appropriate chart.   The buckling stress is determined from Fig. 15-19 or 15-20.

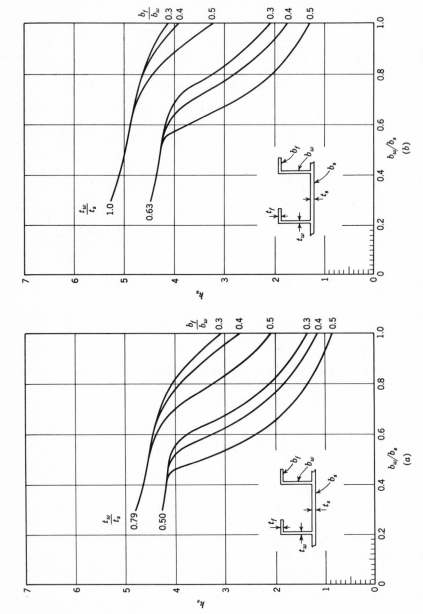

Fig. 16-15 Buckling coefficients for Z-stiffener panels (Ref. 28). (a) $t_w/t_s = 0.50$ and 0.79; (b) $t_w/t_s = 0.63$ and 1.0.

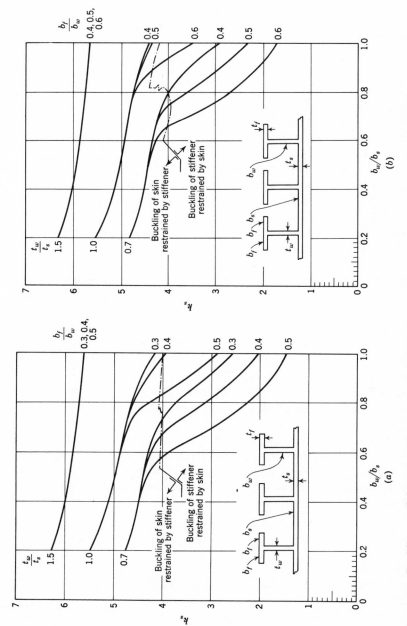

Fig. 16-16 Buckling coefficients for T-stiffener panels (Ref. 27). (a) $t_w/t_f = 1.0$, $b_f/t_f > 10$, $b_w/b_s > 10$, $b_w/b_s > 0.25$; (b) $t_w/t_f = 0.7$, $b_f/t_f > 10$, $b_w/b_s > 0.25$.

Argyris and Dunne[8] have given a comprehensive discussion of the various buckling modes for Z-stiffened panels. The results are given in the form of design charts.

16-6 CRIPPLING OF STIFFENED PANELS

Gerard has demonstrated that, with minor changes, his method for predicting the crippling stress of plates (Sec. 15-9) and thin-walled columns (Sec. 16-3) can also be used[5,17] to determine the local failing stress of longitudinally stiffened panels when $L'/\rho < 20$. The method is applicable to monolithic construction or built-up panels when the inter-fastener buckling and wrinkling stresses (Sec. 16-7) are greater than the crippling stress.

The method is applied to a section of the panel consisting of a stiffener and a width of skin equal to the stiffener spacing. The section is obtained by envisioning a cut in the skin midway between the lines of attachment of the stiffeners to the skin. If the stiffener has two lines of attachment, as in the case of a hat section, a cut is imagined between each of the lines of attachment.

It is assumed that the attachment of the skin to the stiffener provides the same edge restraint to the skin that a corner gives to the flanges of an angle. Each of the skin sections that remains after the cuts is considered to be a flange. The parameter g is therefore equal to the g of the stiffener plus the number of cuts in the skin and the number of equivalent flanges of the skin. For a stiffener with a single line of attachment, the contribution of the skin to g is 3 (1 cut plus 2 flanges); for a stiffener with a double line of attachment, it is 6 (2 cuts plus 4 flanges). Examples are shown in Fig. 16-17. Note that the cut to the left of a stiffener is not counted because it is assigned to the stiffener that is to the left of the cut.

When the skin thickness t_s is different from the stiffener thickness t_w, the parameter gt^2/A is replaced by $gt_w t_s/A$. If the stiffener thickness is not constant, a mean thickness $\bar{t}_w$ computed from Eq. (16-5) is used instead of t_w. The compressive yield stress σ_{cys} of the skin may be different from the yield stress σ_{cyw} of the stiffener. In this case σ_{cy} is replaced by a weighted yield stress $\bar{\sigma}_{cy}$ computed from

$$\bar{\sigma}_{cy} = \frac{\sigma_{cys} + \sigma_{cyw}(\bar{t}/t_s - 1)}{\bar{t}/t_s} \tag{16-14}$$

where $\bar{t}$ is the average thickness of the panel. The general form of Eq. (16-3) then becomes

$$\frac{\bar{\sigma}_f}{\bar{\sigma}_{cy}} = \beta_g \left[\frac{g\bar{t}_w t_s}{A} \left(\frac{E}{\bar{\sigma}_{cy}} \right)^{1/2} \right]^m \tag{16-15}$$

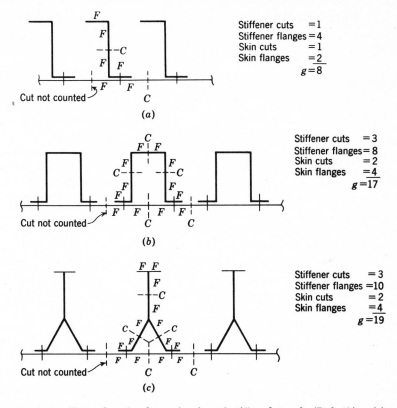

Fig. 16-17 Examples of g determination of stiffened panels (Ref. 17). (a) Z-stiffened panel; (b) hat-stiffened panel; (c) Y-stiffened panel.

The constants β_g and m depend upon the type of stiffeners, and in some cases β_g is also a function of $\bar{t}_w/t_s$. Values of β_g, m, and the cutoff stress for Eq. (16-15) are listed in Table 16-4 for various types of stiffeners. The general angle-type stiffener in this table is one that consists of a series of angles after the cuts are made, and the general T-type stiffener is one that is made up of T elements after the cuts (such as an H section). The Y-section stiffener, which contains both angle and T elements, was found to correlate with the angle-type sections.

Gerard[17] has also been able to correlate short-time creep–crippling-strength data for plates, thin-walled columns, and stiffened panels by his method. The crippling stress is time-dependent at elevated temperatures, and in these cases σ_{cy} is replaced by the stress to attain a 0.002 strain for the given time of exposure to the temperature.

Example 16-5   The formed hat-section stiffeners of the panel in Fig. 16-18 are 7075-T6 aluminum alloy ($\sigma_{cy} = 67$ ksi, $E = 10.5 \times 10^6$ psi), and the skin is 2024-T3 aluminum alloy ($\sigma_{cy} = 40$ ksi).   Determine the crippling stress.

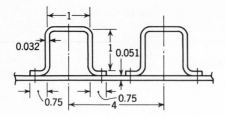

Fig. 16-18   Example 16-5.

The total area of the stiffener and its skin is

$$A = 0.051 \times 4 + 0.032(1 + 2 \times 1 + 2 \times 0.75) = 0.348 \text{ in.}^2$$

and $\bar{t} = 0.348/4 = 0.0871$ in.   From Eq. (16-14)

$$\bar{\sigma}_{cy} = \frac{40 + 67[(0.0871/0.051) - 1]}{0.0871/0.051} = 51.1 \text{ ksi}$$

With $t_w/t_s = 0.032/0.051 = 0.63$, we find from Table 16-4 that $g = 17$, $m = 0.85$, and $\beta_g = 0.50$.   Substituting into Eq. (16-15), we obtain

$$\frac{\bar{\sigma}_f}{\bar{\sigma}_{cy}} = 0.50 \left[\frac{17 \times 0.032 \times 0.051}{0.348} \left(\frac{10.5 \times 10^6}{51.1 \times 10^3} \right)^{1/2} \right]^{0.85} = 0.598$$

This is less than the cutoff value of 0.8; therefore $\bar{\sigma}_f = 0.598 \times 51.1 = 30.6$ ksi.

16-7   INTERFASTENER BUCKLING AND WRINKLING

The method for predicting crippling in the preceding section is for monolithic panels.   While this form of construction is efficient, it is also

Table 16-4   Summary of values of β_g and m for stiffened panels[5,17]

Type of panel	g	m	$\dfrac{\bar{t}_w}{t_s}$	β_g	Cutoff
1. Formed Z stiffeners	8	0.85	—	0.56	σ_{cy}
2. Formed hat stiffeners	17	0.85	1.25	0.59	$0.8\sigma_{cy}$
			1.00	0.56	$0.8\sigma_{cy}$
			0.63	0.50	$0.8\sigma_{cy}$
			0.39	0.48	$0.8\sigma_{cy}$
3. Extruded Y stiffeners	19	0.85	1.16	0.56	
			0.732	0.51	
			0.464	0.48	
4. General angle-type stiffeners	—	0.85	—	0.56	
5. General T-type stiffeners	—	0.40	—	0.67	$0.75\sigma_{cy}$

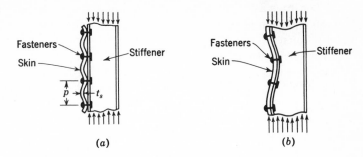

Fig. 16-19 Comparison of (*a*) interfastener buckling and (*b*) wrinkling modes.

expensive, and as a result panels are often built up by riveting or spot-welding the stiffeners to the skin.   In these cases, the panel may experience interfastener buckling or wrinkling failure at a stress lower than the monolithic crippling stress.   In an optimum design the stiffener, skin, and fasteners are proportioned so that all these modes of failure occur at the same stress.

Interfastener (or *interrivet*) *buckling* is defined as a wide-column buckling of the skin between the fasteners that join it to the stiffeners. The wavelength of the buckle is equal to the *pitch* (or spacing) of the fasteners.   The overall appearance of wrinkling is somewhat similar to interfastener buckling, but the wavelength of the buckles and the interaction with the stiffeners are different.   The appearance of the skin and stiffeners in the two types of instability is shown in Fig. 16-19.

Wrinkling is defined as a buckling of the skin in which the stiffener provides an elastic line support, as shown in Fig. 16-20.   The buckle length of the wrinkle is greater than the fastener pitch; and while the skin separates from the stiffener between rivets in interfastener buckling, it maintains contact and deforms the stiffener in wrinkling.   For this reason, wrinkling is often referred to as *forced crippling*.

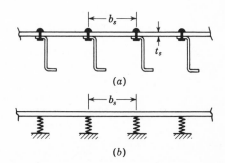

Fig. 16-20 Wrinkling idealization of a longitudinally stiffened panel. (*a*) Actual stiffened panel; (*b*) idealization.

The interfastener buckling stress σ_i can be determined from the wide-column equation (15-10) by letting $a = p$, the fastener pitch. Equation (15-10) must be modified when buckling occurs in the inelastic range. The wide plate behaves like a column, so that $\eta E = E_t$. It is also necessary to account for the fact that the loaded edges of the plate in interfastener buckling are closer to being clamped than simply supported, as assumed in the derivation of Eq. (15-10). Taking these factors into account, we have

$$\sigma_i = \frac{c\pi^2 E_t}{12(1 - \nu_e^2)} \left(\frac{t_s}{p}\right)^2 \tag{16-16}$$

where c is the coefficient of end fixity that is provided by the fasteners. Except for the $1 - \nu_e^2$ factor, Eq. (16-16) can be derived from the column equation (14-74). As a result, σ_i can be found from Fig. 14-22 if $\dfrac{L'}{\rho}\left(\dfrac{\sigma_{0.7}}{E}\right)^{1/2}$ is replaced by $(p/t_s)[12\sigma_{0.7}(1 - \nu_e^2)/cE]^{1/2}$. The end fixity the skin receives depends upon the type of fastener used. Values of c for different fasteners are given in Table 16-5.

The preceding method gives a nominal buckling stress when the skin material is clad and the core properties are used. The interfastener buckling stress $\bar{\sigma}_i$ for the clad material is then determined from

$$\bar{\sigma}_i = \bar{\eta}\sigma_i \tag{16-17}$$

where $\bar{\eta}$ is found from Eqs. (15-47).

The postbuckling behavior of a wide column is similar to that of a column (Fig. 14-15). While the wide column cannot develop loads greater than the buckling load, it can continue to carry the buckling load. Letting b_{ei} be the effective width of the skin at $\sigma_e = \sigma_i$, the load that is carried by the skin after interfastener buckling is $2\sigma_i b_{ei} t_s$. The panel will continue to carry additional load until the crippling stress $\bar{\sigma}_{fst}$ of the stiffener alone is reached, whereupon failure occurs. The load at failure

Table 16-5 End-fixity coefficients for interfastener buckling[5]

Type of fastener	c
Flat-head rivet	4
Brazier-head rivet	3
Machine-countersunk rivet	1
Dimpled-countersunk rivet	1
Spot welds	3.5

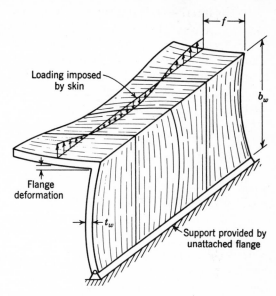

Fig. 16-21 Deformation of the flange and web in the wrinkling idealization of a stiffener.

is therefore $\bar{\sigma}_{fst}A_{st} + 2\sigma_i b_{ei} t_s$, where A_{st} is the stiffener area, and the average stress at failure is

$$\bar{\sigma}_{fr} = \frac{\bar{\sigma}_{fst}A_{st} + 2\sigma_i b_{ei} t_s}{A_{st} + 2b_{ei} t_s} \tag{16-18}$$

In wrinkling, the spring constant of the elastic line support provided by the stiffeners is a function of the flexibilities of the stiffener flange and web. It is also dependent upon the size and spacing of the fasteners. Semonian and Peterson[29] have derived curves for k which can be used with Eq. (15-45) to predict the critical strain for wrinkling buckling. Their analysis is based upon the stiffener idealization shown in Fig. 16-21, where the *effective fastener offset f* is defined as the effective distance from the stiffener web to the line along which the fasteners clamp the flange to the skin. The value of f can be determined from the empirical curves of Fig. 16-22, and k can be found from theoretical results in Fig. 16-23. With k known, ϵ_{cr} is computed from Eq. (15-45) with $t/b = t_s/b_s$, and the wrinkling-buckling stress is obtained from Fig. 15-19. When the material is clad, the critical stress computed from the core properties must be multiplied by $\bar{\eta}$.

Semonian and Peterson also treat the wrinkling failure of the skin, which occurs through buckle growth and plastic deformation. The

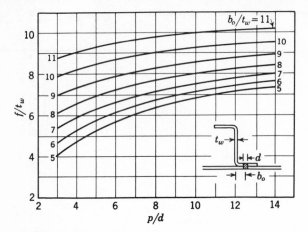

Fig. 16-22 Experimental values for effective rivet offset f (Ref. 29).

wrinkling-failure strain ϵ_w is computed from Eq. (15-45) with $\epsilon_{cr} = \epsilon_w$, $t/b = t_s/b_s$, and $k = k_w$, which is found from the semiempirical curves of Fig. 16-24.   The wrinkling-failure stress $\bar{\sigma}_w$ is determined from Fig. 15-19. If the plate is clad, the nominal failing stress determined by using the Ramberg-Osgood properties of the core material must be multiplied by $\bar{\eta}$ to obtain the failing stress of the clad plate.   A comparison of Figs. 16-23 and 16-24 shows that unless f/b_w is small, there is little difference between the wrinkling instability and failing stresses of the skin.

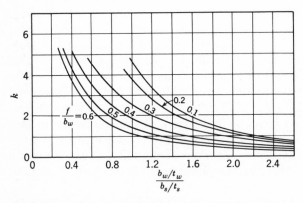

Fig. 16-23 Wrinkling-buckling coefficients for riveted panels (Ref. 29).

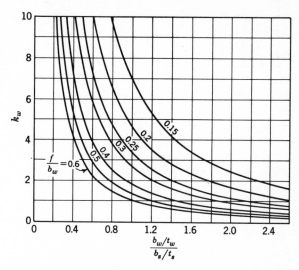

Fig. 16-24 Experimental wrinkling-failure coefficients for riveted panels (Ref. 29).

Wrinkling distorts the flange of the stiffener attached to the skin and usually precipitates failure of the panel.   However, the stiffener will continue to carry load if it is unusually sturdy relative to the skin.   To determine $\bar{\sigma}_{fr}$, the failing stress of the built-up panel, it is necessary to determine whether the stiffener-failing stress is larger or smaller than the wrinkling-failure stress.   This is determined from the relative magnitudes of the crippling stress of the stiffener alone $\bar{\sigma}_{fst}$, the wrinkling-failure stress $\bar{\sigma}_w$, and the crippling stress of the equivalent monolithic panel $\bar{\sigma}_f$.

If $\bar{\sigma}_w \leq \bar{\sigma}_{fst}$, wrinkling occurs before or at stiffener crippling, and $\bar{\sigma}_{fr}$ is the larger of the values obtained from the equations

$$\bar{\sigma}_{fr} = \bar{\sigma}_w \tag{16-19}$$

$$\bar{\sigma}_{fr} = \frac{\bar{\sigma}_{fst}A_{st}}{A_{st} + b_s t_s} \tag{16-20}$$

The first equation assumes that wrinkling leads to forced crippling, and the second accounts for the case of an unusually sturdy stiffener which carries the entire load.   When $\bar{\sigma}_w > \bar{\sigma}_{fst}$, stiffener crippling occurs before wrinkling, in which case it is assumed that the stiffener continues to carry the crippling load.   Failure occurs when the skin reaches the wrinkling stress.   The average stress at failure is therefore

$$\bar{\sigma}_{fr} = \frac{\bar{\sigma}_{fst}A_{st} + \bar{\sigma}_w b_s t_s}{A_{st} + b_s t_s} \tag{16-21}$$

In no case may $\bar{\sigma}_{fr}$ be greater than $\bar{\sigma}_f$, the crippling stress of the equivalent monolithic panel.

Elasticity in the fasteners increases the flexibility of the line support and reduces $\bar{\sigma}_w$. Figures 16-22 and 16-24 are based upon tests in which 2117-T4 aluminum-alloy rivets with diameters greater than $0.9t_s$ were used. The results should be used with caution when smaller-diameter rivets are used. An approximate criterion for the required tensile strength of the rivets is[5,29]

$$P_r > 0.55Eb_s p\epsilon_w{}^2 \qquad (16\text{-}22)$$

Allowable tensile strengths of protruding and flush-head rivets are given in Refs. 30 and 31, respectively.

The results in Fig. 16-22 were obtained from tests on panels with formed stiffeners. The use of extruded stiffeners usually eliminates wrinkling as a mode of failure because of the support supplied by the sharp exterior corner and the fillet at the interior corner. In addition, the fillet radius of an extruded section is usually smaller than the bend radius of a formed section, so that a smaller fastener-offset distance b_0 may be used.

Example 16-6 Determine the interfastener buckling stress, the failing stress, and the required rivet strength for the 2024-T3 aluminum alloy ($\sigma_{cy} = 40$ ksi, $\sigma_{0.7} = 39$ ksi, $E = 10.7 \times 10^6$ psi, $n = 11.5$) Z-stiffener panel shown in Fig. 16-25 (this is used as a numerical example in Ref. 29).

To determine the interfastener buckling stress we evaluate

$$\frac{p}{t_s} \left[\frac{12\sigma_{0.7}(1 - \nu_e{}^2)}{cE} \right]^{1/2} = \frac{1}{0.102} \left[\frac{12 \times 39 \times 10^3(1 - 0.3^2)}{3 \times 10.7 \times 10^6} \right]^{1/2} = 1.129$$

where from Table 16-5, $c = 3$. Using this parameter in place of $\dfrac{L'}{\rho} \left(\dfrac{\sigma_{0.7}}{E} \right)^{1/2}$ in Fig. 14-22, we find $\sigma_i/\sigma_{0.7} = \sigma_{cr}/\sigma_{0.7} = 1.04$, so that $\sigma_i = 1.04 \times 39 = 40.6$ ksi.

The stiffener area $A_{st} = 0.262$ in.², and, from Eq. (16-4) and Table 16-2,

$$\frac{\bar{\sigma}_{fst}}{\sigma_{cy}} = 3.2 \left[\frac{0.064^2}{0.262} \left(\frac{10.7 \times 10^6}{40 \times 10^3} \right)^{1/3} \right]^{3/4} = 0.576$$

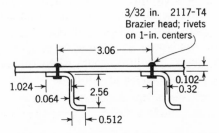

Fig. 16-25 Example 16-6.

for the two-corner Z-section stiffener. From Table 16-2, the cutoff stress is $\bar{\sigma}_{fst}/\sigma_{cy} = 2(0.064/2.56)^{1/3} = 0.586$. This is greater than the previously determined result; therefore $\bar{\sigma}_{fst} = 0.576 \times 40 = 23$ ksi.

The total area of the panel per stiffener spacing is

$$A = A_{st} + b_s t_s = 0.262 + 3.06 \times 0.102 = 0.574 \text{ in.}^2$$

The crippling stress of the equivalent monolithic panel is found from Eq. (16-15) and Table 16-4. Applying the constants for a formed Z-stiffened panel, we have

$$\frac{\bar{\sigma}_f}{\sigma_{cy}} = 0.56 \left[\frac{8 \times 0.064 \times 0.102}{0.574} \left(\frac{10.7 \times 10^6}{40 \times 10^3} \right)^{1/2} \right]^{0.85} = 0.786$$

This is less than the cutoff value of $\bar{\sigma}_f/\sigma_{cy} = 1$; therefore $\bar{\sigma}_f = 0.786 \times 40 = 31.5$ ksi.

To determine $\bar{\sigma}_w$ we first compute the parameters $b_0/t_w = 0.32/0.064 = 5.0$ and $p/d = 3\frac{2}{3} = 10.67$. From Fig. 16-22, $f/t_w = 6.98$, so that $f = 6.98 \times 0.064 = 0.447$ in. Using

$$\frac{f}{b_w} = \frac{0.447}{2.56} = 0.1748 \qquad \frac{b_w t_s}{b_s t_w} = \frac{2.56 \times 0.102}{3.06 \times 0.064} = 1.333$$

we obtain $k_w = 3.64$ from Fig. 16-24. With $t/b = t_s/b_s$ we find from Eq. (15-45) that

$$\frac{\epsilon_w E}{\sigma_{0.7}} = \frac{3.64 \times \pi^2 \times 10.7 \times 10^6}{39 \times 10^3 \times 12(1 - 0.3^2)} \left(\frac{0.102}{3.06} \right)^2 = 1.01$$

Using Fig. 15-19, we obtain $\sigma_{cr}/\sigma_{0.7} = \bar{\sigma}_w/\sigma_{0.7} = 0.86$, which gives $\bar{\sigma}_w = 0.86 \times 39 = 33.6$ ksi.

Noting that $\bar{\sigma}_w > \bar{\sigma}_{fst}$, we find from Eq. (16-21)

$$\bar{\sigma}_{fr} = \frac{23 \times 0.266 + 33.6 \times 3.06 \times 0.102}{0.262 + 3.06 \times 0.102} = 28.8 \text{ ksi}$$

This is less than σ_i and $\bar{\sigma}_f$, and so the failing stress is 28.8 ksi if tensile failure of the rivets does not occur.

Applying Eq. (16-22), the rivets must be capable of withstanding a tensile force of

$$P_r = 0.55 \times 10.7 \times 10^6 \times 3.06 \times 1 \times 0.00365^2 = 240 \text{ lb}$$

From Ref. 30, the allowable tensile strength of a $\frac{3}{32}$-in. brazier-head rivet is 268 lb, so that rivet failure does not occur, and the failing stress of the panel is 28.8 ksi.

16-8 FAILURE OF STIFFENED PANELS

The material in Secs. 16-6 and 16-7 can be used to predict the failing stresses of panels when $L'/\rho \leq 20$. As panel length increases, the failing stress decreases, until at large L'/ρ the panel fails in flexure as a long column. The failing stress in the long-column range can be predicted by Eq. (14-17), where ρ is the radius of gyration of the combination of the skin and its effective width of skin.

At intermediate lengths there is a transition from the pure local to the pure flexural mode of failure. As in thin-walled columns, failure in this region occurs through a combination of the primary and flexural modes and may involve twisting of the stiffeners. There is no current theory that satisfactorily predicts failure in this range, and it is necessary to rely on test data and empirical procedures.

Several methods have been used to predict the failure of stiffened panels. One of these[32] is a modification of the method for thin-walled columns described in Sec. 16-4. It is assumed that the stiffened panel has the same column curve as the column alone when the column buckles about an axis parallel to the skin. Therefore, the column curve is similar to the solid line in Fig. 16-1, where the Euler equation (14-17) is used when $L'/\rho \geq (L'/\rho)_{tr}$ and the Johnson's parabola equation (16-7) is applied when L'/ρ is less than this value.

In using Eq. (14-17) or (16-7), ρ is taken as the radius of gyration of the stiffener and an effective width of skin equal to $2b_e$, which acts with the stiffener as an effective column. It is easily shown that ρ can be determined from the equation[32]

$$\left(\frac{\rho}{\rho_{st}}\right)^2 = \frac{1 + [1 + (s/\rho_{st})^2]2b_e t_s/A_{st}}{(1 + 2b_e t_s/A_{st})^2} \tag{16-23}$$

where ρ_{st} and A_{st} are the radius of gyration and area of the stiffener alone and s is the distance from the stiffener centroid to the midplane of the skin. The skin is fully effective below its buckling stress, so that $2b_e = b_s$, and the failing stress can be found directly from Eq. (16-23) and Eq. (14-17) or (16-7).

When the stress exceeds the critical stress of the skin, b_e depends upon the edge stress σ_e, which is equal to the stiffener stress when the effective column fails (if the skin and stiffener are of the same material). However, the failing stress of the effective column depends upon ρ, which is a function of b_e. As a result, the failing stress cannot be determined directly from Eq. (14-17) or (16-7), and the following method of successive approximations must be used:

1. The value of ρ_{st} is taken as a first approximation of ρ, $(L'/\rho)_{tr}$ is computed, and a first approximation of the failing stress σ_c of the effective column is calculated from Eq. (14-17) or (16-7).
2. The value of b_e is determined from one of the methods of Sec. 15-7, and an improved estimate of ρ is calculated from Eq. (16-23).
3. Steps 1 and 2 are repeated, using the improved approximation for ρ in place of ρ_{st}. The process is repeated until the values at the end of a cycle of iteration are the same as those at the beginning of the cycle.
4. The average failing stress of the panel is computed from

$$\bar{\sigma}_c = \frac{\sigma_c(A_{st} + 2b_e t_s)}{A_{st} + b_s t_s} \tag{16-24}$$

The effective width must be modified when interfastener buckling occurs. By assuming that the skin continues to carry the same load that it resisted at σ_i we find

$$b_e = b_{ei} \frac{\sigma_c}{\sigma_i} \tag{16-25}$$

where b_{ei} is the effective width at $\sigma_e = \sigma_i$, and σ_c is the failing stress of the effective column.

Gerard[5] has suggested a modification of the preceding method which uses the column curve of Fig. 16-26 in place of that in Fig. 16-1. For $L'/\rho \leq 20$, it is assumed that the failing stress is constant and equal to $\bar{\sigma}_f$ for a monolithic panel or $\bar{\sigma}_{fr}$ for a built-up panel. At stresses below the local buckling stress σ_{cr} and the proportional limit σ_{pl}, the Euler equation (14-17) is used. A parabola is utilized in the intermediate range. For the monolithic panel, the equation of a parabola which satisfies the conditions $d\sigma_c/d(L'/\rho) = 0$ and $\sigma_c = \bar{\sigma}_f$ at $L'/\rho = 20$ and which joins the Euler curve at $\sigma_c = \sigma_{cr}$ is[5]

$$\sigma_c = \bar{\sigma}_f \left[1 - \frac{\sigma_{cr}}{\sigma_E} \left(1 - \frac{\sigma_{cr}}{\bar{\sigma}_f} \right) \frac{\sigma_{20}^{1/2} - \sigma_E^{1/2}}{(\sigma_{20}^{1/2} - \sigma_{cr}^{1/2})^2} \right] \tag{16-26}$$

In this equation, $\sigma_E = \pi^2 E/(L'/\rho)^2$ is the Euler stress, and σ_{20} is the Euler stress evaluated at $L'/\rho = 20$. If $\sigma_{pl} < \sigma_{cr}$, σ_{cr} is replaced by σ_{pl}. For a built-up panel, $\bar{\sigma}_{fr}$ is used in place of $\bar{\sigma}_f$ if σ_i or σ_w is less than $\bar{\sigma}_f$.

Current practice is to use direct-reading column curves for predicting panel failure rather than the curves of Fig. 16-1 or 16-26. The type of chart in Fig. 16-27 has found widespread acceptance. In it the average failing stress of the panel $\bar{\sigma}_c$ is plotted against N/L', where N is the compressive force per unit width of the loaded edges. The parameter N/L' is known by the designer from the applied loads and geometry. This type of chart is advantageous because it does not require successive approximations to find the failing stress. In addition, the failing stresses of alternative designs can be compared directly for a given value of the loading parameter N/L'. Since $\bar{\sigma}_c$ is inversely proportional to the cross-

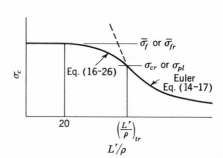

Fig. 16-26 Gerard's column curve for stiffened panels.

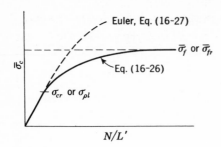

Fig. 16-27 Direct-reading chart for column failure of a stiffened panel.

sectional area of the panel, the lightest panel will have the smallest value of $\bar{\sigma}_c$ for the given N/L'.

When $\bar{\sigma}_c$ is less than σ_{cr} and σ_{pl}, the failing stress is the Euler stress σ_E. To express σ_E in terms of N/L' we multiply and divide the right side of Eq. (14-17) by N and obtain $\sigma_E = \pi^2 E \left(\dfrac{\rho}{N}\right)^2 \left(\dfrac{N}{L'}\right)^2$. At failure $N = \sigma_E \bar{t}$, where $\bar{t}$ is the average thickness of the panel given by $\bar{t} = (A_{st} + b_s t_s)/b_s$. Substituting this into the $(\rho/N)^2$ term and solving for σ_E, we find

$$\sigma_E = \left[\pi^2 E \left(\frac{\rho}{\bar{t}}\right)^2 \left(\frac{N}{L'}\right)^2 \right]^{1/3} \tag{16-27}$$

which is used in Fig. 16-27 when $\bar{\sigma}_c$ is less than σ_{cr} and σ_{pl}. At higher stresses, Eq. (16-26) is used with σ_E obtained from Eq. (16-27).

After a comprehensive study of the failure of hat-, Z-, and Y-section stiffened panels the NACA gave the results in direct-reading charts similar to Fig. 16-27. The results are too extensive to reproduce here; however, a bibliography of this work is given in Ref. 5. The NACA charts are useful in arriving at the optimum design of a stiffened panel where all modes of failure occur at the same stress level. The skin thickness is often fixed by torsional-rigidity requirements associated with aeroelastic criteria. The NACA has also given charts for the minimum-weight design of panels with this additional condition. In these cases, the results are given as a family of curves of $\bar{\sigma}_c$ versus N/t_s, with N/L' as a parameter.

REFERENCES

1. Gerard, G., and H. Becker: Handbook of Structural Stability, pt. I: Buckling of Flat Plates, *NACA Tech. Note* 3781, 1957.
2. Becker, H.: Handbook of Structural Stability, pt. II: Buckling of Composite Elements, *NACA Tech. Note* 3782, 1957.
3. Gerard, G., and H. Becker: Handbook of Structural Stability, pt. III: Buckling of Curved Plates and Shells, *NACA Tech. Note* 3783, 1957.
4. Gerard, G., and H. Becker: Handbook of Structural Stability, pt. IV: Failure of Plates and Composite Elements, *NACA Tech. Note* 3784, 1957.
5. Gerard, G.: Handbook of Structural Stability, pt. V: Compressive Strength of Flat Stiffened Panels, *NACA Tech. Note* 3785, 1957.

6. Becker, H.: Handbook of Structural Stability, pt. VI: Strength of Stiffened Curved Plates and Shells, *NACA Tech. Note* 3786, 1957.

7. Gerard, G., and H. Becker: Handbook of Structural Stability, pt. VII: Strength of Thin-wing Construction, *NACA Tech. Note* D-162, 1959.

8. Argyris, D. E., and P. C. Dunne: "Handbook of Aeronautics, pt. 2: Structural Analysis," 4th ed., Pitman Publishing Corporation, New York, 1952.

9. Heller, C. O.: Behavior of Stiffened Plates, vol. 1, Analysis, U.S. Naval Acad. *Eng. Dept. Rept.* 67-1, 1967.

10. Roy, J. A., and E. H. Schuette: The Effect of Angle of Bend between Plate Elements on the Local Instability of Formed Z-sections, *NACA Wartime Report* L-268, 1944.

11. Lundquist, E. E., E. Z. Stowell, and E. H. Schuette: Principles of Moment Distribution Applied to the Stability of Structures Composed of Bars or Plates, *NACA Wartime Report* L-326, 1943.

12. Kroll, W. D., G. P. Fisher, and G. P. Heimerl: Charts for the Calculation of the Critical Stress for Local Instability of Columns with I-, Z-, Channel, and Rectangular Tube Section, *NACA Wartime Report* L-429, 1943.

13. van der Mass, C. J.: Charts for the Calculations of the Critical Compressive Stress for Local Instability of Columns with Hat Sections, *J. Aeron. Sci.*, **21**(6): 399–403 (June, 1954).

14. Stowell, E. Z.: Compressive Strength of Flanges, *NACA Tech. Note* 2020, 1950.

15. Crockett, H. B.: Predicting Stiffener and Stiffened Panel Crippling Stresses *J. Aeron. Sci.*, **9**(13): 501–509 (November, 1942).

16. Needham, R. A.: The Ultimate Strength of Aluminum-alloy Formed Structural Shapes in Compression, *J. Aeron. Sci.*, **21**(4): 217–229 (April, 1954).

17. Gerard, G.: The Crippling Strength of Compression Elements, *J. Aeron. Sci.*, **25**(1): 37–52 (January, 1958).

18. Bleich, F.: "Buckling Strength of Metal Structures," McGraw-Hill Book Company, New York, 1952.

19. Niles, A. S., and J. S. Newell: "Airplane Structures," vol. 2, 3d ed., John Wiley & Sons, Inc., New York, 1946.

20. "Metallic Materials and Elements for Flight Vehicle Structures," *Military Handbook* MIL-HDBK-5A, Feb. 8, 1966.

21. Seide, P., and M. Stein: Compressive Buckling of Simply Supported Plates with Longitudinal Stiffeners, *NACA Tech. Note* 1825, 1949.

22. Budiansky, B., and P. Seide: Compressive Buckling of Simply Supported Plates with Transverse Stiffeners, *NACA Tech. Note* 1557, 1948.

23. Seide, P.: The Effect of Longitudinal Stiffeners Located on One Side of a Plate on the Compressive Buckling Stress of the Plate-Stiffener Combination, *NACA Tech. Note* 2873, 1953.

24. Wittrick, W. H.: Correlation between Some Stability Problems for Orthotropic and Isotropic Plates under Bi-axial and Uniaxial Direct Stress, *Aeron. Quart.*, **4**(1): 83–92 (August, 1952).

25. Dow, N. F., C. Libove, and R. E. Hubka: Formulas for Elastic Constants of Plates with Integral Waffle-like Stiffening, *NACA Rept.* 1195, 1954.

26. Timoshenko, S. P., and S. Woinowsky-Krieger: "Theory of Plates and Shells," 2d ed., McGraw-Hill Book Company, New York, 1959.

27. Boughan, R. B., and G. H. Baab: Charts for Calculation of the Critical Compressive Stress for Local Instability of Idealized Web- and T-stiffened Panels, *NACA Wartime Report* L-204, 1944.

28. Gallaher, G. L., and G. L. Boughan: A Method of Calculating the Compressive Strength of Z-stiffened Panels that Develop Local Instability, *NACA Tech. Note* 1482, 1947.

29. Semonian, J. W., and J. P. Peterson: An Analysis of the Stability of Short Sheet-stringer Panels with Special Reference to the Influence of Riveted Connection between Sheet and Stringer, *NACA Tech. Note* 3431, 1955.
30. Schuette, E. H., L. M. Bartone, and M. W. Mandel: Tensile Tests of Round-head, Flat-head, and Brazier-head Rivets, *NACA Tech. Note* 930, 1944.
31. Mandel, M. W., and L. M. Bartone: Tensile Tests of NACA and Conventional Machine-countersunk Flush Rivets, *NACA Wartime Report* L-176, 1944.
32. Sechler, E. E., and L. G. Dunn: "Airplane Structural Analysis and Design," John Wiley & Sons, Inc., New York, 1942.

PROBLEMS

16-1. Determine the local buckling stress of the extruded section shown.   The material is 7075-T6 aluminum alloy ($E = 10.5 \times 10^6$ psi, $\sigma_{cy} = 70$ ksi, $\sigma_{0.7} = 72$ ksi, $n = 16.6$).

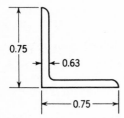

0.75

0.63

0.75   **Fig. P16-1**

16-2. Determine the local buckling stress of the extruded section shown.   The material is the same as in Prob. 16-1.

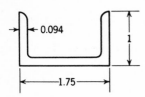

0.094

1

1.75   **Fig. P16-2**

16-3. Determine the local buckling stress of the formed section shown.   The material is 7075-T6 aluminum alloy sheet ($E = 10.5$ psi, $\sigma_{cy} = 67$ ksi, $\sigma_{0.7} = 70$ ksi, $n = 9.2$).

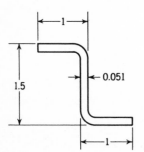

1

0.051

1.5

1   **Fig. P16-3**

16-4. What diameter bulb is required on the outstanding leg of the section of Prob. 16-1 to make the leg behave like a web instead of a flange?

16-5. What length of lip is required on the outstanding legs of the section of Prob. 16-3 to make them behave like webs?

16-6. Determine the crippling stress of the section in Prob. 16-1.

16-7. Determine the crippling stress of the section in Prob. 16-2.

16-8. Determine the crippling stress of the section in Prob. 16-3.

16-9. The minimum value of ρ of the section in Probs. 16-2 and 16-7 is 0.304.   Determine the failing stress of the column if $c = 2$ and the length is (a) 10 in. and (b) 60 in.

16-10. Determine the local buckling stress of a panel with a 0.051-in. skin and the stiffeners of Prob. 16-1 if the stiffener spacing is 2.5 in.

16-11. Determine the local buckling stress of a panel with a 0.064-in. skin and the stiffeners of Prob. 16-3.   The stiffener spacing is 3 in.

16-12. Determine the buckling stress of the panel of Example 16-4 if the stiffeners are in the transverse instead of the longitudinal direction.

16-13. Determine the monolithic crippling stress of the section of Prob. 16-11.

16-14. The stiffeners in Prob. 16-11 are attached to the skin by $\frac{1}{8}$-in. rivets with a 1-in. pitch.   Determine the interfastener-buckling and wrinkling-failure stresses. Determine the panel failing stress for $L'/\rho \leq 20$.

16-15. Construct a direct-reading column chart for the panel of Prob. 16-14 using the results of Probs. 16-11, 16-13, and 16-14.

Index

Index